THE HUMAN BODY COLORING BOOK

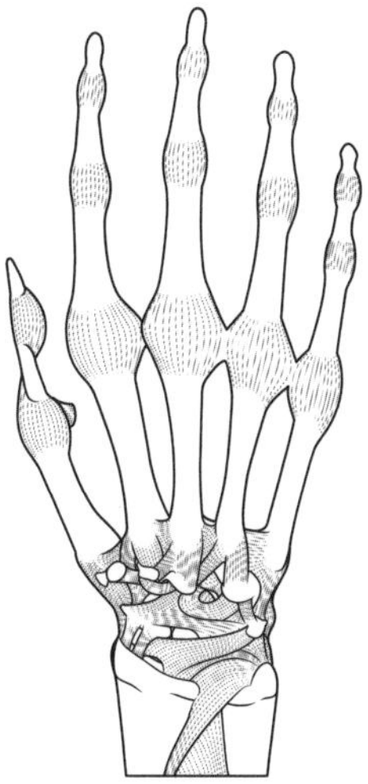

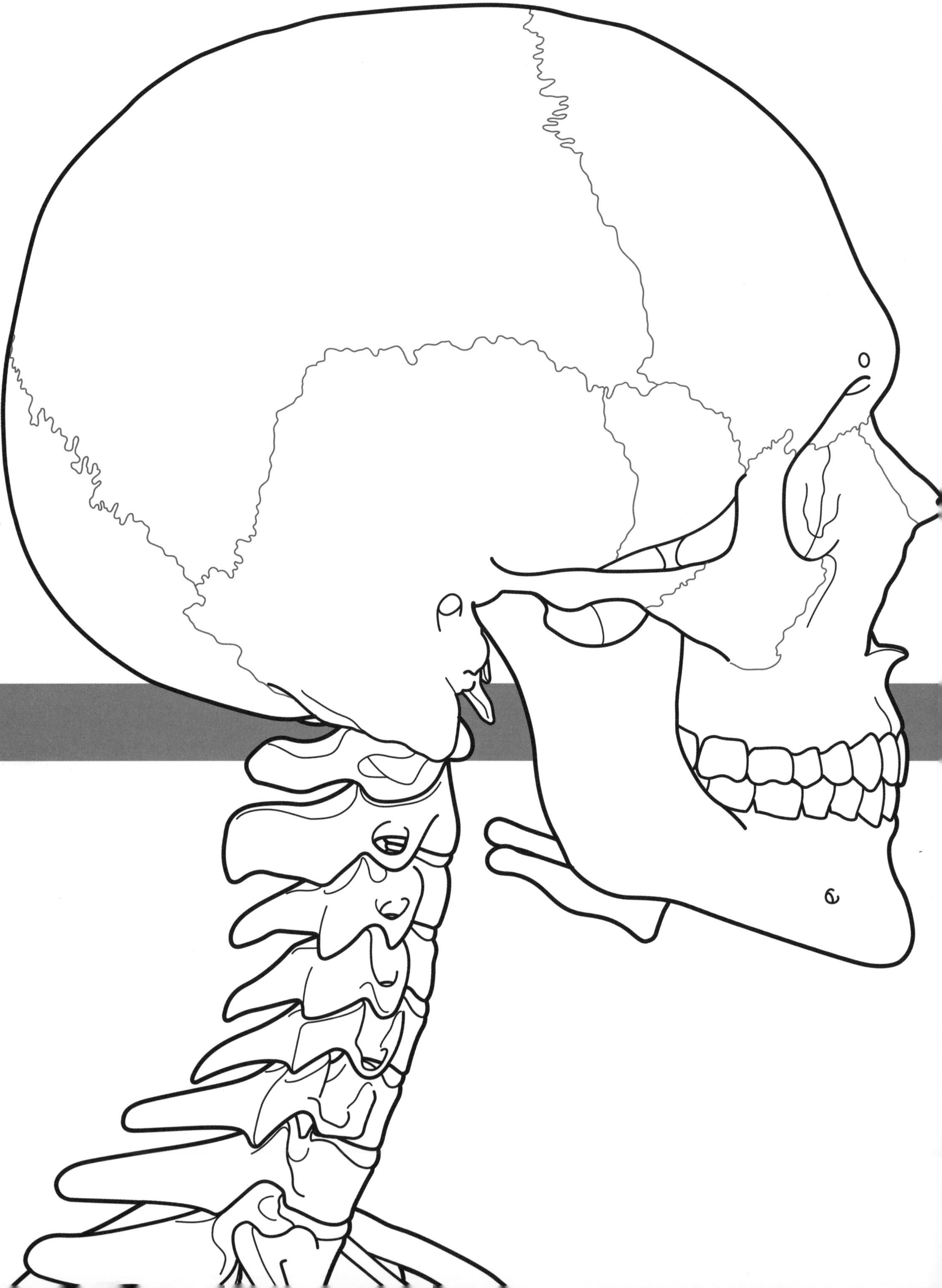

THE HUMAN BODY COLORING BOOK

DK

DK | Penguin Random House

Editorial Consultant
Professor Susan Standring

DK LONDON

Senior Designer Clare Marshall
Senior Editor Martyn Page
Jacket Designer Silke Spinges
Production Editor Tony Phipps
Illustrations Peter Bull Art Studio
Production Controller Erika Pepe
Managing Art Editor Owen Peyton Jones
Managing Editor Julie Ferris
US Editor Rebecca Warren
Associate Publishing Director Liz Wheeler
Art Director Philip Ormerod
Publishing Director Jonathan Metcalf

DK DELHI

Senior Art Editor Mitun Banerjee
Senior Editor Alka Ranjan
Designers Pooja Pipil, Amit Malhotra, Pallavi Narain,
Editors Megha Gupta, Antara Moitra, Pallavi Singh
Additional design support Arup Giri, Niyati Gosain, Payal Rosalind Malik, Nidhi Mehra, Mahipal Singh, Zaurin Thoidingjam, Deep Shikha Walia
Additional editorial support Sreshtha Bhattacharya, Suefa Lee
Managing Editor Rohan Sinha
Deputy Managing Art Editor Priyabrata Roy Chowdhury
DTP Designer Anita Yadav
Managing Art Editor Arunesh Talapatra
DTP Manager/CTS Balwant Singh

First American Edition, August 2011
Published in the United States by DK Publishing
1450 Broadway, Suite 801, New York, NY 10018

21 22 23 24 20 19 18 17 16
018–179803–August 2011

Published in Great Britain by Dorling Kindersley Limited

A catalog record for this book is available from the Library of Congress.
ISBN 978-0-7566-8234-7

DK books are available at special discounts when purchased in bulk for sales promotions, premium, fund-raising, or educational use. For details, contact: DK Publishing Special Markets, 1450 Broadway, Suite 801, New York, NY 10018
SpecialSales@dk.com

Printed and bound in China

For the curious
www.dk.com

MIX Paper from responsible sources FSC™ C018179 FSC www.fsc.org

This book was made with Forest Stewardship Council ™ certified paper – one small step in DK's commitment to a sustainable future. For more information go to www.dk.com/our-green-pledge

CONTENTS

HOW TO USE THIS BOOK

This book can be used as a learning tool in two different ways. Either color in and/or label the diagrams, using the accompanying keys, or cover up the keys and label the diagrams from memory. The book is organized by body system but can also be navigated by body region.
Look for the » cross-references at the bottom of every spread.

THE BODY

01

SURFACE ANATOMY

DIFFERENT AREAS OF THE BODY'S SURFACE HAVE BEEN GIVEN ANATOMICAL NAMES SO THAT LOCATIONS MAY BE DESCRIBED ACCURATELY AND UNAMBIGUOUSLY. THE ILLUSTRATIONS HERE SHOW THE MAIN REGIONS OF THE ANTERIOR AND POSTERIOR SURFACES OF THE BODY.

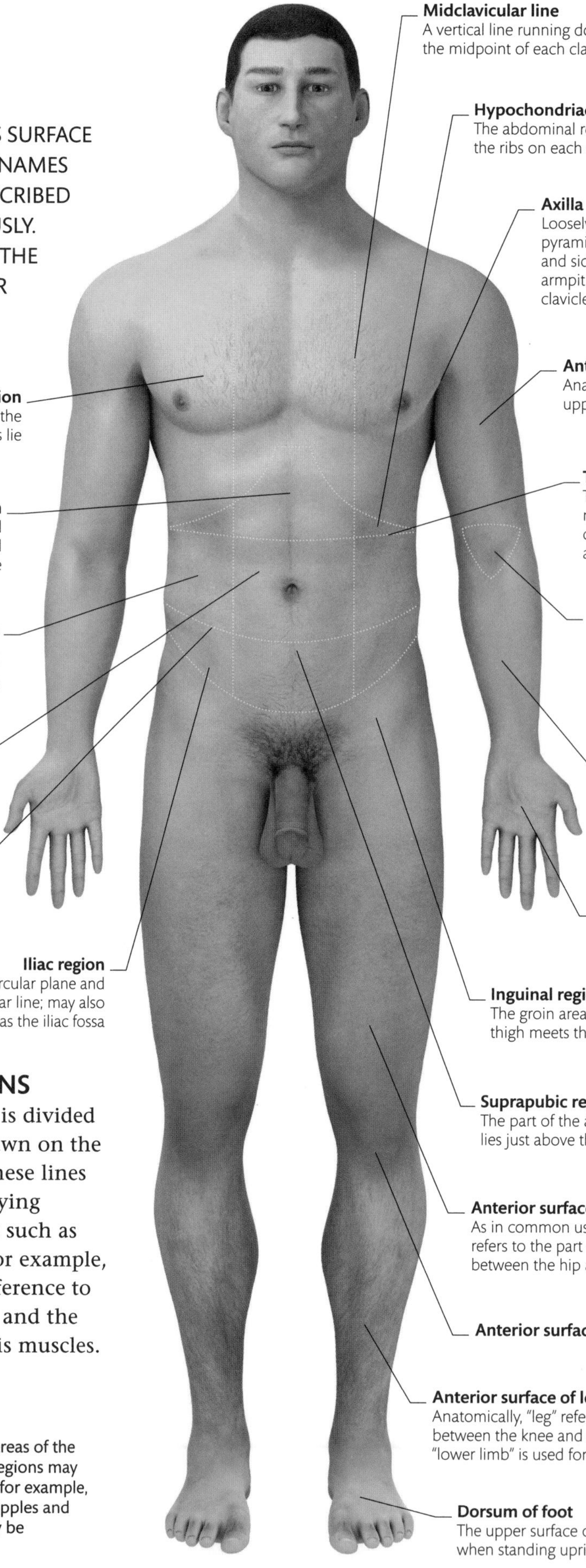
Pectoral region
The chest; sometimes used to refer to just the upper chest, where the pectoral muscles lie
Epigastric region
The area of the abdominal wall above the transpyloric plane and framed by the diverging margins of the rib cage
Lumbar region
The sides of the abdominal wall, between the transpyloric and transtubercular planes
Umbilical region
Central region of the abdominal wall, around the umbilicus
Transtubercular plane
This plane passes through the iliac tubercles and lies at the level of the fifth lumbar vertebra
Iliac region
The area below the transtubercular plane and lateral to the midclavicular line; may also be referred to as the iliac fossa
Midclavicular line
A vertical line running down from the midpoint of each clavicle
Hypochondriac region
The abdominal region under the ribs on each side
Axilla
Loosely, the armpit; more precisely, the pyramid-shaped area between the upper arm and side of the thorax. Its floor is the skin of the armpit, and it reaches up to the level of the clavicle, top of the scapula, and first rib
Anterior surface of arm
Anatomically, "arm" refers just to the part of the upper limb between the shoulder and the elbow
Transpyloric plane
The horizontal plane joining the tips of the ninth costal cartilage, at the margins of the rib cage; also level with the first lumbar vertebra and the pylorus of the stomach
Cubital fossa
The triangular area anterior to the elbow, bounded by a line between the bony epicondyles of the humerus on each side, and framed below by the pronator teres and brachioradialis muscles
Anterior surface of forearm
Anatomically (and colloquially) the forearm is the part of the upper limb between the elbow and wrist
Palmar surface of hand
The anterior surface of the hand
Inguinal region
The groin area, where the thigh meets the trunk
Suprapubic region
The part of the abdomen that lies just above the pubic bones
Anterior surface of thigh
As in common usage, the term "thigh" refers to the part of the lower limb between the hip and knee
Anterior surface of knee
Anterior surface of leg
Anatomically, "leg" refers just to the part between the knee and ankle; the term "lower limb" is used for the whole limb
Dorsum of foot
The upper surface of the foot, when standing upright

ANTERIOR SURFACE REGIONS

The anterior surface of the body is divided into areas by imaginary lines drawn on the body. The location of many of these lines is defined by reference to underlying anatomical structures or features such as muscles or bony prominences; for example, the cubital fossa is defined by reference to the epicondyles of the humerus, and the pronator teres and brachioradialis muscles.

ANTERIOR ANATOMICAL TERMS
This illustration shows the main anatomical areas of the anterior surface of the body. Some of these regions may sometimes be subdivided into smaller areas; for example, in the pectoral region, the area around the nipples and extending a short distance into the axilla may be referred to as the mammary region.

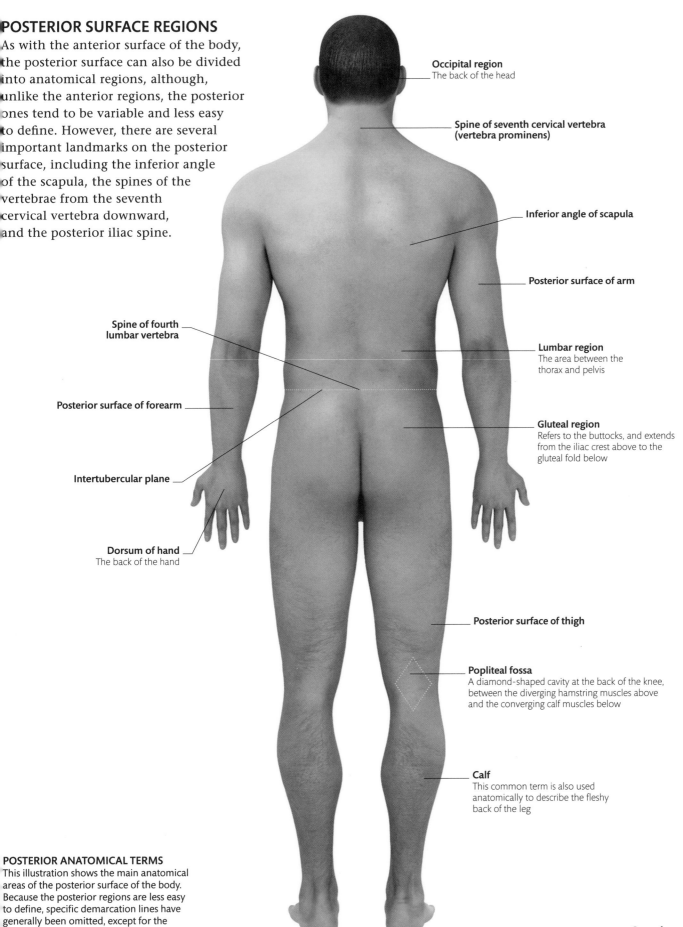

POSTERIOR SURFACE REGIONS

As with the anterior surface of the body, the posterior surface can also be divided into anatomical regions, although, unlike the anterior regions, the posterior ones tend to be variable and less easy to define. However, there are several important landmarks on the posterior surface, including the inferior angle of the scapula, the spines of the vertebrae from the seventh cervical vertebra downward, and the posterior iliac spine.

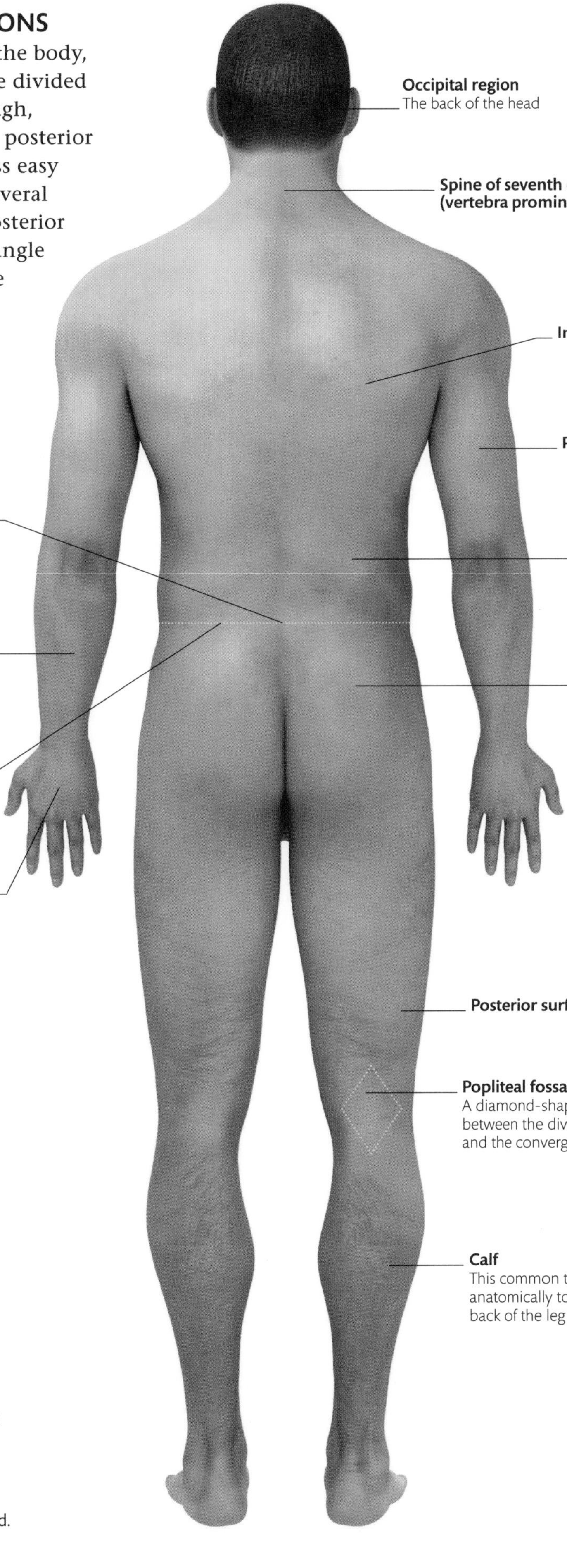

Occipital region
The back of the head
Spine of seventh cervical vertebra (vertebra prominens)
Inferior angle of scapula
Posterior surface of arm
Spine of fourth lumbar vertebra
Lumbar region
The area between the thorax and pelvis
Posterior surface of forearm
Gluteal region
Refers to the buttocks, and extends from the iliac crest above to the gluteal fold below
Intertubercular plane
Dorsum of hand
The back of the hand
Posterior surface of thigh
Popliteal fossa
A diamond-shaped cavity at the back of the knee, between the diverging hamstring muscles above and the converging calf muscles below
Calf
This common term is also used anatomically to describe the fleshy back of the leg

POSTERIOR ANATOMICAL TERMS
This illustration shows the main anatomical areas of the posterior surface of the body. Because the posterior regions are less easy to define, specific demarcation lines have generally been omitted, except for the popliteal fossa, which can be clearly defined.

See also p. 10 »

PLANES, MOVEMENTS, AND DIRECTIONS

AS WITH SURFACE REGIONS, PLANES THROUGH THE BODY, BODY MOVEMENTS, AND DIRECTIONS ALSO HAVE SPECIFIC ANATOMICAL TERMS. THESE TERMS ASSUME THAT THE BODY IS IN A STANDARD POSITION: STANDING UPRIGHT; FACING FORWARD; UPPER LIMBS BY THE SIDES WITH THE PALMS FACING FORWARD; LOWER LIMBS TOGETHER WITH THE FEET FACING FORWARD.

PLANES AND MOVEMENTS

There are three main anatomical planes through the body. The coronal plane cuts vertically down through the body, through or parallel to the shoulders. The sagittal plane cuts vertically down through the body, through or parallel to the sternum. The transverse plane cuts horizontally through the body, dividing it into upper and lower parts. The main movements are flexion, extension, abduction, and adduction. Some joints, such as the hip and shoulder, also allow rotation of a limb along its axis. A special type of rotation between the forearm bones allows the palm to be moved from a forward or upward-facing position (supination) to a backward or downward-facing position (pronation).

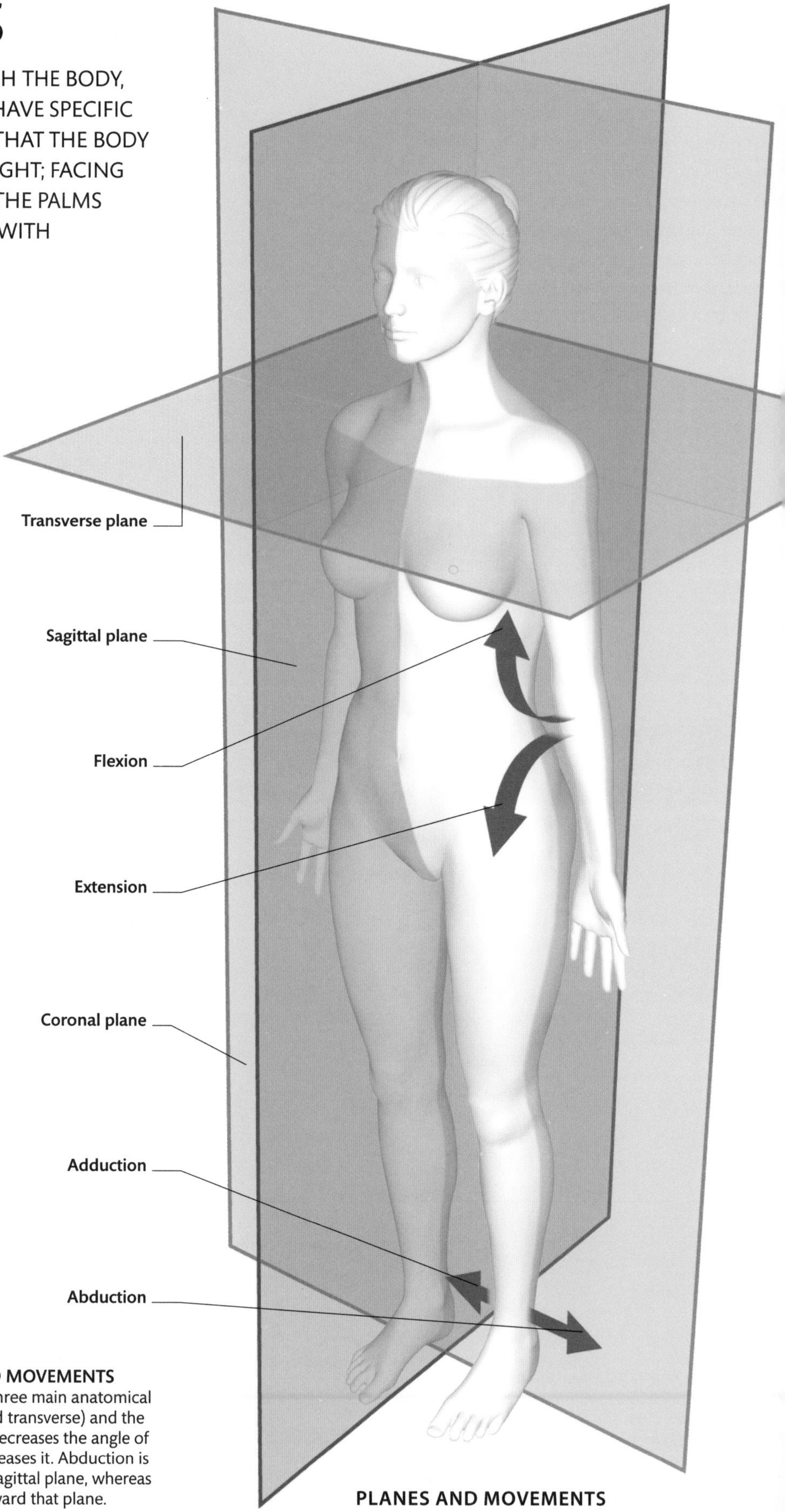

TERMS FOR PLANES AND MOVEMENTS
This illustration shows the three main anatomical planes (coronal, sagittal, and transverse) and the main movements. Flexion decreases the angle of a joint, while extension increases it. Abduction is movement away from the sagittal plane, whereas adduction is movement toward that plane.

PLANES AND MOVEMENTS

DIRECTIONS

As well as movements and planes, there are also anatomical terms for directions and relative positions of the body. Medial and lateral describe positions of structures toward or away from the midline, respectively. Superior and inferior refer to vertical position—toward the top or bottom of the body. Proximal and distal describe a relative position toward the center or periphery of the body; these two terms are most commonly applied to structures in the limbs, describing their position relative to the attachment of the limb to the body.

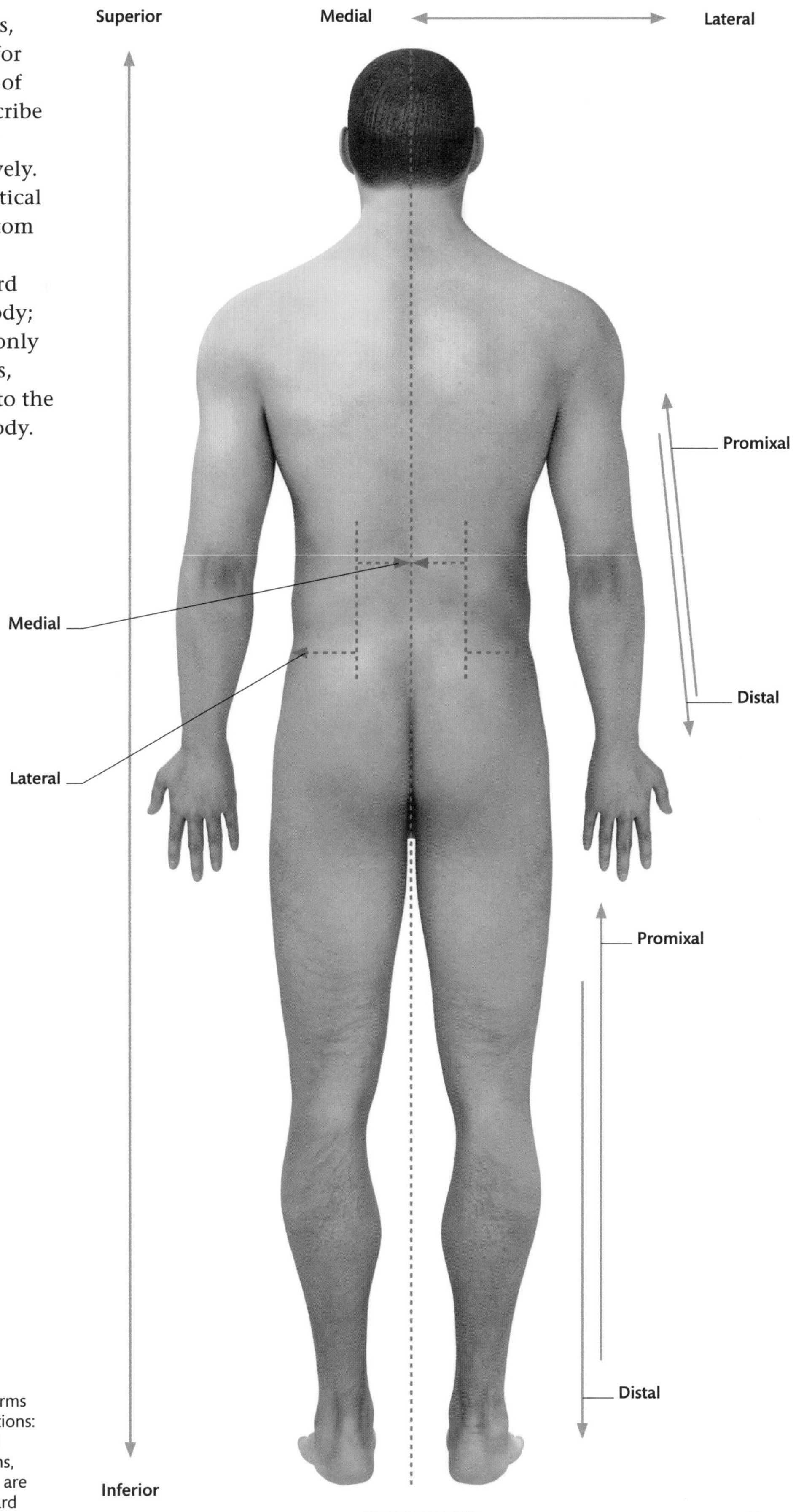

TERMS FOR DIRECTIONS
This image illustrates the main anatomical terms used to describe directions and relative positions: superior and inferior; medial and lateral; and proximal and distal. In addition to these terms, anterior (or ventral) and posterior (or dorsal) are also used, referring to relative positions toward the front or back or the body, respectively.

DIRECTIONS

See also p. 8 »

BODY COMPOSITION

THE HUMAN BODY IS COMPOSED OF ABOUT 75 TRILLION CELLS, WHICH ARE ORGANIZED HIERARCHICALLY TO FORM TISSUES, ORGANS, AND BODY SYSTEMS.

CELLS AND TISSUES

The overall organization of the body can be visualized in the form of a hierarchy of levels, from cells (which are made up of chemicals), through tissues and organs to systems. As the hierarchy ascends, the number of components in each of its levels decreases, culminating in a single organism.

Cells are the smallest living units. They are created from molecules, which shape their outer covering and inner structures and drive the metabolic reactions that keep them alive. There are more than 200 types of cell in the human body, each adapted to carry out a specific role. Groups of cells form communities called tissues. The body has four basic tissue types: epithelial, connective, muscular, and nervous tissue. Epithelial tissue covers surfaces and lines cavities. Connective tissue supports and protects body structures. Muscular tissue produces movement. Nervous tissue facilitates rapid internal communication.

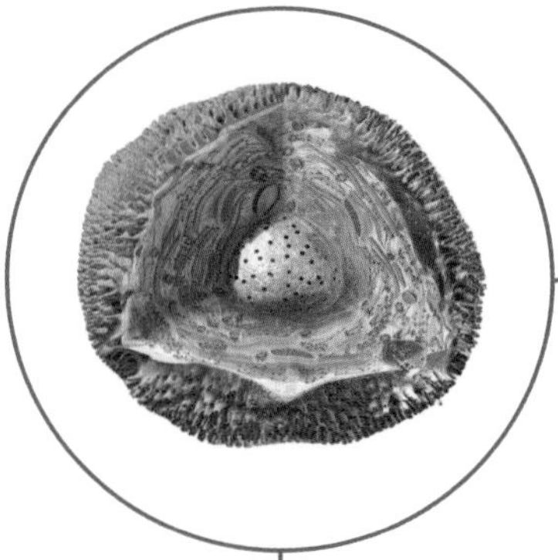

CELLS

TISSUES

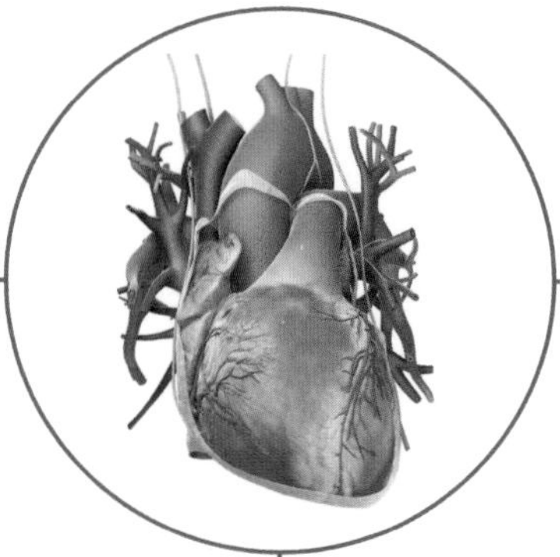

ORGANS

CELLS
The body's most basic living components, cells differ in size and shape according to their function, but all (except red blood cells) possess the same basic features: an outer membrane; organelles in the jellylike cytoplasm; and a nucleus, which contains DNA. Red blood cells are highly specialized and lack a nucleus.

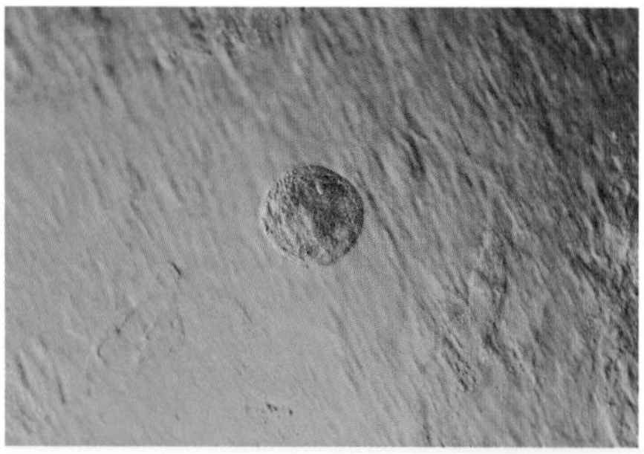

Stem cells
These unspecialized cells have the unique ability to differentiate into a wide range of specialized tissue cells, such as muscle, brain, or blood cells.

TISSUES
Tissues are made up of groups of similar cells. For example, cardiac muscle tissue (one of three types of muscle tissue) consists of specialized cardiac muscle cells that contract together to make the heart pump and, working as a network, conduct the nerve signals that ensure the heart's pumping is precisely coordinated.

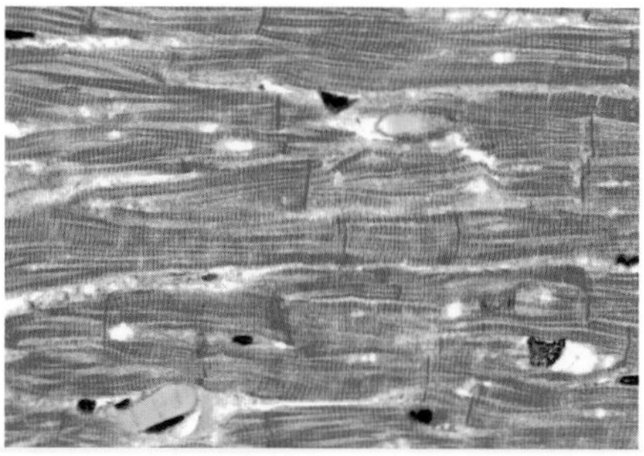

Cardiac muscle fibers
The cells, or fibers, in cardiac tissue are long and cylindrical and have branches that form junctions with other cells to create an interconnected network.

ORGANS
Organs consist of different tissues grouped together in a single structure to perform a unique function or group of functions. For example, the heart includes cardiac muscle tissue; connective tissues, which protect the heart and hold other tissues together; and epithelial tissues, which line the heart's chambers and cover its valves.

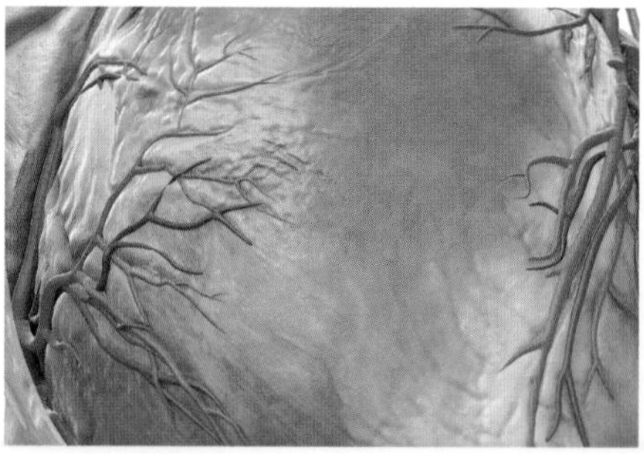

Heart structure
The heart has a complex structure and is connected to a vast network of veins and arteries. Internally, it has four chambers with muscular walls.

ORGANS AND SYSTEMS

Organs are discrete structures consisting of at least two types of tissue. Each has a specialized role or roles that no other organ can perform. When organs collectively have a common purpose, they are linked together within a system, such as the cardiovascular system, which transports oxygen, nutrients, and waste products around the body. Integrated and interdependent, the systems combine to produce a complete human being.

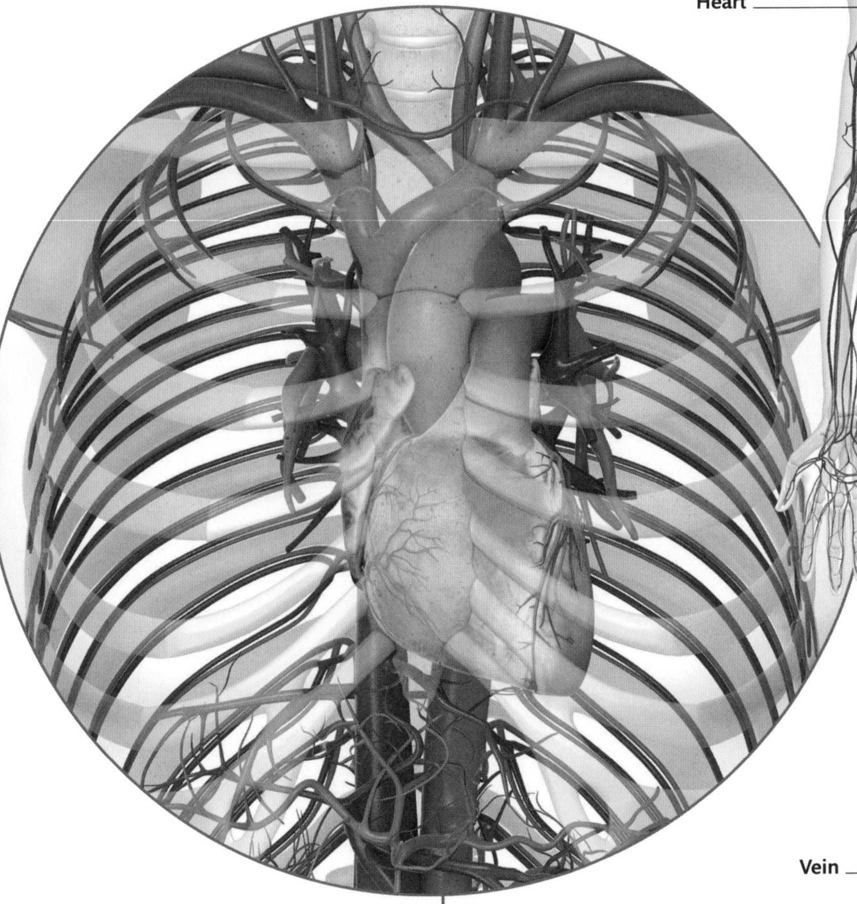

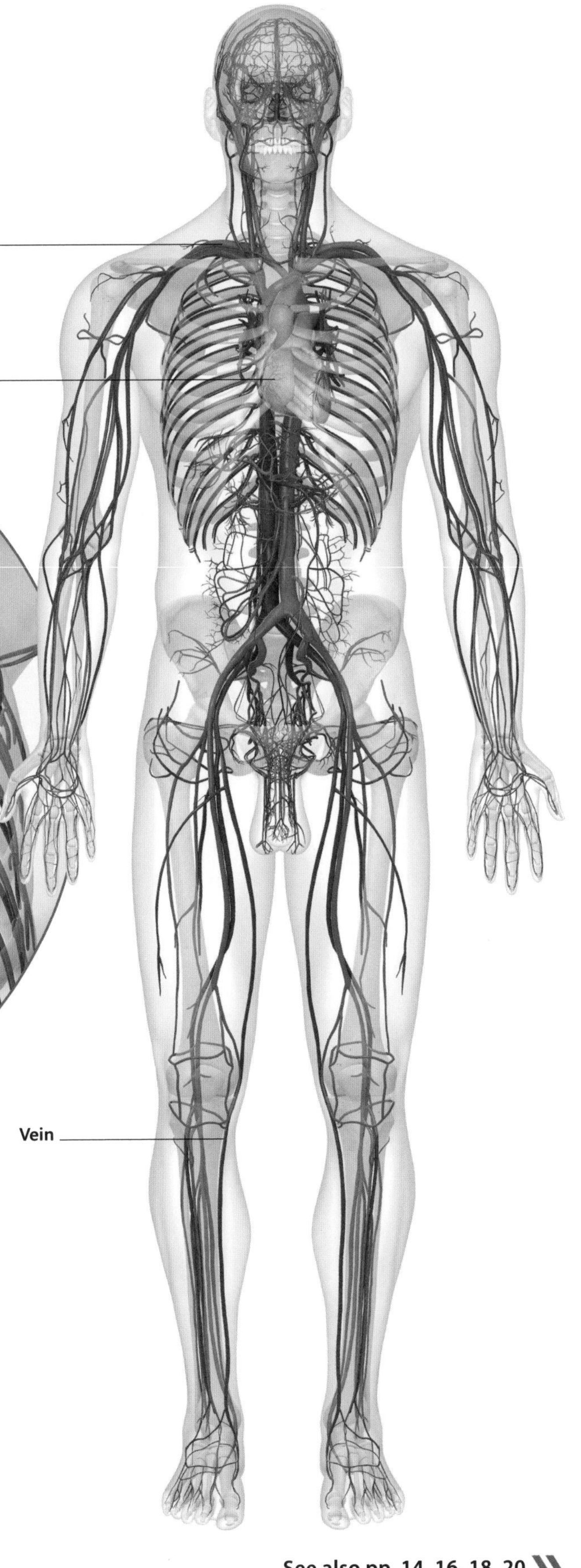

BODY SYSTEMS
A body system is a group of organs that have a common function or functions. The cardiovascular system, for instance, comprises the heart, blood and lymphatic vessels, and blood, which function together to form the body's key transportation system. Its basic function is to pump blood around the body, delivering oxygen and nutrients to, and removing wastes from, the cells of the body's other systems (skeletal, muscular, integumentary, nervous, endocrine, respiratory, digestive, urinary, and reproductive). At the same time, the cardiovascular system also depends on the other systems to function properly.

See also pp. 14, 16, 18, 20 »

CELLS

THE CELL IS THE BASIC FUNCTIONAL UNIT OF THE HUMAN BODY. IT IS THE SMALLEST PART CAPABLE OF THE PROCESSES THAT DEFINE LIFE, SUCH AS REPRODUCTION, RESPIRATION, AND MOVEMENT, ALTHOUGH NOT EVERY CELL HAS THESE ABILITIES.

CELL ANATOMY

Most cells are microscopic—typically only about 0.0004 in (0.01 mm) across—although some specialized cells, such as certain neurons, may be more than 12 in (30 cm) long. Cells are also diverse, with specialized features that facilitate their functions. Nevertheless, there are certain features that almost all cells have in common, such as an outer membrane, and organelles such as a nucleus and mitochondria. The exception is mature red blood cells, which lack organelles.

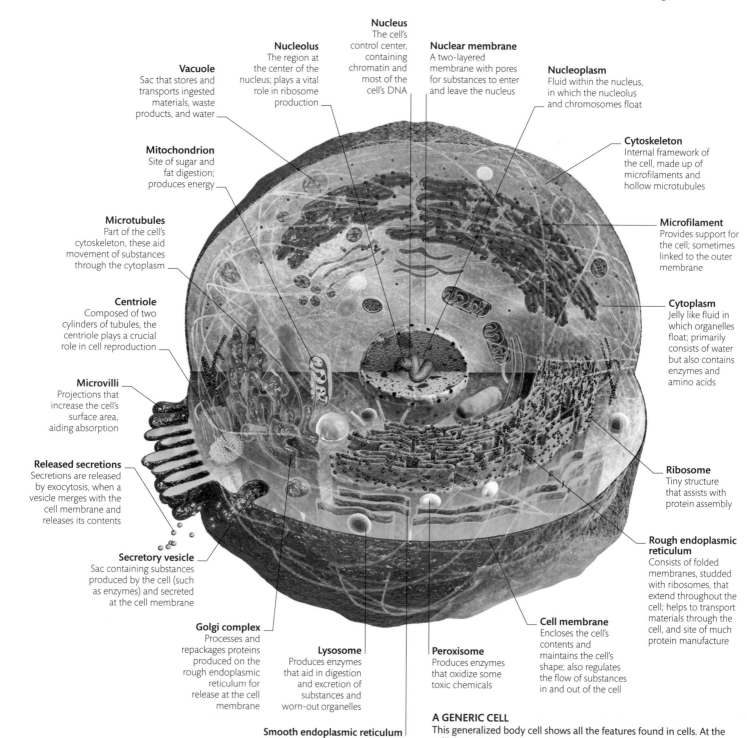

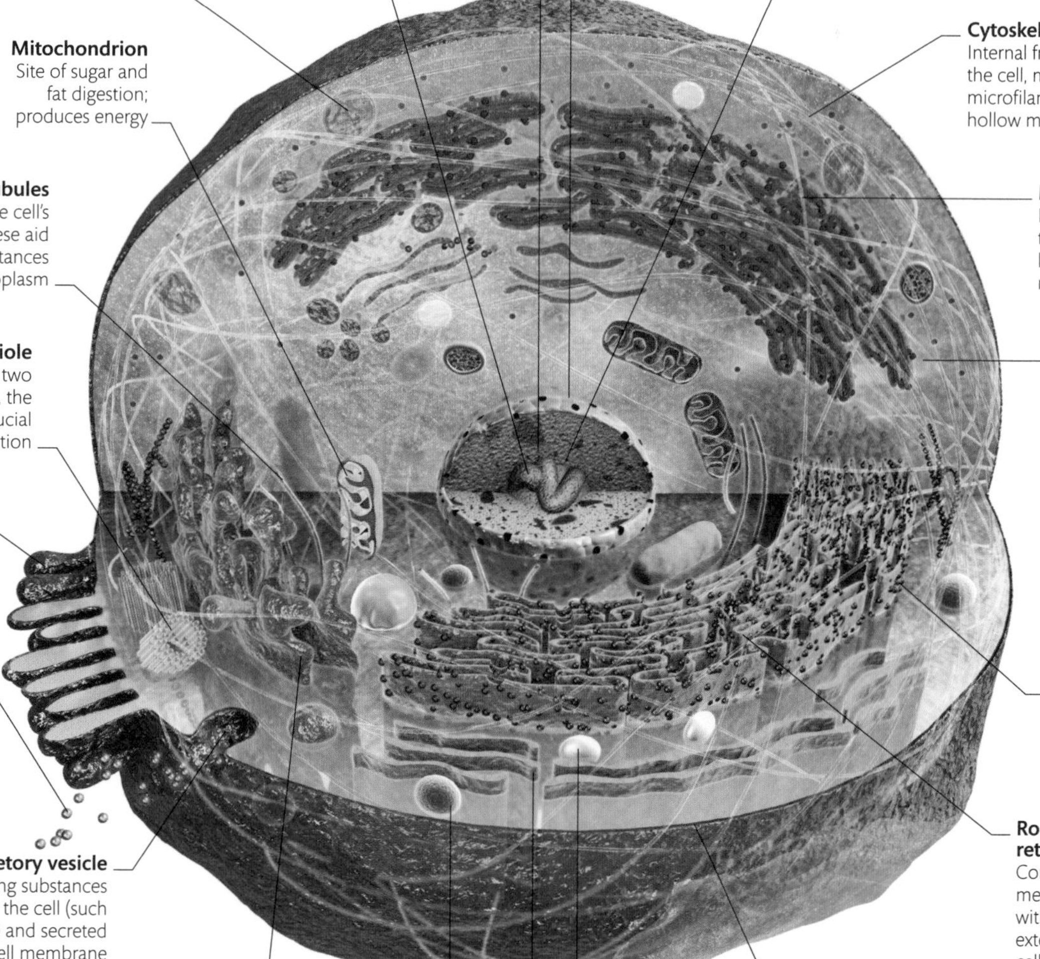
Nucleus
The cell's control center, containing chromatin and most of the cell's DNA
Nucleolus
The region at the center of the nucleus; plays a vital role in ribosome production
Vacuole
Sac that stores and transports ingested materials, waste products, and water
Nuclear membrane
A two-layered membrane with pores for substances to enter and leave the nucleus
Nucleoplasm
Fluid within the nucleus, in which the nucleolus and chromosomes float
Mitochondrion
Site of sugar and fat digestion; produces energy
Cytoskeleton
Internal framework of the cell, made up of microfilaments and hollow microtubules
Microtubules
Part of the cell's cytoskeleton, these aid movement of substances through the cytoplasm
Microfilament
Provides support for the cell; sometimes linked to the outer membrane
Centriole
Composed of two cylinders of tubules, the centriole plays a crucial role in cell reproduction
Cytoplasm
Jelly like fluid in which organelles float; primarily consists of water but also contains enzymes and amino acids
Microvilli
Projections that increase the cell's surface area, aiding absorption
Released secretions
Secretions are released by exocytosis, when a vesicle merges with the cell membrane and releases its contents
Ribosome
Tiny structure that assists with protein assembly
Secretory vesicle
Sac containing substances produced by the cell (such as enzymes) and secreted at the cell membrane
Rough endoplasmic reticulum
Consists of folded membranes, studded with ribosomes, that extend throughout the cell; helps to transport materials through the cell, and site of much protein manufacture
Golgi complex
Processes and repackages proteins produced on the rough endoplasmic reticulum for release at the cell membrane
Lysosome
Produces enzymes that aid in digestion and excretion of substances and worn-out organelles
Peroxisome
Produces enzymes that oxidize some toxic chemicals
Cell membrane
Encloses the cell's contents and maintains the cell's shape; also regulates the flow of substances in and out of the cell
Smooth endoplasmic reticulum
Network of tubes and sacs that helps to transport materials through the cell; also a site of calcium storage and main location of fat metabolism

A GENERIC CELL
This generalized body cell shows all the features found in cells. At the cell's heart is the nucleus, where genetic material is stored and the first stages of protein synthesis occur. Cells also contain other structures for assembling proteins, including ribosomes, the endoplasmic reticulum, and the Golgi apparatus. Mitochondria provide the cell with energy.

CELL TYPES

Although every cell contains the same genetic information (except mature red blood cells, which lack DNA, and germ cells, which have only half the normal complement of genes), not all the genes are "switched on" in every cell. It is this pattern of gene expression that determines the structure and function of a cell. A cell's fate is largely determined before birth, influenced by its position in the body and the cocktail of chemicals to which it is exposed. Early during development, stem cells begin to differentiate into three layers of more specialized cells: ectoderm, endoderm, and mesoderm. Ectodermal cells will form the skin and nails, the epithelium of the nose, mouth, and anus, the eyes, and the brain and spinal cord. Endodermal cells become the linings of the digestive and respiratory tracts, and glandular organs including the liver and pancreas. Mesodermal cells develop into the muscles, circulatory system, and the excretory system, including the kidneys.

CELL VARIETIES

This illustration shows just a few of the more than 200 different types of cell in the body. Despite the variety of their appearance, structure, and function, all are based on the same basic template, as illustrated in the generic cell opposite.

Red blood cells
Unlike other human cells, mature red blood cells lack a nucleus and organelles. Instead, they are packed with hemoglobin, the iron-containing molecule that binds with oxygen and which gives blood its red color. Red blood cells develop in the bone marrow and circulate for about 120 days before being broken down and recycled.

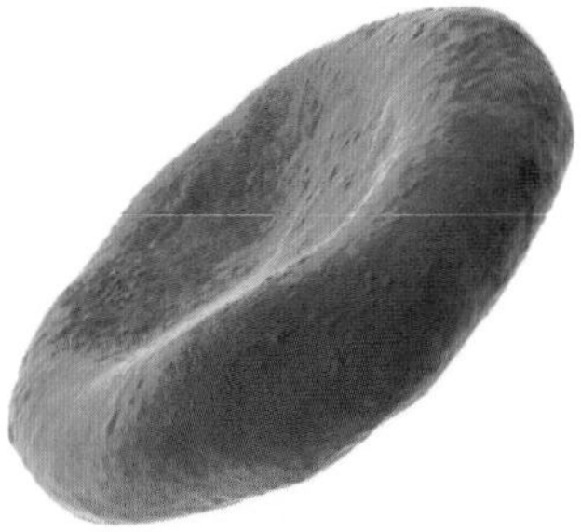

Epithelial cells
These are barrier cells lining the cavities and surfaces of the body, such as the skin and lining of the lungs and reproductive tract. Some epithelial cells have cilia that project from their surface, to waft eggs down the Fallopian tubes or move mucus out of the lungs, for example.

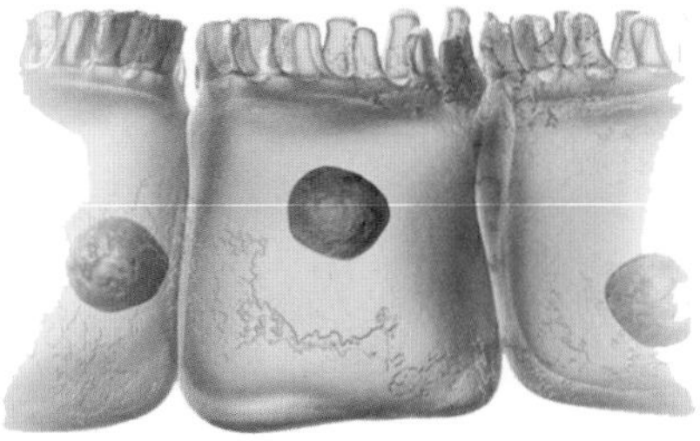

Adipose cells
These cells are highly adapted for fat storage, and most of the interior is taken up by a large droplet of semiliquid fat. When we gain weight, the adipose cells fill with even more fat and swell, although eventually they also start to increase in number.

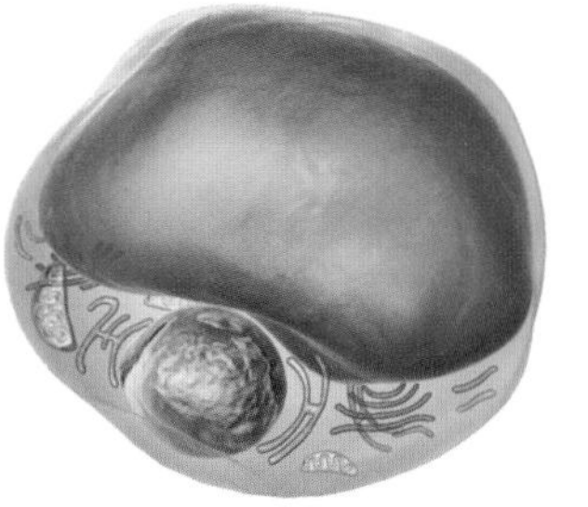

Nerve cells
These electrically excitable cells transmit action potentials down the axon. Found throughout the body, they act as detectors and as information conveyors and processors. They communicate with each other across connections called synapses.

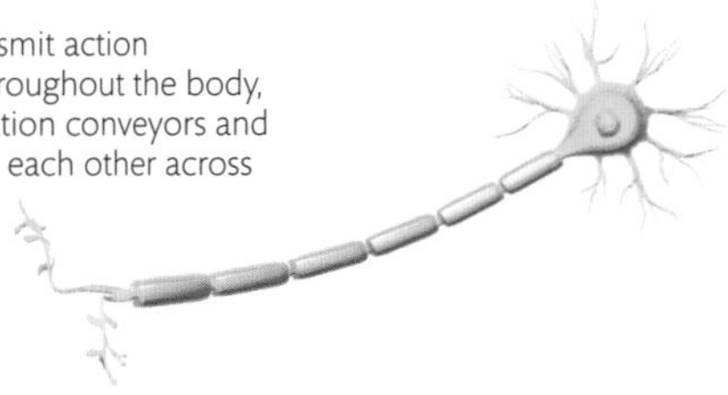

Sperm cells
Sperm have a tail that enables them to swim up the female reproductive tract to fertilize an egg. Each sperm contains only 23 chromosomes; at fertilization, these pair up with an egg's 23 chromosomes to create an embryo with the full complement of 46 chromosomes.

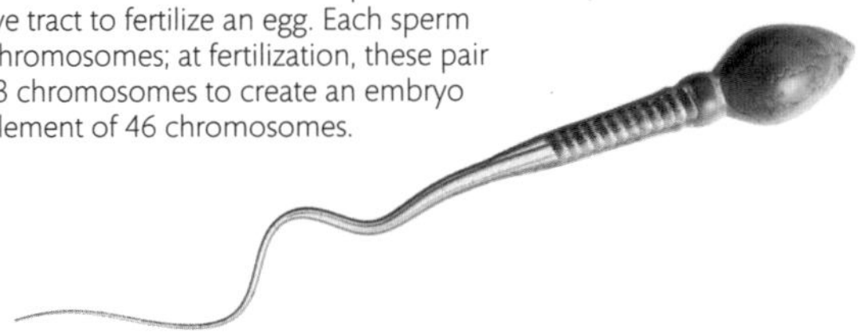

Egg cells (ova)
One of the largest cells in the human body, an ovum is just visible to the naked eye. Like sperm, each egg contains only 23 chromosomes. Every woman is born with a finite number of eggs, which decreases as she ages.

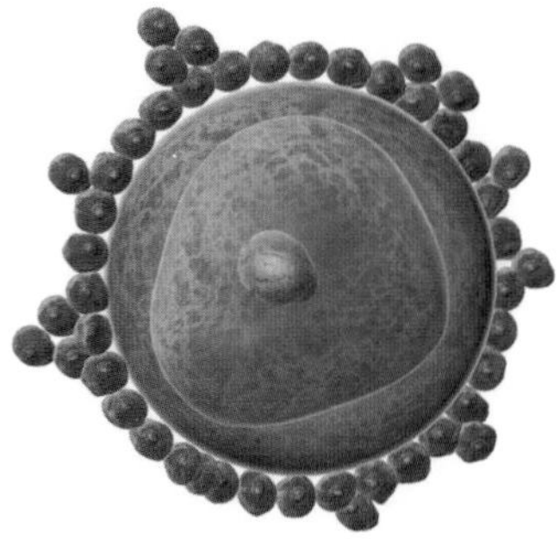

Photoreceptor cells
Found at the back of the eye, these are specialized nerve cells that contain a light-sensitive pigment and generate a nerve impulse when struck by light. There are two main types of photoreceptor: rods (right), which work well in low light, but enable us to see only in black and white; and cones, which work better in bright light and enable us to detect colors.

Smooth muscle cells
One of three types of muscle cell, smooth muscle cells are spindle-shaped cells found in the arteries and digestive tract that produce long, wavelike contractions. To do this, they are packed with contractile filaments and large numbers of mitochondria.

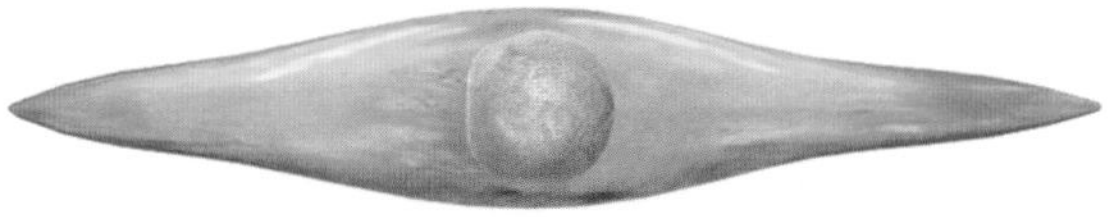

See also pp. 12, 16 »

TISSUES

TISSUES ARE GROUPS OF SIMILAR CELLS THAT CARRY OUT A SIMILAR FUNCTION. THERE ARE FOUR BASIC TISSUE TYPES IN THE HUMAN BODY: MUSCLE, CONNECTIVE, NERVOUS, AND EPITHELIAL.

TISSUE TYPES

Although the individual cells in a tissue carry out the same function and tend to be structurally similar, not all cells within a tissue are necessarily identical. In addition, the different forms of the basic tissue types can have very different appearances and functions. For example, blood, bone, and cartilage are all types of connective tissue, but so are fat layers, tendons, ligaments, and the fibrous tissue that holds organs and epithelial layers in place. Nervous tissue forms nerves, and individual nerve cells can be classified according to their structure into unipolar, bipolar, and multipolar neurons. There are three main types of muscle tissue: smooth (involuntary), skeletal (voluntary), and cardiac. Epithelial tissue forms the epidermis and the tissues that line almost every organ in the body. There are various types of epithelial tissue, classified according to the shape of the constituent cells and number of cell layers; for example, simple columnar epithelium consists of one layer of column-shaped cells.

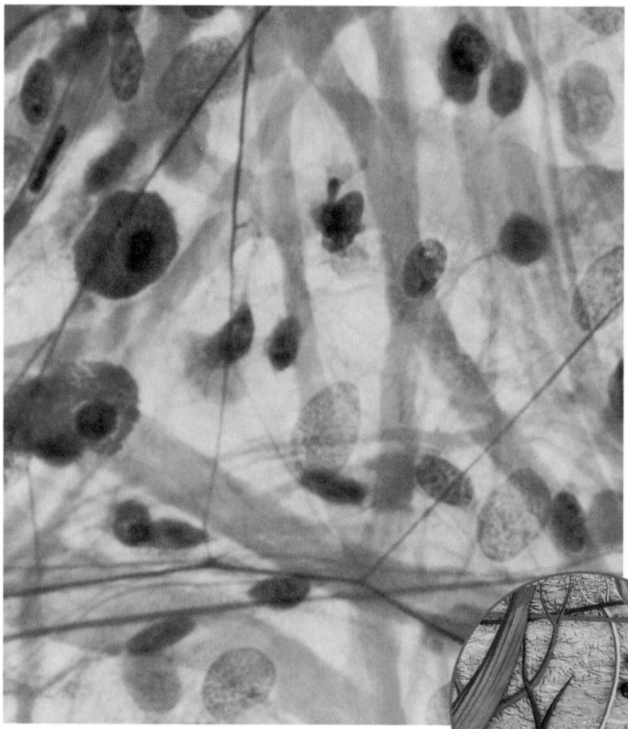

Dermal tissue

Loose connective tissue
This type of tissue contains cells called fibroblasts. The fibers they secrete are loosely organized and run in random directions, making the tissue quite pliable. Loose connective tissue holds organs in place, and provides cushioning and support.

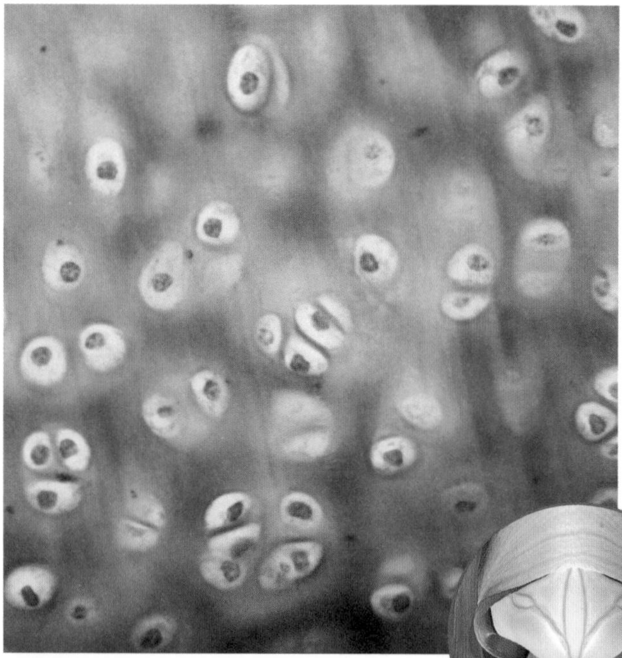

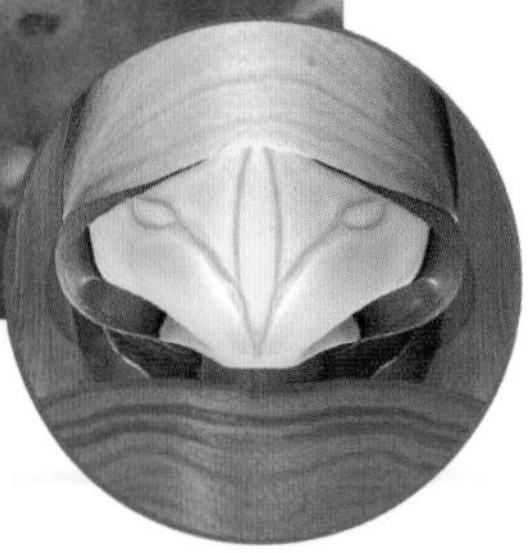

Nasal cartilage

Cartilage
This stiff, rubbery connective tissue is composed of cells called chondrocytes embedded in a matrix of gel-like material, which the cells secrete. Cartilage is found in the joints between bones, and in the ear and nose. The high water content of cartilage makes it tough but flexible.

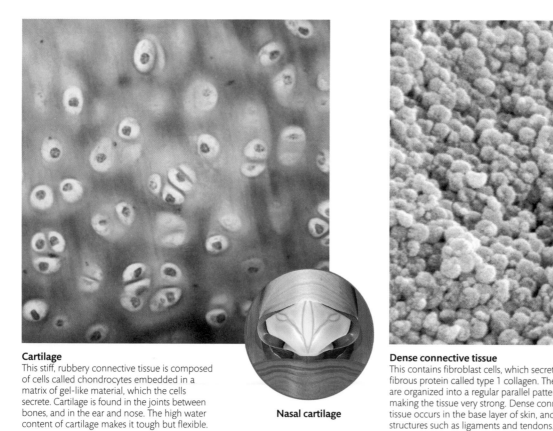

Ligaments at the knee

Dense connective tissue
This contains fibroblast cells, which secrete a fibrous protein called type 1 collagen. The fibers are organized into a regular parallel pattern, making the tissue very strong. Dense connective tissue occurs in the base layer of skin, and forms structures such as ligaments and tendons.

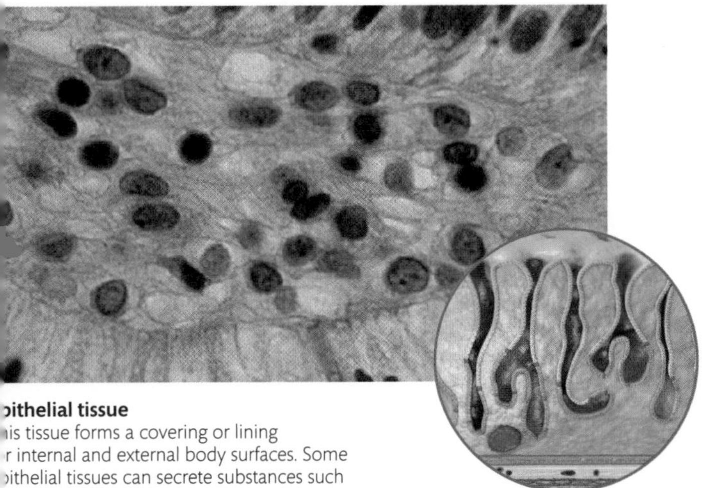

›ithelial tissue
›is tissue forms a covering or lining
r internal and external body surfaces. Some
›ithelial tissues can secrete substances such
 digestive enzymes; others can absorb
ıbstances, such as nutrients or water.

Stomach wall

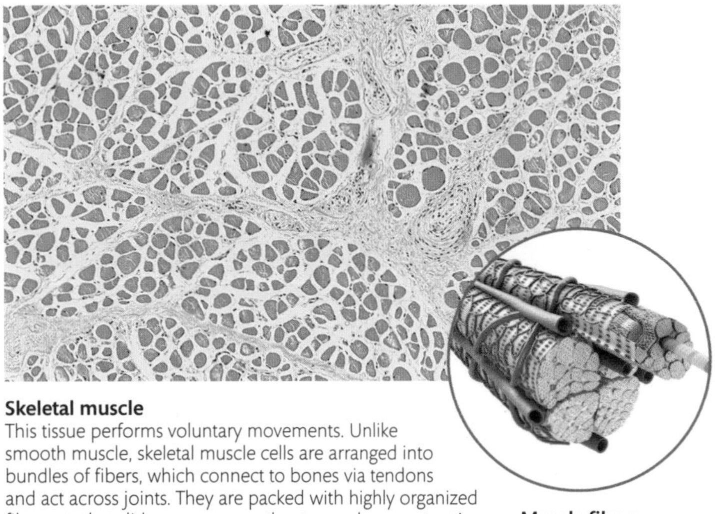

Skeletal muscle
This tissue performs voluntary movements. Unlike smooth muscle, skeletal muscle cells are arranged into bundles of fibers, which connect to bones via tendons and act across joints. They are packed with highly organized filaments that slide over one another to produce contractions.

Muscle fibers

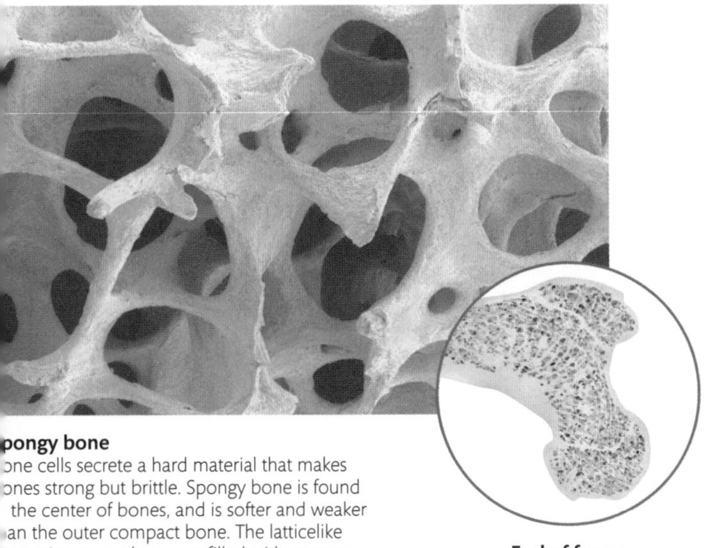

›pongy bone
›one cells secrete a hard material that makes
›ones strong but brittle. Spongy bone is found
 the center of bones, and is softer and weaker
›an the outer compact bone. The latticelike
›aces in spongy bone are filled with marrow
 connective tissue.

End of femur

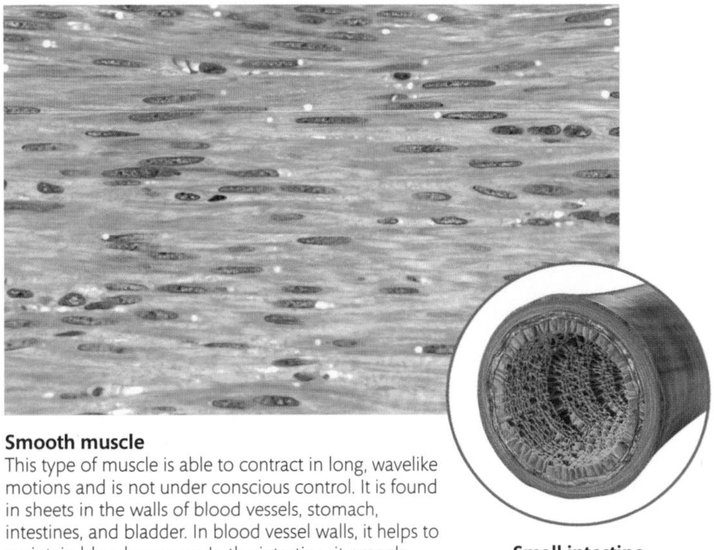

Smooth muscle
This type of muscle is able to contract in long, wavelike motions and is not under conscious control. It is found in sheets in the walls of blood vessels, stomach, intestines, and bladder. In blood vessel walls, it helps to maintain blood pressure. In the intestine, it propels food through the digestive tract.

Small intestine

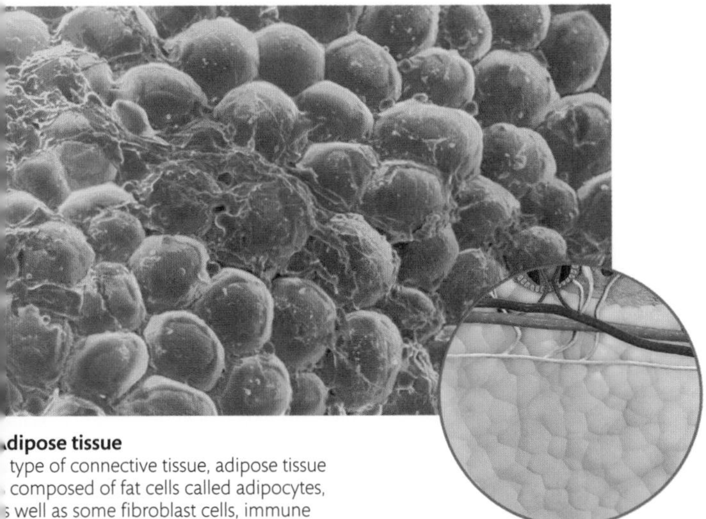

›dipose tissue
 type of connective tissue, adipose tissue
 composed of fat cells called adipocytes,
s well as some fibroblast cells, immune
›ells, and blood vessels. Its main function
 to act as an energy store, and to cushion,
›rotect, and insulate the body.

Subcutaneous fat

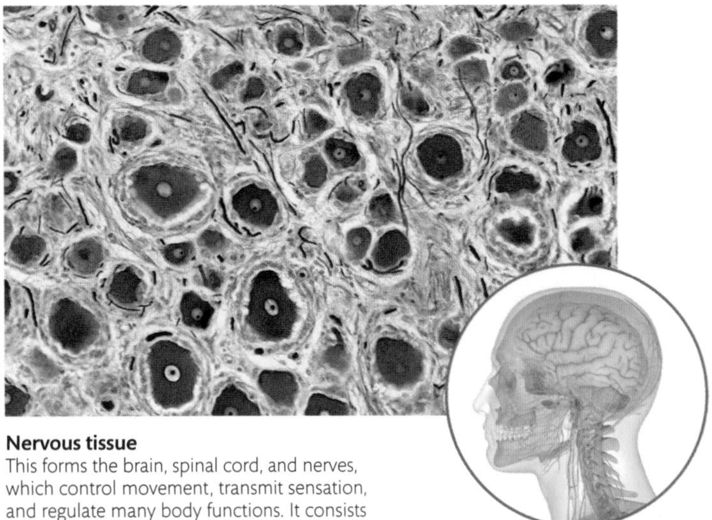

Nervous tissue
This forms the brain, spinal cord, and nerves, which control movement, transmit sensation, and regulate many body functions. It consists of networks of neurons supported by glial cells.

Brain and upper spinal cord

See also pp. 12, 18, 20 ››

BODY SYSTEMS 1

THE BODY'S ORGANS ARE ARRANGED INTO SYSTEMS, EACH WITH A PRIMARY FUNCTION, SUCH AS DIGESTION OR TRANSPORTING SUBSTANCES AROUND THE BODY. HOWEVER, THE SYSTEMS ARE INTERDEPENDENT: ALL HAVE TO WORK TOGETHER TO MAINTAIN THE HEALTH AND EFFICIENCY OF THE BODY AS A WHOLE.

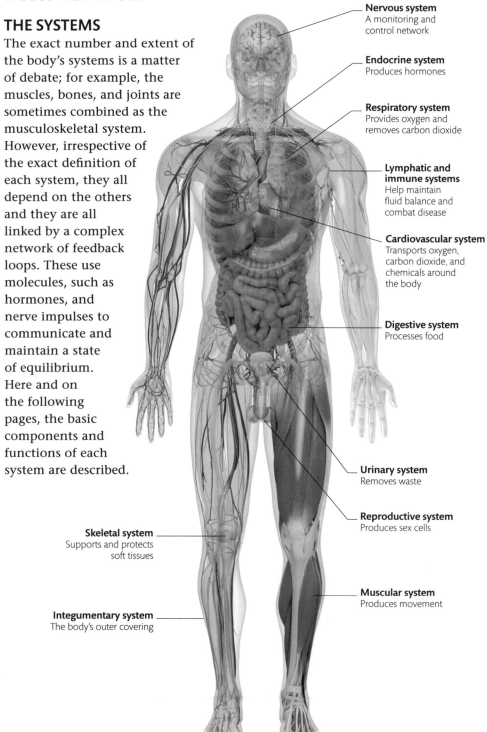

THE SYSTEMS

The exact number and extent of the body's systems is a matter of debate; for example, the muscles, bones, and joints are sometimes combined as the musculoskeletal system. However, irrespective of the exact definition of each system, they all depend on the others and they are all linked by a complex network of feedback loops. These use molecules, such as hormones, and nerve impulses to communicate and maintain a state of equilibrium. Here and on the following pages, the basic components and functions of each system are described.

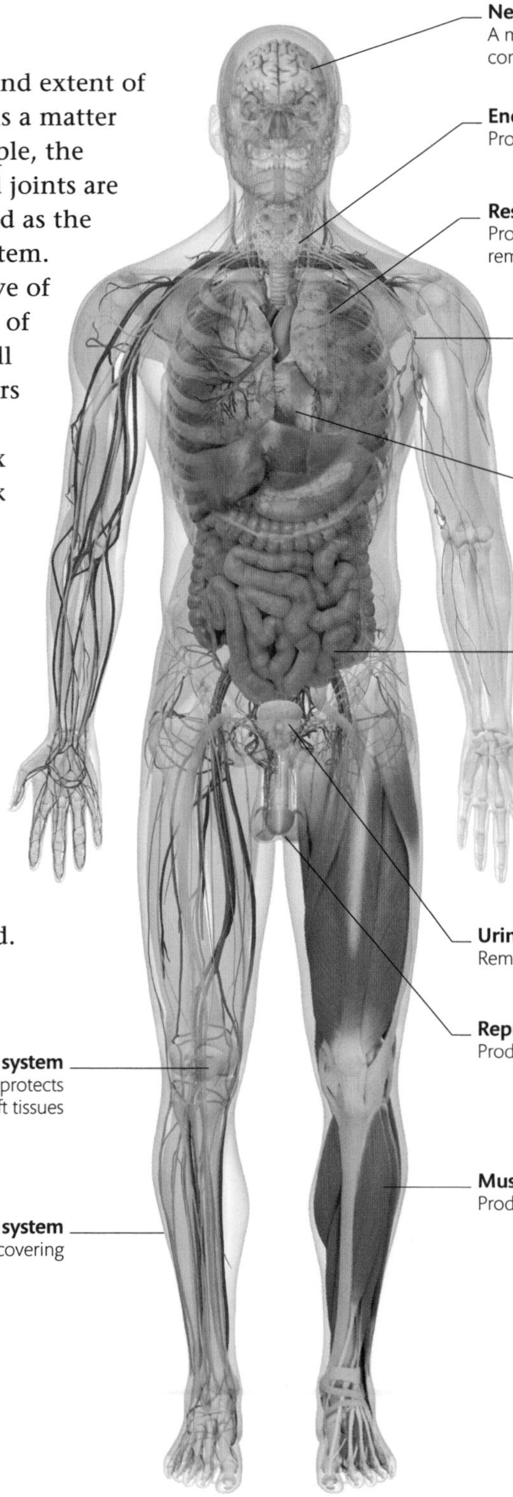
Nervous system
A monitoring and control network
Endocrine system
Produces hormones
Respiratory system
Provides oxygen and removes carbon dioxide
Lymphatic and immune systems
Help maintain fluid balance and combat disease
Cardiovascular system
Transports oxygen, carbon dioxide, and chemicals around the body
Digestive system
Processes food
Urinary system
Removes waste
Reproductive system
Produces sex cells
Skeletal system
Supports and protects soft tissues
Muscular system
Produces movement
Integumentary system
The body's outer covering

SKELETAL SYSTEM

This system uses bones, cartilage, and ligaments to provide the body with structural support and protection. It encases much of the nervous system within a protective skull and vertebrae, and the vital organs of the respiratory and circulatory systems within the rib cage. The skeletal system also supports the circulatory and immune systems by manufacturing red and white blood cells.

Main components

- Skull, vertebral column, ribs, and sternum (axial skeleton)
- Limb bones, shoulder girdle, and pelvic girdle (appendicular skeleton)
- Ligaments

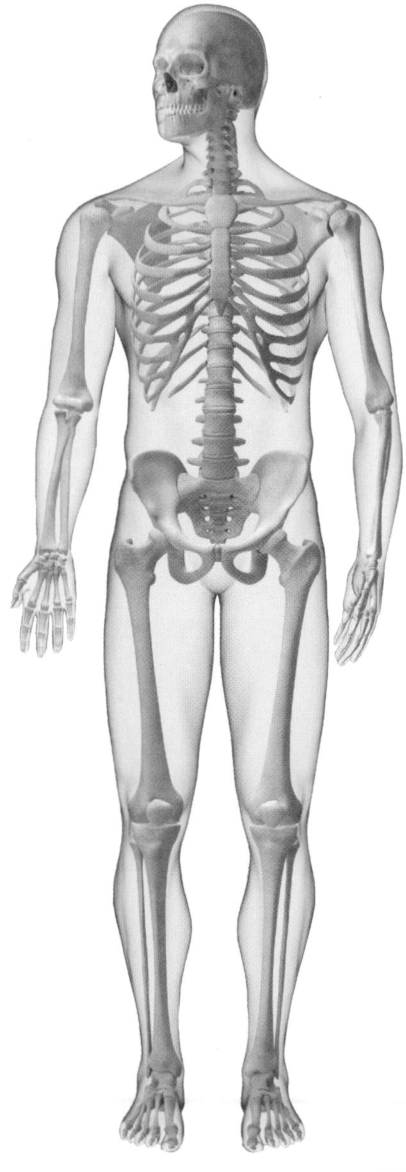

See also pp. 22–51 »

MUSCULAR AND INTEGUMENTARY SYSTEMS

The muscular system is responsible for generating movement, both of the limbs and within other body systems, such as the digestive system and respiratory system. The integumentary system comprises the skin, hair, and nails, which form the body's outer covering. In addition to providing physical protection, the skin also helps to regulate body temperature and, in its layer of fat, acts as an energy store.

Main components
- Skeletal muscles (attached to bones)
- Smooth muscle within organs
- Cardiac muscle of heart
- Tendons
- Skin, hair, and nails

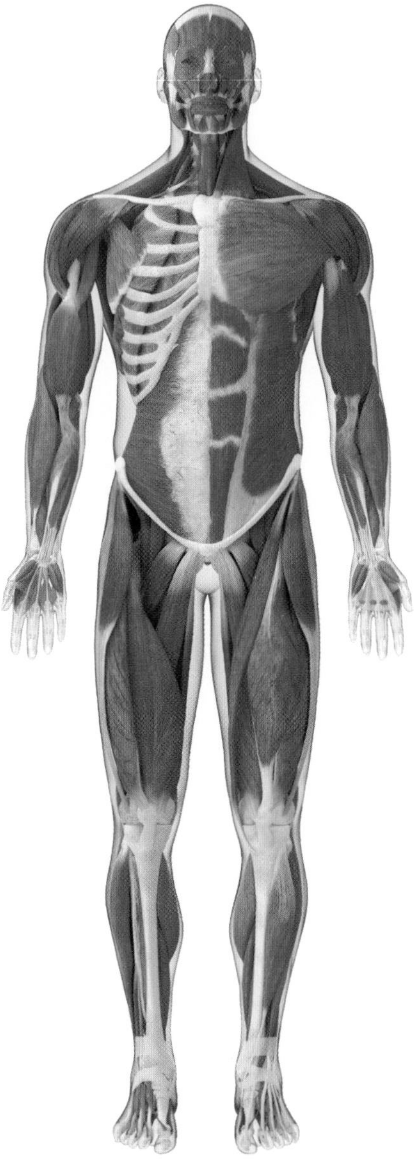

See also pp. 52–79 »

NERVOUS SYSTEM

The brain, spinal cord, and nerves work together to collect, process, and disseminate information from the body's internal and external environments. The nervous system communicates via networks of nerve cells, which connect with other body systems. The brain monitors and controls these systems.

Main components
- Brain
- Spinal cord
- Peripheral nerves
- Sense organs

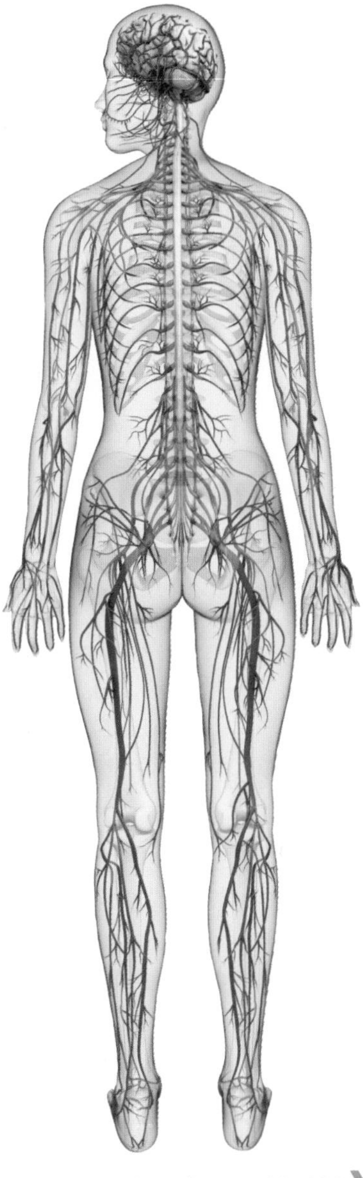

See also pp. 80–113 »

RESPIRATORY SYSTEM

Every cell in the body needs oxygen and must get rid of the waste product of metabolism—carbon dioxide—in order to function. The respiratory system allows this to happen. Air is inhaled into the lungs, where passive exchange of oxygen and carbon dioxide molecules occurs between the air and blood, and then carbon dioxide is exhaled from the lungs.

Main components
- Nasal and other air passages in the skull
- Pharynx
- Trachea
- Lungs
- Airways in lungs (bronchi and bronchioles)
- Diaphragm and other respiratory muscles

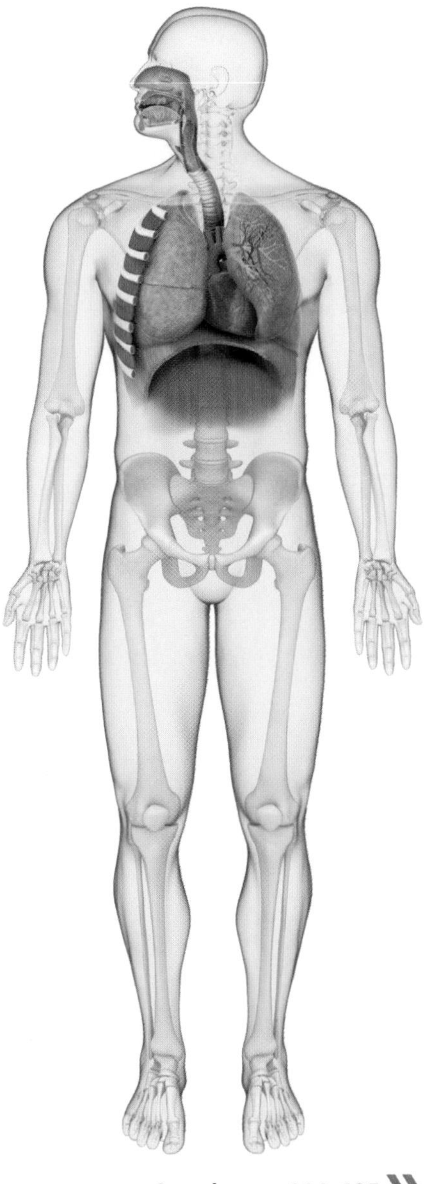

See also pp. 114–125 »

BODY SYSTEMS 2

CARDIOVASCULAR SYSTEM

The cardiovascular system uses blood to carry oxygen from the lungs and nutrients from the digestive system to all the body's cells. It also carries waste products away from the cells. At the center of the cardiovascular system is the heart, which pumps blood through the blood vessels.

Main components
- Heart
- Major blood vessels (arteries and veins)
- Minor blood vessels (arterioles and venules)
- Capillaries
- Blood

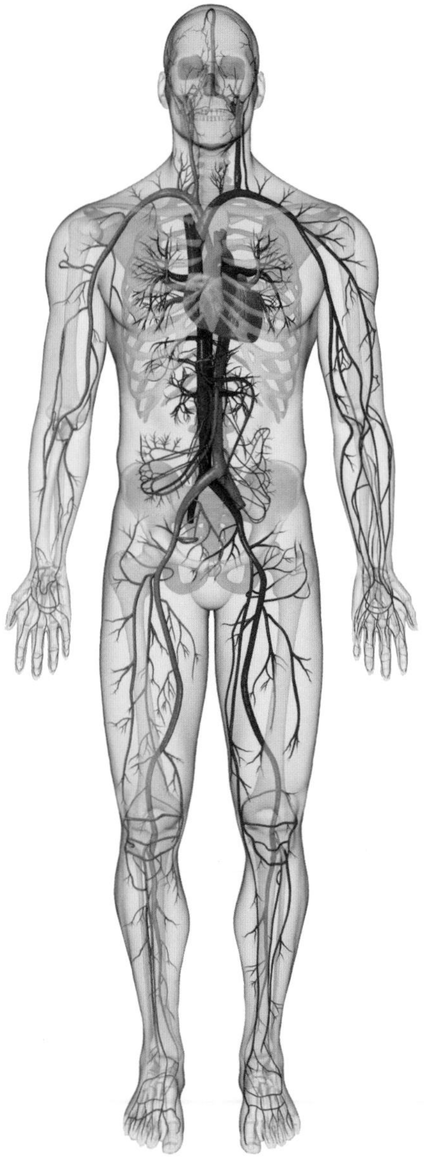

See also pp. 126–153 »

LYMPHATIC AND IMMUNE SYSTEMS

The lymphatic system consists of a network of vessels and nodes that drain fluid from capillaries and return it to the veins. Its main functions are to maintain fluid balance in the cardiovascular system and to distribute immune cells. The immune system consists of physical, cellular, and chemical defenses that provide resistance to threats such as infectious diseases and internal pathological processes such as cancer.

Main components
- Lymph nodes, ducts, and vessels
- Tonsils and adenoid
- Spleen
- Thymus gland
- Lymph fluid
- White blood cells
- Antibodies

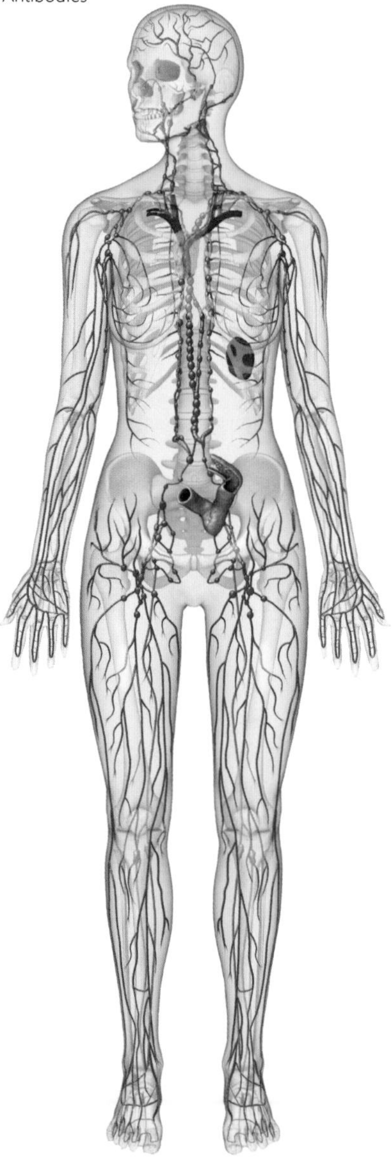

See also pp. 154–167 »

DIGESTIVE SYSTEM

The digestive system—the mouth, pharynx, esophagus, stomach, intestines, and organs such as the liver, gallbladder, and pancreas—breaks down and processes the food we eat so that vital nutrients can be absorbed from the intestines into the circulatory system and transported to the cells.

Main components
- Mouth, pharynx, and esophagus
- Stomach
- Small intestine (duodenum, jejunum, and ileum)
- Large intestine (appendix, colon, and rectum)
- Anus
- Liver, pancreas, and gallbladder

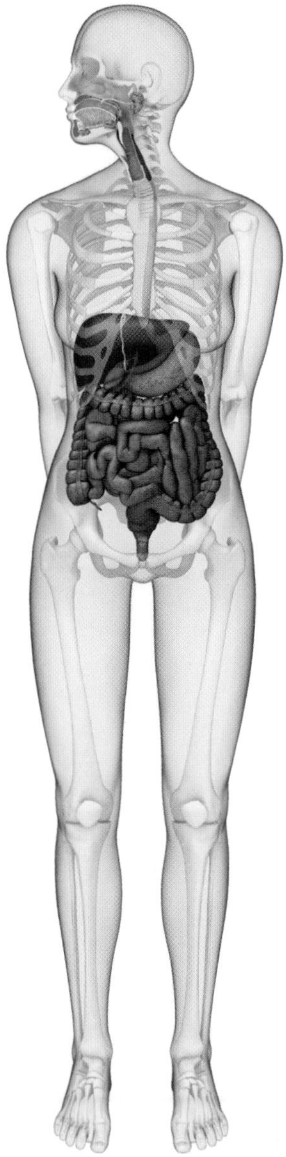

See also pp. 168–185 »

ENDOCRINE SYSTEM

Like the nervous system, the endocrine system communicates messages between the body's systems, enabling them to be monitored and controlled. To do this, it uses hormones, which are usually secreted into the blood from specialized endocrine glands.

Main components
- Pituitary gland
- Hypothalamus
- Thyroid gland
- Pancreas
- Suprarenal (adrenal) glands
- Heart
- Stomach and intestines

Male:
- Testes

Female:
- Ovaries

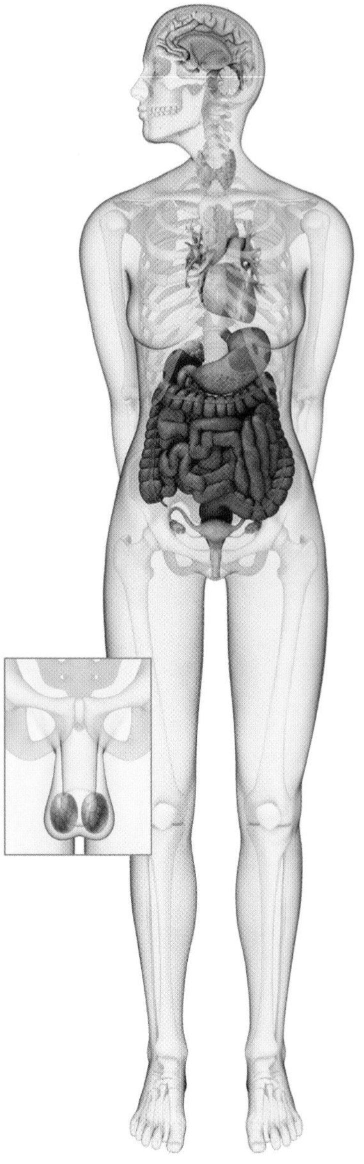

See also pp. 186–199 »

URINARY SYSTEM

This system filters and removes many of the waste products generated by other body systems. It does this by filtering blood through the kidneys and producing urine, which is excreted through the urethra. The kidneys also help to maintain blood pressure by ensuring that the correct amount of water is re-absorbed by the blood.

Main components
- Kidneys
- Ureters
- Bladder
- Urethra

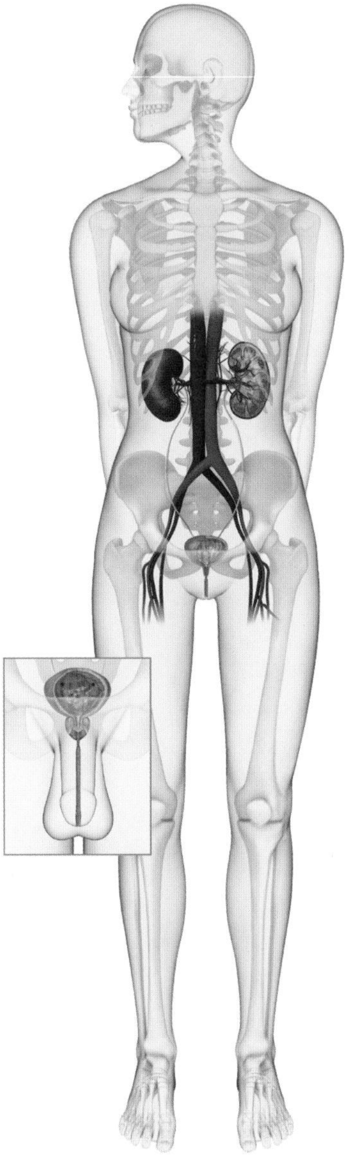

See also pp. 200–209 »

REPRODUCTIVE SYSTEM

The reproductive system is not essential for maintaining life but is essential for propagating it. The anatomy and physiology of the system differ significantly between the sexes. The production of sperm in the male is continual, whereas the production of ripe eggs in the female is cyclical. The ovaries and testes also produce hormones, and therefore also form part of the endocrine system.

Main components

Female:
- Ovaries, Fallopian tubes, and uterus
- Vagina and external genitalia
- Breasts

Male:
- Testes, spermatic ducts, seminal vesicles, urethra, and penis
- Prostate and bulbourethral (Cowper's) glands

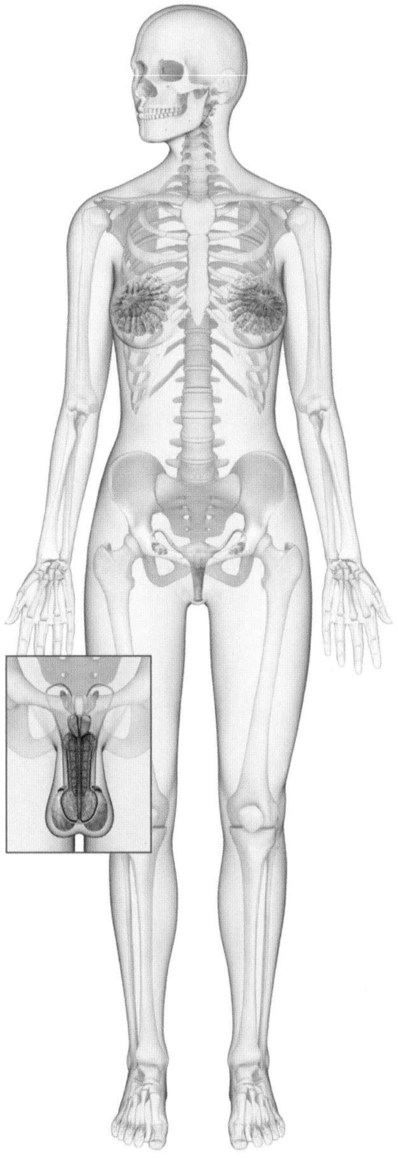

See also pp. 210–245 »

SKELETAL SYSTEM

02

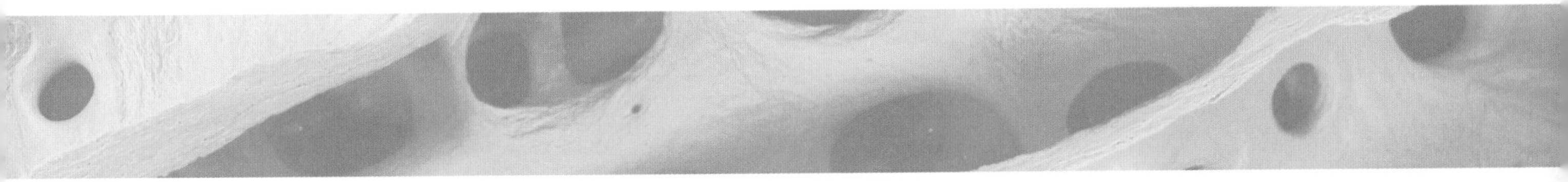

SKELETAL SYSTEM 1

THE SKELETON IS THE INNER FRAMEWORK THAT GIVES THE BODY SHAPE AND SUPPORTS THE WEIGHT OF ALL OTHER TISSUES. ITS INTERLINKING BONES ALLOW COORDINATED MOVEMENT, PROVIDE STABILITY, AND PROTECT SOME OF THE MOST DELICATE AND VITAL PARTS OF THE BODY.

STRUCTURE OF THE SKELETON

The average adult skeleton has 206 bones. There are natural variations: about one individual in 20 has an extra rib, for example. The skeleton also differs in shape between the sexes. In males, bones tend to be heavier and more robust. The difference is most obvious in the pelvis, which in females is usually wider and shallower than in males.

There are four main bone types: long, short, flat, and irregular. Long bones occur in the limbs and consist of a central shaft between two epiphyses. Short bones, such as the calcaneus in the heel, are roughly spherical or cuboidal in shape. Flat bones are constructed in thin sheets and include the bones of the vault of the skull. Irregular bones, such as the tiny ossicles of the middle ear and the sphenoid bone in the skull, have highly varied shapes. A further type, sesamoid bones, are usually small and rounded, and occur within tendons. They include the patella.

The bones are arranged symmetrically on either side of the body and are grouped into two main divisions, the axial and appendicular skeletons, which have different roles.

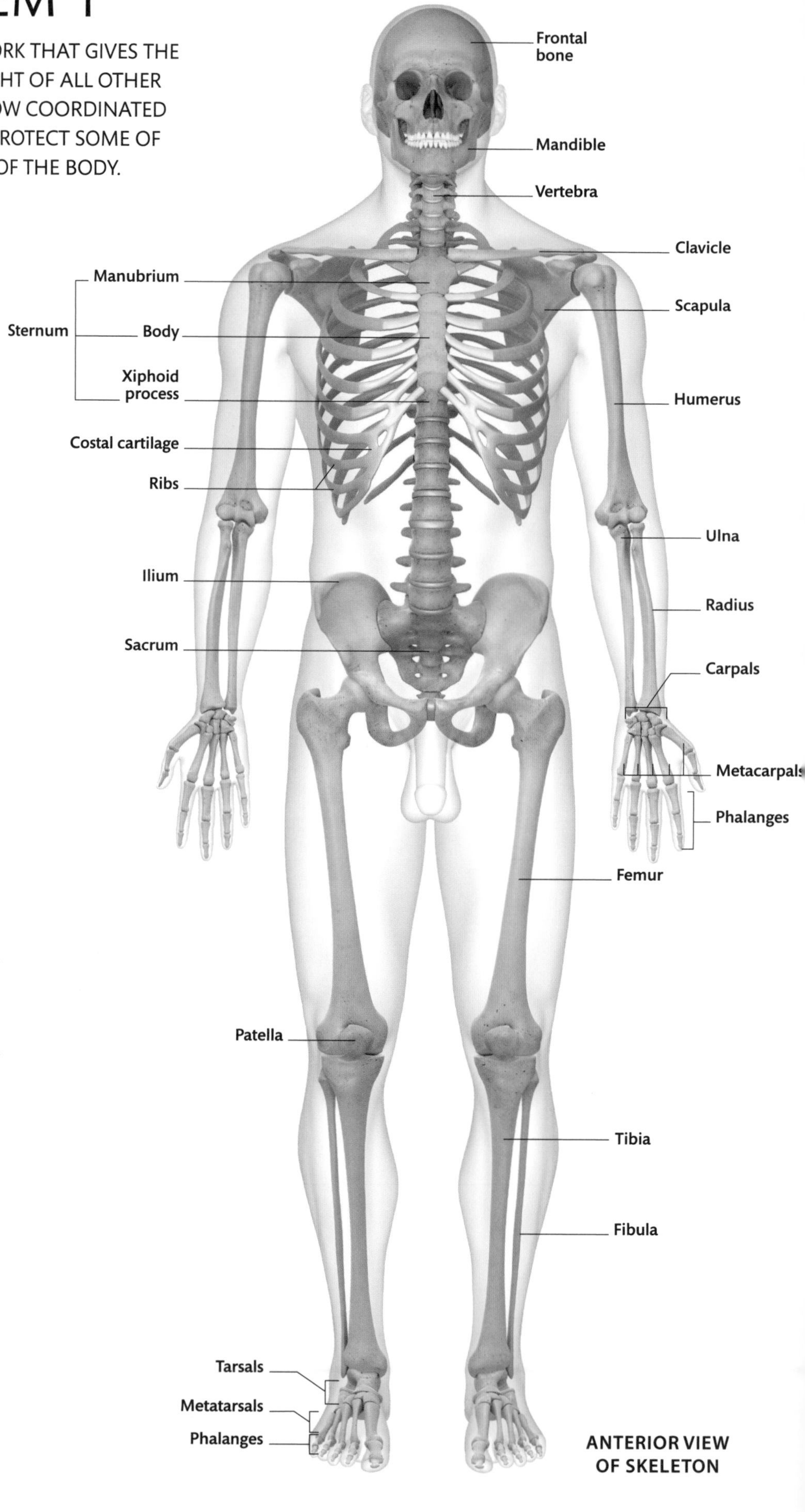

ANTERIOR VIEW OF SKELETON

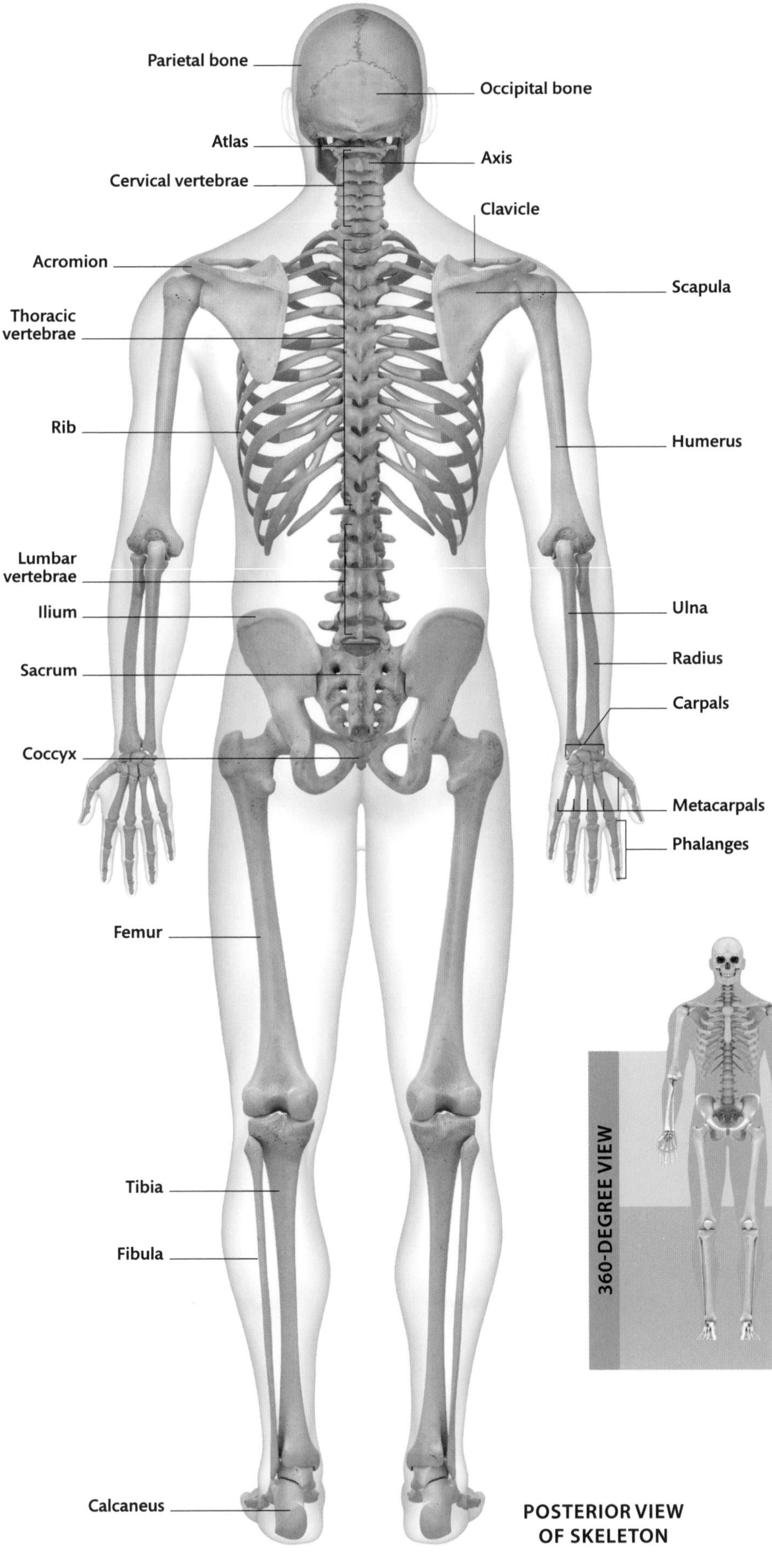

POSTERIOR VIEW
OF SKELETON

AXIAL SKELETON

This part of the skeleton comprises the bones of the central part of the body. The axial skeleton includes the skull, vertebral column, rib cage, and sternum—80 bones in all. Such bones have a predominantly protective role, surrounding some of the body's most vulnerable parts: the brain, spinal cord, heart, and lungs.

APPENDICULAR SKELETON

This part of the skeleton consists of the bones of the upper and lower limbs and the pectoral and pelvic girdles—some 126 bones. The pectoral girdles, each made up of a scapula and a clavicle, attach the upper arm bones to the axial skeleton. The pelvic girdle articulates with the sacrum and the femurs on either side.

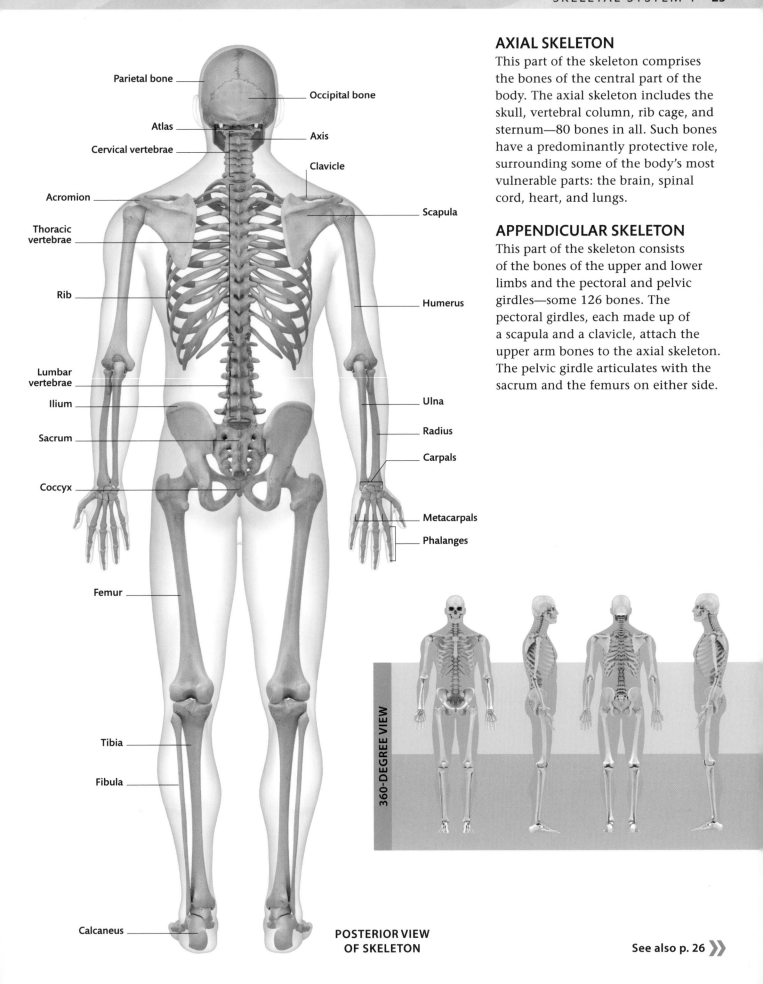

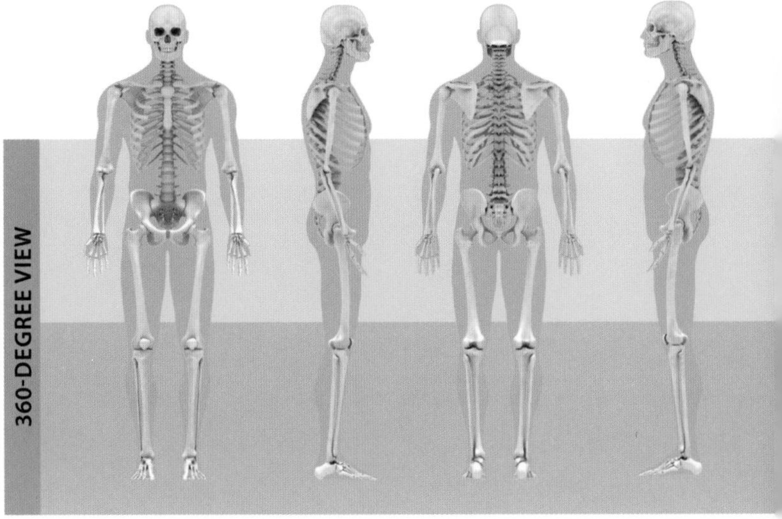

See also p. 26 »

SKELETAL SYSTEM 2

MOST OF THE SKELETON DEVELOPS FIRST AS CARTILAGE, NEARLY ALL OF WHICH HAS BEEN TRANSFORMED INTO BONE BY ADULTHOOD. BONE IS NOT AN INERT MATERIAL BUT CONSTANTLY BREAKS DOWN AND REMODELS ITSELF IN A LIFELONG PROCESS.

BONE STRUCTURE

The main constituents of bone are specialized cells, protein fibers (chiefly collagen), mineral salts, and water. Bone cells are of three main types: osteoblasts, which calcify new bone as it forms; osteoclasts, which break down and absorb unwanted bone tissue; and osteocytes, which maintain healthy bone structure.

Each bone, regardless of shape or size, has an outer layer of cortical (compact) bone, which is a dense, heavy tissue, and an inner layer of cancellous (spongy) bone. Cortical bone is made up of osteons, rod-shaped units consisting of concentric layers of hard tissue. Tiny spaces between the layers house bone cells and nutrients. Cancellous bone comprises an open network of interconnecting struts, or trabeculae, the gaps of which are filled with marrow. In a long bone, a central cavity filled with marrow (the medullary canal) runs along the length of the shaft.

At birth, all bones contain red marrow, which produces blood cells. By adulthood, the red marrow in long bones has become yellow marrow, which is mostly fat. Blood-producing red marrow persists in the skull, vertebrae, ribs, and pelvis.

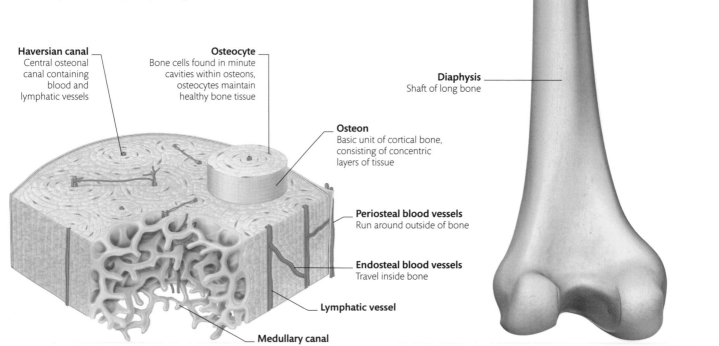

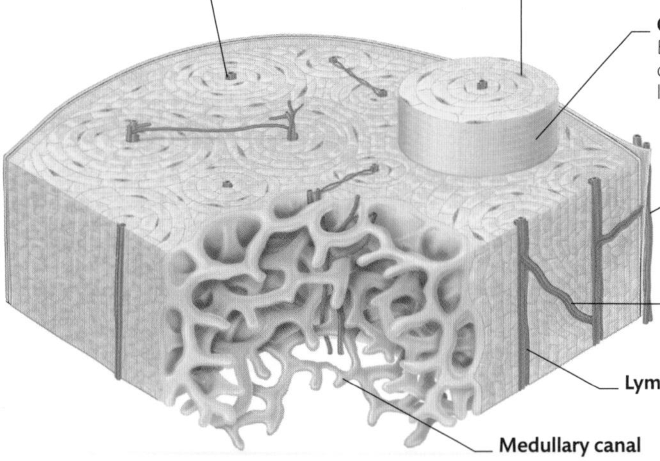
Haversian canal
Central osteonal canal containing blood and lymphatic vessels
Osteocyte
Bone cells found in minute cavities within osteons, osteocytes maintain healthy bone tissue
Osteon
Basic unit of cortical bone, consisting of concentric layers of tissue
Periosteal blood vessels
Run around outside of bone
Endosteal blood vessels
Travel inside bone
Lymphatic vessel
Medullary canal

STRUCTURE OF CORTICAL BONE
Cortical bone is composed of osteons, which are concentric cylinders of bone tissue, each around $^{1}/_{250}$–$^{4}/_{250}$ in (0.1–0.4 mm) in diameter, with a central (Haversian) canal. Blood vessels in the osteons connect to blood vessels within the medullary canal of the bone, as well as to vessels in the outer layer of periosteum.

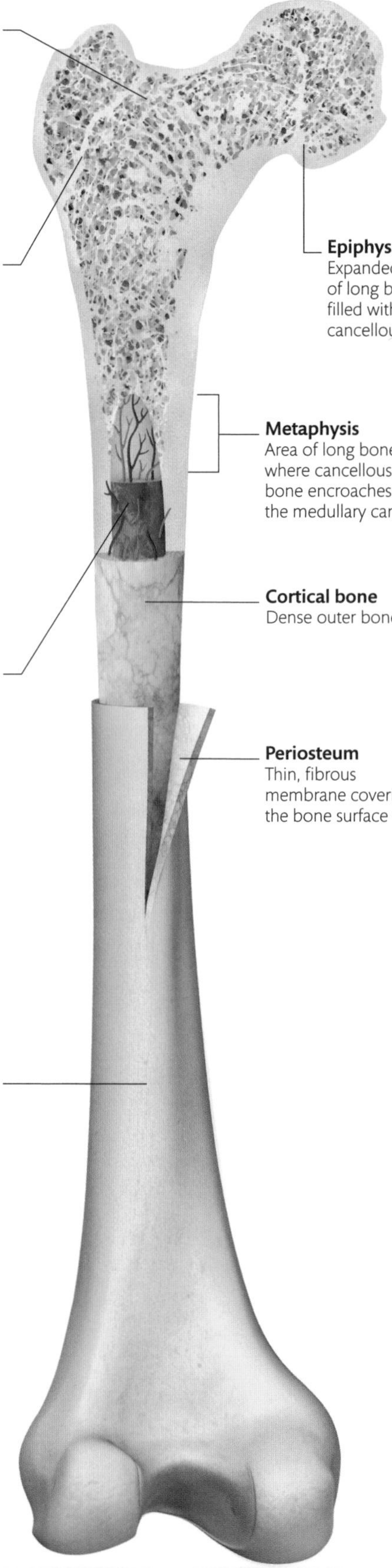
Cancellous bone
Latticework structure of trabeculae filled with bone marrow
Line of fusion of growth plate
A cartilage plate allows long bones to grow rapidly in length during childhood. This growth plate fuses by adulthood, but the line of fusion may remain evident for several years
Epiphysis
Expanded head of long bone, filled with cancellous bone
Metaphysis
Area of long bone, where cancellous bone encroaches on the medullary canal
Cortical bone
Dense outer bone layer
Medullary canal
Contains blood-cell-forming red marrow at birth, which becomes yellow marrow by adulthood and loses this capacity
Periosteum
Thin, fibrous membrane covering the bone surface
Diaphysis
Shaft of long bone

INSIDE A LONG BONE
A typical long bone such as the femur has a central cavity, the medullary canal, filled with blood vessels and bone marrow. This cavity is surrounded by a layer of cancellous bone, which is surrounded by a layer of tough cortical bone. A thin membrane, the periosteum, covers the bone's outer surface.

JOINTS

A joint, or an articulation, is the site at which two bones meet. Joints are classified according to their structure and by the types of movement they allow. The body has more than 300 different joints.

SYNOVIAL JOINTS

The body's most numerous, versatile, and freely moving joints are known as synovial joints. They are enclosed by a protective outer covering of connective tissue—the joint capsule. The delicate inner lining of this capsule, the synovial membrane, continually secretes viscous synovial fluid that keeps the joint well lubricated, allowing the joint surfaces in contact to slide with minimal friction and wear. The bone ends within a synovial joint are covered with hyaline cartilage, a smooth, shock-absorbing tissue. There are around 250 synovial joints in the body. A synovial joint's range of movement is determined by the shape and fit of its articular surfaces. Ligaments and muscles surrounding the joint provide stability.

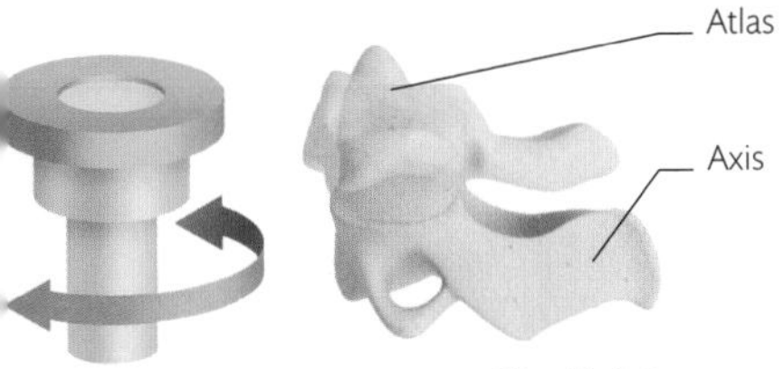

Pivot joint
A peglike projection from one bone turns in the ring-shaped socket of another or, conversely, the ring turns around the peg. The pivot joint between the top (atlas) and second (axis) cervical vertebrae enables the skull to rotate on the spinal axis.

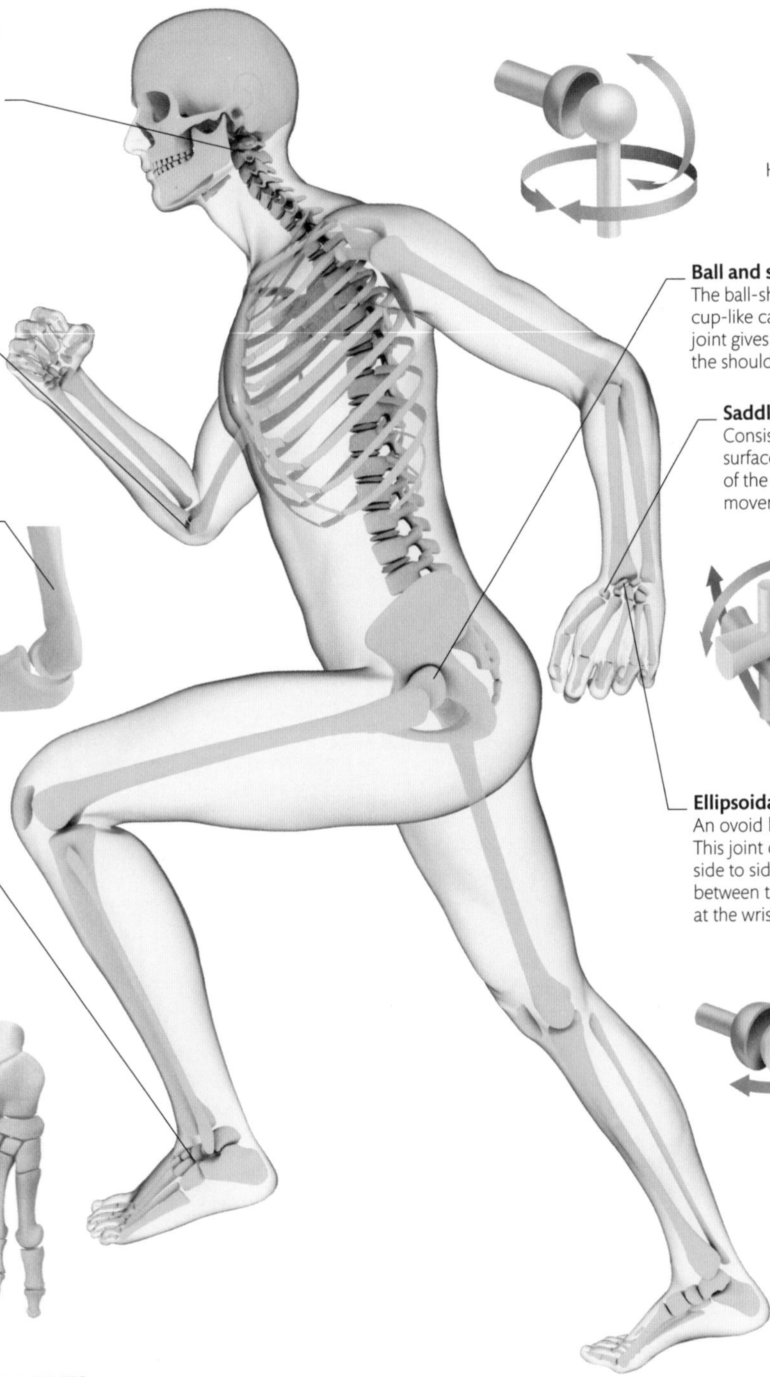

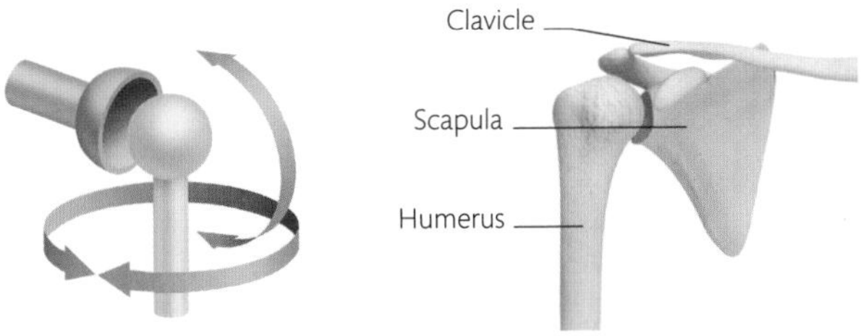

Ball and socket joint
The ball-shaped head of one bone fits into the cup-like cavity of another bone. This type of joint gives the widest range of movement—the shoulder and hip are examples.

Hinge joint
The convex surface of one bone fits into the concave surface of another to allow to-and-fro movement, mainly in one plane. The elbow is a modified hinge joint that permits limited rotation of the arm bones.

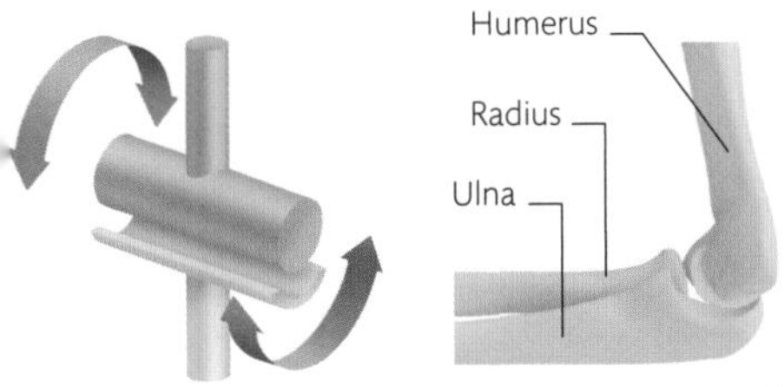

Saddle joint
Consisting of two U-shaped articular surfaces, and found only at the base of the thumb, this joint permits movement in two planes.

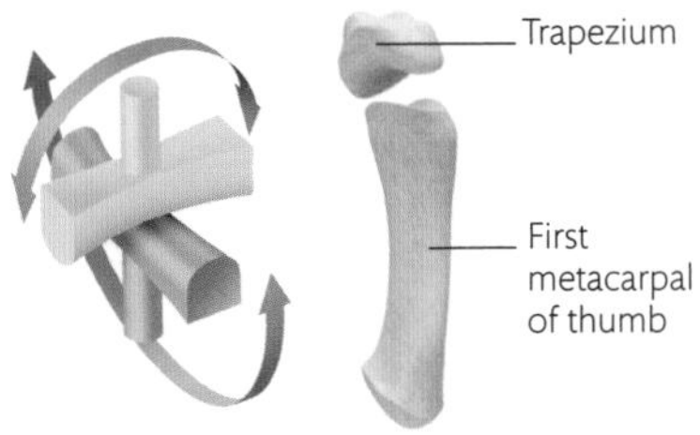

Ellipsoidal joint
An ovoid bone fits in an ellipsoidal cavity. This joint can be flexed and moved from side to side, but rotation is limited. The joint between the radius and the scaphoid bone at the wrist is an example.

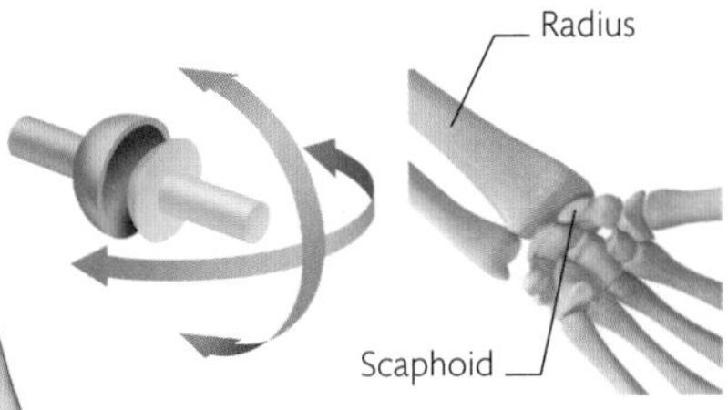

Gliding joint
The bone joint surfaces are almost flat and slide over one another. Some joints between the tarsals of the ankle and between the carpals of the wrist move in this way.

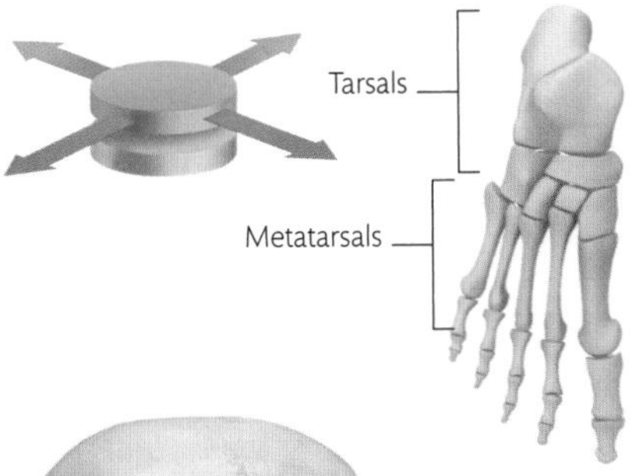

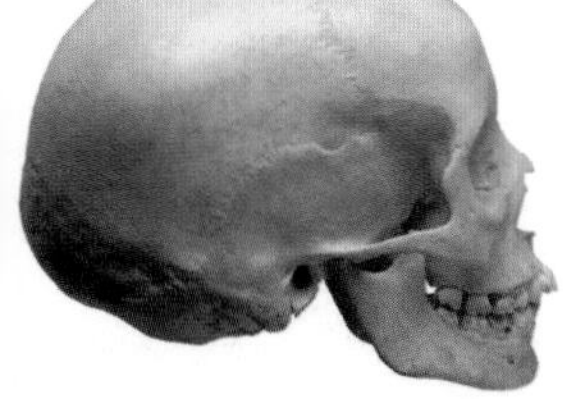

FIXED JOINTS

In infancy, the joints of the separate bones of the vault of the skull are loosely attached to allow for expansion of the rapidly growing brain. Once growth is complete, these joints fuse.

See also pp. 24, 36, 38, 40, 42, 44, 46, 48, 50 »

HEAD AND NECK

THE MAIN SKELETAL ELEMENTS OF THE HEAD AND NECK ARE THE SKULL AND CERVICAL SPINE. THE SKULL CONSISTS OF THE CRANIUM AND MANDIBLE. THE CERVICAL SPINE CONSISTS OF SEVEN CERVICAL VERTEBRAE.

STRUCTURE

The cranium is made up of more than 20 bones, many of which meet at joints called sutures. The mandible in a newborn is in two halves, joined by a fibrous joint; this joint fuses during early infancy so that the mandible becomes a single bone. The mandible articulates with the cranium at the temporomandibular joint. The top two bones of the cervical spine have specific names. The first cervical vertebra is called the atlas. The second cervical vertebra is the axis. The hyoid bone provides an anchor for muscles that form the tongue and floor of the mouth, as well as muscles that attach to the larynx and pharynx.

ANTERIOR VIEW

Color and/or label the structures indicated on the diagram using the key below.

1. Frontal bone
2. Supraorbital foramen
3. Zygomatic process of frontal bone
4. Superior orbital fissure
5. Inferior orbital fissure
6. Infraorbital foramen
7. Nasal crest
8. Maxilla
9. Cervical vertebra
10. First rib
11. Mental protuberance
12. Mental foramen
13. Mandible
14. Inferior nasal concha
15. Piriform aperture
16. Frontal process of maxilla
17. Orbit
18. Nasal bone

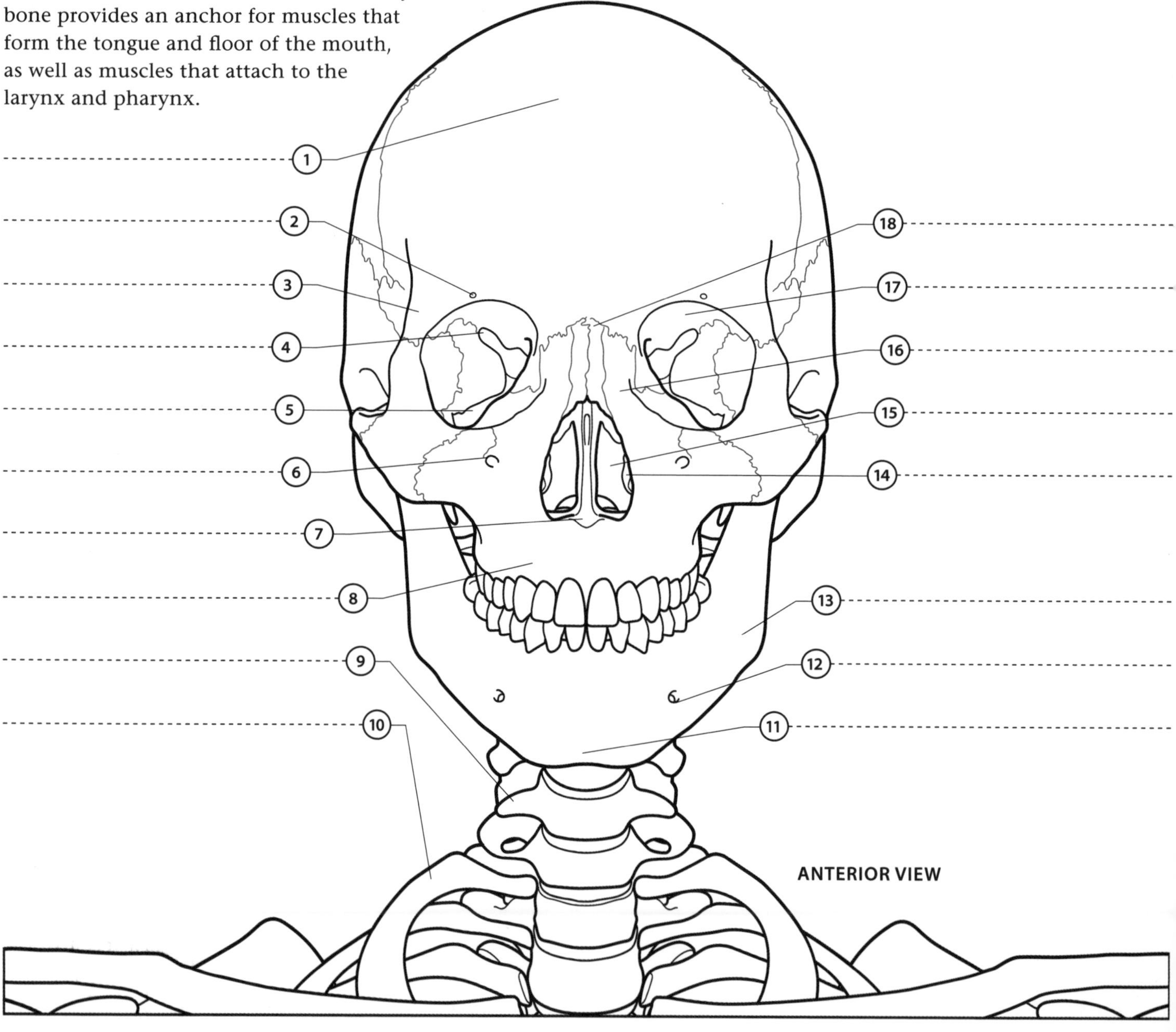

LATERAL VIEW

Color and/or label the structures indicated on the diagram using the key below.

1. Zygomatic arch
2. Condyle of mandible
3. Parietal bone
4. Squamosal suture
5. Lambdoid suture
6. Occipital bone
7. Occipitomastoid suture
8. Temporal bone
9. Styloid process
10. Mastoid process
11. Hyoid bone
12. Ramus of mandible
13. Body of mandible
14. Mental foramen
15. Alveolar process of mandible
16. Maxilla
17. Zygomatic bone
18. Nasal bone
19. Lacrimal bone
20. Coronoid process of mandible
21. Greater wing of sphenoid bone
22. Pterion
23. Frontal bone
24. Coronal suture

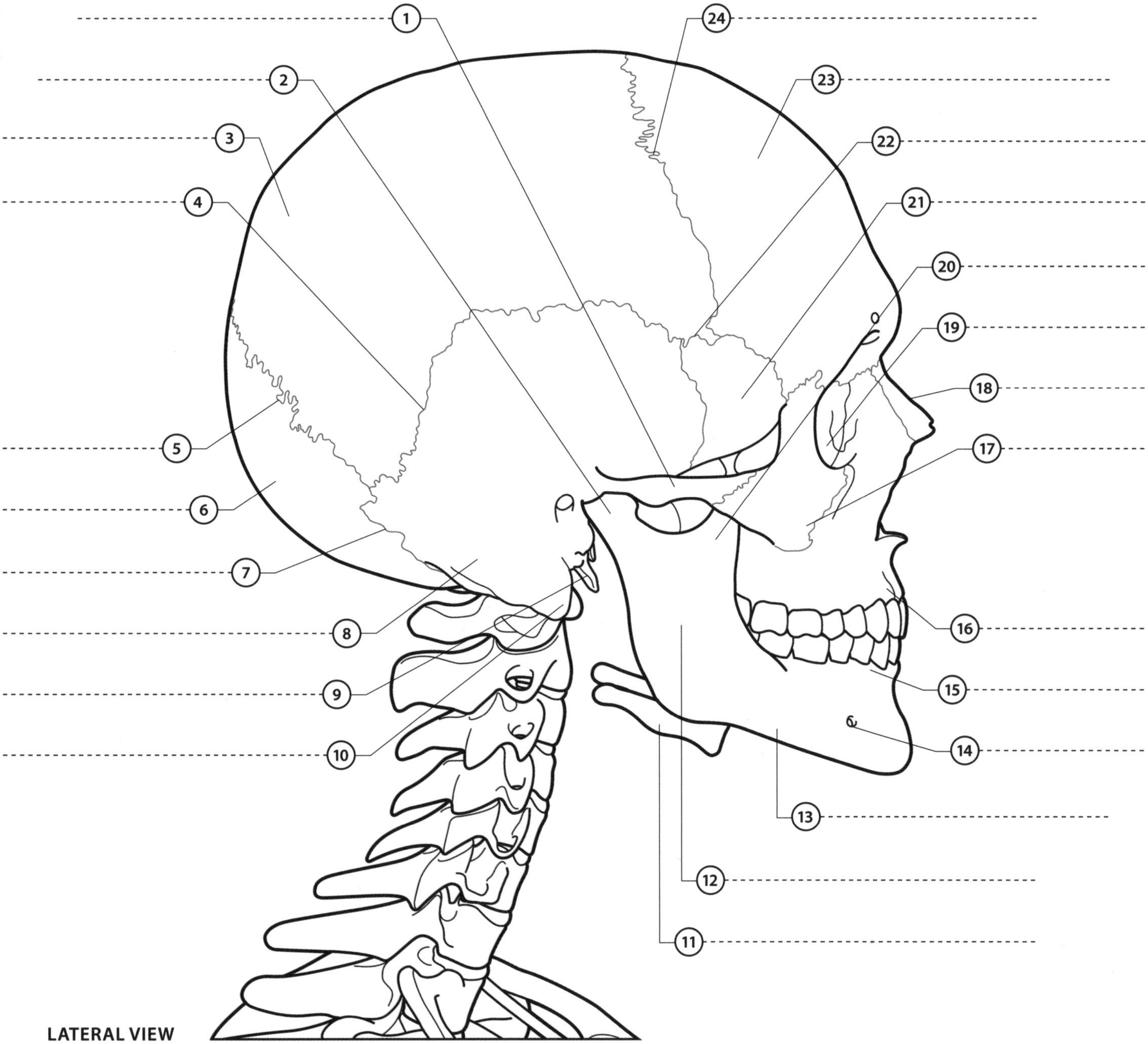

LATERAL VIEW

See also pp. 30, 32, 58, 60, 92, 118, 134, 136, 160, 174 »

SKULL 1

THE SKULL IS MADE UP OF 29 BONES (INCLUDING THE HYOID BONE AND SIX OSSICLES), MOST OF WHICH ARE FUSED TOGETHER.

SAGITTAL SECTION THROUGH SKULL

The skull may be divided into the bones of the cranial vault, facial skeleton, and skull base. The trabecular bone (or diploe) of the skull contains red marrow. Some skull bones also contain air-filled spaces, chiefly around the nasal cavity, such as the sphenoidal and maxillary sinuses.

SAGITTAL SECTION THROUGH SKULL

Color and/or label the structures indicated on the diagram using the key below.

1. Pituitary fossa
2. Sphenoidal sinus
3. Frontal bone
4. Frontal sinus
5. Nasal bone
6. Superior nasal concha
7. Anterior nasal crest
8. Middle nasal concha
9. Inferior nasal concha
10. Palatine bone
11. Medial pterygoid process
12. Styloid process
13. Hypoglossal canal
14. Internal acoustic meatus
15. Occipital bone
16. Lambdoid suture
17. Squamosal suture
18. Parietal bone
19. Temporal bone

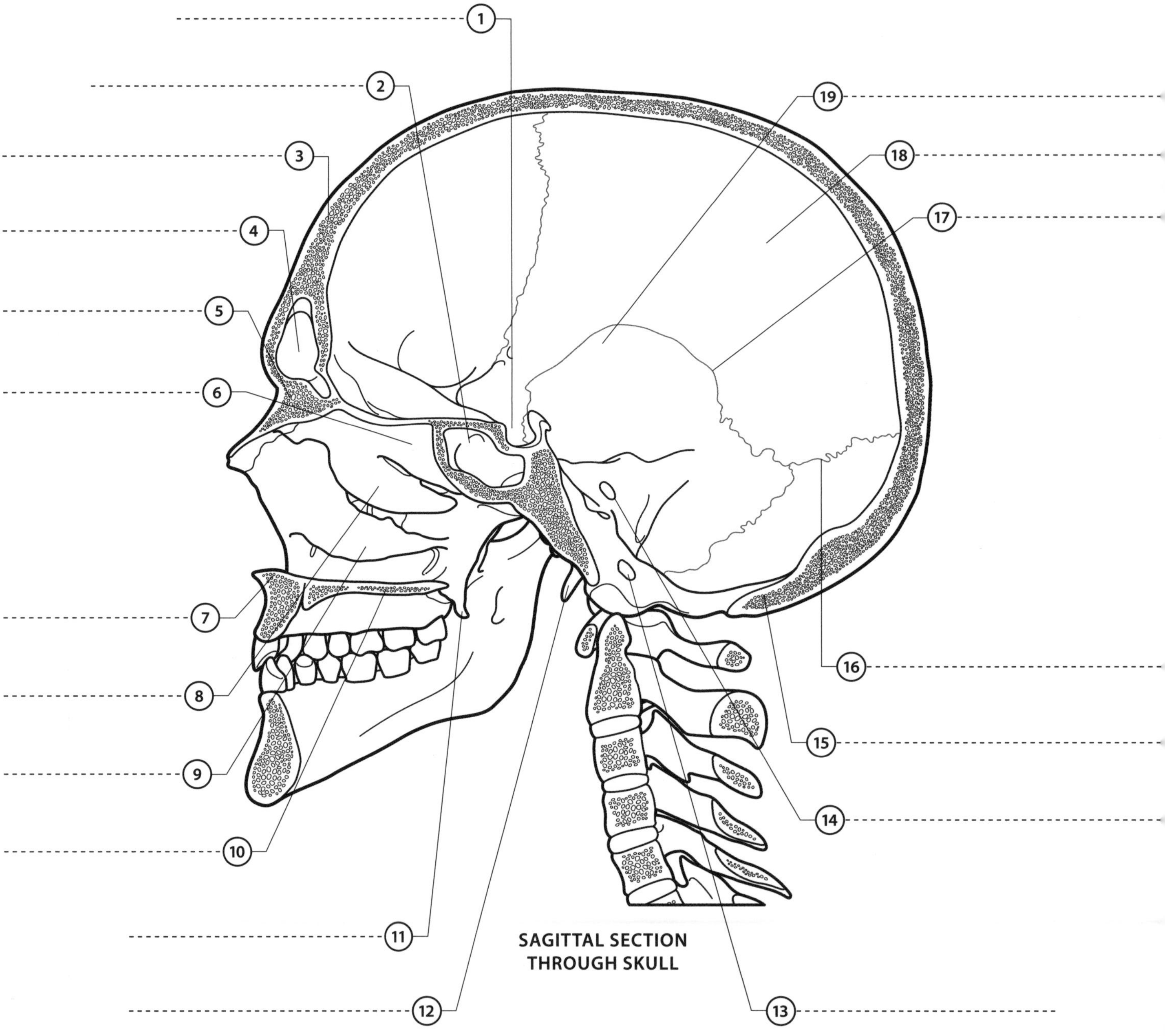

SAGITTAL SECTION THROUGH SKULL

BONES OF THE SKULL

This "exploded" view shows the individual skull bones (not to scale). The butterfly-shaped sphenoid bone forms part of the skull's base, the orbits, and the side walls of the skull; it articulates with several of the other skull bones. The temporal bones also form part of the skull's base and side walls. The dense petrous parts of the temporal bones contain structures in the external, middle, and inner ear, including the three ossicles (malleus, incus, and stapes).

SKULL BONES

Color and/or label the structures indicated on the diagram using the key below.

1. Occipital bone
2. Frontal bone
3. Ossicles
4. Ethmoid bone
5. Nasal bone
6. Mandible
7. Maxilla
8. Inferior nasal concha
9. Palatine bone
10. Vomer
11. Lacrimal bone
12. Zygomatic bone
13. Sphenoid bone
14. Temporal bone
15. Parietal bone

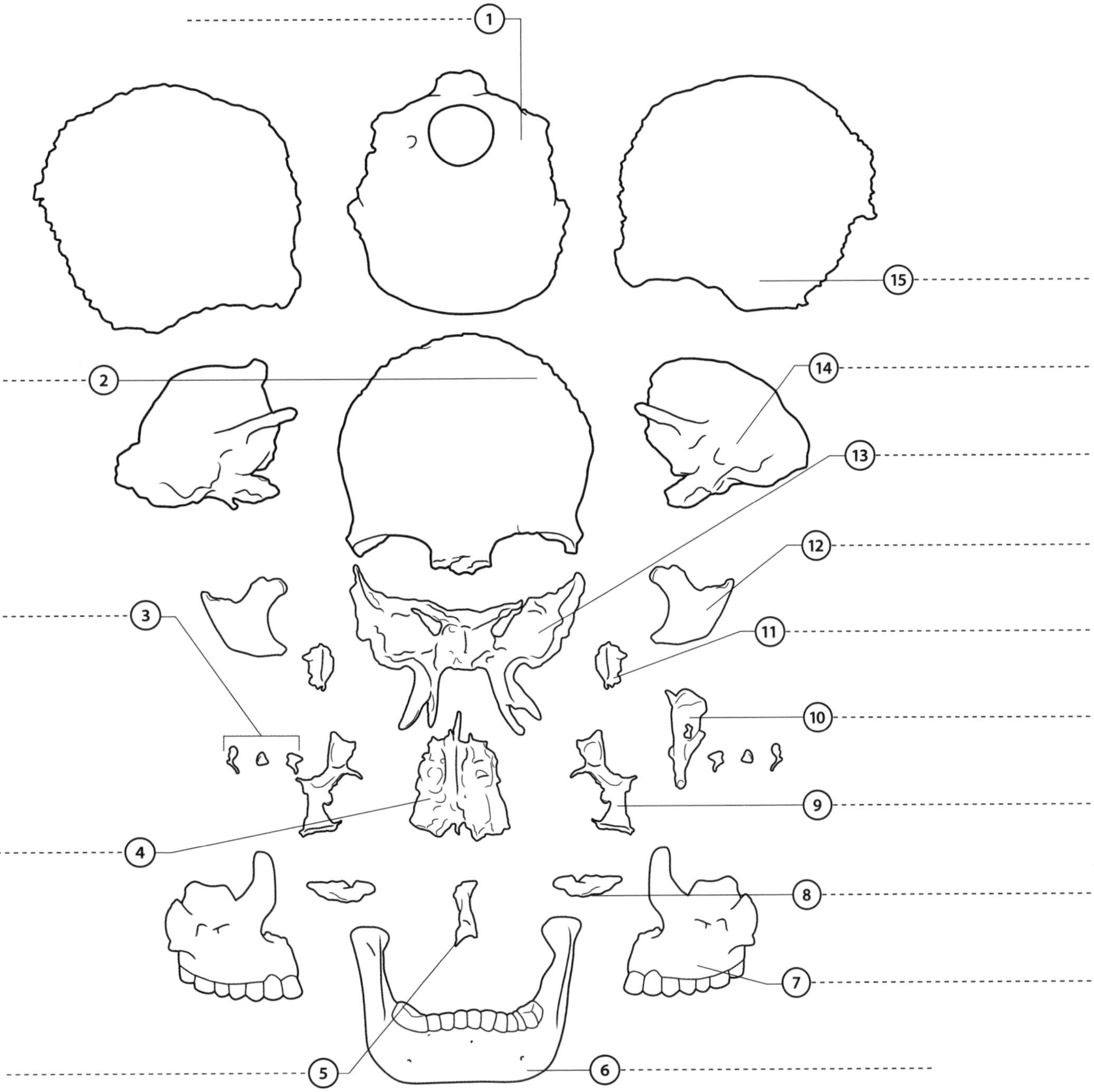

SKULL BONES

See also pp. 28, 32

SKULL 2

THESE IMAGES OF THE SKULL SHOW TWO VIEWS OF ITS BASE: AN INTERNAL VIEW FROM ABOVE AND AN EXTERNAL VIEW FROM BELOW. IN BOTH CASES, THE ANTERIOR OF THE SKULL IS AT THE BOTTOM.

BASE OF SKULL

The most notable features of the base of the skull are the holes in it. The largest (foramen magnum) is where the brain stem emerges to become part of the spinal cord. Most of the smaller holes are paired and provide a passageway for nerves from the brain (notably the cranial nerves) and blood vessels to and from the brain. The upper teeth in their sockets in the maxilla, and the hard palate, can be seen in the external view anteriorly.

INTERNAL SURFACE OF BASE OF SKULL

Color and/or label the structures indicated on the diagram using the key below.

1. Foramen magnum
2. Hypoglossal canal
3. Basiocciput
4. Foramen spinosum
5. Optic canal
6. Cribriform plate of ethmoid
7. Foramen cecum
8. Crista galli
9. Orbital part of frontal bone
10. Foramen rotundum
11. Pituitary fossa
12. Foramen ovale
13. Foramen lacerum
14. Jugular foramen
15. Internal occipital protuberance

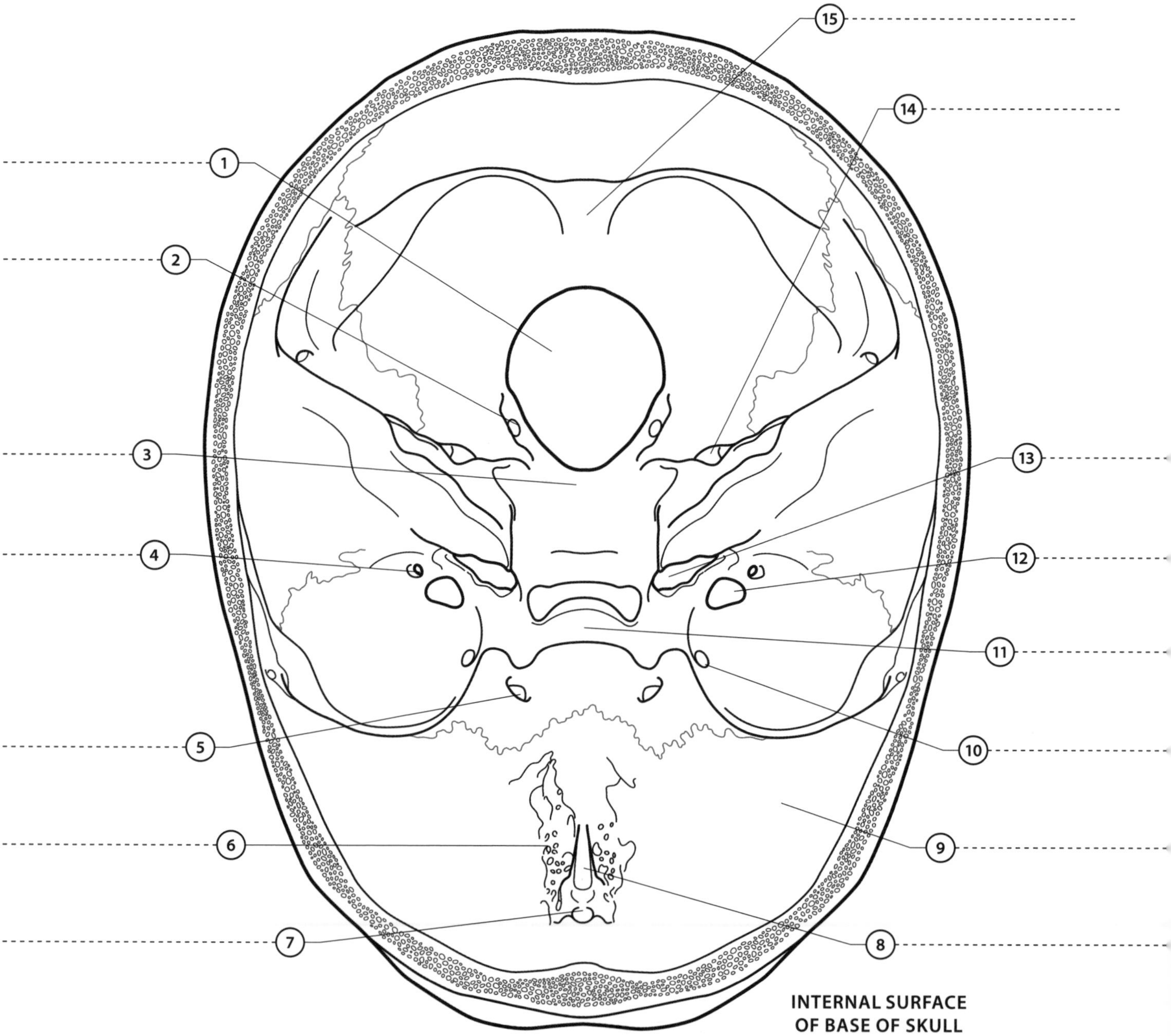

INTERNAL SURFACE OF BASE OF SKULL

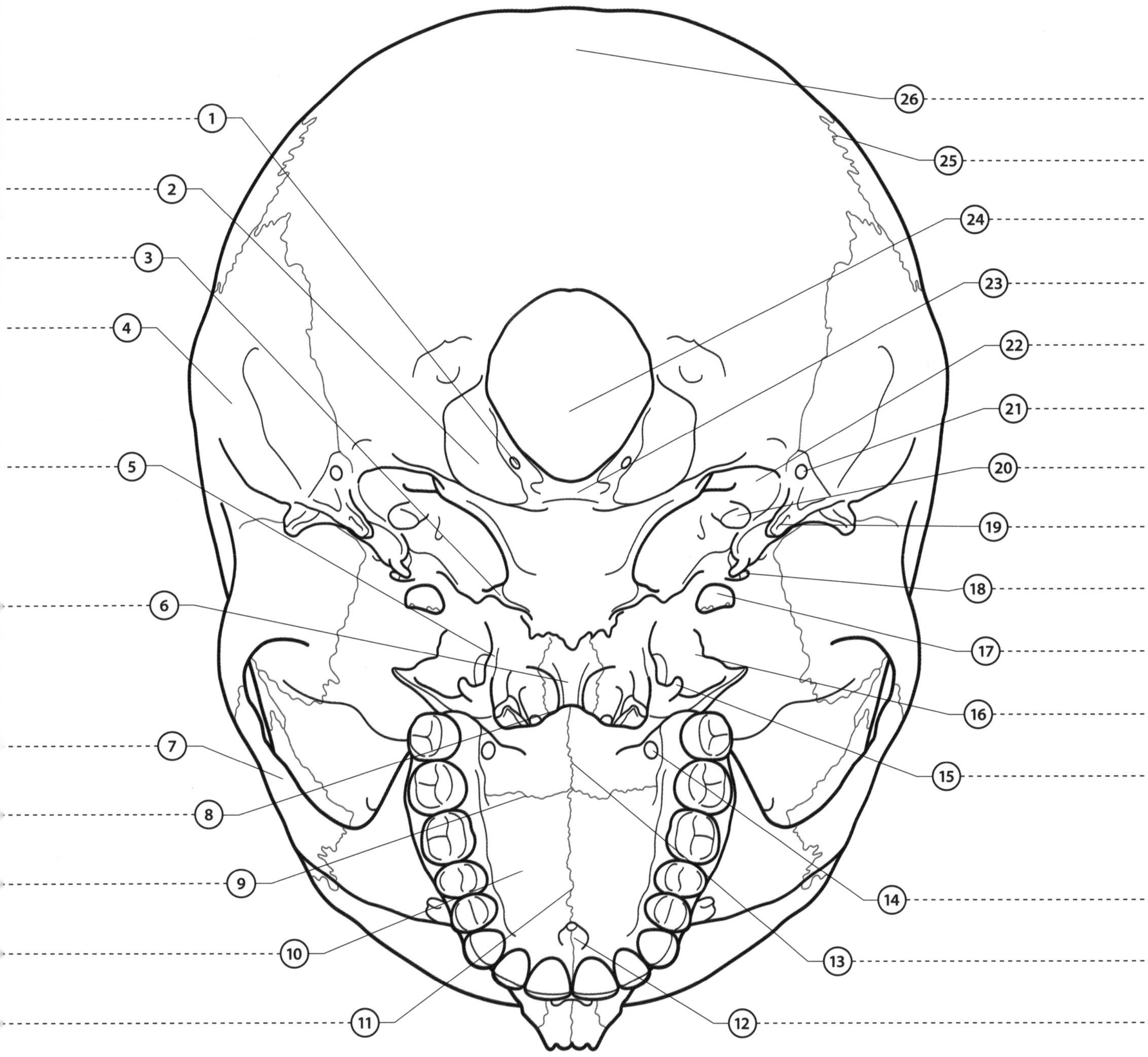

UNDERSIDE OF SKULL

UNDERSIDE OF SKULL

Color and/or label the structures indicated on the diagram using the key below.

1. Hypoglossal canal
2. Occipital condyle
3. Foramen lacerum
4. Mastoid process
5. Medial pterygoid plate
6. Vomer
7. Zygomatic bone
8. Posterior nasal spine
9. Palatomaxillary suture
10. Hard palate
11. Intermaxillary suture
12. Incisive fossa
13. Interpalatine suture
14. Greater palatine foramen
15. Pterygoid hamulus
16. Lateral pterygoid plate
17. Foramen ovale
18. Foramen spinosum
19. Styloid process
20. Carotid canal
21. Stylomastoid foramen
22. Jugular foramen
23. Pharyngeal tubercle
24. Foramen magnum
25. Lambdoid suture
26. Occipital bone

See also pp. 28, 30, 92, 94, 96 »

SPINE

ALSO CALLED THE SPINAL OR VERTEBRAL COLUMN, THE SPINE RUNS FROM THE BASE OF THE SKULL TO THE COCCYX.

STRUCTURE

The entire vertebral column is about 28 in (70 cm) long in men and 24 in (60 cm) in women. About a quarter of this length is made up by the cartilaginous intervertebral disks between the bones. The number of vertebrae varies from 32 to 35, because of variation in the number of small vertebrae that make up the coccyx. Although there is a general pattern for a vertebra—most have a body, a neural arch, and spinous and transverse processes—there are specific features that differentiate the vertebrae of each section of the spine.

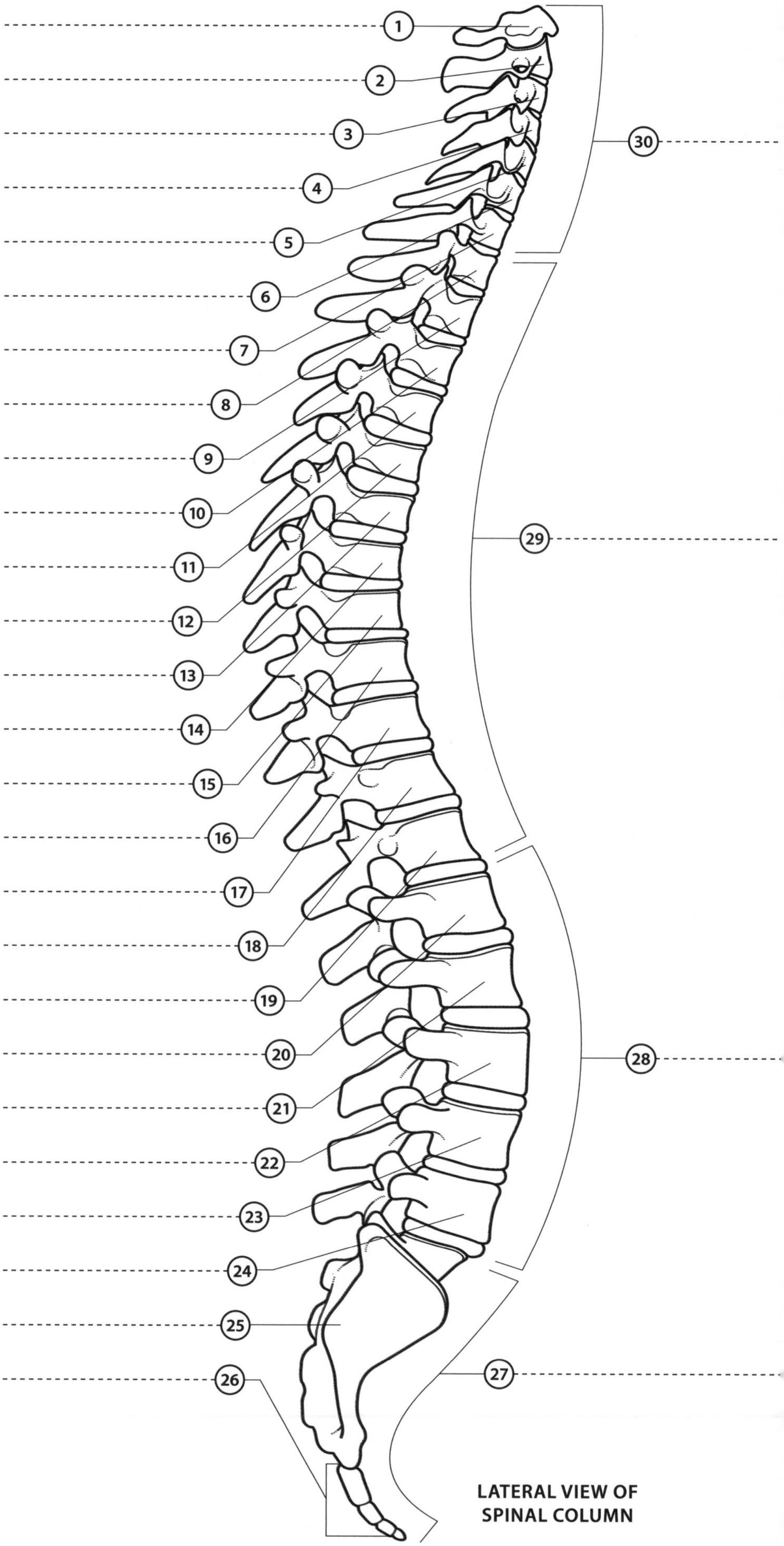

LATERAL VIEW OF SPINAL COLUMN

LATERAL VIEW OF SPINAL COLUMN

Color and/or label the structures indicated on the diagram using the key below.

1. Atlas (C1)
2. Axis (C2)
3. C3
4. C4
5. C5
6. C6
7. C7
8. T1
9. T2
10. T3
11. T4
12. T5
13. T6
14. T7
15. T8
16. T9
17. T10
18. T11
19. T12
20. L1
21. L2
22. L3
23. L4
24. L5
25. Sacrum
26. Coccyx
27. Sacral curvature
28. Lumbar curvature
29. Thoracic curvature
30. Cervical curvature

VERTEBRAE

The cervical spine consists of seven vertebrae. The atlas (first cervical vertebra), which articulates with the base of the skull, has no body but has an anterior arch that forms a joint with the dens (an upward-projecting process, also called the odontoid peg) of the axis (second cervical vertebra). The other five cervical vertebrae have a more obvious body and a vertebral foramen that is large relative to the size of the body. The 12 thoracic vertebrae are characterized by having facets for articulation with the ribs. The five lumbar vertebrae have kidney-shaped bodies that are large relative to the size of the vertebral foramen. The sacrum consists of five fused vertebrae; the lateral part articulates with the pelvis at the sacroiliac joints. The coccyx consists of between three and five bones.

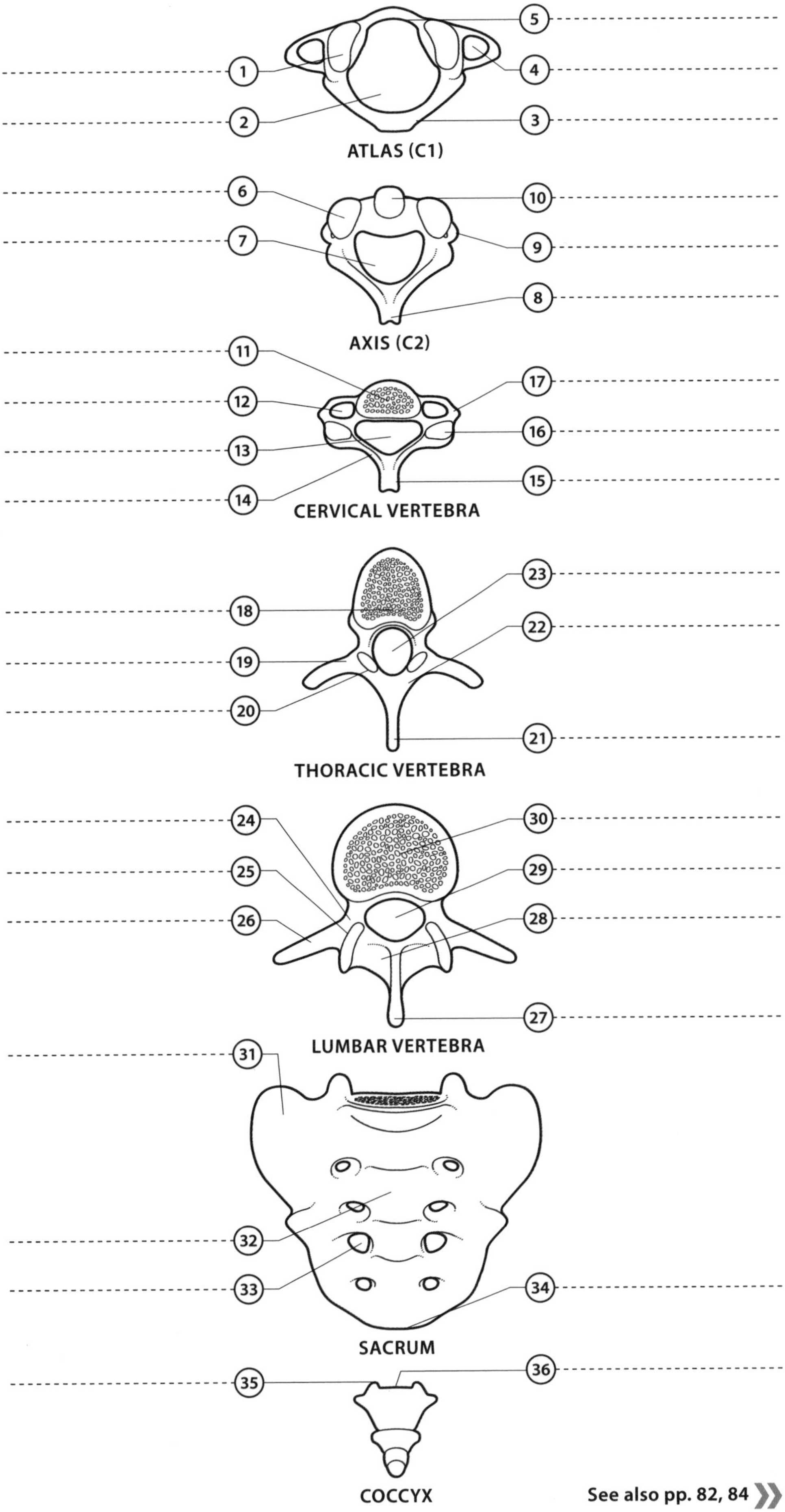

TYPES OF VERTEBRAE

Color and/or label the structures indicated on the diagram using the key below.

1. Superior articular facet
2. Vertebral foramen
3. Posterior arch
4. Transverse foramen
5. Anterior arch
6. Superior articular facet
7. Vertebral foramen
8. Spinous process
9. Transverse process
10. Dens (odontoid peg)
11. Body
12. Transverse foramen
13. Vertebral foramen
14. Lamina
15. Spinous process
16. Superior articular facet
17. Transverse process
18. Body
19. Transverse process
20. Superior articular facet
21. Spinous process
22. Lamina
23. Vertebral foramen
24. Pedicle
25. Superior articular facet
26. Transverse process
27. Spinous process
28. Lamina
29. Vertebral foramen
30. Body
31. Ala of sacrum
32. Body
33. Anterior sacral foramen
34. Apex of sacrum
35. Coccygeal cornu
36. Facet for apex of sacrum

See also pp. 82, 84 »

THORAX

THE SKELETON OF THE THORAX HAS SEVERAL IMPORTANT ROLES: IT PROTECTS THE HEART AND LUNGS; IT ACTS AS AN ANCHOR FOR MUSCLE ATTACHMENT; AND THE RIBS AND STERNUM MOVE TO CHANGE THE VOLUME OF THE THORAX DURING BREATHING.

ANTERIOR VIEW OF THORAX

Color and/or label the structures indicated on the diagram using the key below.

1. T1 (first thoracic vertebra)
2. Transverse process of T1
3. Head of first rib
4. Clavicle
5. Scapula
6. First rib
7. Manubrium sterni
8. Second costal cartilage
9. Third rib
10. Fourth rib
11. Body of sternum
12. Fifth rib
13. Sixth rib
14. Seventh rib
15. Xiphoid process
16. Eighth to tenth ribs
17. Eleventh and twelfth ribs

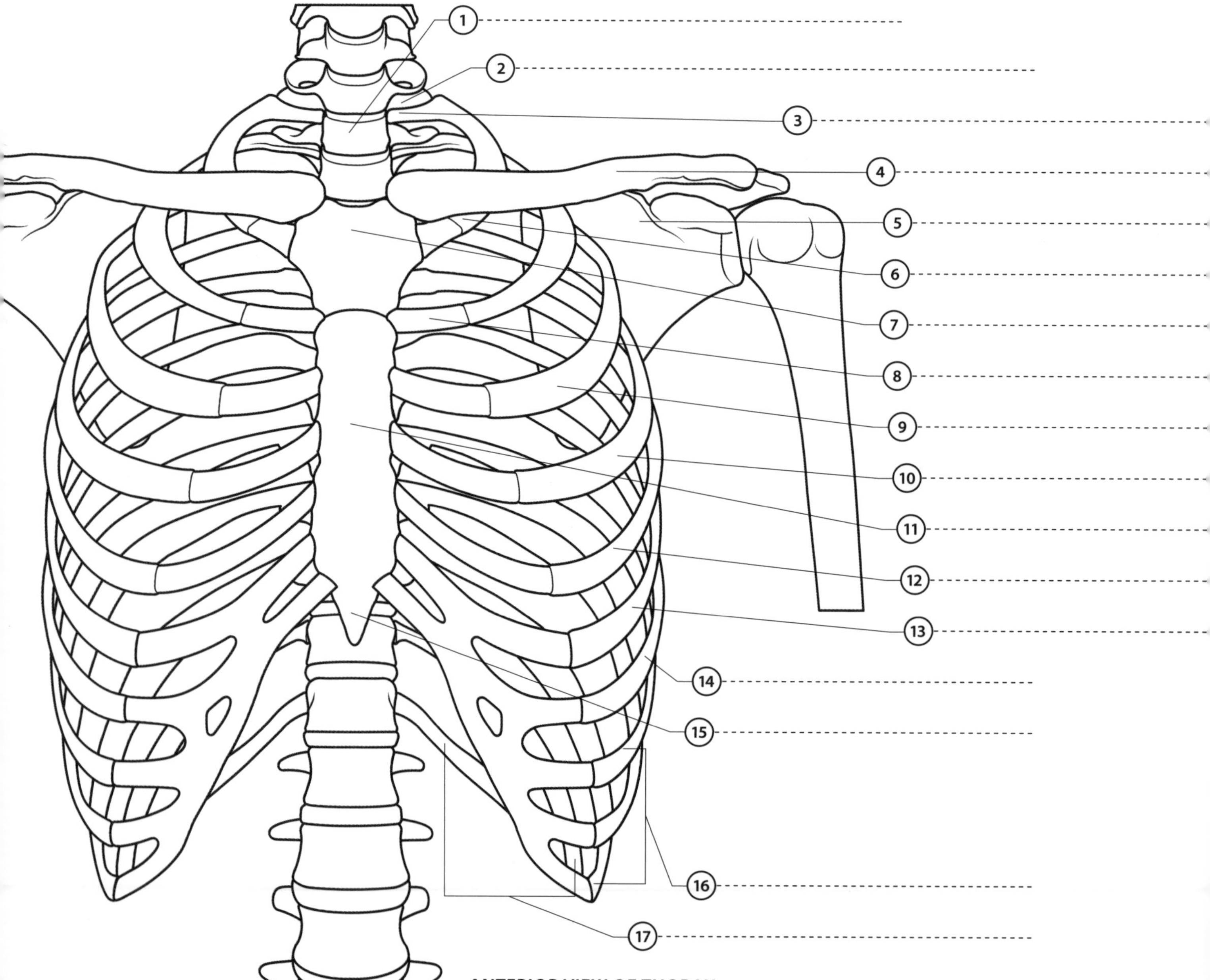

ANTERIOR VIEW OF THORAX

STRUCTURE

The bones of the thorax include 12 thoracic vertebrae, 12 pairs of ribs and costal cartilages, and the sternum. The upper seven ribs articulate with the sternum via their costal cartilages. The eighth to tenth costal cartilages each join to the cartilage above, creating the sweeping curve of the rib cage below the sternum on each side. The eleventh and twelfth ribs are short and do not join any other ribs; they are sometimes known as free or floating ribs. The vertebrae articulate via intervertebral disks. The body, manubrium, and xiphoid process of the sternum articulate via secondary cartilaginous joints. The joints between the ribs and vertebrae at the back are synovial, allowing the ribs to move during breathing.

POSTERIOR VIEW OF THORAX

Color and/or label the structures indicated on the diagram using the key below.

1. C7 (seventh cervical vertebra)
2. Transverse process of T1
3. First rib
4. Third rib
5. Fifth rib
6. Seventh rib
7. Ninth rib
8. Tenth rib
9. Eleventh rib
10. Twelfth rib

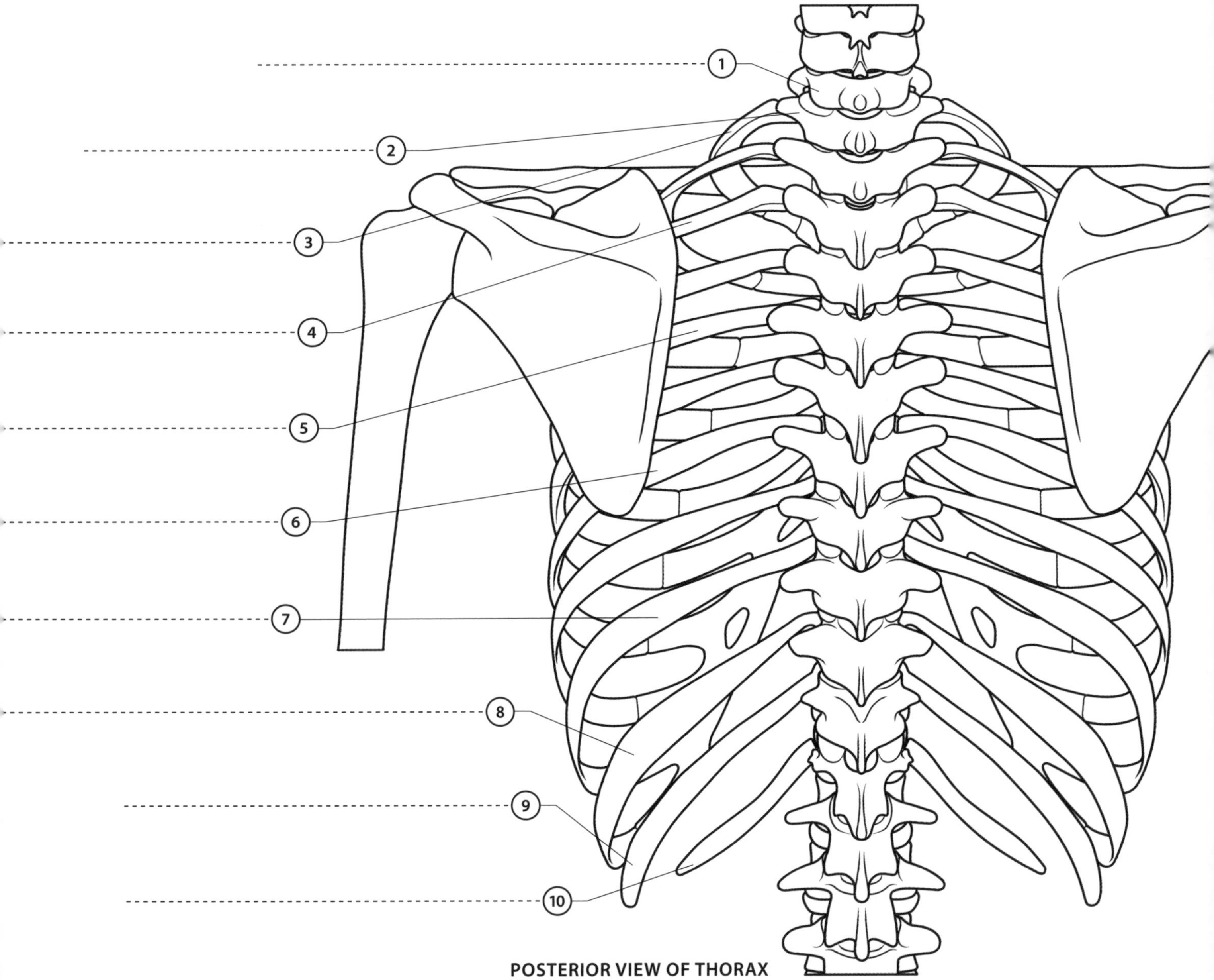

POSTERIOR VIEW OF THORAX

See also pp. 62, 64, 100, 120, 122, 124, 138, 162, 178 »

ABDOMEN AND PELVIS

THE BONES OF THE ABDOMEN AND PELVIS PROTECT INTERNAL ORGANS SUCH AS THOSE OF THE DIGESTIVE, URINARY, AND REPRODUCTIVE SYSTEMS. THE PELVIS ALSO PROVIDES ATTACHMENT FOR THE BONES AND MUSCLES OF THE LEGS.

ANTERIOR VIEW

Color and/or label the structures indicated on the diagram using the key below.

1. Sacroiliac joint
2. Iliac crest
3. Ilium
4. Sacrum
5. Coccyx
6. Superior pubic ramus
7. Body of ischium
8. Ischiopubic ramus
9. Obturator foramen
10. Femur
11. Pubic tubercle
12. Pubic symphysis
13. Anterior sacral foramina
14. Ala of sacrum
15. Anterior superior iliac spine
16. Lumbar vertebrae
17. Twelfth rib

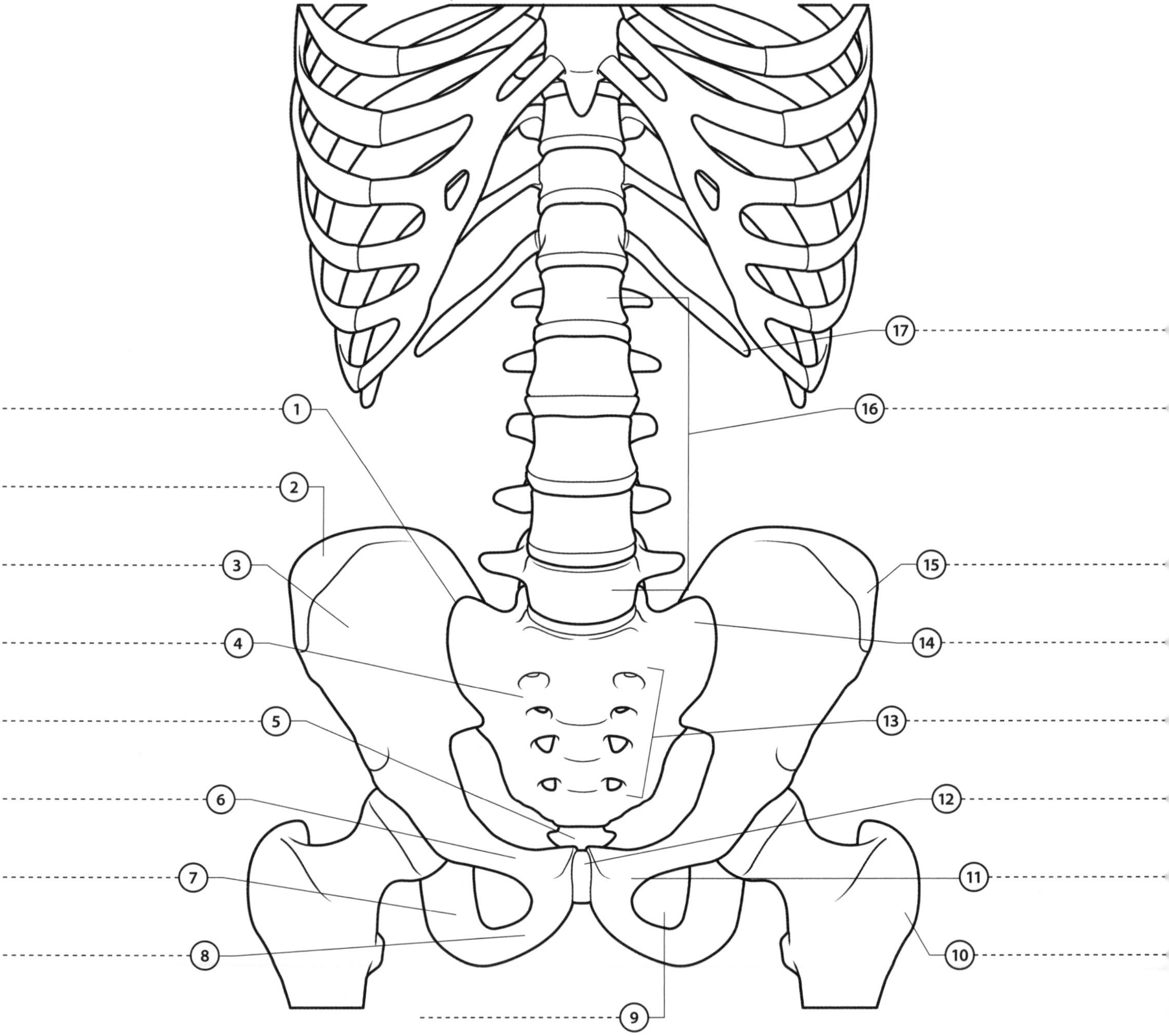

ANTERIOR VIEW

STRUCTURE

The skeletal boundaries of the abdomen include the five lumbar vertebrae at the back, the lower margins of the ribs above, and the iliac crest of the pelvic bones below. The abdominal cavity extends up under the rib cage, as high as the gap between the fifth and sixth ribs, due to the domed shape of the diaphragm. The pelvis is a basin shape, and is enclosed by the two pelvic (innominate) bones at the front and sides, and the sacrum at the back. Each pelvic bone is made of three fused bones: the ilium at the rear, the ischium at the lower front, and the pubis above it.

POSTERIOR VIEW

Color and/or label the structures indicated on the diagram using the key below.

1. Lumbar vertebrae
2. Iliac crest
3. Posterior superior iliac spine
4. Gluteal surface of ilium
5. Sacroiliac joint
6. Sacrum
7. Greater trochanter
8. Body of pubis
9. Lesser trochanter
10. Femur
11. Ischiopubic ramus
12. Ischial tuberosity
13. Coccyx
14. Obturator foramen
15. Superior pubic ramus
16. Posterior sacral foramina
17. Lumbosacral joint
18. Twelfth rib

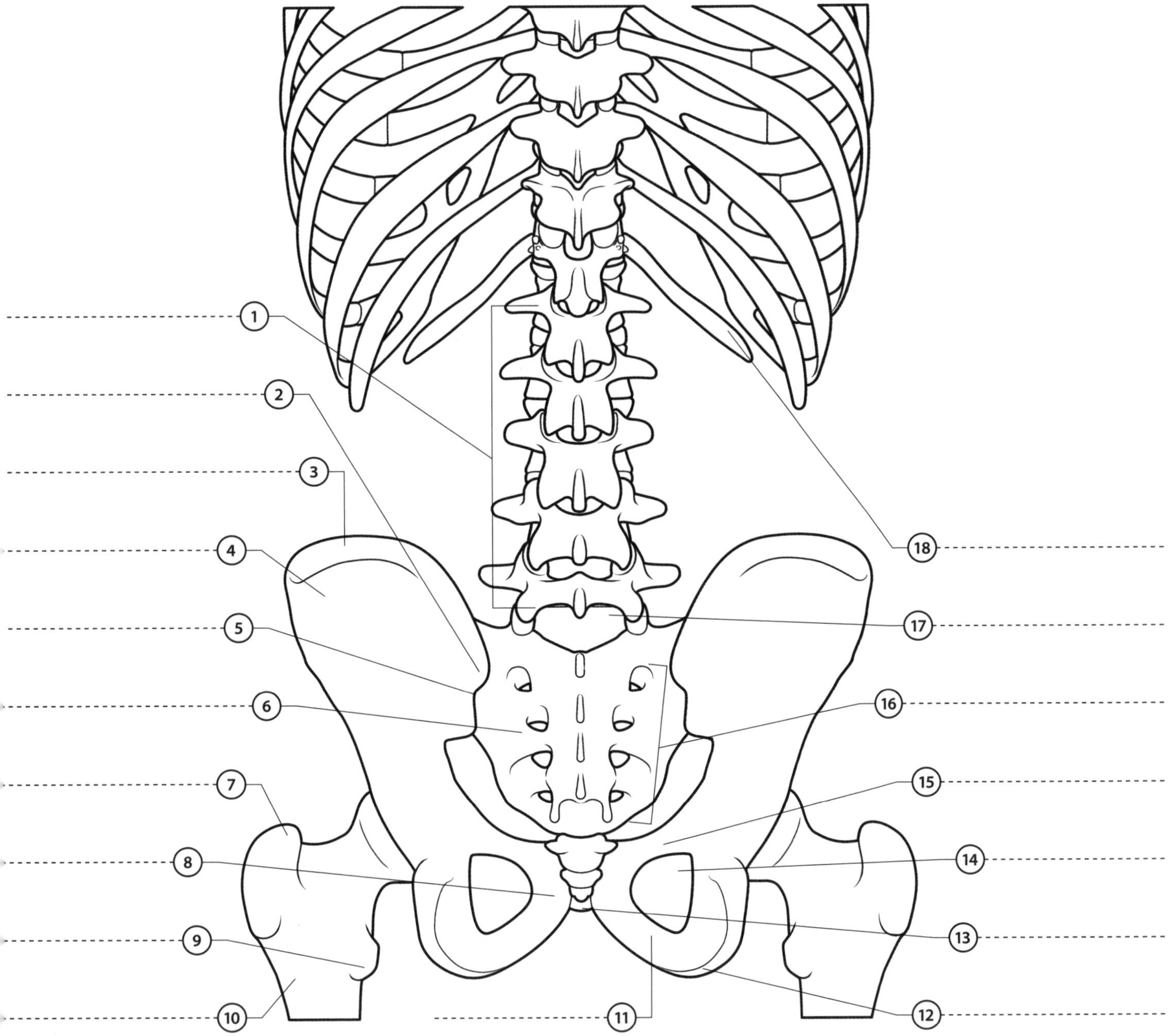

POSTERIOR VIEW See also pp. 66, 68, 100, 144, 162, 178, 198, 204, 218, 224

SHOULDER AND ARM

THE SHOULDER AND ARM CONSIST OF 32 INDIVIDUAL BONES: TWO IN THE SHOULDER GIRDLE; ONE IN THE UPPER ARM; TWO IN THE LOWER ARM; EIGHT WRIST BONES (CARPALS); AND 19 BONES (PHALANGES AND METACARPALS) IN THE HAND.

STRUCTURE

The scapula and clavicle make up the shoulder girdle, which anchors the arm to the thorax. This is a very mobile attachment—the scapula “floats” on the rib cage, attached to it by muscles only. The shoulder joint—a ball and socket joint—is the most mobile joint in the body. The back of the scapula is divided into two sections by its spine, which runs to the side and projects out above the shoulder to form the acromion. The clavicle articulates with the acromion of the scapula laterally and with the sternum medially. The humerus articulates at its upper end with the scapula and at its lower end with the radius and ulna, forming the elbow joint. The lower ends of the radius and ulna articulate with some of the carpals to form the wrist joint. The carpals articulate with five metacarpals in the hand, which, in turn, articulate with the proximal phalanges of the fingers.

POSTERIOR VIEW

Color and/or label the structures indicated on the diagram using the key below.

1. Supraspinous fossa
2. Clavicle
3. Acromion
4. Spine of scapula
5. Glenoid cavity
6. Infraspinous fossa
7. Humerus
8. Olecranon fossa
9. Lateral epicondyle of humerus
10. Interosseus border of ulna
11. Interosseus border of radius
12. Ulna
13. Radius
14. Styloid process of radius
15. Phalanges
16. Metacarpals
17. Carpals
18. Olecranon of ulna
19. Medial epicondyle of humerus

ANTERIOR VIEW

Color and/or label the structures indicated on the diagram using the key below.

1. Scapula
2. Acromion
3. Clavicle
4. Coracoid process
5. Glenoid fossa
6. Shaft of humerus
7. Coronoid fossa
8. Radial fossa
9. Lateral epicondyle
10. Capitulum of humerus
11. Trochlea of humerus
12. Phalanges
13. Metacarpals
14. Carpals
15. Ulna
16. Interosseus border of ulna
17. Radius
18. Interosseus border of radius
19. Medial epicondyle

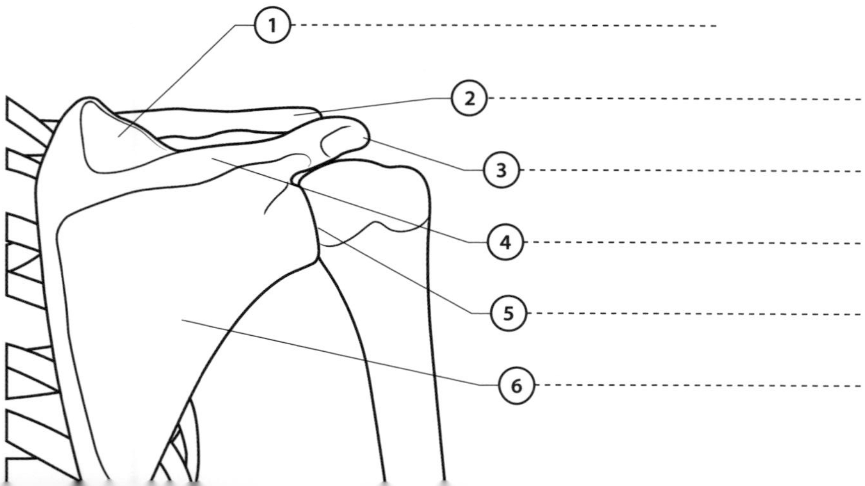

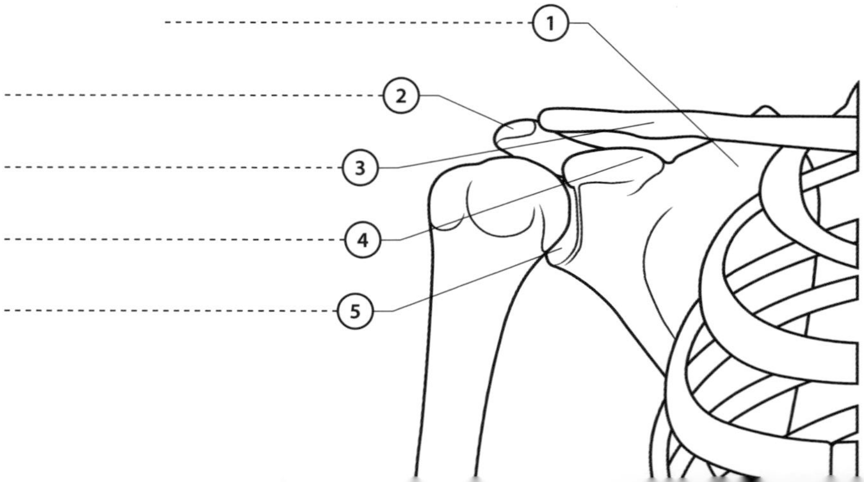

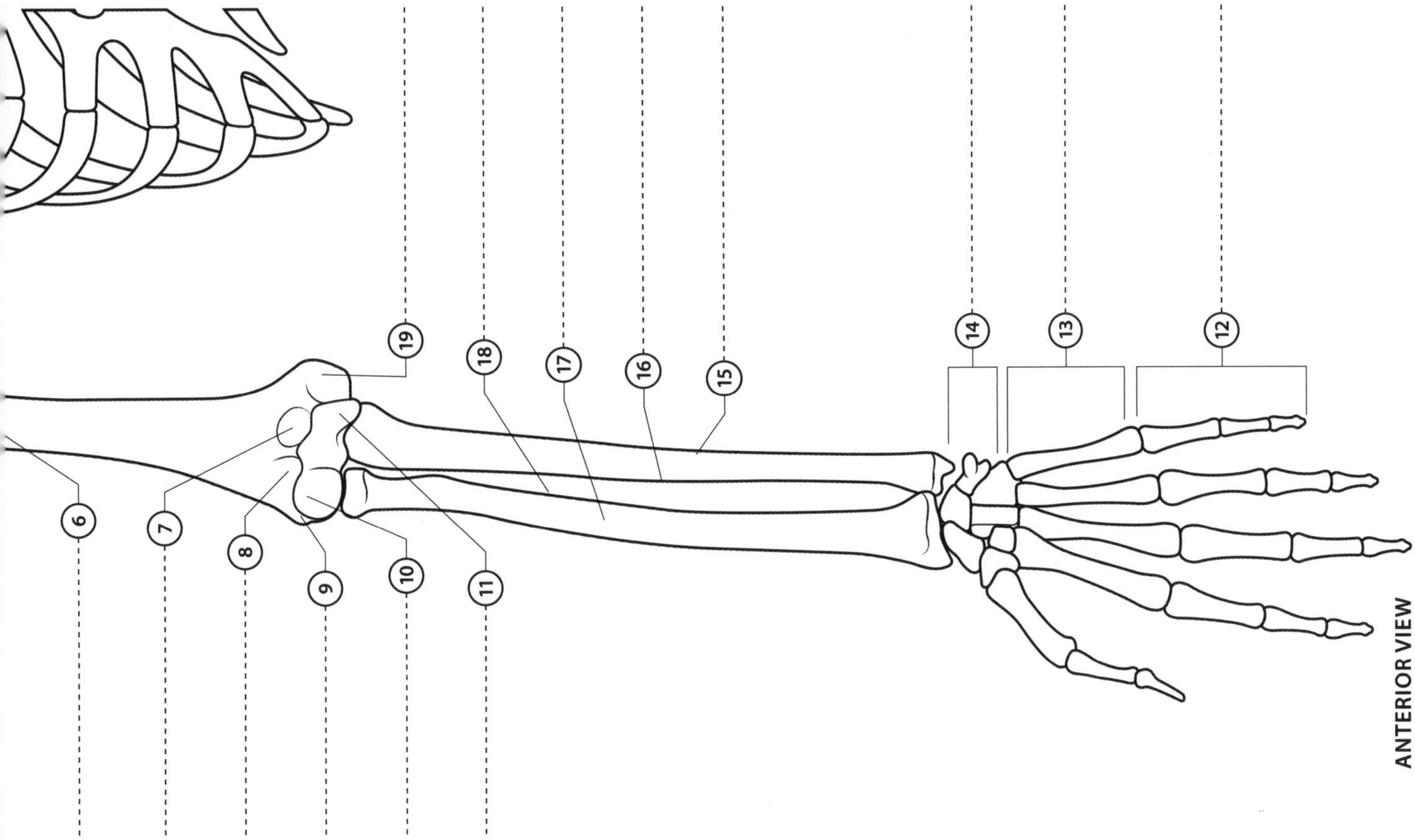

ANTERIOR VIEW

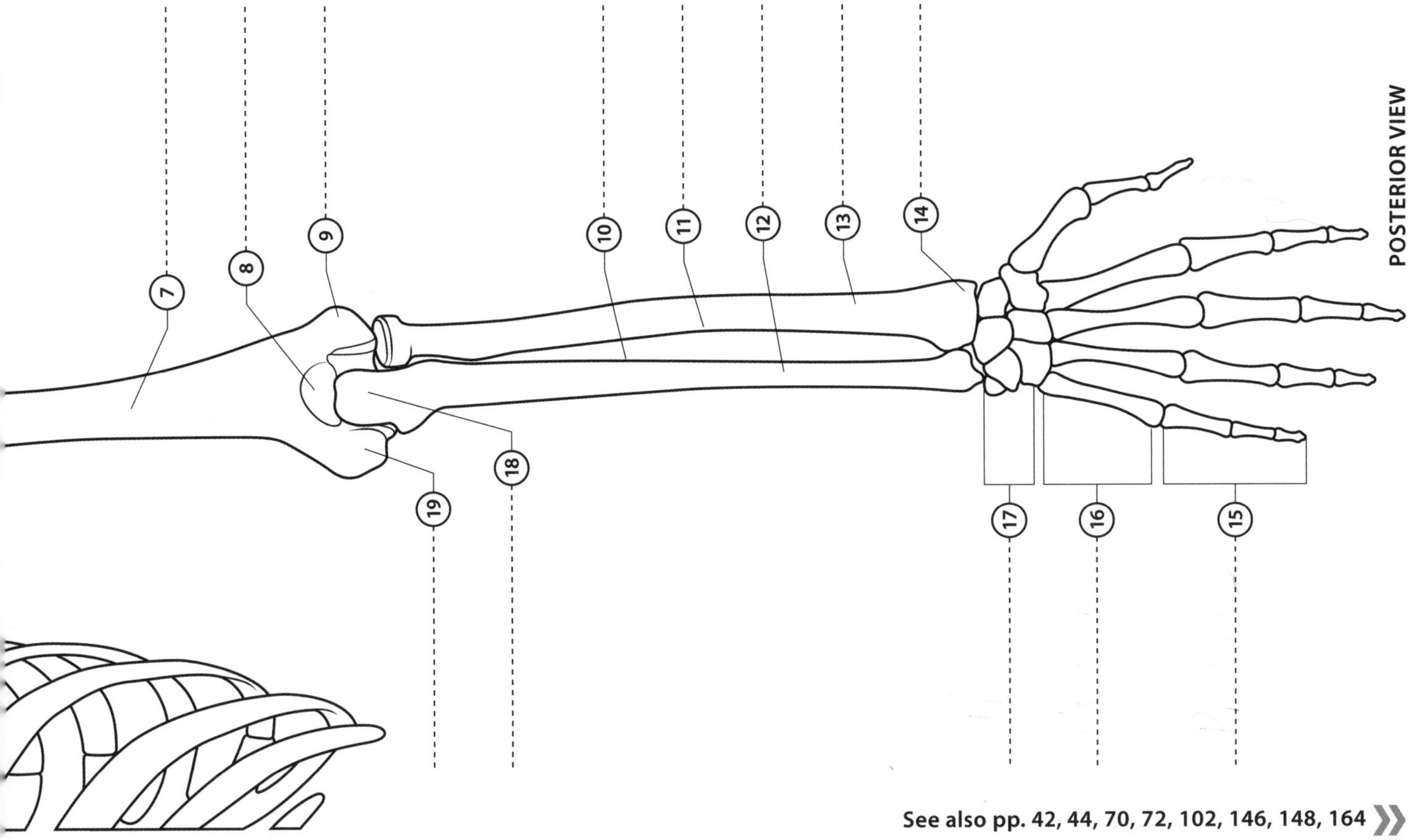

POSTERIOR VIEW

See also pp. 42, 44, 70, 72, 102, 146, 148, 164 »

ELBOW

THE ELBOW IS THE HINGE JOINT BETWEEN THE UPPER AND LOWER ARM, ENABLING FLEXION AND EXTENSION OF THE LOWER ARM.

STRUCTURE

The elbow joint is formed by the articulation of the humerus with the forearm bones. The trochlea of the humerus articulates with the ulna; the capitulum of the humerus articulates with the head of the radius. The elbow joint is stabilized by collateral ligaments on each side.

ELBOW (MEDIAL SIDE)

Color and/or label the structures indicated on the diagram using the key below.

1. Humerus
2. Medial epicondyle
3. Olecranon of ulna
4. Ulnar collateral ligament
5. Ulna
6. Radius
7. Biceps tendon
8. Anular ligament of the radius

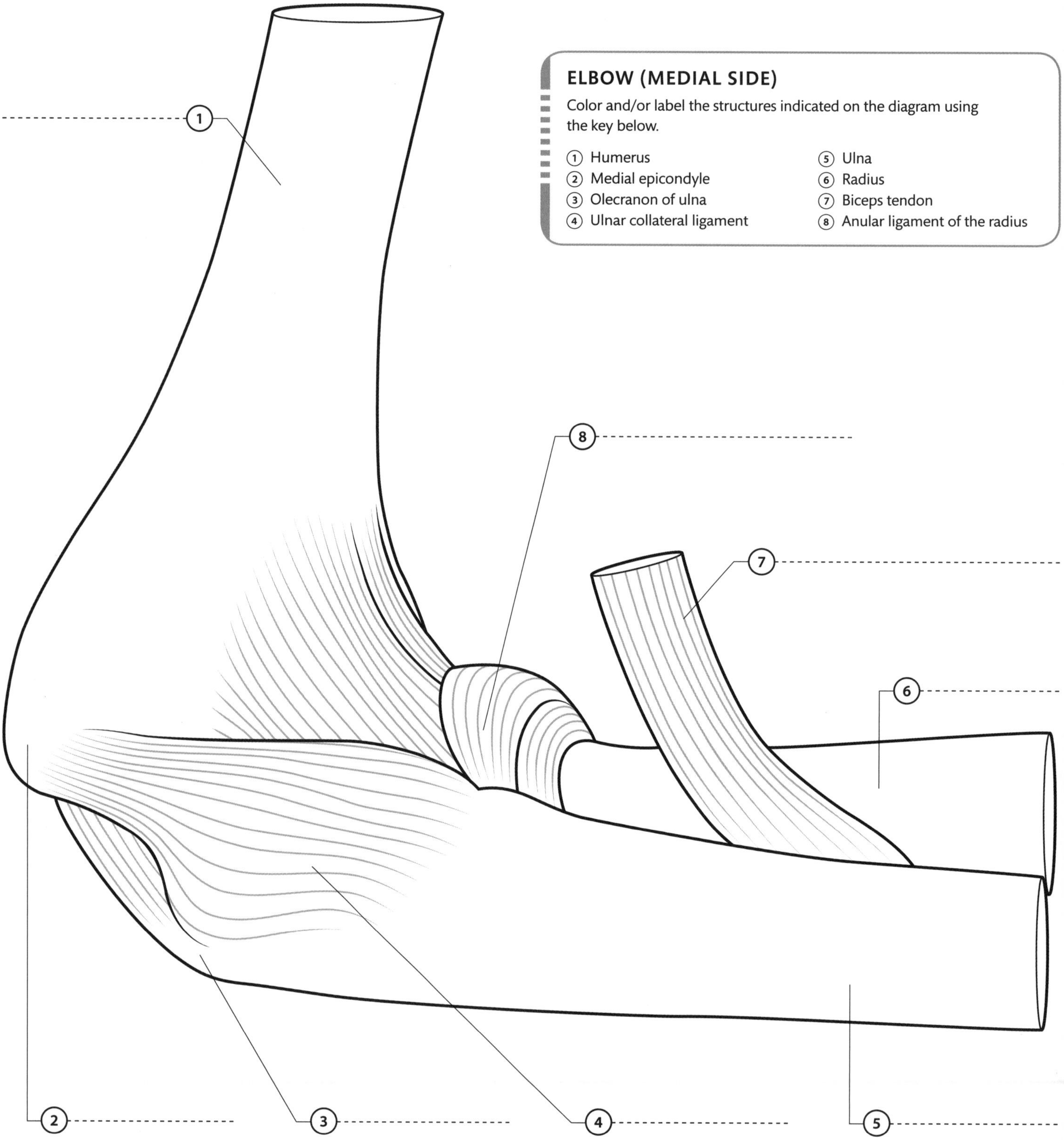

ELBOW (MEDIAL SIDE)

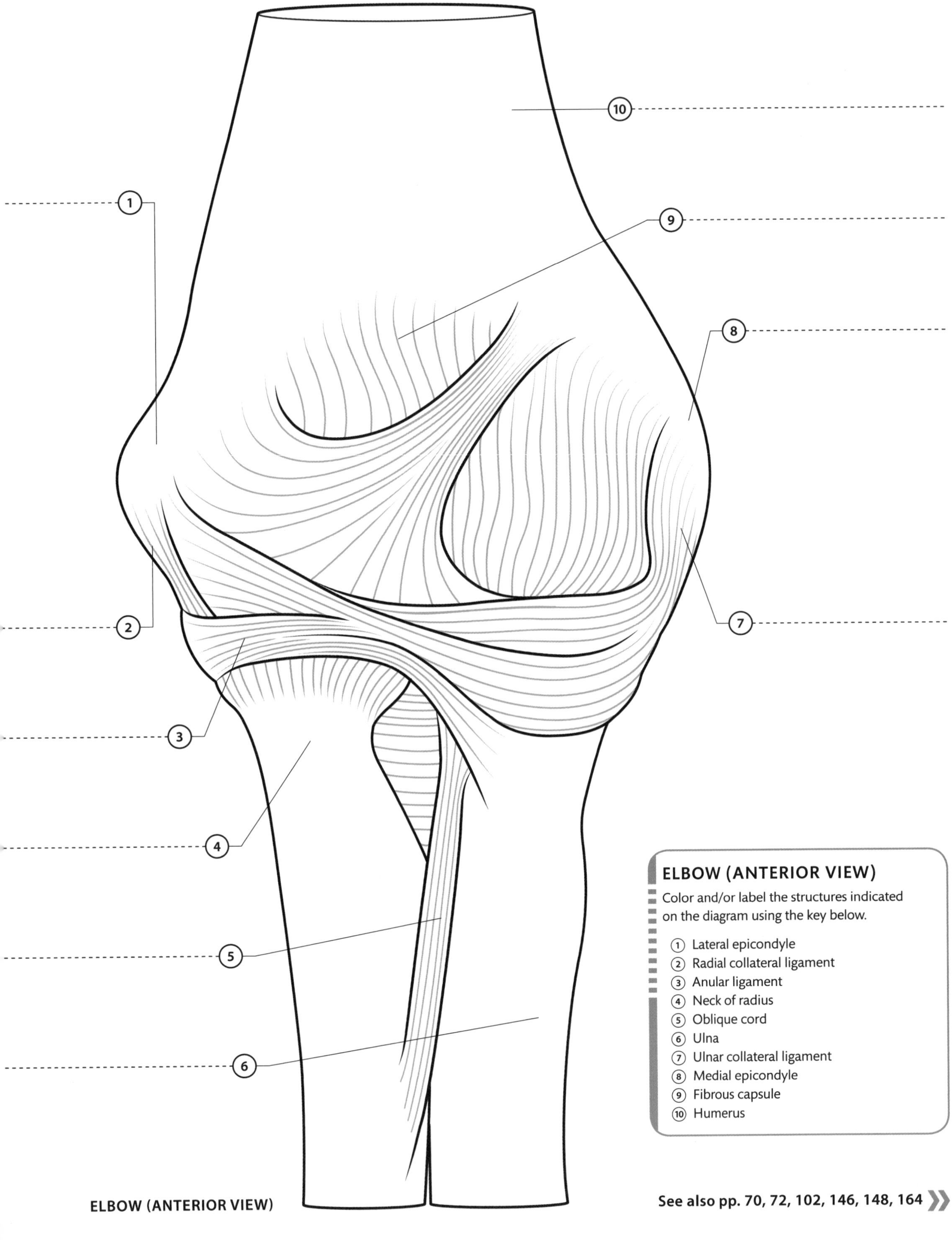

ELBOW (ANTERIOR VIEW)

ELBOW (ANTERIOR VIEW)

Color and/or label the structures indicated on the diagram using the key below.

1. Lateral epicondyle
2. Radial collateral ligament
3. Anular ligament
4. Neck of radius
5. Oblique cord
6. Ulna
7. Ulnar collateral ligament
8. Medial epicondyle
9. Fibrous capsule
10. Humerus

See also pp. 70, 72, 102, 146, 148, 164 »

WRIST AND HAND

THE WRIST IS A COMPLEX JOINT COMPRISING THE DISTAL ENDS OF THE RADIUS AND ULNA AND EIGHT CARPAL BONES. THE HAND CONSISTS OF FIVE METACARPAL BONES, WITH 14 PHALANGES IN THE FINGERS.

HAND AND WRIST JOINTS

The radius widens out at its distal end to form the wrist joint with the nearest two carpals, the lunate and scaphoid. This joint allows flexion, extension, adduction, and abduction. There are also synovial joints between the carpals in the wrist, which increase the range of motion during wrist flexion and extension. Synovial joints between the metacarpals and phalanges make it possible to spread or close the fingers, as well as to flex or extend the whole finger. Joints between the individual phalanges enable the fingers to bend and straighten. Humans, like many other primates, have opposable thumbs. The joints at the base of the thumb are shaped differently from those of the fingers. The joint between the metacarpal of the thumb and the wrist bones is especially mobile and allows the tip of the thumb to touch the other fingertips. The wrist and finger joints are stabilized by a complex series of ligaments.

PALMAR (ANTERIOR) VIEW

Color and/or label the structures indicated on the diagram using the key below.

1. Palmar metacarpal ligament
2. Hook of hamate bone
3. Pisiform bone
4. Palmar ulnocarpal ligament
5. Ulna
6. Radius
7. Lunate bone
8. Palmar radiocarpal ligament
9. Radiate carpal ligament
10. Capitate bone
11. First metacarpal
12. Proximal phalanx
13. Distal phalanx
14. Palmar ligament
15. Middle phalanx
16. Distal phalanx
17. Deep transverse metacarpal ligament

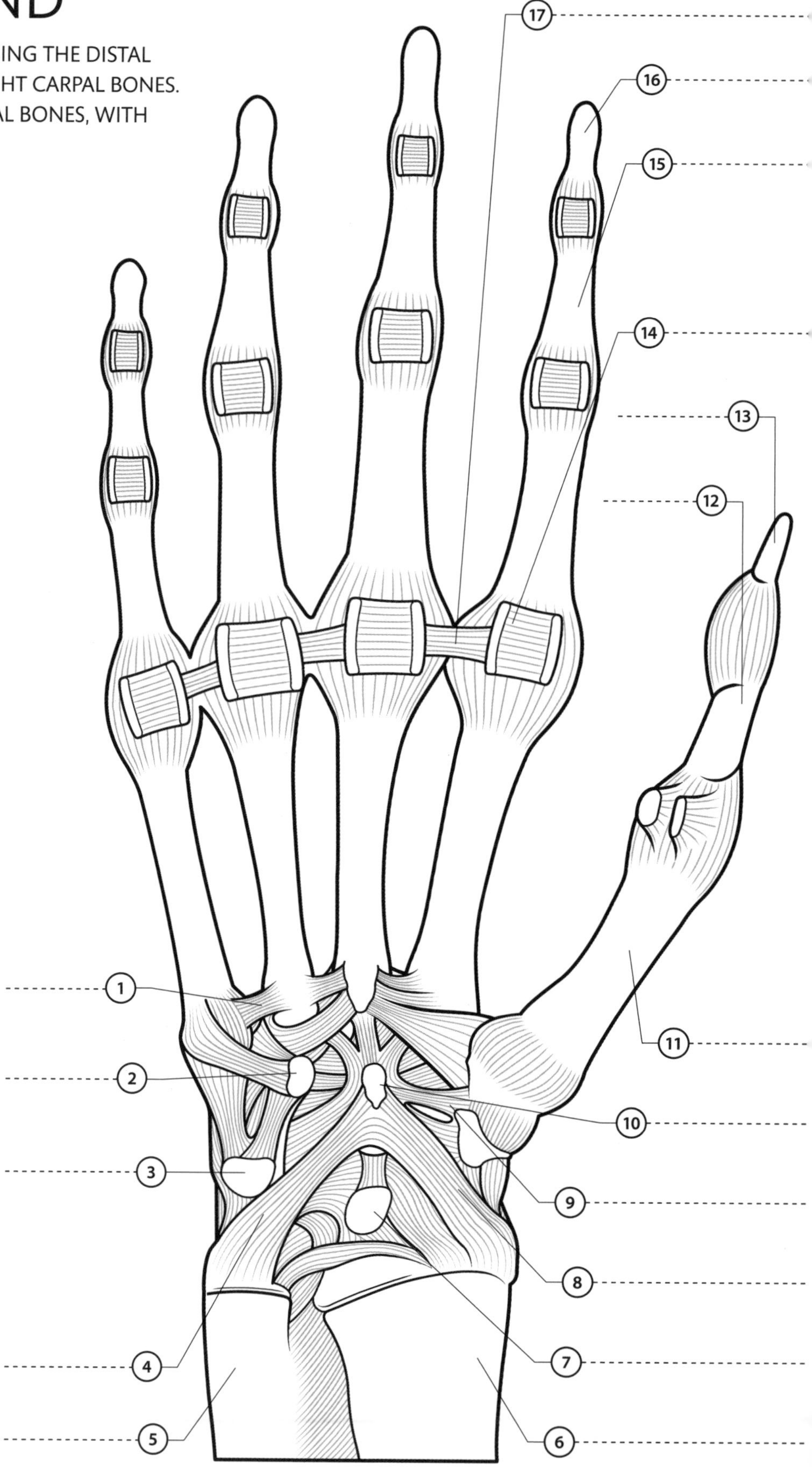

PALMAR (ANTERIOR) VIEW

DORSAL (POSTERIOR) VIEW

Color and/or label the structures indicated on the diagram using the key below.

1. Middle phalanx
2. Proximal phalanx
3. First metacarpal
4. Dorsal intercarpal ligament
5. Styloid process of radius
6. Scaphoid bone
7. Radius
8. Ulna
9. Dorsal radiocarpal ligament
10. Styloid process of ulna
11. Triquetrum bone
12. Capitate bone
13. Hamate bone
14. Dorsal carpometacarpal ligament
15. Fifth metacarpal
16. Fibrous capsule of metacarpophalangeal joint
17. Fibrous capsule of distal interphalangeal joint
18. Distal phalanx

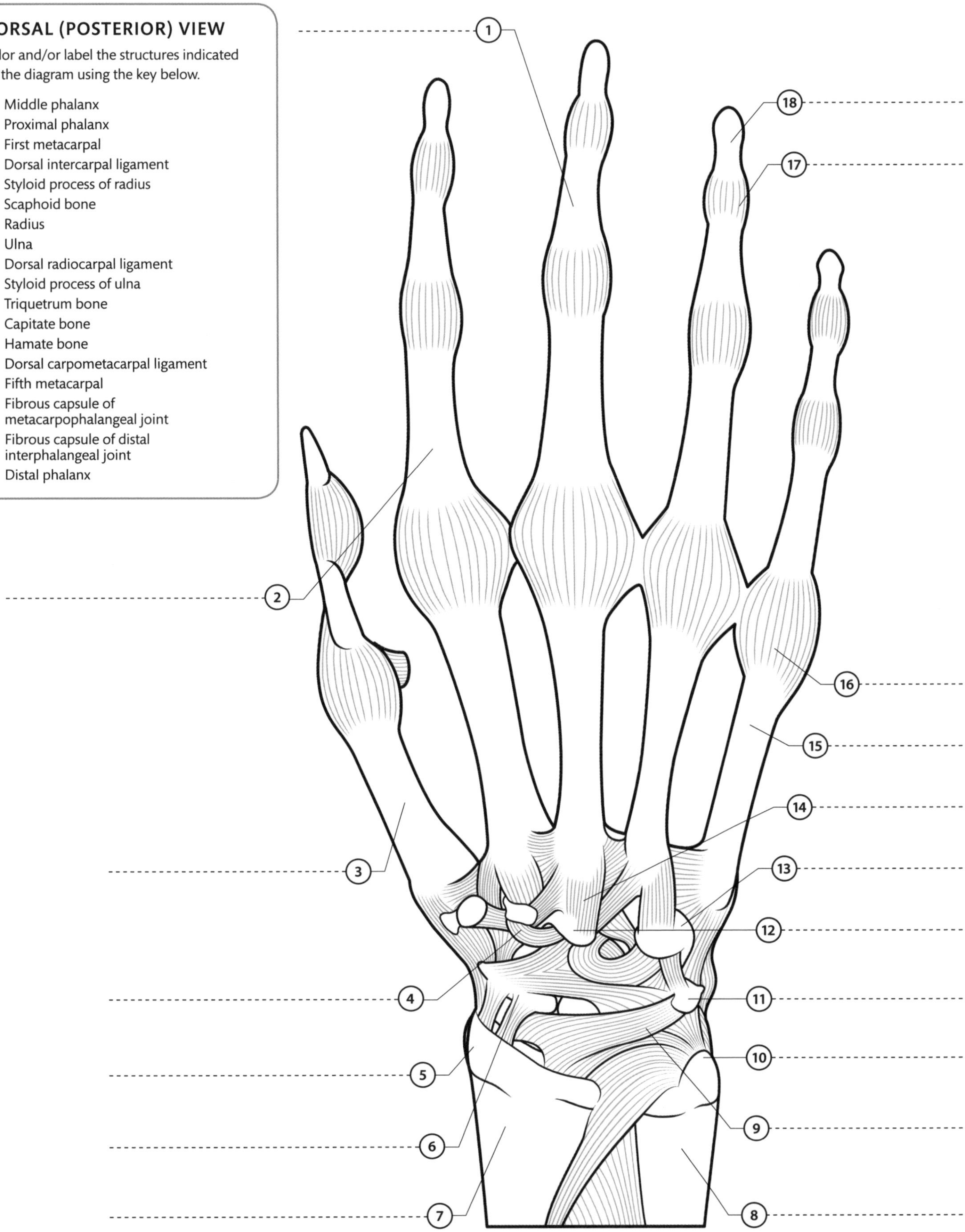

DORSAL (POSTERIOR) VIEW

See also pp. 40, 70, 72, 102, 148

HIP AND LEG

THE LEG CONSISTS OF 30 INDIVIDUAL BONES: ONE IN THE THIGH; THE PATELLA AT THE KNEE; TWO LOWER LEG BONES; SEVEN ANKLE BONES (TARSALS); AND 19 BONES IN THE FOOT.

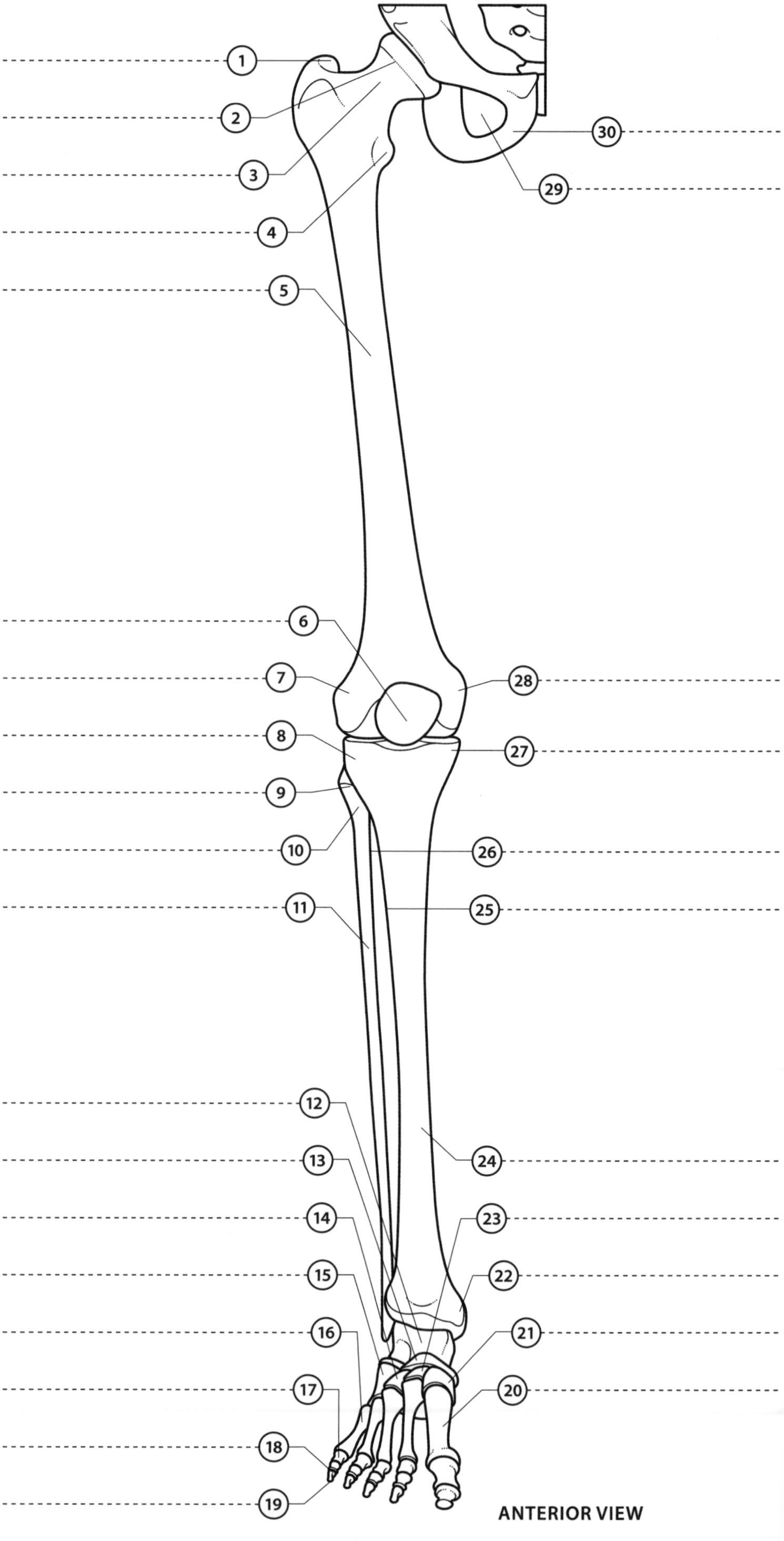

ANTERIOR VIEW

ANTERIOR VIEW

Color and/or label the structures indicated on the diagram using the key below.

1. Greater trochanter
2. Head of femur
3. Neck of femur
4. Lesser trochanter
5. Shaft of femur
6. Patella
7. Lateral epicondyle of femur
8. Lateral condyle of tibia
9. Head of fibula
10. Neck of fibula
11. Shaft of fibula
12. Talus
13. Navicular
14. Lateral cuneiform
15. Cuboid
16. Fifth metatarsal
17. Proximal phalanx
18. Middle phalanx
19. Distal phalanx
20. First metatarsal
21. Medial cuneiform
22. Medial malleolus
23. Intermediate cuneiform
24. Shaft of tibia
25. Interosseus border of tibia
26. Interosseus border of fibula
27. Medial condyle of tibia
28. Medial epicondyle of femur
29. Obturator foramen
30. Ischiopubic ramus

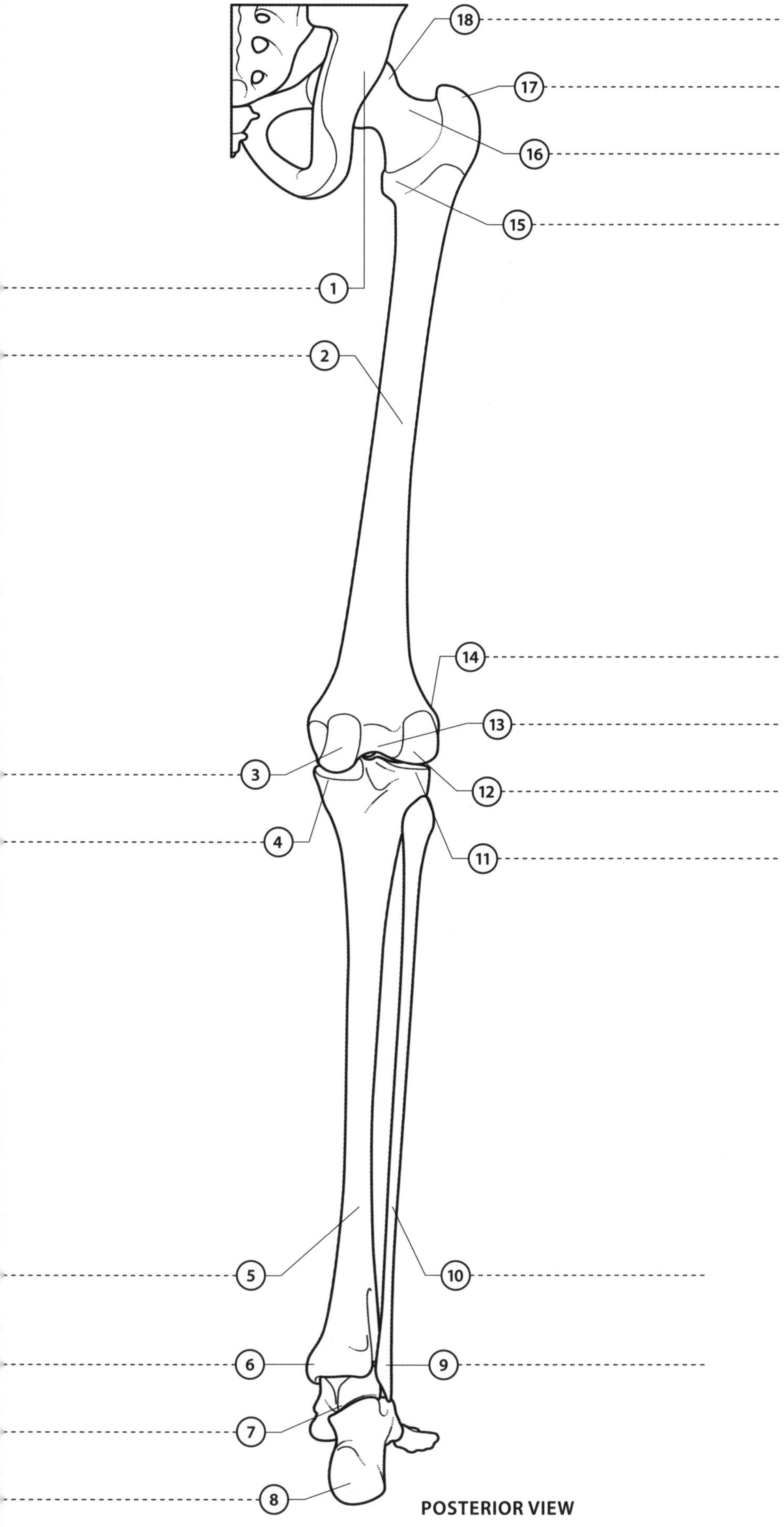

POSTERIOR VIEW

STRUCTURE

The leg is attached to the spine by the pelvic bones. This is a more stable arrangement than that of the shoulder girdle because the legs and pelvis must bear the body weight. The sacroiliac joint provides a strong attachment between the ilium of the pelvis and the sacrum, and the hip joint is a much deeper and more stable joint than that in the shoulder. The neck of the femur joins the head at an obtuse angle. The shaft of the femur is cylindrical, and at the distal end widens to form the knee joint with the tibia and patella. The tibia is the main weight-bearing bone of the lower leg, and also forms part of the ankle joint. The fibula, which articulates with the tibia below the knee joint, provides extra areas for the attachment of muscles and also forms part of the ankle joint. The tarsals of the ankle articulate with the metatarsals of the foot, which in turn articulate with the proximal phalanges of the toes—an arrangement that is similar to that of the carpals, metacarpals, and phalanges in the hand.

POSTERIOR VIEW

Color and/or label the structures indicated on the diagram using the key below.

1. Posterior lip of acetabulum
2. Shaft of femur
3. Medial condyle of femur
4. Medial condyle of tibia
5. Shaft of tibia
6. Medial malleolus
7. Talus
8. Calcaneus
9. Lateral malleolus
10. Shaft of fibula
11. Lateral condyle of tibia
12. Lateral condyle of femur
13. Intercondylar fossa
14. Lateral epicondyle
15. Lesser trochanter
16. Neck of femur
17. Greater trochanter
18. Head of femur

See pp. 48, 50, 74, 76, 104, 150, 152, 165 »

KNEE

THE KNEE JOINT IS FORMED BY THE ARTICULATION OF THE FEMUR IN THE THIGH WITH THE TIBIA OF THE LOWER LEG AND THE PATELLA (KNEECAP).

FUNCTION

The knee is primarily a hinge joint but also permits some rotation to occur. These complex movements are reflected in the complexity of the joint: crescent-shaped articular disks (menisci) inside the joint, powerful collateral ligaments on either side of the joint, and crossed-over cruciate ligaments binding the femur to the tibia. Numerous extra pockets of synovial fluid (bursae) lubricate tendons around the joint.

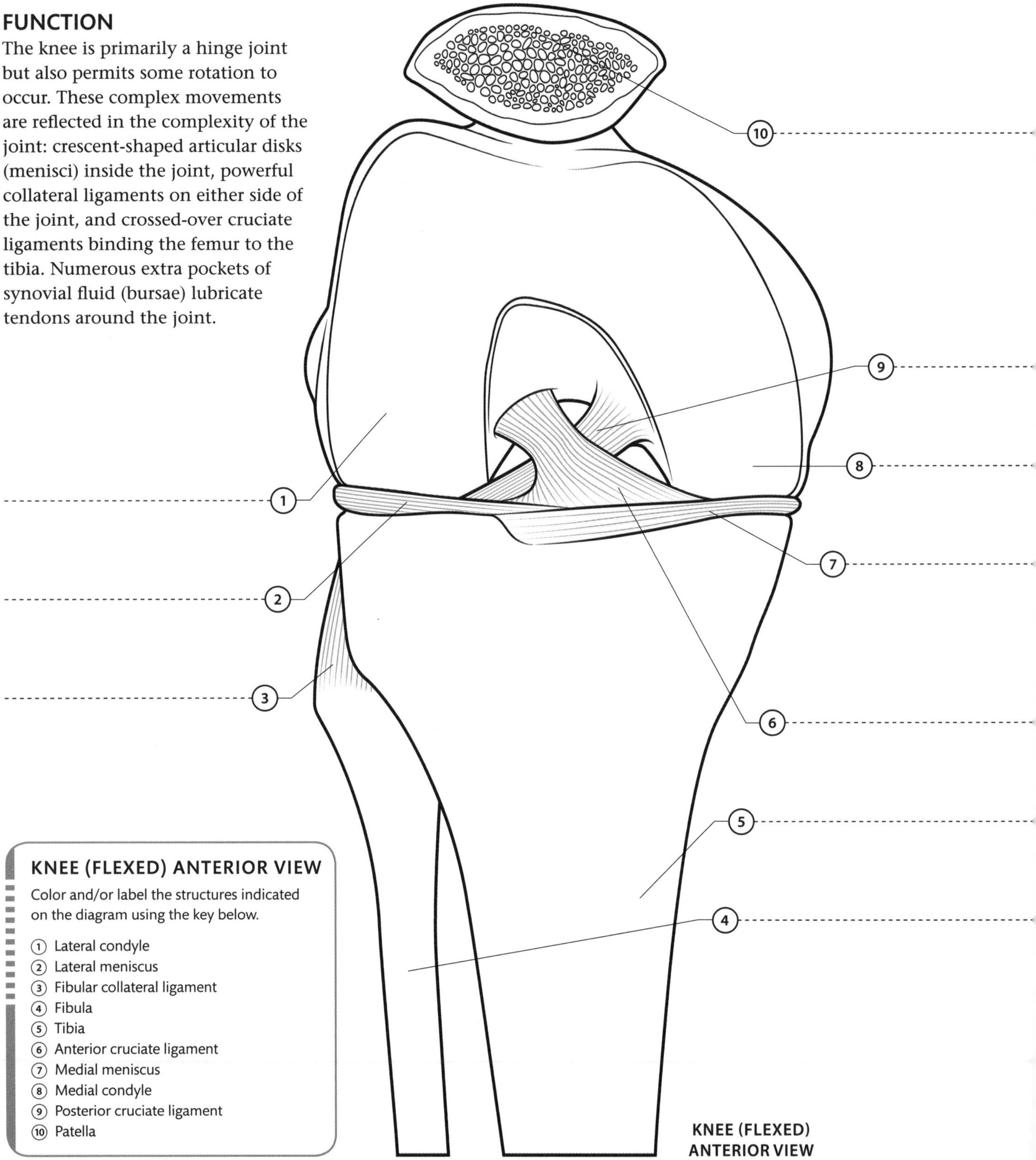

KNEE (FLEXED) ANTERIOR VIEW

KNEE (FLEXED) ANTERIOR VIEW

Color and/or label the structures indicated on the diagram using the key below.

1. Lateral condyle
2. Lateral meniscus
3. Fibular collateral ligament
4. Fibula
5. Tibia
6. Anterior cruciate ligament
7. Medial meniscus
8. Medial condyle
9. Posterior cruciate ligament
10. Patella

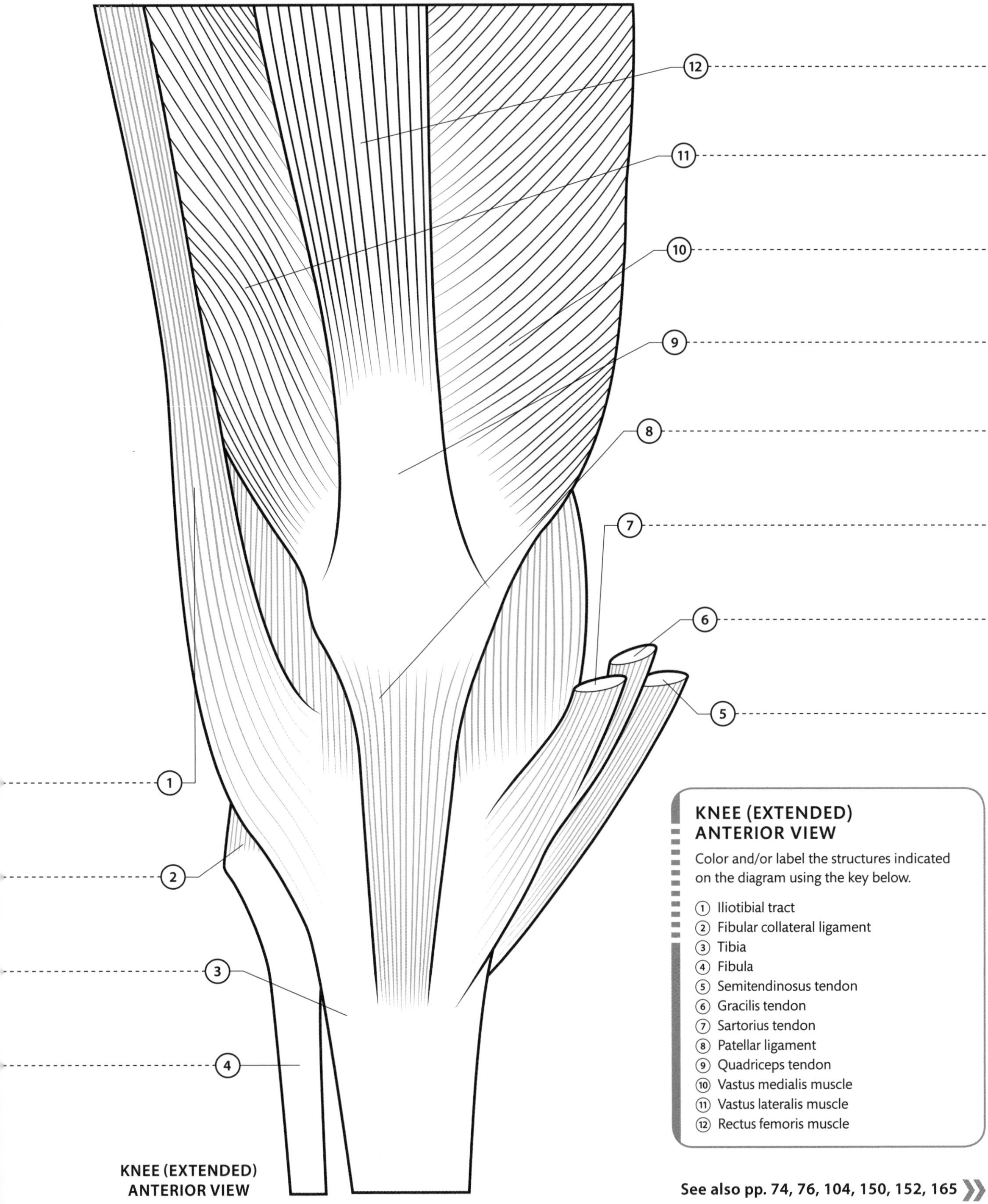

KNEE (EXTENDED)
ANTERIOR VIEW

KNEE (EXTENDED) ANTERIOR VIEW

Color and/or label the structures indicated on the diagram using the key below.

1. Iliotibial tract
2. Fibular collateral ligament
3. Tibia
4. Fibula
5. Semitendinosus tendon
6. Gracilis tendon
7. Sartorius tendon
8. Patellar ligament
9. Quadriceps tendon
10. Vastus medialis muscle
11. Vastus lateralis muscle
12. Rectus femoris muscle

See also pp. 74, 76, 104, 150, 152, 165 »

ANKLE AND FOOT

THE ANKLE JOINT COMPRISES THE DISTAL ENDS OF THE TIBIA AND FIBULA AND SEVEN TARSAL BONES. THE FOOT CONSISTS OF FIVE METATARSAL BONES, WITH 14 PHALANGES IN THE TOES.

STRUCTURE

The ankle joint is a simple hinge joint. The distal ends of the tibia and fibula are firmly bound together by ligaments, forming a strong fibrous joint, and making a wrench shape that sits around the "nut" of the talus. The joint is stabilized by strong collateral ligaments on either side. The talus forms synovial joints with the calcaneus beneath it and the navicular bone in front of it. Level with the joint between the talus and navicular is a joint between the calcaneus and cuboid. These joints together allow the foot to be angled inward (inversion) or outward (eversion). The skeleton of the foot is a sprung structure, with the bones forming arches, held together by ligaments and also supported by tendons.

LATERAL VIEW

Color and/or label the structures indicated on the diagram using the key below.

1. Fibula
2. Anterior tibiofibular ligament
3. Lateral malleolus
4. Calcaneofibular ligament
5. Calcaneus
6. Calcaneal (Achilles) tendon
7. Long plantar ligament
8. Short plantar (calcaneocuboid) ligament
9. Calcaneonavicular ligament
10. Dorsal calcaneocuboid ligament
11. Cuboid
12. Fibularis brevis tendon
13. Deep transverse metatarsal ligaments
14. Distal phalanx
15. Middle phalanx
16. Proximal phalanx
17. First metatarsal
18. Dorsal metatarsal ligaments
19. Dorsal tarsometatarsal ligaments
20. Navicular
21. Talus
22. Anterior talofibular ligament
23. Tibia

LATERAL VIEW

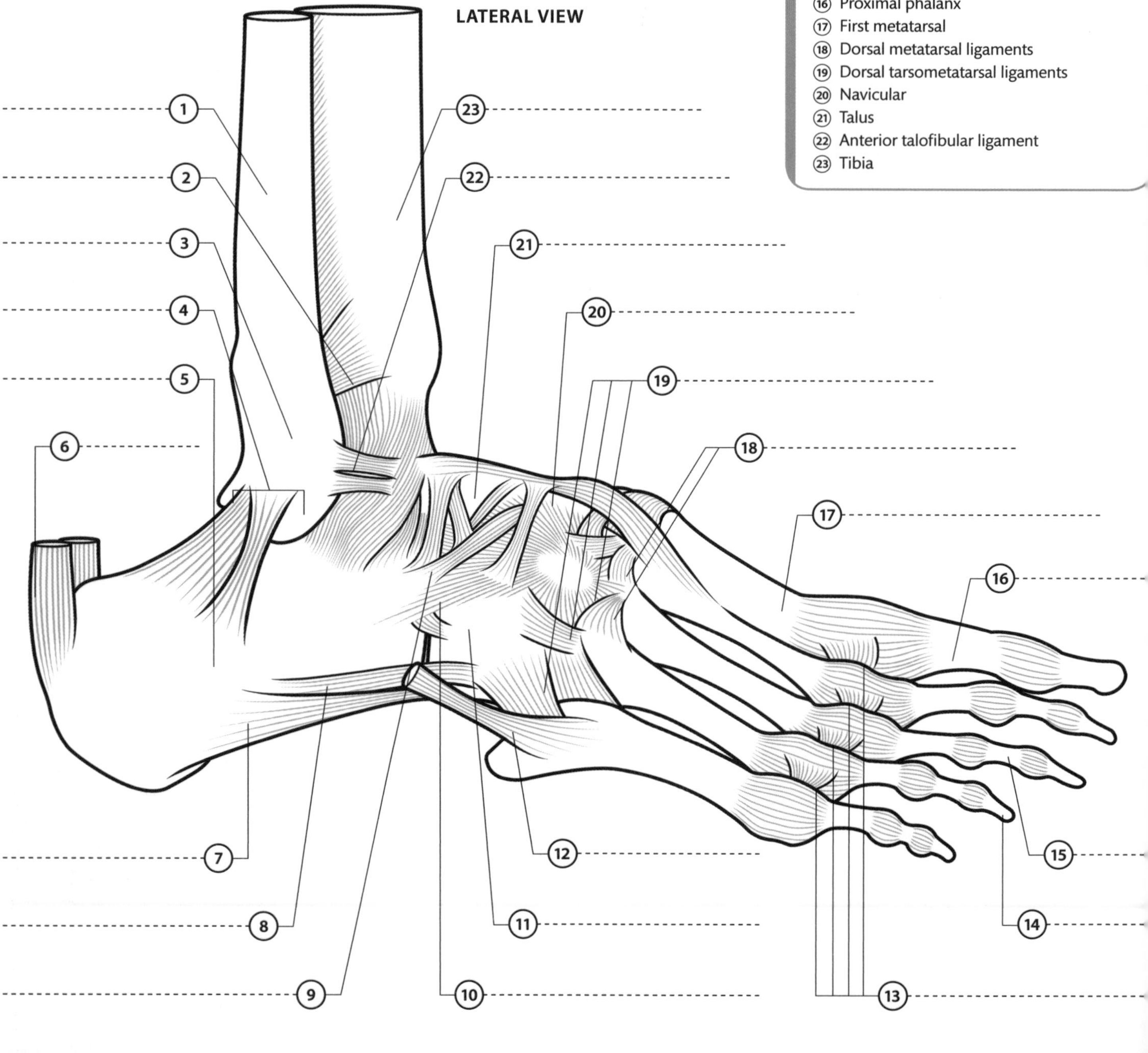

POSTERIOR VIEW

Color and/or label the structures indicated on the diagram using the key below.

1. Fibula
2. Tibia
3. Medial malleolus
4. Posterior tibiofibular ligament
5. Lateral malleolus
6. Posterior talofibular ligament
7. Calcaneofibular ligament
8. Deltoid ligament (posterior tibiotalar part)
9. Calcaneal (Achilles) tendon
10. Calcaneus

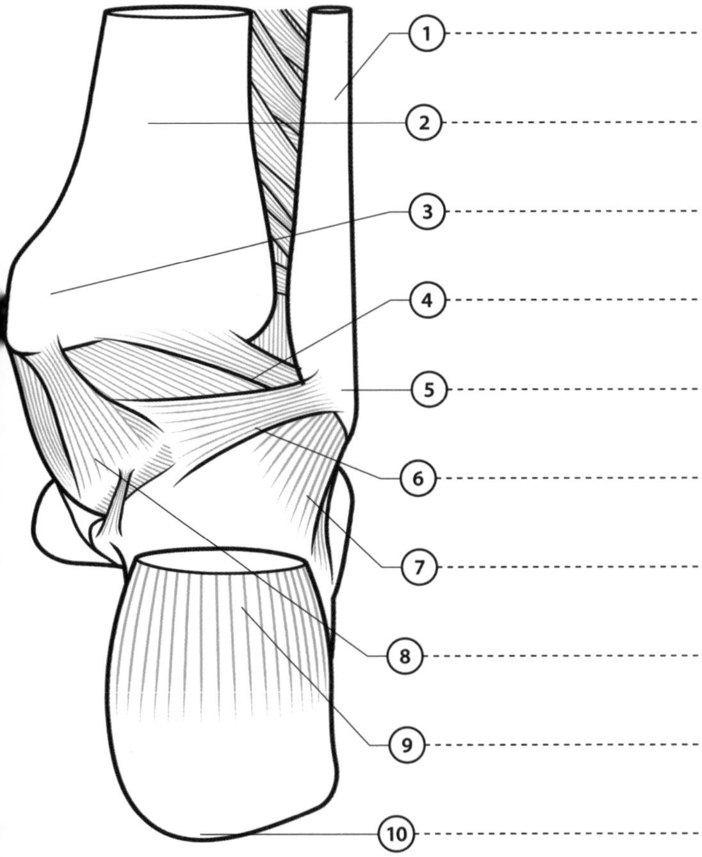

POSTERIOR VIEW

MEDIAL VIEW

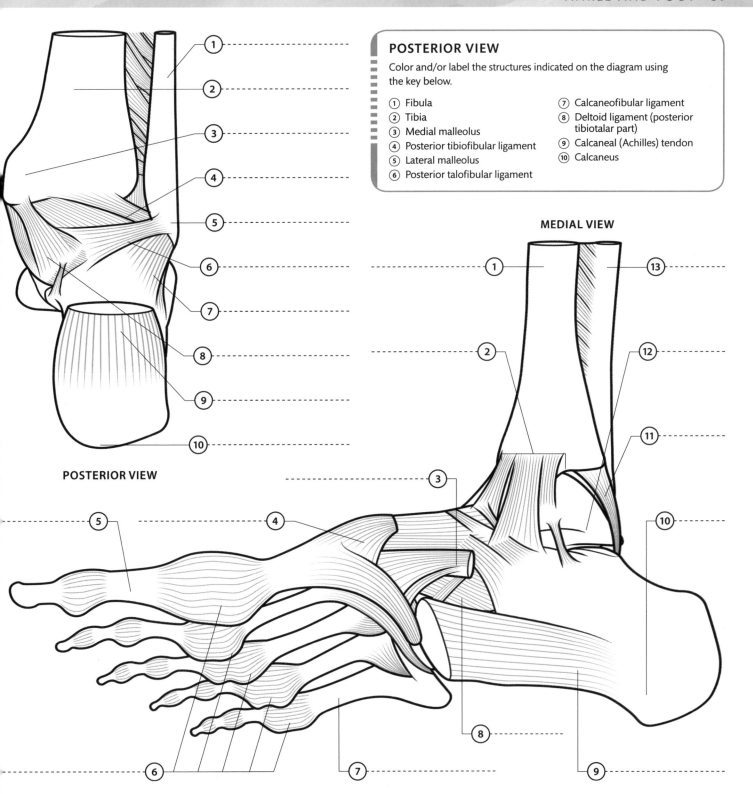

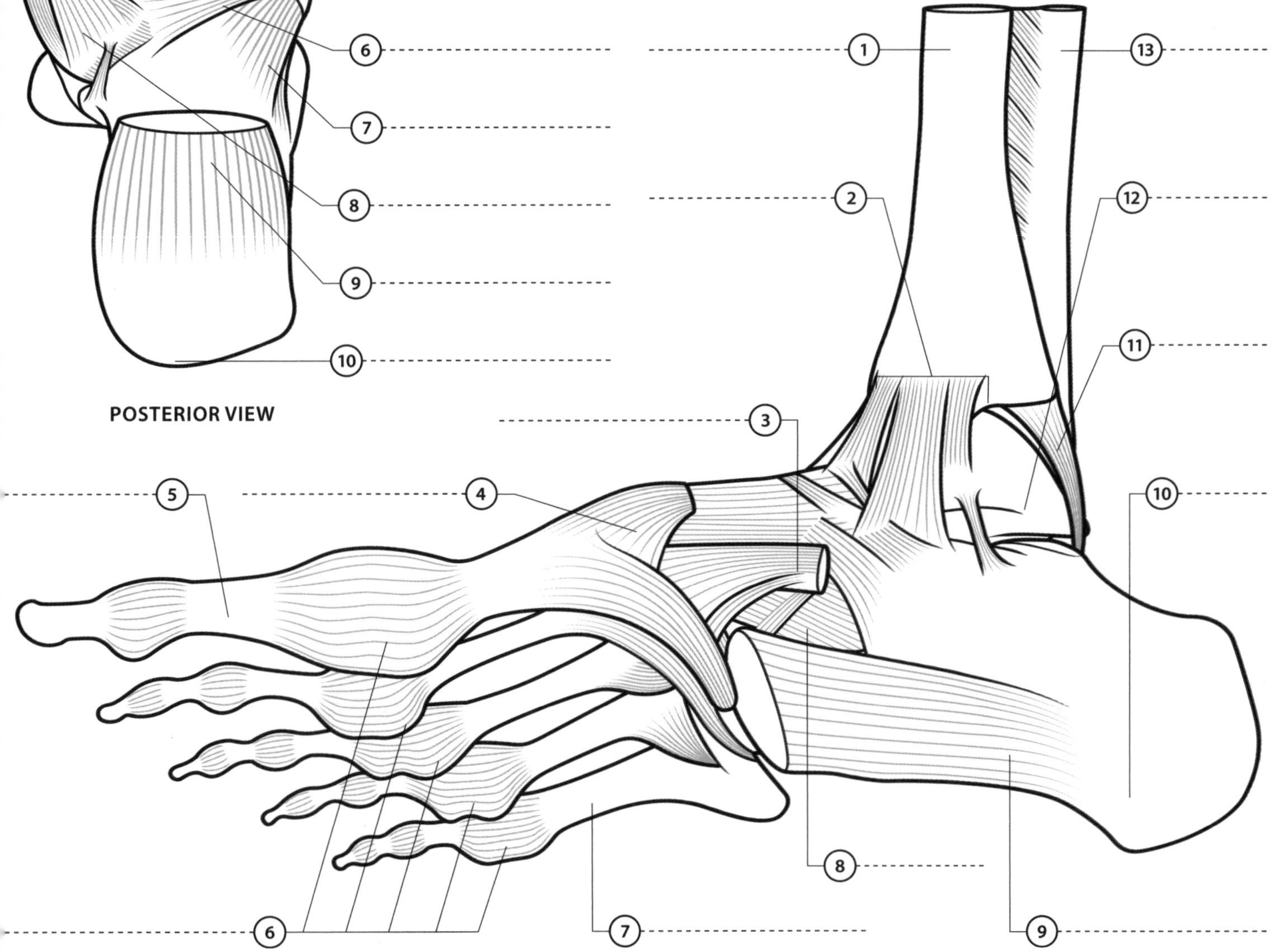

MEDIAL VIEW

Color and/or label the structures indicated on the diagram using the key below.

1. Tibia
2. Deltoid ligament
3. Tibialis posterior tendon
4. Tibialis anterior tendon
5. Proximal phalanx
6. Fibrous capsules of metatarsophalangeal joints
7. Fifth metatarsal
8. Plantar calcaneonavicular ligament (spring ligament)
9. Long plantar ligament
10. Calcaneus
11. Posterior tibiofibular ligament
12. Talus
13. Fibula

See also pp. 46, 74, 76, 104, 152, 165

MUSCULAR AND INTEGUMENTARY SYSTEMS

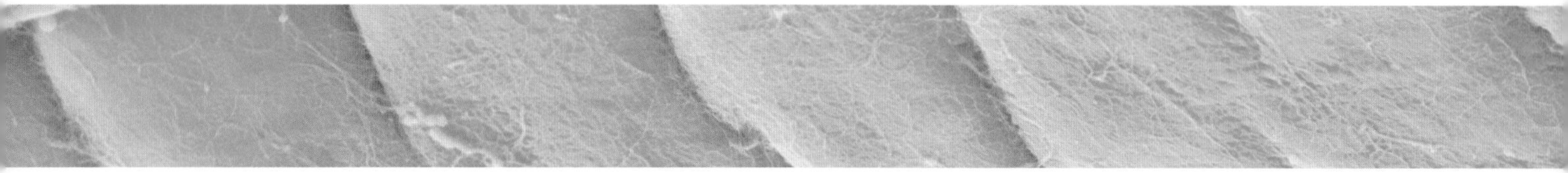

03

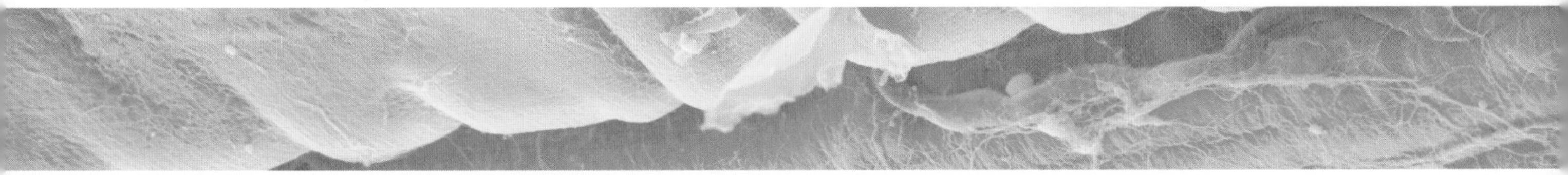

MUSCULAR SYSTEM 1

THESE ANTERIOR AND POSTERIOR VIEWS OF THE MALE BODY ILLUSTRATE THE SUPERFICIAL MUSCLES JUST BELOW THE SKIN AND THE DEEP MUSCLES UNDERLYING THEM.

MUSCLES OF THE BODY

A typical body contains approximately 640 striated muscles, making up about two-fifths of its total weight in a male and a slightly smaller proportion in a female. A typical muscle spans a joint and tapers at each end into a fibrous tendon anchored to bone. Some muscles divide to attach to different bones. Muscles achieve movement by contracting and pulling on the bones to which they are attached; they usually work in groups.

Sometimes the name of a muscle derives from its shape (e.g. the rhomboid major), whereas other muscles are named according to their bone attachments (e.g. the intercostals). Flexor muscles act to bend a limb at the joint they span, while extensors straighten the joint. Abductor muscles cause movement away from the midline of the body, as when holding an arm out to the side. Their adductor partners are responsible for the reverse movement back toward the midline.

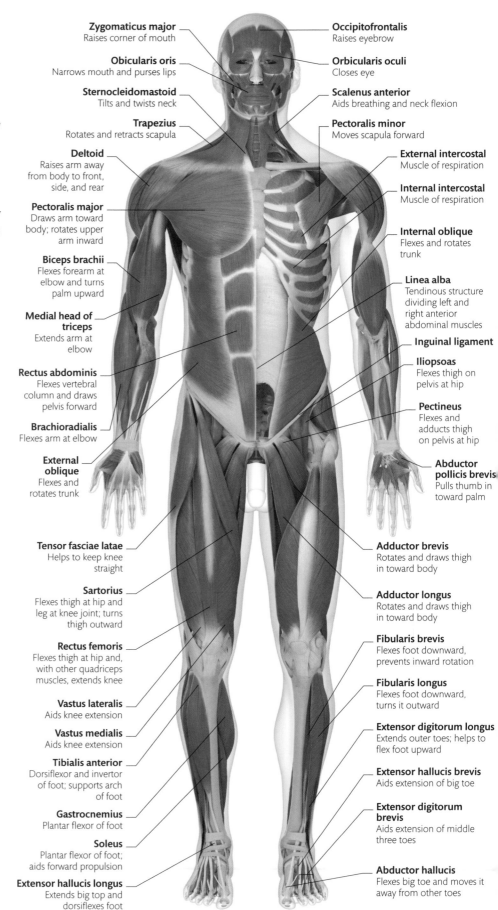

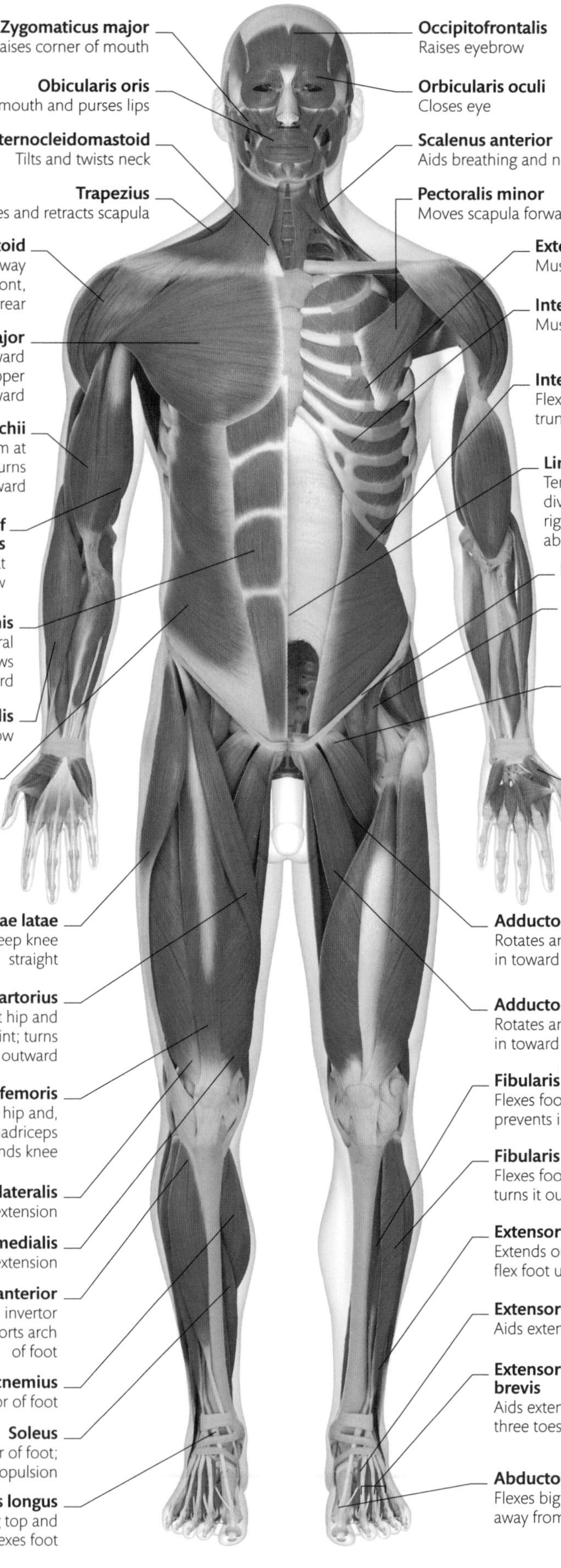

ANTERIOR VIEW

ANTERIOR VIEW
In this illustration the main superficial muscles are on the right side of the body, and some of the intermediate and deep muscles are on the left.

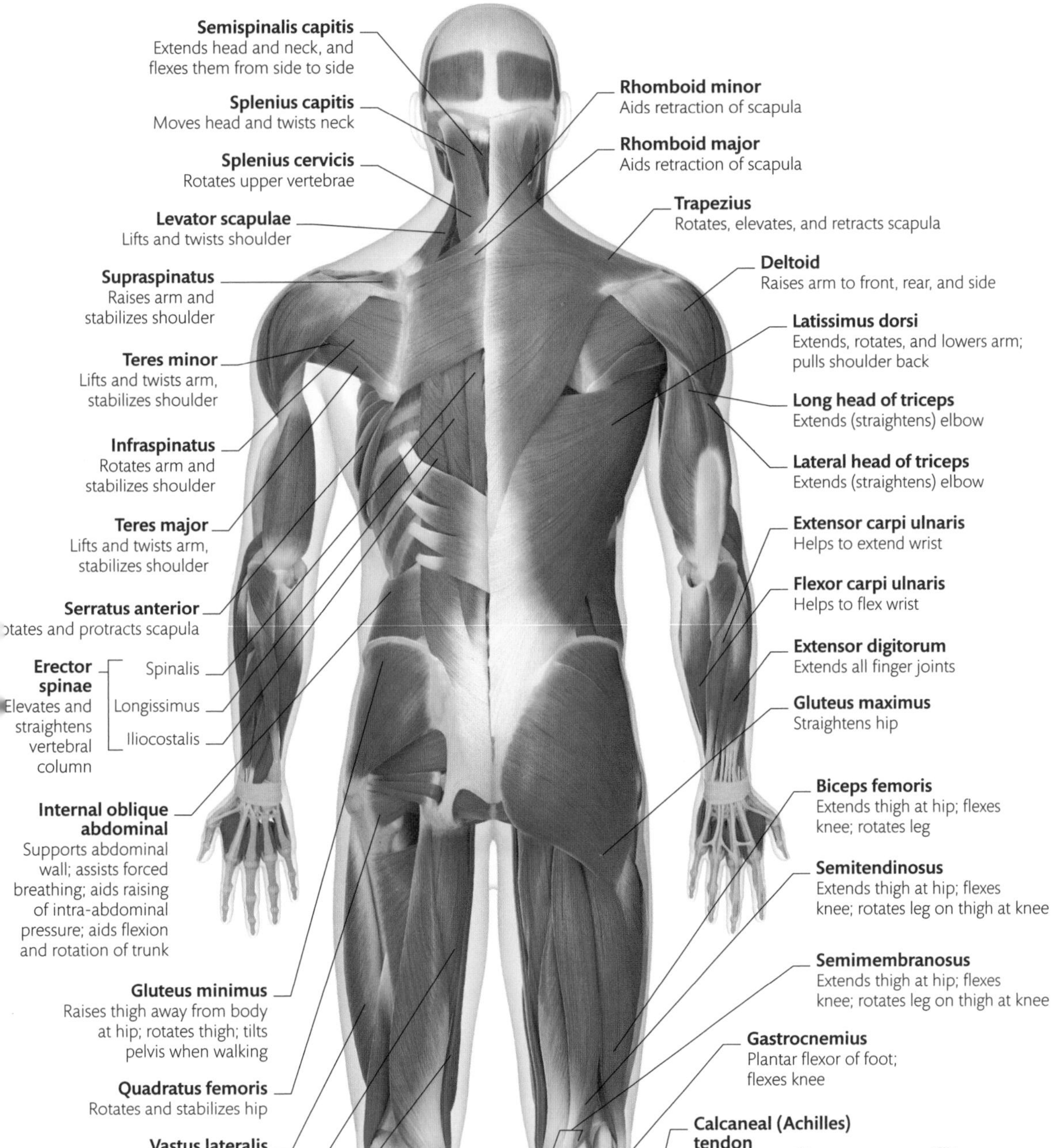
Semispinalis capitis
Extends head and neck, and flexes them from side to side
Splenius capitis
Moves head and twists neck
Splenius cervicis
Rotates upper vertebrae
Levator scapulae
Lifts and twists shoulder
Supraspinatus
Raises arm and stabilizes shoulder
Teres minor
Lifts and twists arm, stabilizes shoulder
Infraspinatus
Rotates arm and stabilizes shoulder
Teres major
Lifts and twists arm, stabilizes shoulder
Serratus anterior
otates and protracts scapula
Erector spinae
Elevates and straightens vertebral column
Spinalis
Longissimus
Iliocostalis
Internal oblique abdominal
Supports abdominal wall; assists forced breathing; aids raising of intra-abdominal pressure; aids flexion and rotation of trunk
Gluteus minimus
Raises thigh away from body at hip; rotates thigh; tilts pelvis when walking
Quadratus femoris
Rotates and stabilizes hip
Vastus lateralis
Extends and stabilizes knee
Adductor magnus
Extends thigh on pelvis at hip
Gracilis
Moves thigh toward body; flexes and rotates thigh
Popliteus
Flexes and turns leg to unlock extended knee
Tibialis posterior
Main muscle in turning sole of foot inward (inversion)
Rhomboid minor
Aids retraction of scapula
Rhomboid major
Aids retraction of scapula
Trapezius
Rotates, elevates, and retracts scapula
Deltoid
Raises arm to front, rear, and side
Latissimus dorsi
Extends, rotates, and lowers arm; pulls shoulder back
Long head of triceps
Extends (straightens) elbow
Lateral head of triceps
Extends (straightens) elbow
Extensor carpi ulnaris
Helps to extend wrist
Flexor carpi ulnaris
Helps to flex wrist
Extensor digitorum
Extends all finger joints
Gluteus maximus
Straightens hip
Biceps femoris
Extends thigh at hip; flexes knee; rotates leg
Semitendinosus
Extends thigh at hip; flexes knee; rotates leg on thigh at knee
Semimembranosus
Extends thigh at hip; flexes knee; rotates leg on thigh at knee
Gastrocnemius
Plantar flexor of foot; flexes knee
Calcaneal (Achilles) tendon

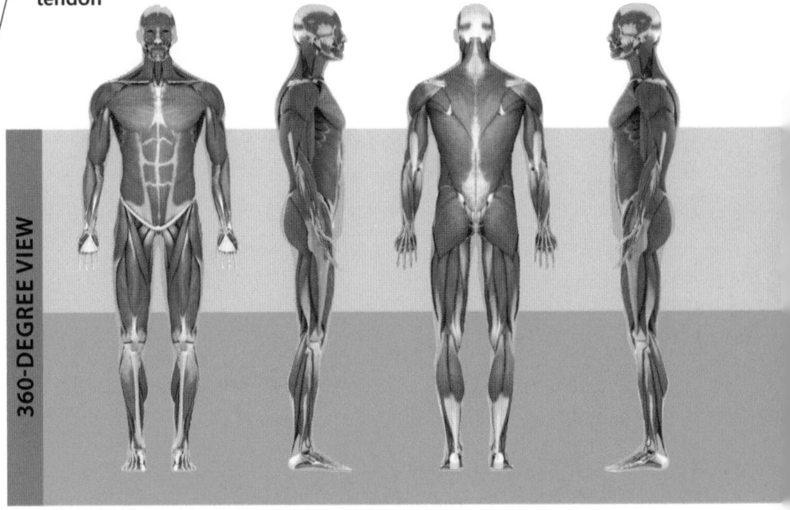
360-DEGREE VIEW

POSTERIOR VIEW
In this illustration the superficial muscles are on the right side of the body, and the deeper ones are shown on the left.

POSTERIOR VIEW

See also p. 56 »

MUSCULAR SYSTEM 2

THE BODY HAS THREE MAIN TYPES OF MUSCLE—SKELETAL, SMOOTH, AND CARDIAC MUSCLE. EACH TYPE HAS A DISTINCTIVE MICROSCOPIC STRUCTURE.

SKELETAL MUSCLE

Skeletal muscle, also known as striated or voluntary muscle, is usually attached to bone at both ends. It consists of densely packed groups of elongated cells called myofibers. Each myofiber is made up of smaller structures called myofibrils, within which are thick and thin myofilaments composed mainly of the proteins actin and myosin. Skeletal muscles are innervated by peripheral or cranial nerves and contract to produce bodily movement, usually under conscious control. These muscles automatically adjust their tension to maintain the body's posture.

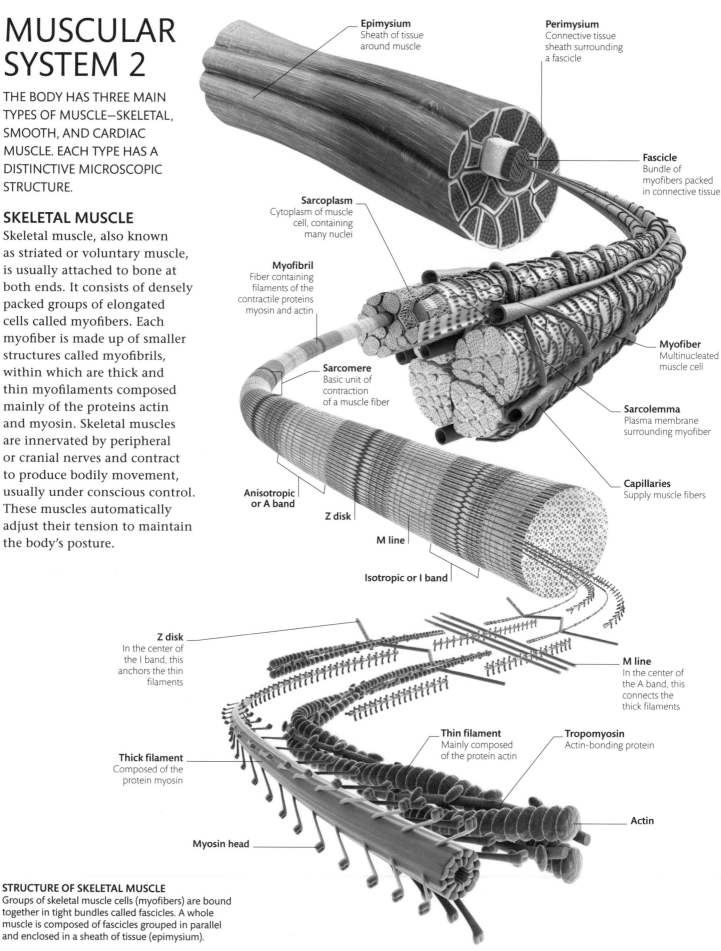

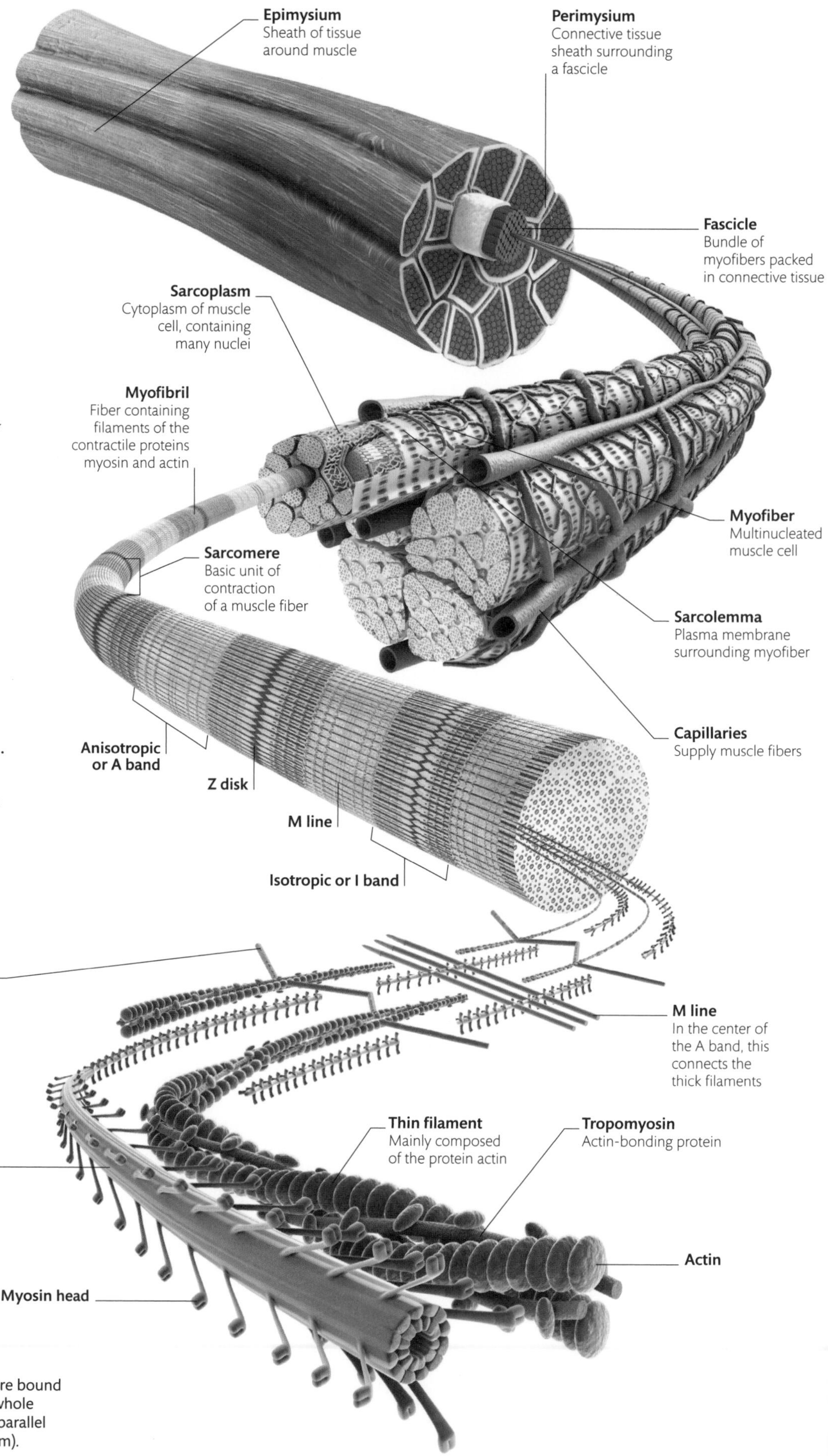

STRUCTURE OF SKELETAL MUSCLE
Groups of skeletal muscle cells (myofibers) are bound together in tight bundles called fascicles. A whole muscle is composed of fascicles grouped in parallel and enclosed in a sheath of tissue (epimysium).

SMOOTH MUSCLE

This type of muscle is found in the organs of the body, particularly in the walls of tubes, such as the gut, blood vessels, and the respiratory tract. Smooth muscle tissue is made of individual, tapering cells and is supplied by autonomic motor nerves, which control the operation of body systems, usually without conscious awareness.

STRUCTURE OF SMOOTH MUSCLE
The short spindle-shaped fibers of smooth muscle form sheets of muscle. This tissue can contract for long periods of time, usually under involuntary control.

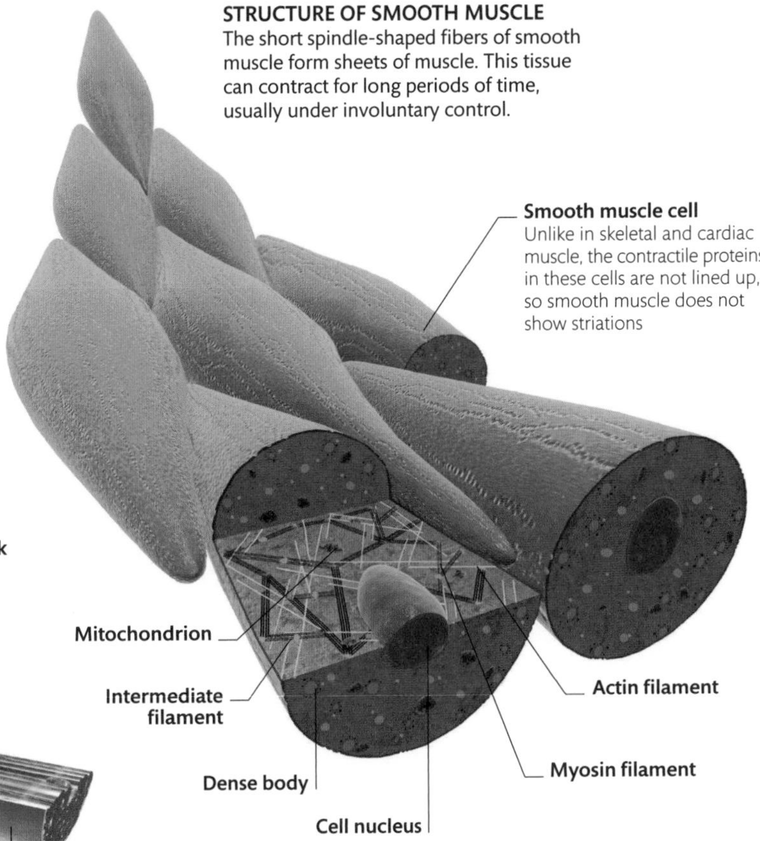

STRUCTURE OF CARDIAC MUSCLE
The fibers in cardiac muscle are short and branching, usually Y- or V-shaped. They form a network within the wall of the heart.

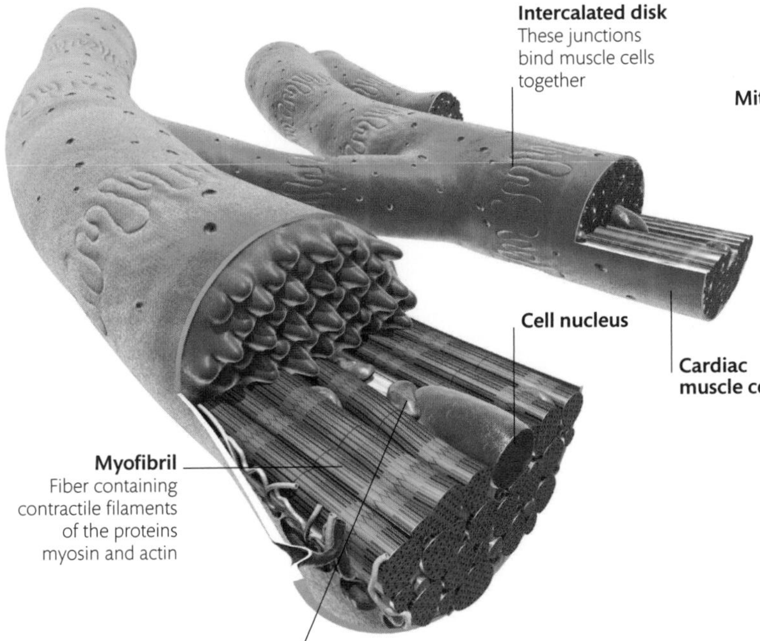

CARDIAC MUSCLE

Also known as myocardium, cardiac muscle is found only in the wall of the heart. This tissue is made up of short interlinked fibers capable of sustained rhythmic movement. Controlled by the autonomic nervous system, cardiac muscle contracts spontaneously, the rate of contraction increasing or decreasing to match the heart's output to the body's needs.

MUSCLE CONTRACTION

Striated muscles contract when nerve impulses trigger the interaction of filaments of the proteins actin and myosin within their cells. These filaments are arranged in overlapping patterns in segments called sarcomeres. Thin actin filaments extend inward from a Z disk, which separates one sarcomere from the next and surround and overlap thick myosin filaments in the center of the sarcomere. When a muscle contracts, small "heads" extending from each myosin filament interact with actin filaments to make the myofibril shorten.

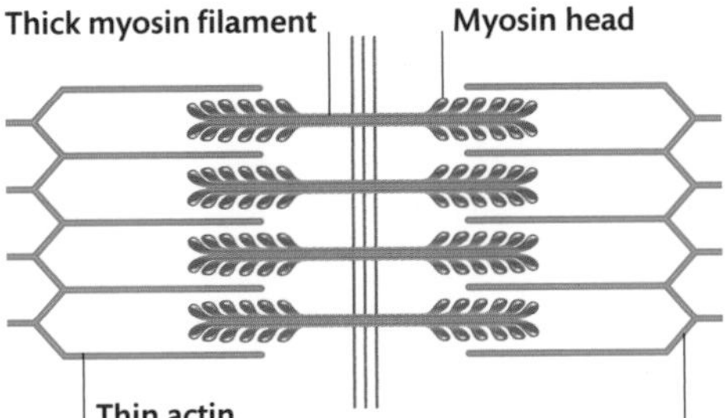

RELAXED MUSCLE
This diagram shows a longitudinal section through a sarcomere (the section between Z disks) in a relaxed muscle. The thick and thin filaments overlap only slightly. The myosin heads are "energized" and ready for action but they do not interact with the actin filaments.

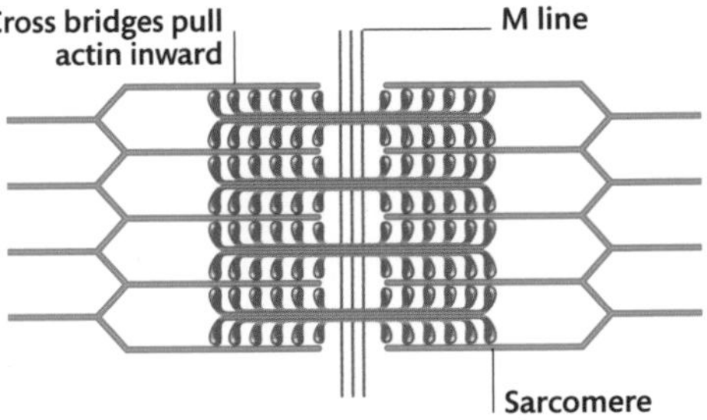

CONTRACTED MUSCLE
During muscle contraction, repeated cycles of cross bridge attachment and detachment pull actin filaments inward so that they slide over the thick filaments, shorten the sarcomere, and increase the overlap between filaments. As a result, muscles become significantly shorter than their resting length.

See also p. 54 »

HEAD AND NECK 1

THE HEAD AND NECK CONTAIN A LARGE NUMBER OF MUSCLES, MANY OF WHICH ARE CONCERNED WITH FACIAL EXPRESSION.

FACIAL MUSCULATURE

The muscles of the face have important functions. They open and close the apertures in the face—the eyes, nostrils, and mouth—and they also play a key role in communication, which is why they are often known as the muscles of facial expression. These muscles are attached to bone at one end and skin and subcutaneous tissue at the other, and they make possible expressions such as frowning, raising the eyebrows, smiling, and pouting.

ANTERIOR VIEW

Color and/or label the structures indicated on the diagram using the key below.

1. Epicranial aponeurosis
2. Frontal belly of occipitofrontalis
3. Temporalis
4. Orbicularis oculi
5. Depressor labii inferioris
6. Mentalis
7. Sternocleidomastoid
8. Sternohyoid
9. Trapezius
10. Scalenus anterior
11. Omohyoid (superior belly)
12. Levator scapulae
13. Depressor anguli oris
14. Risorius
15. Orbicularis oris
16. Masseter
17. Zygomaticus major
18. Zygomaticus minor
19. Levator labii superioris
20. Levator labii superioris alaeque nasi
21. Nasalis

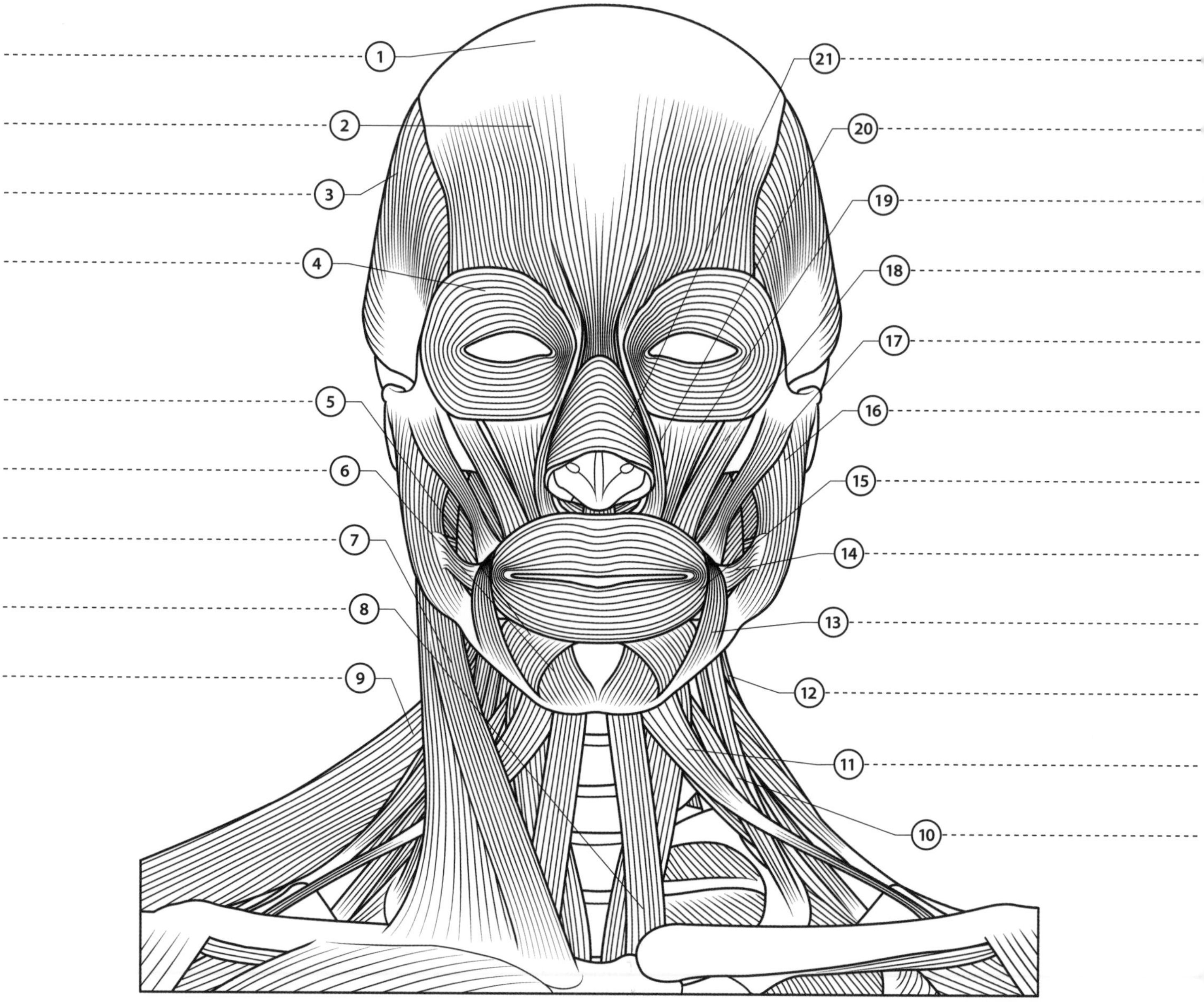

ANTERIOR VIEW

EYE MUSCLES

Color and/or label the structures indicated on the diagram using the key below.

1. Trochlea of superior oblique
2. Superior oblique
3. Medial rectus
4. Superior rectus
5. Lateral rectus
6. Common anular tendon

POSTERIOR VIEW

Color and/or label the structures indicated on the diagram using the key below.

1. Semispinalis capitis
2. Splenius capitis
3. Levator scapulae
4. Rhomboid minor
5. Rhomboid major
6. Spine of scapula
7. Acromion
8. Trapezius
9. Sternocleidomastoid
10. Occipital belly of occipitofrontalis
11. Temporalis

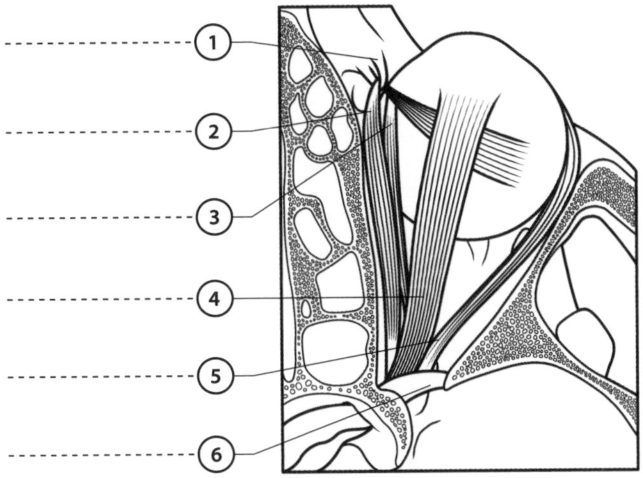

EYE MUSCLES
(VIEW FROM ABOVE)

EYE MUSCLES

The six muscles that move the eye do not move independently. The lateral rectus rotates the eye outward. The medial rectus rotates the eye inward. The superior rectus rotates the eye upward. The superior oblique rotates the eye downward and outward. The inferior oblique (not shown) rotates the eye upward and inward. The inferior rectus (not shown) rotates the eye downward.

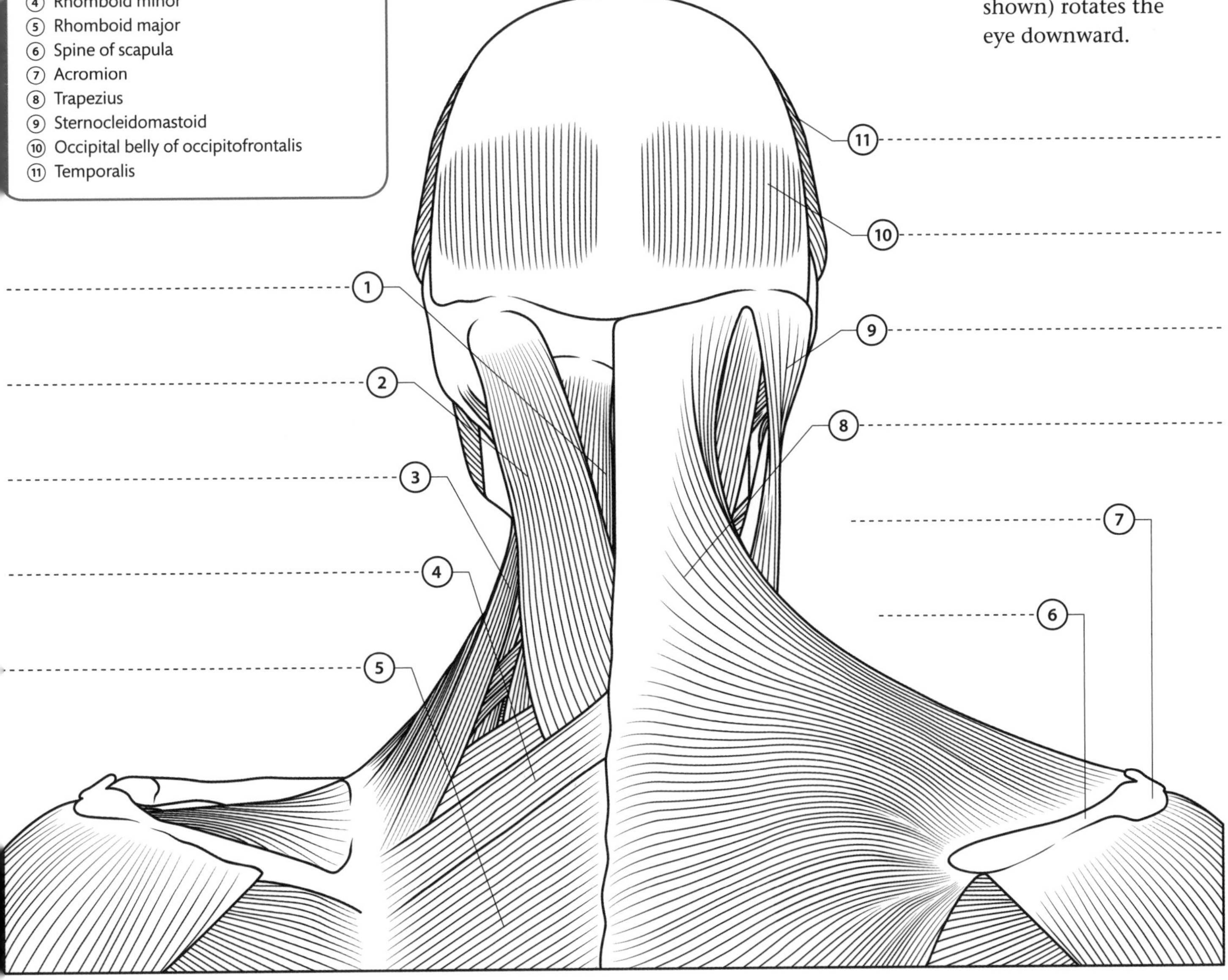

DEEP MUSCLES

POSTERIOR VIEW

SUPERFICIAL MUSCLES

See also pp. 28, 60, 92, 118, 134, 136, 160, 174, 192

HEAD AND NECK 2

AS WELL AS THE MUSCLES OF FACIAL EXPRESSION, THE HEAD AND NECK ALSO CONTAIN OTHER MUSCLES, INCLUDING THOSE THAT MOVE THE JAW, CONTROL SWALLOWING, AND CONTROL THE VOCAL CORDS.

HEAD AND NECK MUSCULATURE

The muscles of mastication attach from the skull to the mandible, operating to open and close the mouth and grind the teeth together for chewing. The two largest muscles of mastication are the temporalis and masseter. The soft palate, tongue, pharynx, and larynx also contain muscles. The soft palate comprises five pairs of muscles. The tongue is a great mass of muscle covered in mucosa. The pharyngeal muscles are important in swallowing, and the laryngeal muscles control the vocal cords.

LATERAL VIEW

Color and/or label the structures indicated on the diagram using the key below.

1. Epicranial aponeurosis
2. Temporalis
3. Occipital belly of occipitofrontalis
4. Posterior belly of digastric
5. Sternocleidomastoid
6. Splenius capitis
7. Inferior constrictor of pharynx
8. Trapezius
9. Levator scapulae
10. Scalenus medius
11. Scalenus anterior
12. Scalenus posterior
13. Inferior belly of omohyoid
14. Sternothyroid
15. Sternohyoid
16. Superior belly of omohyoid
17. Thyrohyoid
18. Masseter
19. Anterior belly of digastric
20. Mentalis
21. Depressor labii inferioris
22. Depressor anguli oris
23. Orbicularis oris
24. Risorius
25. Nasalis
26. Levator labii superioris
27. Levator labii superioris alaeque nasi
28. Zygomaticus major
29. Orbicularis oculi
30. Frontal belly of occipitofrontalis

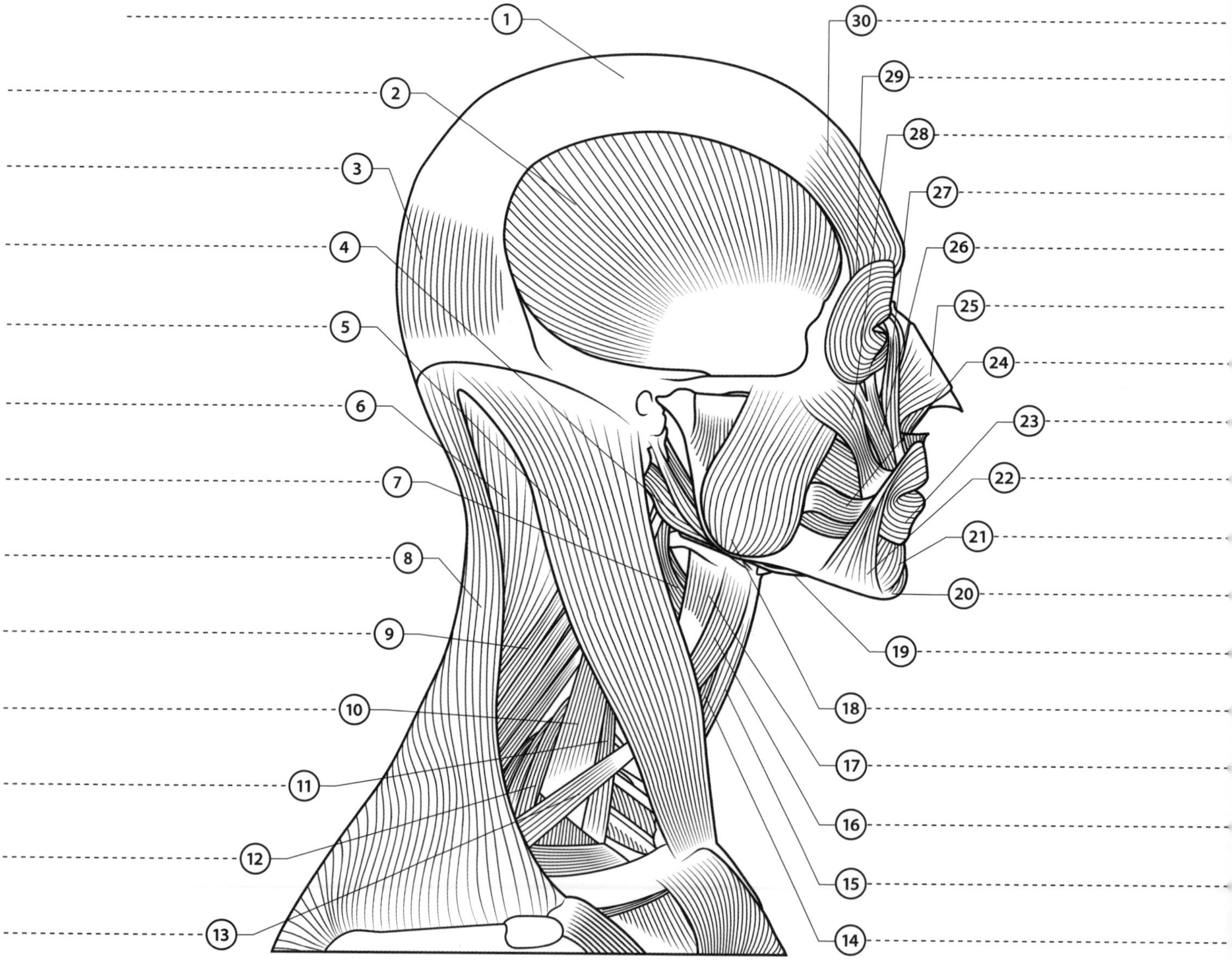

LATERAL VIEW

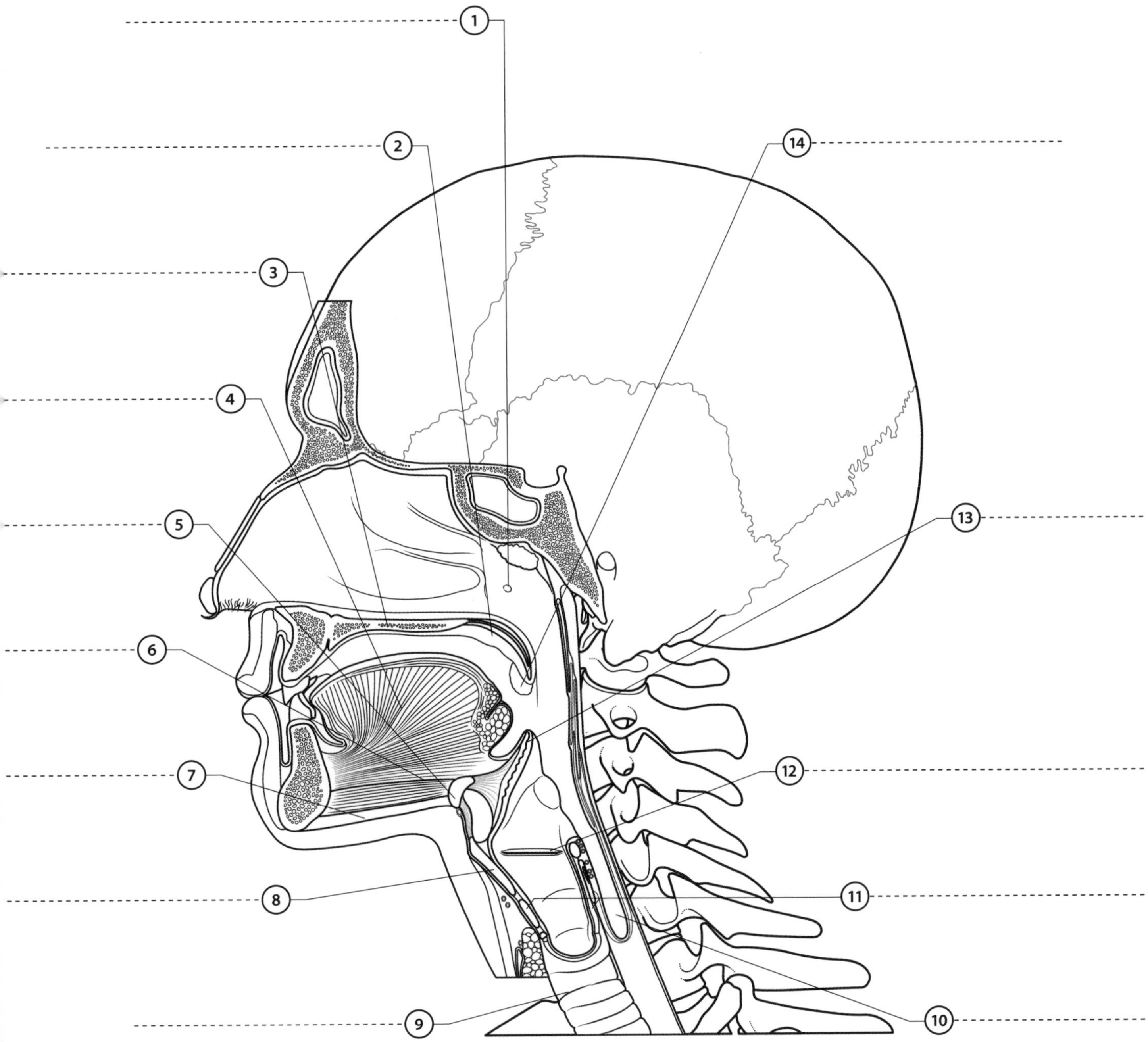

SAGITTAL SECTION

SAGITTAL SECTION

Color and/or label the structures indicated on the diagram using the key below.

1. Eustachian (pharyngotympanic) tube
2. Soft palate
3. Hard palate
4. Genioglossus
5. Hyoid bone
6. Genohyoid
7. Mylohyoid
8. Thyroid cartilage
9. Trachea
10. Esophagus
11. Cricoid cartilage
12. Vocal cord
13. Epiglottis
14. Palatine tonsil

See also pp. 28, 58, 98, 118, 134, 160, 170 »

THORAX 1

THE ANTERIOR MUSCLES OF THE THORAX ARE PRIMARILY CONCERNED WITH MOVING OR STABILIZING THE THORACIC WALL DURING BREATHING.

ANTERIOR MUSCULATURE

The walls of the thorax are filled in, between the ribs, by the intercostal muscles. There are three layers of these muscles, and the muscle fibers of each layer lie in different directions. The main muscle for breathing is the diaphragm (not shown here), which divides the thorax and abdomen. It attaches to the spine and deep muscles in the back, around the margins of the rib cage, and to the sternum at the front. Although the intercostal muscles are active during breathing, their main function seems to be to prevent the spaces between the ribs from being "sucked in." Other muscles may also be recruited to help with deep breathing. The sternocleidomastoid and scalene muscles in the neck can help by pulling the sternum and upper ribs upward. The pectoral muscles can also pull the ribs up and out, if the arm is held in a fixed position.

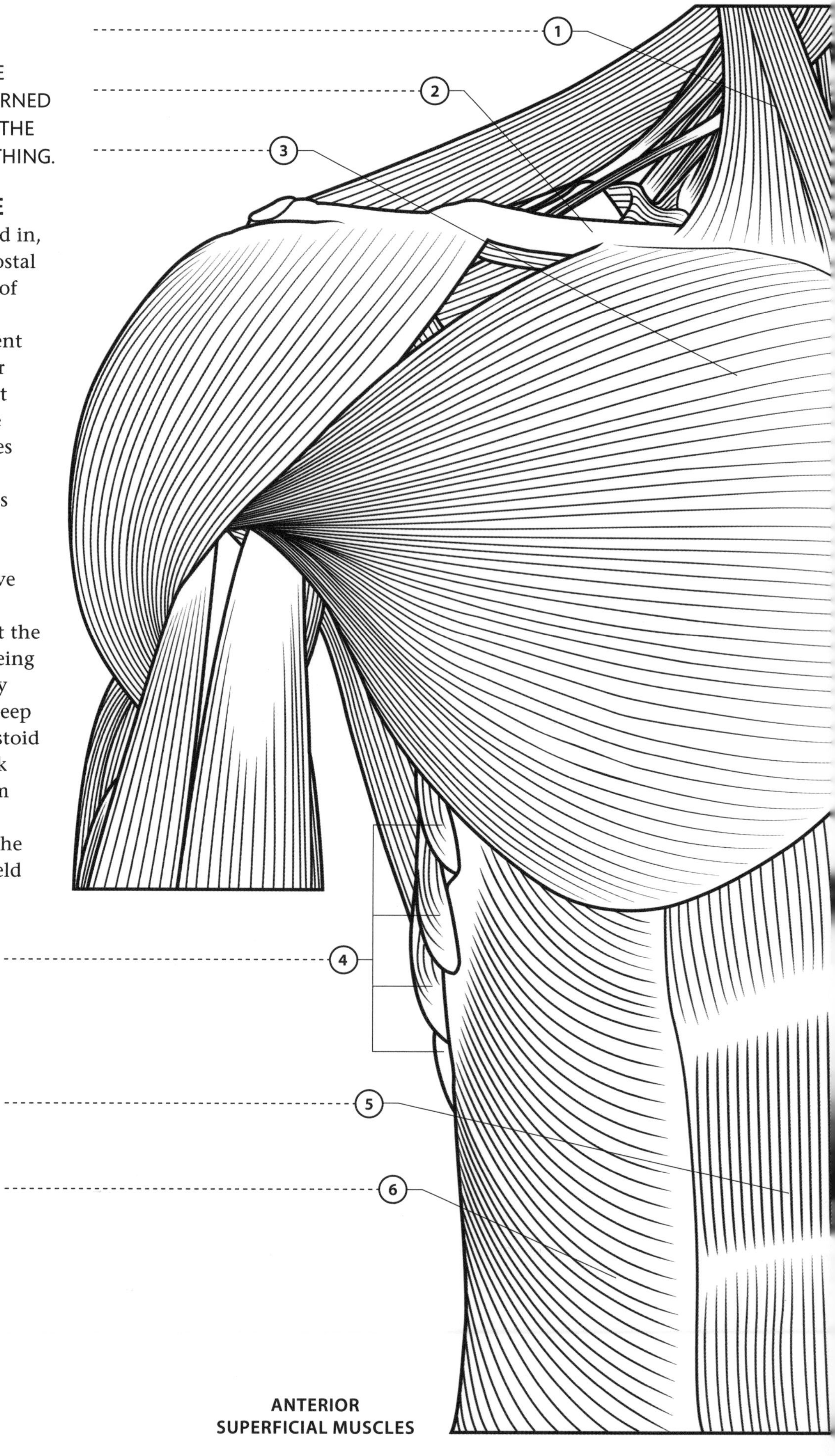

ANTERIOR
SUPERFICIAL MUSCLES

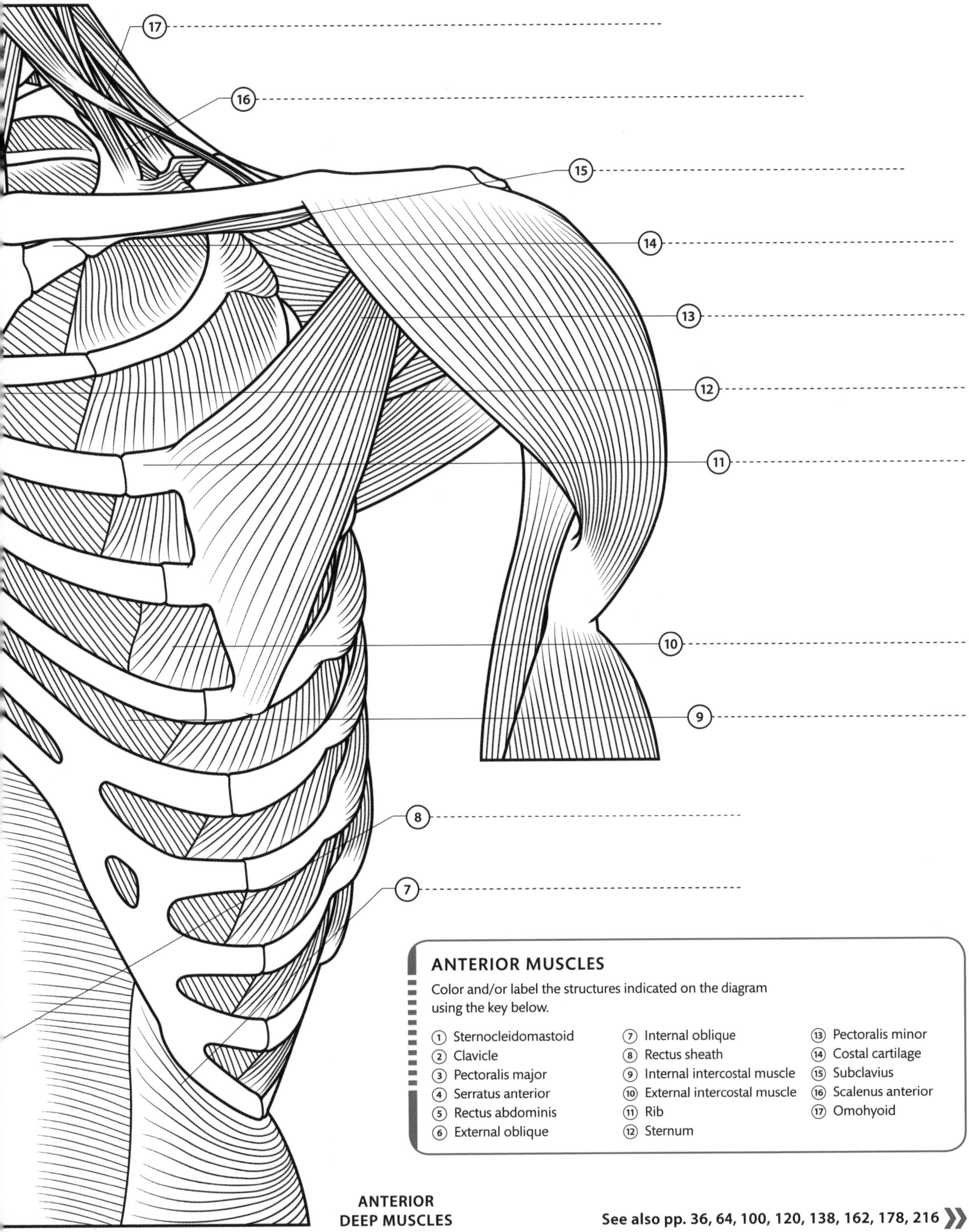

ANTERIOR MUSCLES

Color and/or label the structures indicated on the diagram using the key below.

1. Sternocleidomastoid
2. Clavicle
3. Pectoralis major
4. Serratus anterior
5. Rectus abdominis
6. External oblique
7. Internal oblique
8. Rectus sheath
9. Internal intercostal muscle
10. External intercostal muscle
11. Rib
12. Sternum
13. Pectoralis minor
14. Costal cartilage
15. Subclavius
16. Scalenus anterior
17. Omohyoid

ANTERIOR DEEP MUSCLES

See also pp. 36, 64, 100, 120, 138, 162, 178, 216

THORAX 2

THE POSTERIOR MUSCLES OF THE THORAX ARE MAINLY CONCERNED WITH BODY MOVEMENTS AND WITH MAINTAINING POSTURE.

POSTERIOR MUSCULATURE

The superficial muscles of the back include two large, triangular-shaped muscles: the latissimus dorsi and trapezius. Although the latissimus dorsi is involved in forced expiration, squeezing the lower chest to expel air, it is really a climbing muscle: when hanging by the arms, it is largely the powerful latissimus dorsi that makes it possible to pull the body weight up. Underneath those superficial muscles are the deep extensor muscles of the spine, which can be felt as a distinct ridge on each side of the spine, especially in the lumbar region. The most bulky of these muscles are collectively known as erector spinae and play a vital role in keeping the spine erect or extending a flexed spine.

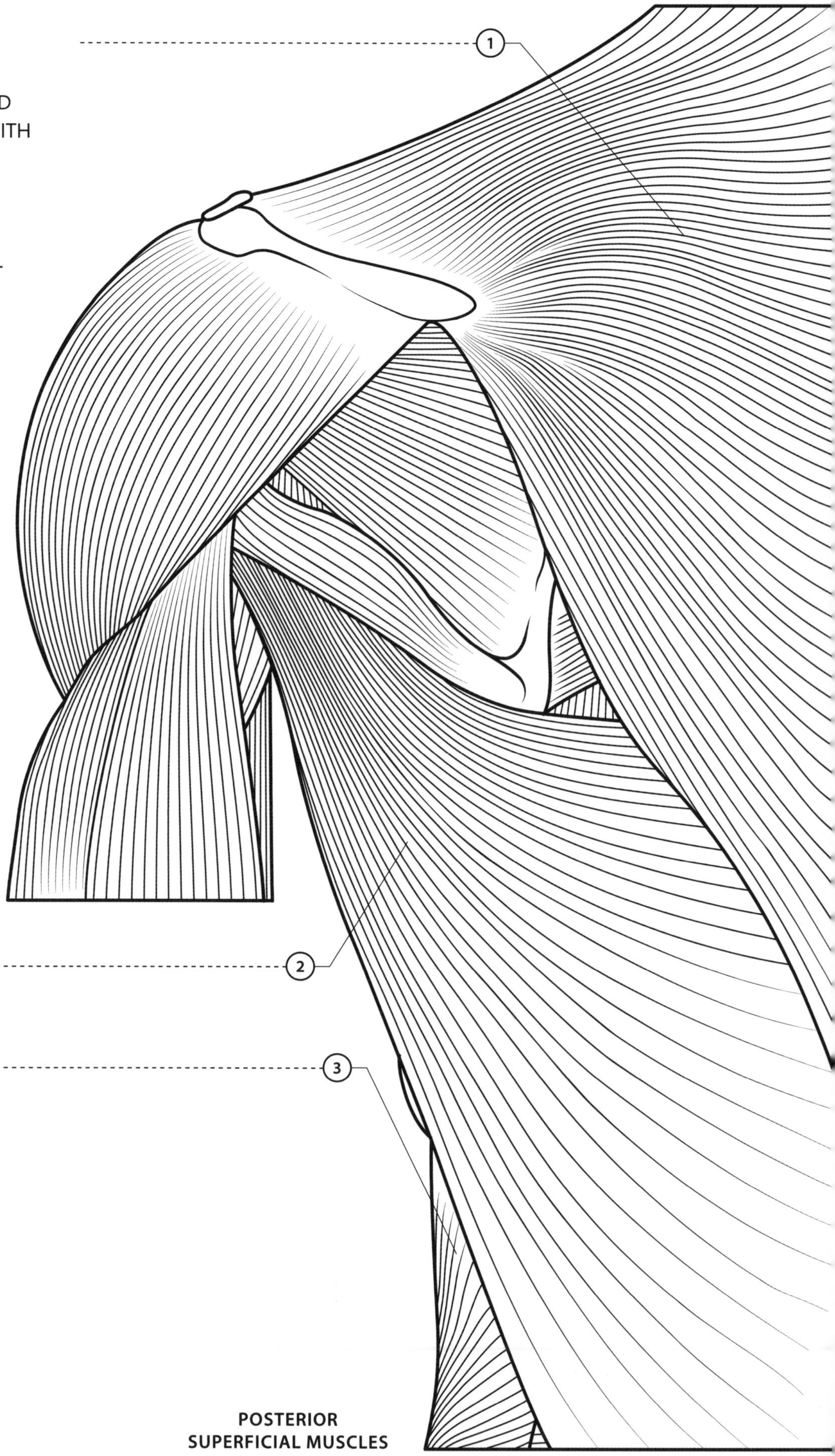

POSTERIOR SUPERFICIAL MUSCLES

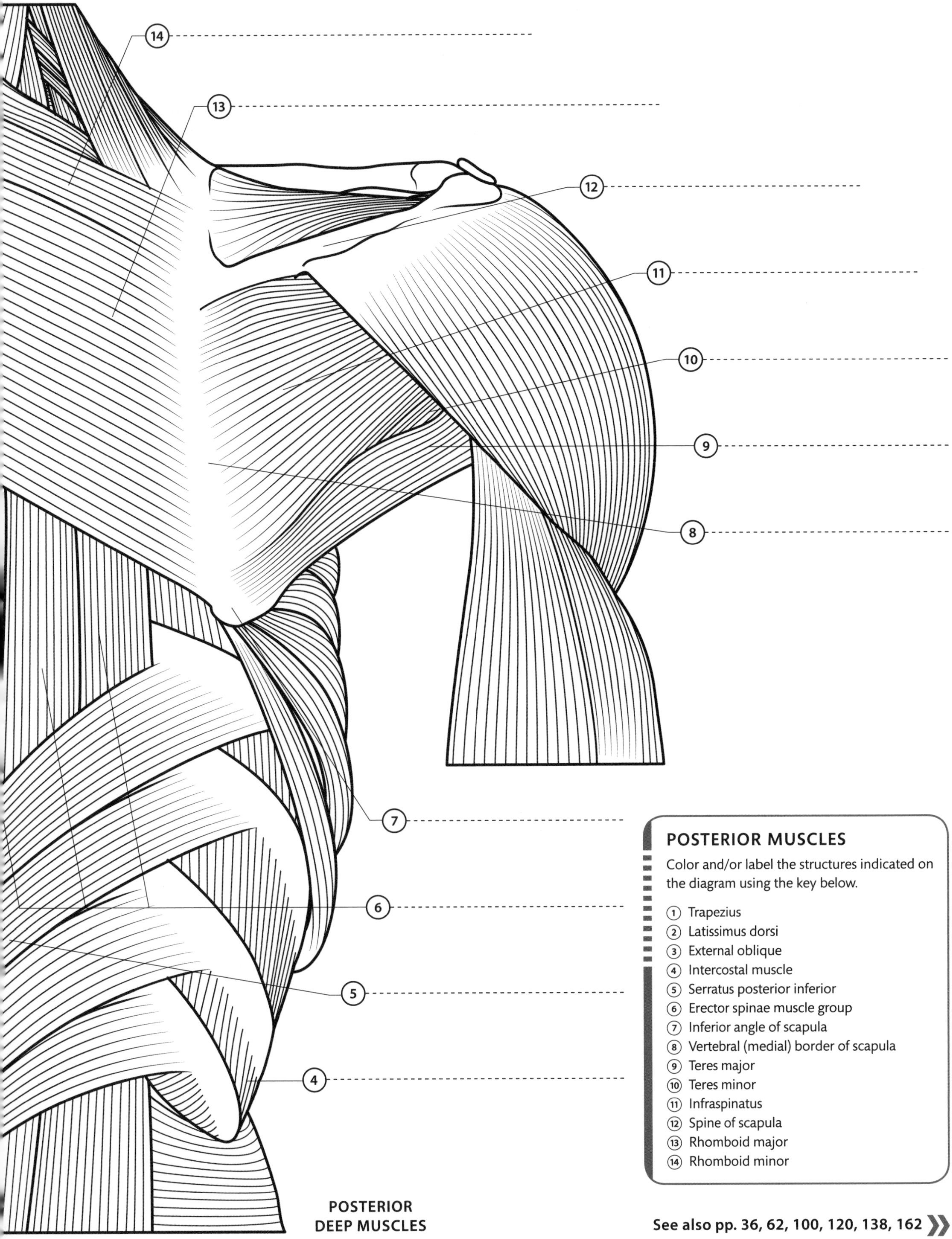

POSTERIOR
DEEP MUSCLES

POSTERIOR MUSCLES

Color and/or label the structures indicated on the diagram using the key below.

1. Trapezius
2. Latissimus dorsi
3. External oblique
4. Intercostal muscle
5. Serratus posterior inferior
6. Erector spinae muscle group
7. Inferior angle of scapula
8. Vertebral (medial) border of scapula
9. Teres major
10. Teres minor
11. Infraspinatus
12. Spine of scapula
13. Rhomboid major
14. Rhomboid minor

See also pp. 36, 62, 100, 120, 138, 162 »

ABDOMEN AND PELVIS 1

THE ANTERIOR MUSCLES OF THE ABDOMEN AND PELVIS ARE INVOLVED NOT ONLY IN POSTURE AND BODY MOVEMENTS BUT ALSO IN FUNCTIONS SUCH AS DEFECATION AND MICTURITION.

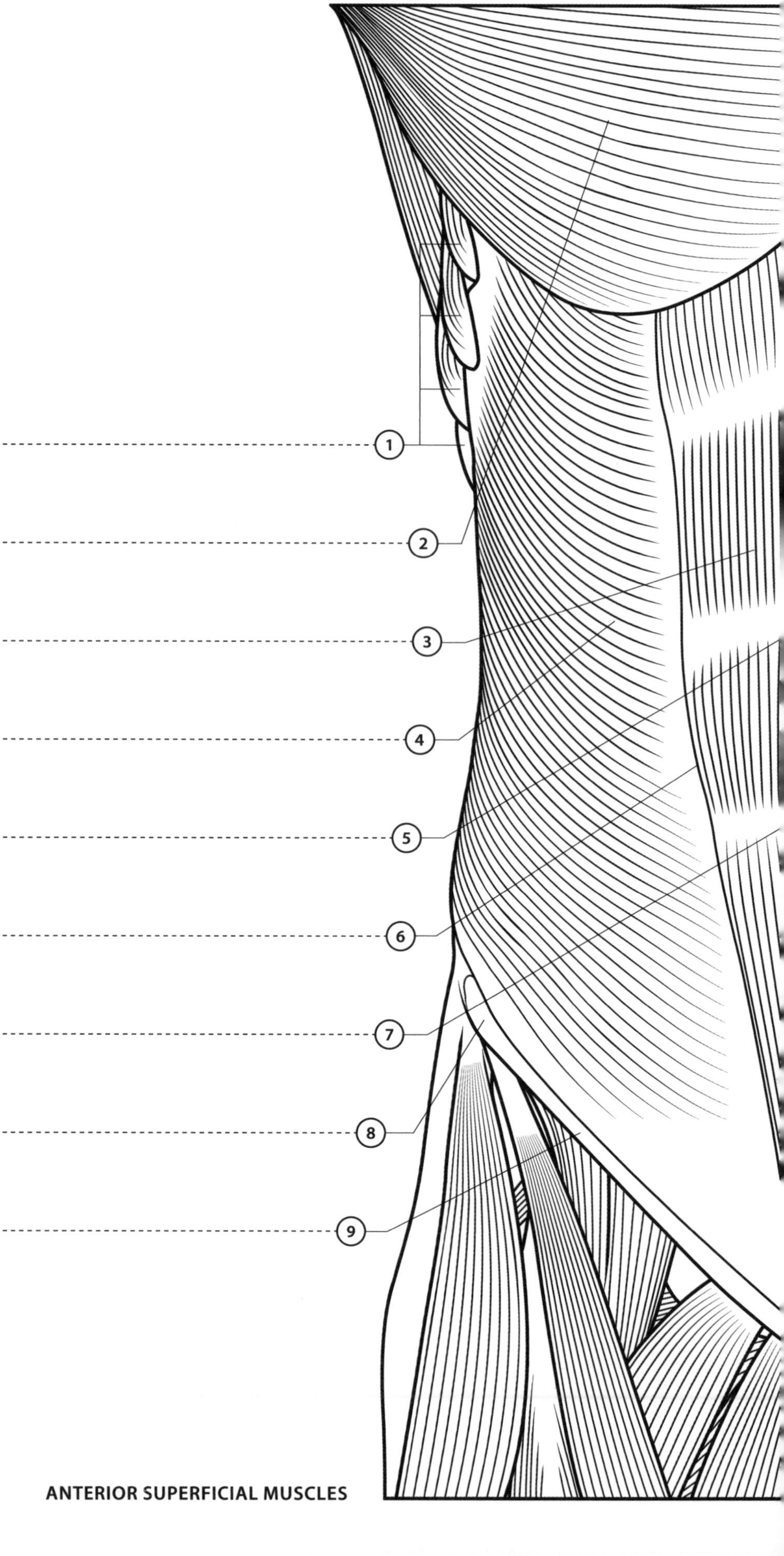

ANTERIOR SUPERFICIAL MUSCLES

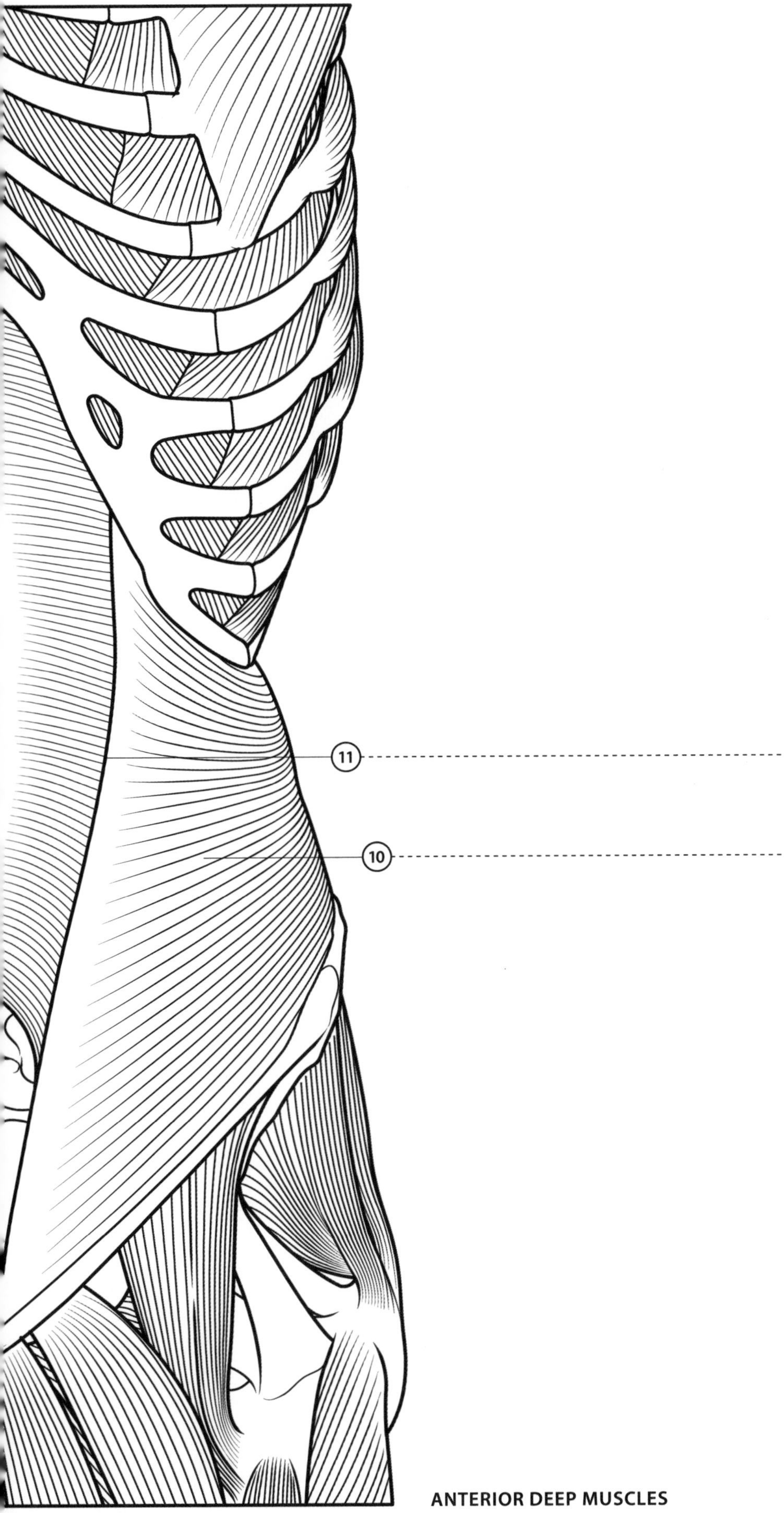

ANTERIOR DEEP MUSCLES

ANTERIOR MUSCULATURE

The abdominal muscles can move the trunk—flexing the spine to the front or side, or twisting the abdomen from side to side. They are important in posture, helping to support the upright spine, and are also brought into action when lifting heavy objects. They are involved during defecation, micturition, and forced expiration, as they compress the abdomen and raise the internal pressure. At the front, lying either side of the midline, there are two straight, straplike rectus abdominis muscles, which are broken up by horizontal tendons; in a well-toned, slim person, this creates the "six-pack" appearance. Flanking the recti muscles on each side are three layers of broad, flat muscles.

ANTERIOR MUSCLES

Color and/or label the structures indicated on the diagram using the key below.

1. Serratus anterior
2. Pectoralis major
3. Rectus abdominis
4. External oblique
5. Linea alba
6. Linea semilunaris
7. Tendinous intersection
8. Anterior superior iliac spine
9. Inguinal ligament
10. Internal oblique
11. Aponeurosis of internal oblique (cut edge)

See also p. 70 »

ABDOMEN AND PELVIS 2

THE POSTERIOR MUSCLES OF THE ABDOMEN AND PELVIS ARE PRIMARILY CONCERNED WITH POSTURE AND BODY MOVEMENTS.

POSTERIOR MUSCULATURE

The most superficial muscle of the lower back is the broad latissimus dorsi. Underneath this, lying along the spine on each side, there is a large bulk of muscle that forms two ridges in the lumbar region in a well-toned person. This muscle mass is collectively known as the erector spinae and, as its name suggests, it is important in keeping the spine upright. When the spine is flexed forward, the erector spinae can pull it back into an upright position, and can even take it further, into extension. The muscle can be divided into three main strips on each side: iliocostalis, longissimus, and spinalis. Most of the muscle bulk of the buttock is due to one muscle: the fleshy gluteus maximus, which extends the hip joint. Beneath the gluteus maximus is a range of smaller muscles that also move the hip.

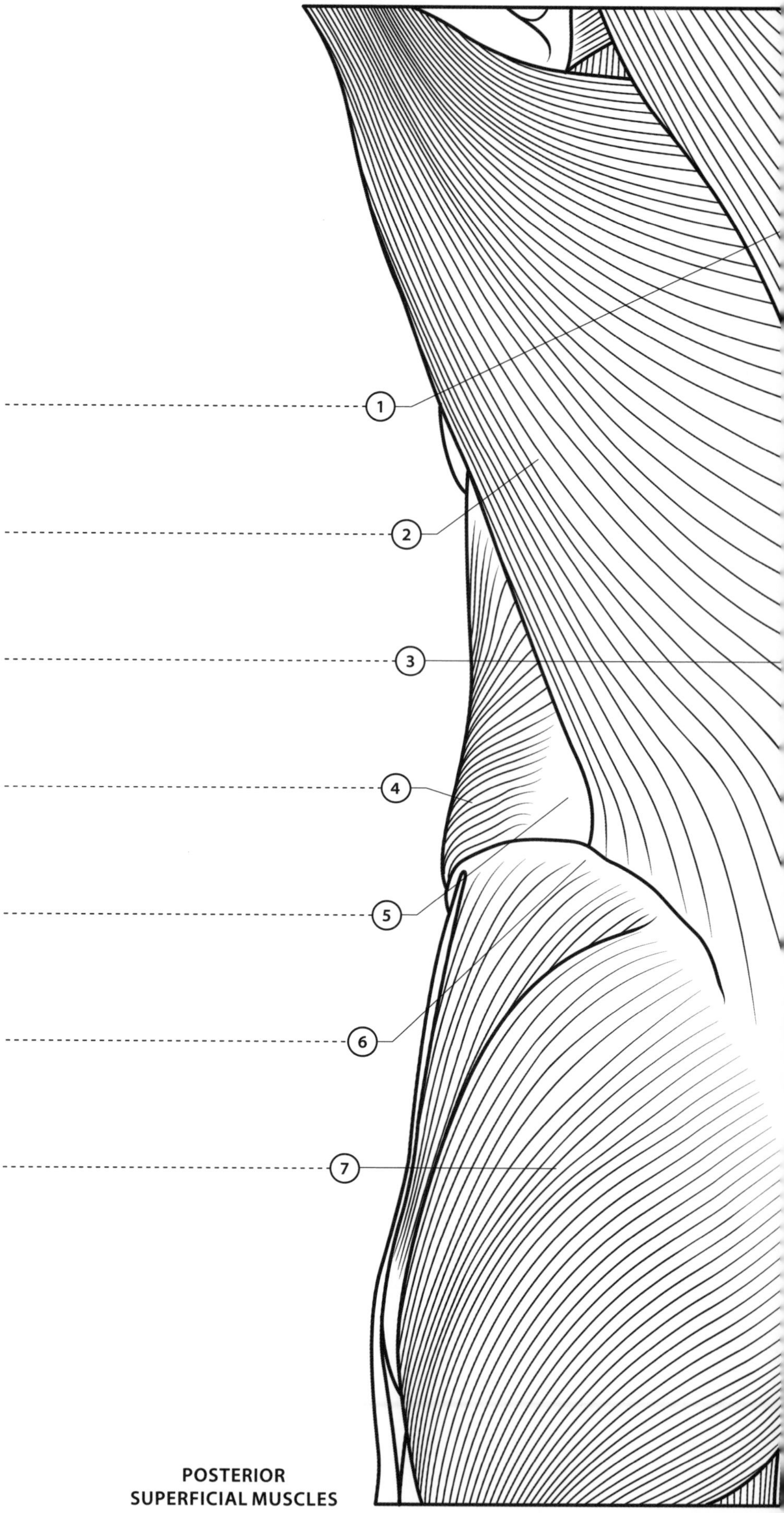

POSTERIOR SUPERFICIAL MUSCLES

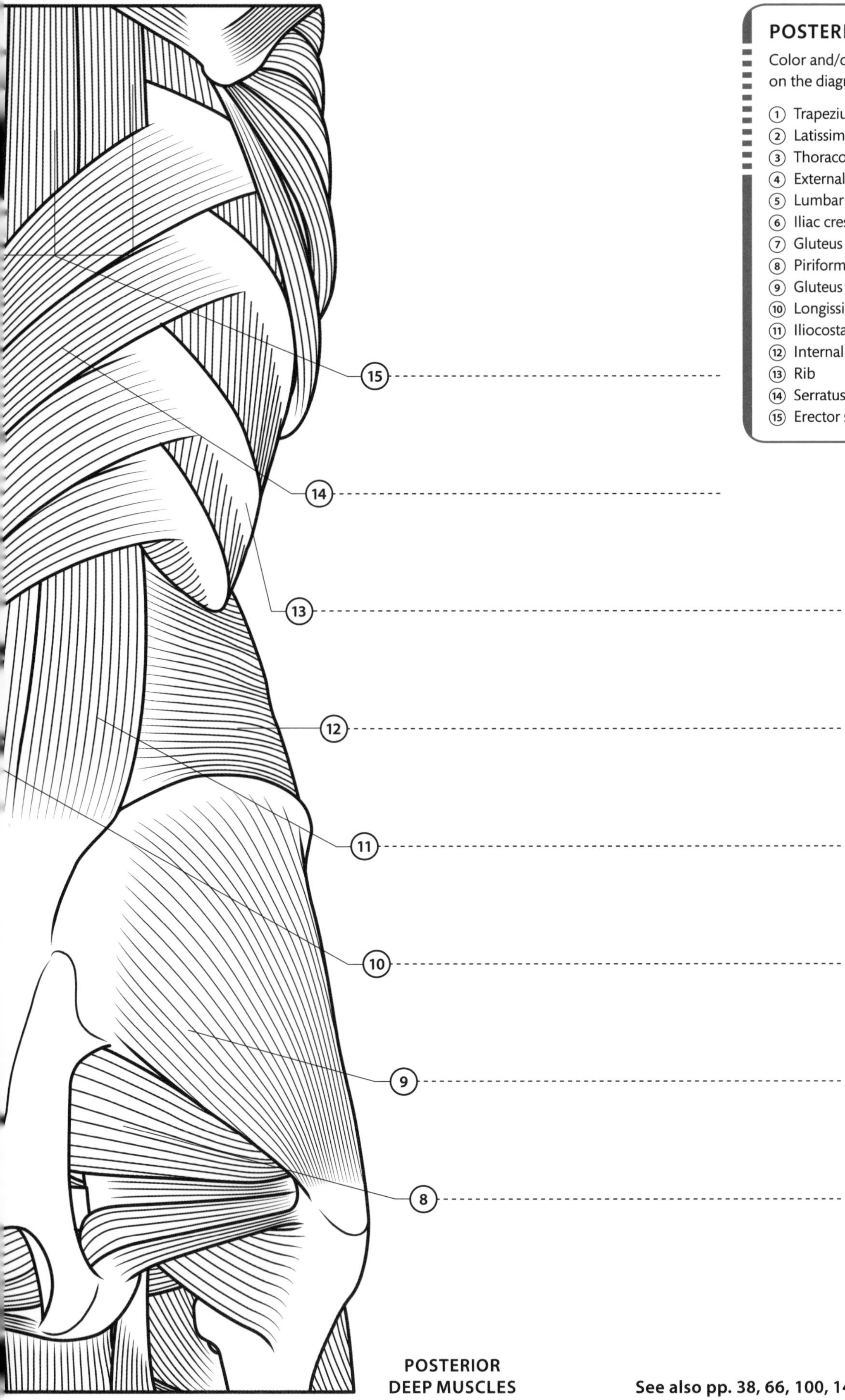

POSTERIOR
DEEP MUSCLES

POSTERIOR MUSCLES

Color and/or label the structures indicated on the diagram using the key below.

1. Trapezius
2. Latissimus dorsi
3. Thoracolumbar fascia
4. External oblique
5. Lumbar triangle
6. Iliac crest
7. Gluteus maximus
8. Piriformis
9. Gluteus medius
10. Longissimus
11. Iliocostalis
12. Internal oblique
13. Rib
14. Serratus posterior inferior
15. Erector spinae muscle group

See also pp. 38, 66, 100, 144, 163, 179, 199, 219, 224 »

SHOULDER AND ARM 1

THE ANTERIOR MUSCLES OF THE SHOULDER AND ARM, IN COMBINATION WITH THE POSTERIOR MUSCLES, ARE RESPONSIBLE FOR THE COMPLEX RANGE OF MOVEMENTS THAT THE ARM, WRIST, AND HAND CAN PERFORM.

ANTERIOR SUPERFICIAL MUSCULATURE

The deltoid lies over the shoulder. Acting as a whole, this muscle abducts the arm, but its fibers that attach to the front of the clavicle can also move the arm forward. The pectoralis major flexes the arm forward and adducts it. Biceps brachii is a flexor of the elbow, and can also rotate the radius so the palm faces upward (supination). There are five superficial muscles on the front of the forearm, all taking their attachment from the medial epicondyle of the humerus. Pronator teres attaches across to the radius and can pull this bone into pronation (so the palm turns downward). The other muscles run further down the forearm, becoming tendons that attach around the wrist or continue into the hand.

ANTERIOR SUPERFICIAL MUSCLES

Color and/or label the structures indicated on the diagram using the key below.

1. Platysma
2. Pectoralis major
3. Deltoid
4. Short head of biceps
5. Long head of biceps
6. Biceps brachii
7. Biceps aponeurosis
8. Pronator teres
9. Flexor carpi radialis
10. Brachioradialis
11. Palmaris longus tendon
12. Flexor digitorum superficialis
13. Flexor retinaculum
14. Thenar muscles
15. Lumbricals
16. Tendons of flexor digitorum profundus
17. Tendons of flexor digitorum superficialis
18. Hypothenar muscles
19. Palmar aponeurosis
20. Flexor carpi ulnaris

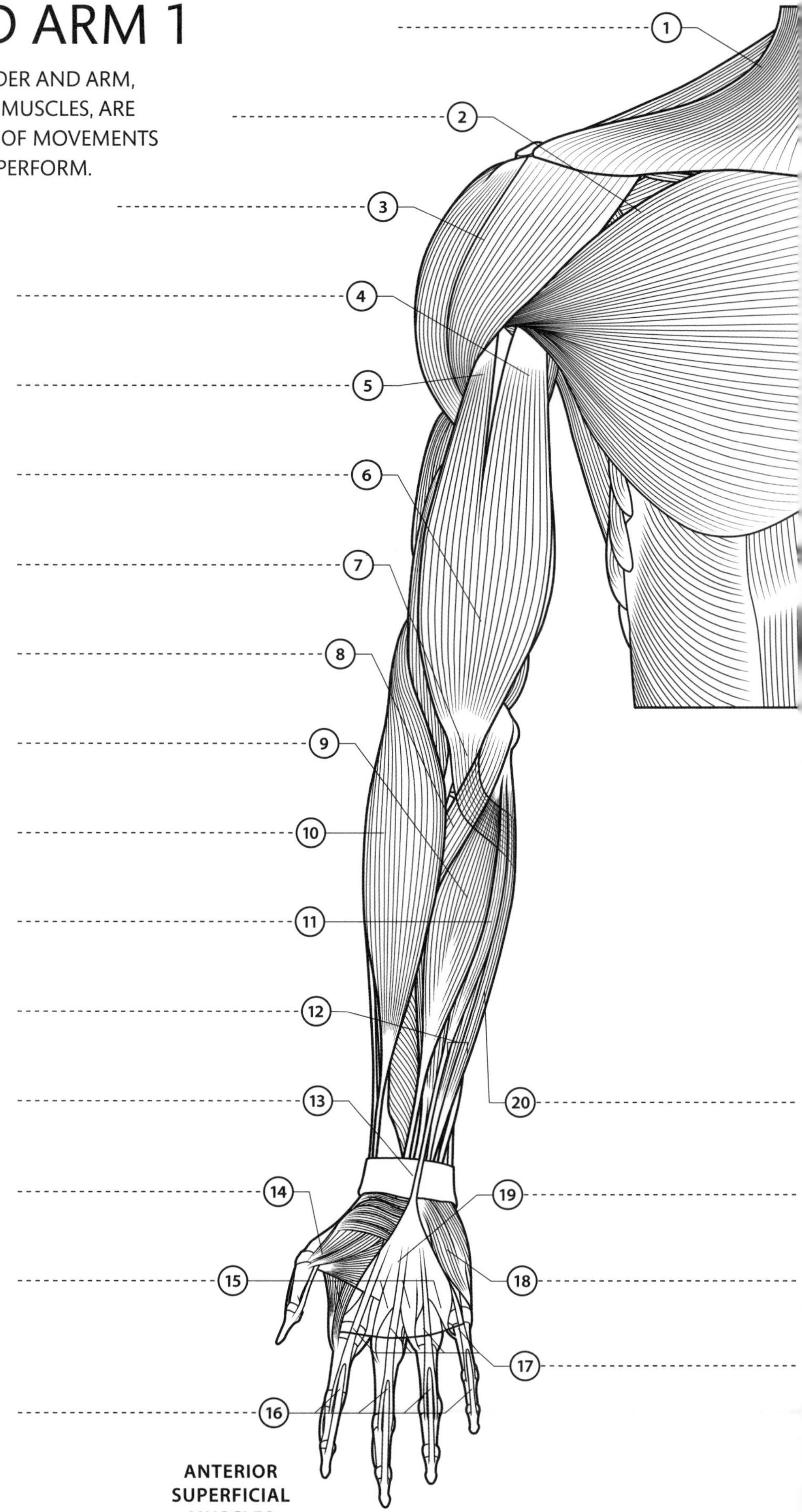

ANTERIOR SUPERFICIAL MUSCLES

ANTERIOR DEEP MUSCULATURE

The deep muscles around the shoulder include four muscles that help move and stabilize the shoulder joint, known as the rotator cuff group (subscapularis, teres major, teres minor, and supraspinatus). The tendon of supraspinatus passes through a narrow gap between the head of the humerus and the acromion of the scapula, making it vulnerable to compression and damage in this gap. On the front of the humerus, the biceps brachii has been removed to reveal the brachialis, which runs from the lower humerus to the ulna. The brachialis is a flexor of the elbow. Under the superficial muscles on the anterior of the forearm several deeper muscle layers attach to the radius and ulna and to the interosseus membrane between the bones. Also visible is the long, quill-like flexor of the thumb, the flexor pollicis longus.

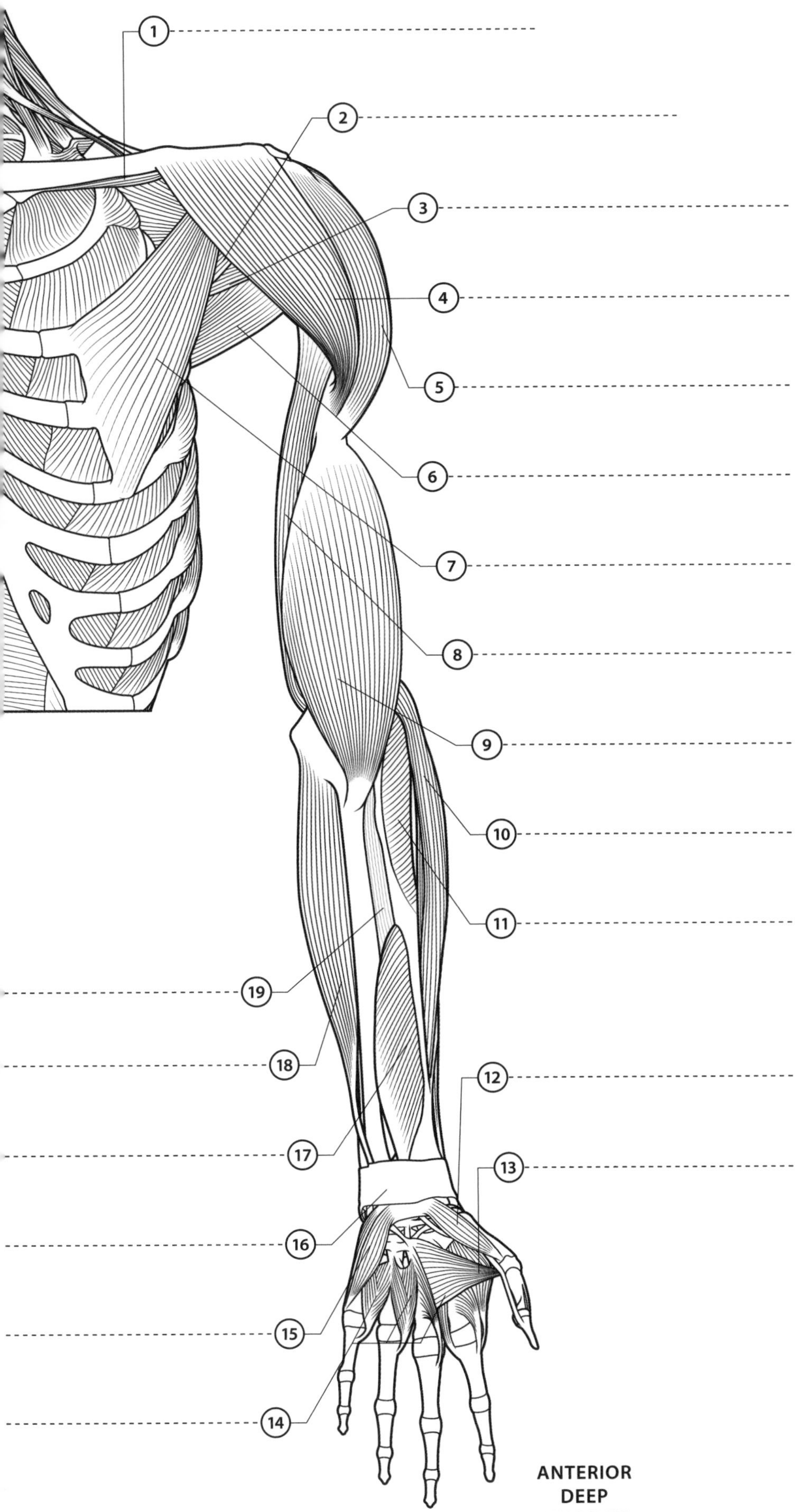

ANTERIOR DEEP MUSCLES

ANTERIOR DEEP MUSCLES

Color and/or label the structures indicated on the diagram using the key below.

1. Subclavius
2. Subscapularis
3. Teres major
4. Anterior fibers of deltoid
5. Middle fibers of deltoid
6. Latissimus dorsi
7. Pectoralis minor
8. Medial head of triceps
9. Brachialis
10. Brachioradialis
11. Supinator
12. Thenar muscles
13. Adductor pollicis
14. Palmar interosseus muscles
15. Hypothenar muscles
16. Flexor retinaculum
17. Flexor pollicis longus
18. Flexor carpi ulnaris
19. Interosseus membrane

See also pp. 40, 72, 102, 146, 148, 164 »

SHOULDER AND ARM 2

THE POSTERIOR MUSCLES OF THE SHOULDER AND ARM WORK IN COMBINATION WITH THE ANTERIOR MUSCLES TO PRODUCE THE WIDE RANGE OF MOVEMENTS THAT THE ARM, WRIST, AND HAND CAN MAKE.

POSTERIOR SUPERFICIAL MUSCULATURE

The posterior fibers of the deltoid attach from the spine of the scapula down to the humerus, and this part of the muscle can draw back the arm or extend it. The latissimus dorsi (a broad muscle attaching from the back of the trunk and ending in a narrow tendon that attaches onto the humerus) can also extend the arm. The triceps brachii is the sole extensor of the elbow. In a superficial dissection (represented in this view) only two of the three heads of the triceps can be seen: the long and lateral heads. The triceps tendon attaches to the lever-like olecranon of the ulna, which forms the bony knob at the back of the elbow. On the back of the forearm, seven superficial extensor muscles attach to the lateral epicondyle of the humerus. Most of their tendons run down to the wrist or into the hand.

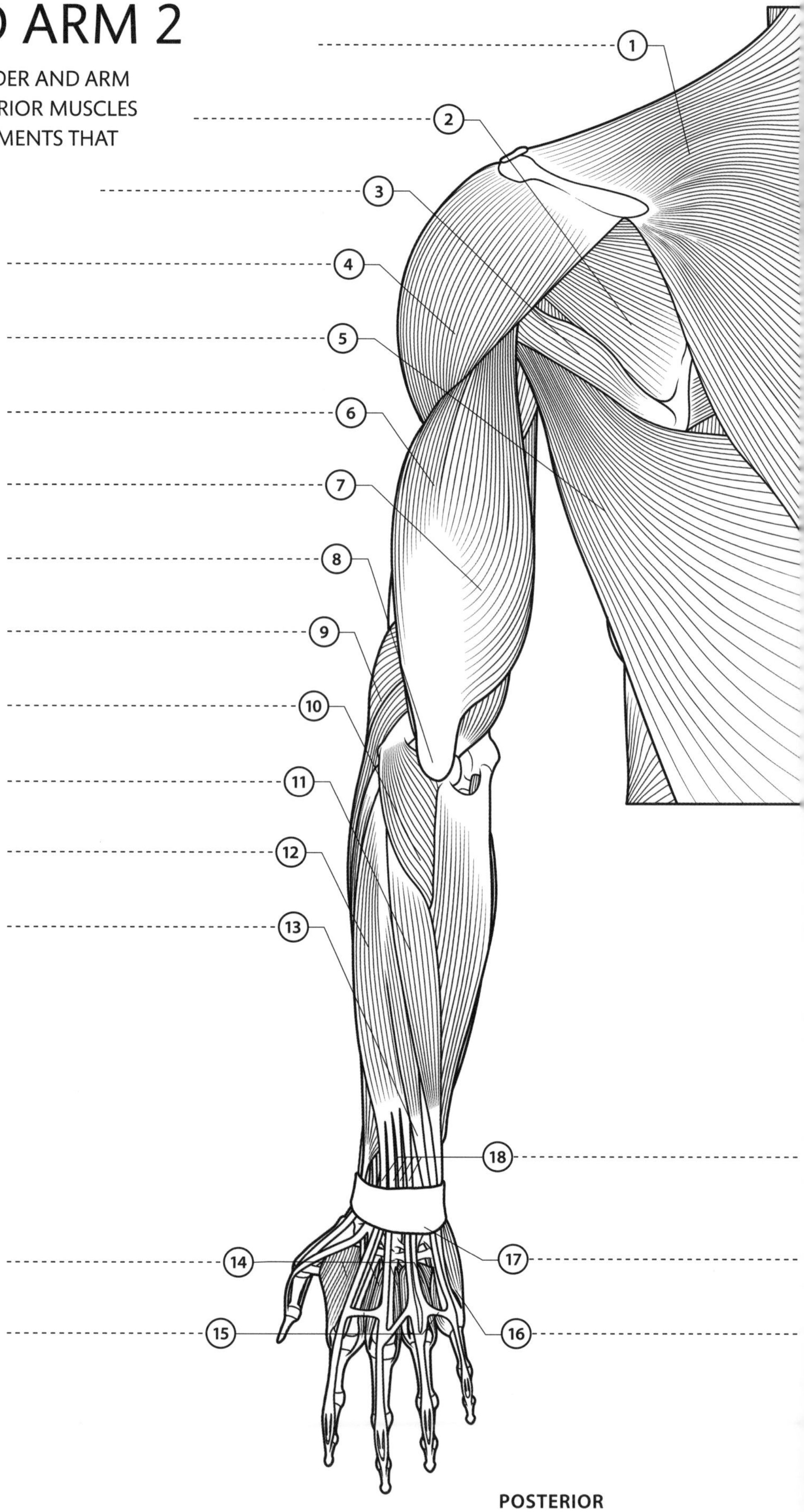

POSTERIOR SUPERFICIAL MUSCLES

POSTERIOR SUPERFICIAL MUSCLES

Color and/or label the structures indicated on the diagram using the key below.

1. Trapezius
2. Infraspinatus
3. Teres major
4. Deltoid
5. Latissimus dorsi
6. Lateral head of triceps
7. Long head of triceps
8. Triceps tendon
9. Brachioradialis
10. Anconeus
11. Extensor carpi ulnaris
12. Extensor digitorum
13. Extensor digiti minimi
14. Dorsal interosseus muscles
15. Intertendinous connections
16. Abductor digiti minimi
17. Extensor retinaculum
18. Tendons of extensor digitorum

POSTERIOR DEEP MUSCULATURE

More of the rotator cuff muscles—the supraspinatus, infraspinatus, and teres minor—are visible from the back. As well as moving the shoulder joint in various directions, including rotation, these muscles are important in stabilizing the shoulder joint. On the back of the arm, a deeper view reveals the third (medial) head of the triceps, which attaches from the back of the humerus. It joins with the lateral and long heads to form the triceps tendon, attaching to the olecranon. Most of the forearm muscles take their attachment from the epicondyles of the humerus, just above the elbow, but the brachioradialis and extensor carpi radialis longus have higher origins from the side of the humerus. Other deep muscles on the back of the forearm include the long extensors of the thumb and index finger, and the supinator, which pulls on the radius to rotate the pronated arm (held with the palm facing downward) into supination (palm facing up). In the hand, a deep dissection reveals the interosseus muscles that act on the metacarpophalangeal joints to either spread or close the fingers.

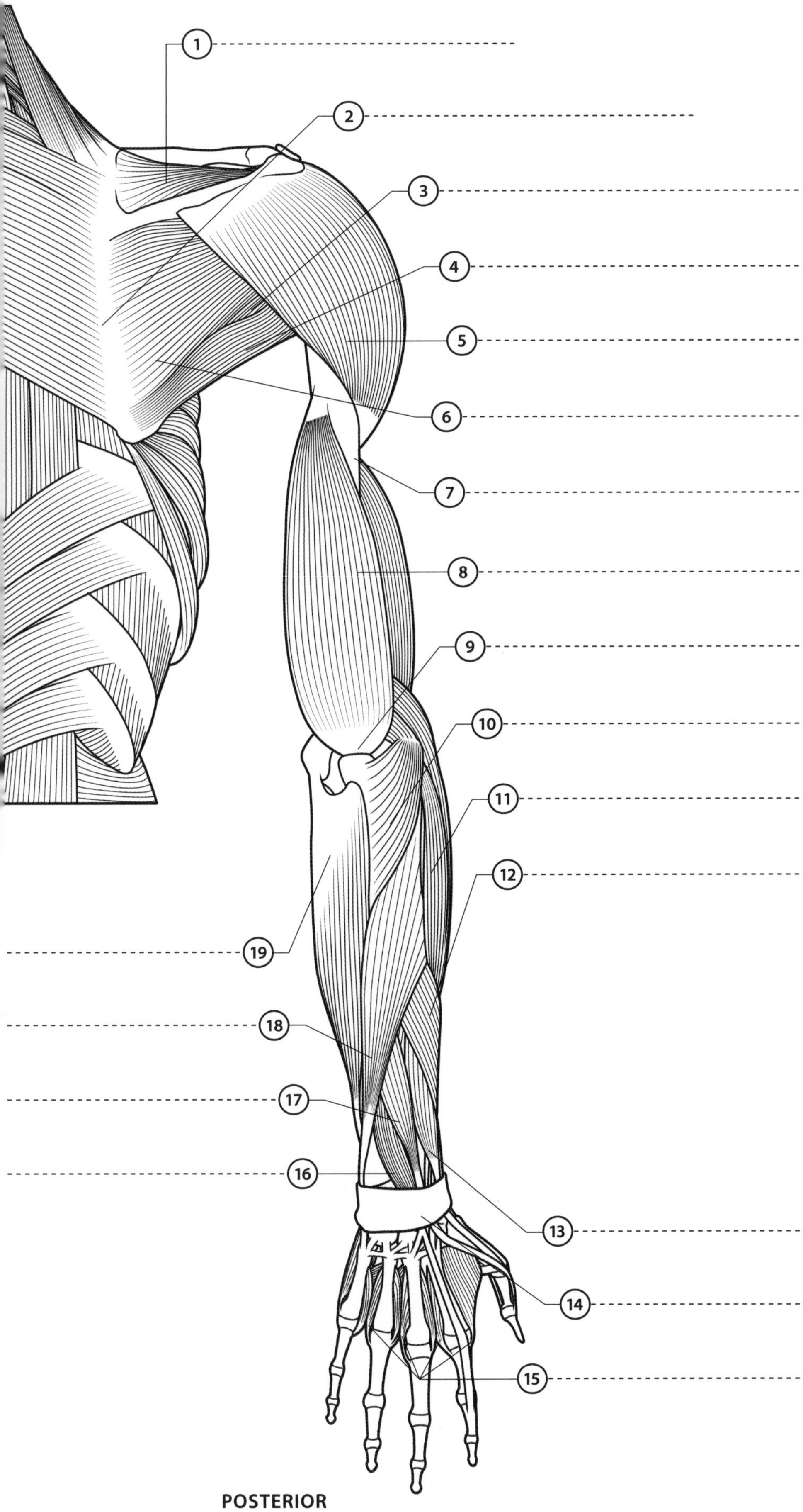

POSTERIOR DEEP MUSCLES

POSTERIOR DEEP MUSCLES

Color and/or label the structures indicated on the diagram using the key below.

1. Supraspinatus
2. Medial border of scapula
3. Teres minor
4. Teres major
5. Posterior fibers of deltoid
6. Infraspinatus
7. Humerus
8. Medial head of triceps
9. Triceps tendon
10. Anconeus
11. Extensor carpi radialis brevis
12. Abductor pollicis longus
13. Extensor pollicis brevis
14. Extensor retinaculum
15. Interossei
16. Extensor indicis
17. Extensor pollicis longus
18. Extensor carpi ulnaris
19. Flexor carpi ulnaris

See also pp. 40, 70, 102, 146, 148, 164 »

HIP AND LEG 1

THE MUSCLES OF THE HIP AND LEG MOVE THE HIP, LEG, ANKLE, AND FOOT.

ANTERIOR SUPERFICIAL MUSCULATURE

Most of the muscle bulk on the front of the thigh is the four-headed quadriceps femoris. Three of its heads are visible in a superficial dissection of the thigh: the rectus femoris, vastus lateralis, and vastus medialis. The patella is embedded in the quadriceps tendon; the part of the tendon below the patella is called the patellar ligament. On the outer (lateral) side of the tibia and running alongside it is a soft wedge of muscles. Their tendons run down to the foot. These muscles can pull the foot upward at the ankle (a movement called dorsiflexion). Some extensor tendons continue all the way to the toes.

ANTERIOR SUPERFICIAL MUSCLES

Color and/or label the diagram using the key below.

1. Iliopsoas
2. Pectineus
3. Adductor longus
4. Tensor fascia latae
5. Iliotibial tract
6. Rectus femoris
7. Vastus lateralis
8. Quadriceps tendon
9. Patellar tendon
10. Fibularis (peroneus) longus
11. Extensor digitorum longus
12. Fibularis (peroneus) brevis
13. Inferior extensor retinaculum
14. Extensor hallucis brevis
15. Extensor digitorum longus tendons
16. Dorsal interossei
17. Extensor hallucis longus tendon
18. Superior extensor retinaculum
19. Soleus
20. Tibialis anterior
21. Medial head of gastrocnemius
22. Vastus medialis
23. Sartorius
24. Gracilis
25. Inguinal ligament

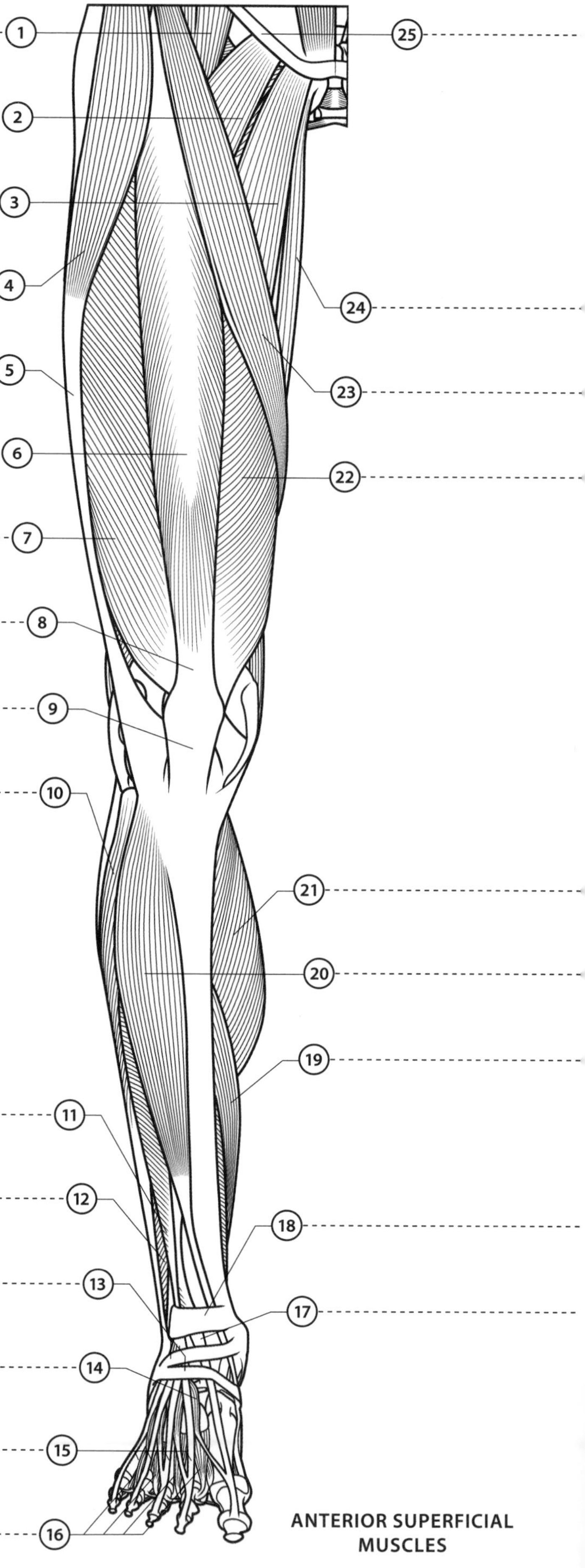

ANTERIOR SUPERFICIAL MUSCLES

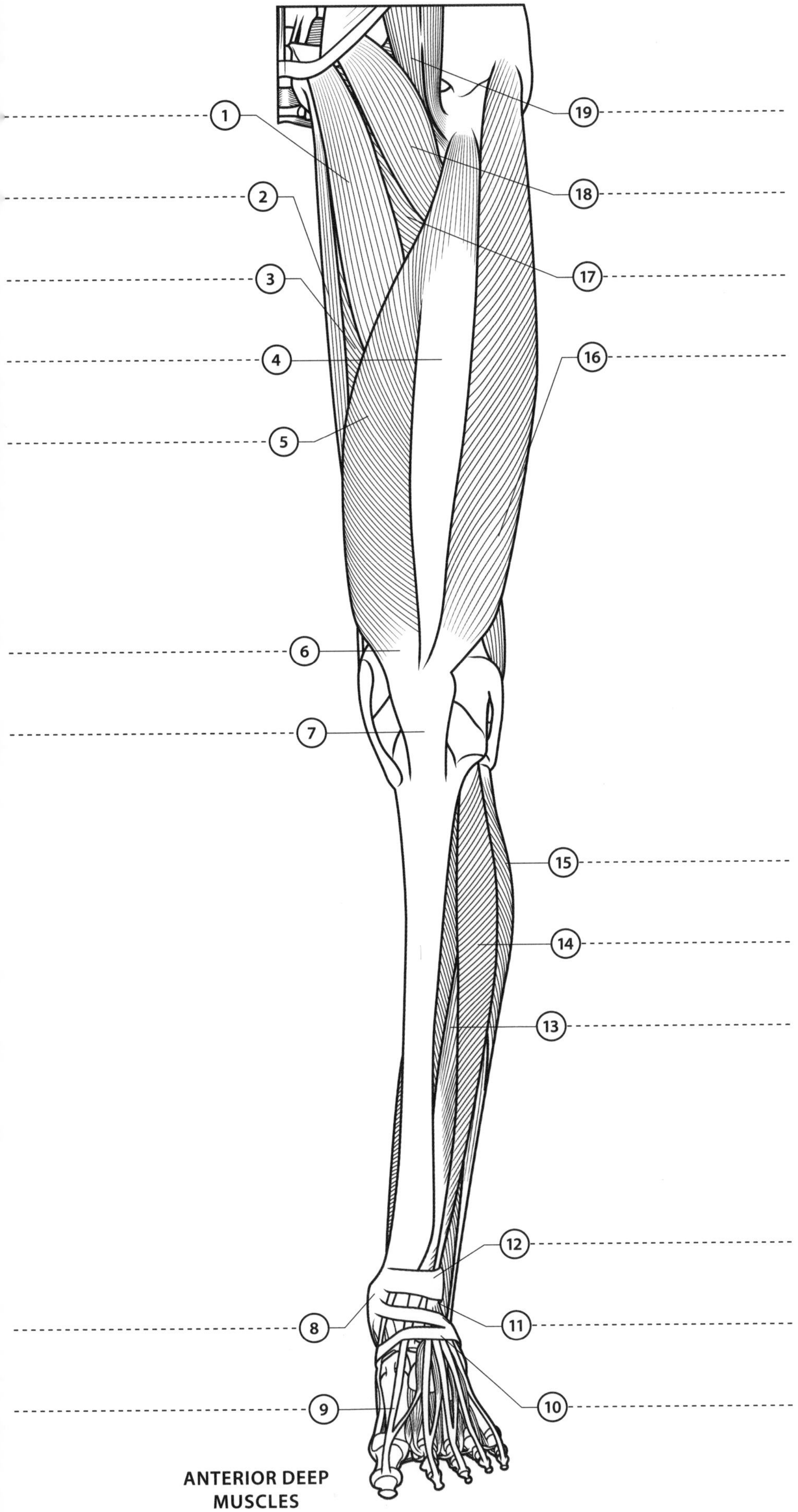

ANTERIOR DEEP MUSCLES

ANTERIOR DEEP MUSCULATURE

With the rectus femoris and sartorius stripped away, the deep, fourth head of the quadriceps (vastus intermedius) is visible. The adductor muscles that bring the thighs together can also be seen clearly, including the long, slender gracilis. The largest adductor muscle—the adductor magnus—has an opening in its tendon through which the leg's main artery (the femoral artery) passes. The adductor tendons attach to the pubis and ischium of the pelvis. Two muscles run along the outer (lateral) side of the leg, down into the foot: the fibularis longus and the fibularis brevis. These muscles pull the outer side of the foot upward (a movement called eversion). The tendon of the fibularis longus runs underneath the foot, to attach to the inner side, and helps to maintain the transverse arch of the foot.

ANTERIOR DEEP MUSCLES

Color and/or label the diagram using the key below.

1. Adductor longus
2. Gracilis
3. Adductor magnus
4. Vastus intermedius
5. Vastus medialis
6. Quadriceps tendon
7. Patellar tendon
8. Medial malleolus
9. Extensor hallucis longus tendon
10. Inferior extensor retinaculum
11. Extensor digitorum longus tendon
12. Superior extensor retinaculum
13. Extensor hallucis longus
14. Extensor digitorum longus
15. Fibularis (peroneus) longus
16. Vastus lateralis
17. Adductor brevis
18. Pectineus
19. Iliopsoas

See also pp. 46, 76, 104, 150, 152, 165 »

HIP AND LEG 2

IN COMBINATION WITH THE ANTERIOR MUSCLES, THE POSTERIOR MUSCLES OF THE HIP AND LEG ARE RESPONSIBLE FOR MOVING THE HIP, LEG, ANKLE, AND FOOT, AND HELP THE LEG TO BEAR THE BODY'S WEIGHT.

POSTERIOR SUPERFICIAL MUSCULATURE

At the back of the hip and thigh are the gluteus maximus and the three hamstring muscles (biceps femoris, semitendinosus, and semimembranosus). The gluteus maximus extends the hip joint, swinging the leg backward. The hamstrings attach to the ischial tuberosity of the pelvis and sweep down the back of the thigh to the tibia and fibula. They are the main flexors of the knee. The lower leg includes the relatively bulky muscles that form the calf. The gastrocnemius, and the soleus underneath it, are large muscles that join together to form the Achilles tendon. They pull up on the calcaneus, pushing the ball of the foot down, and are involved as the foot pushes off from the ground during walking and running.

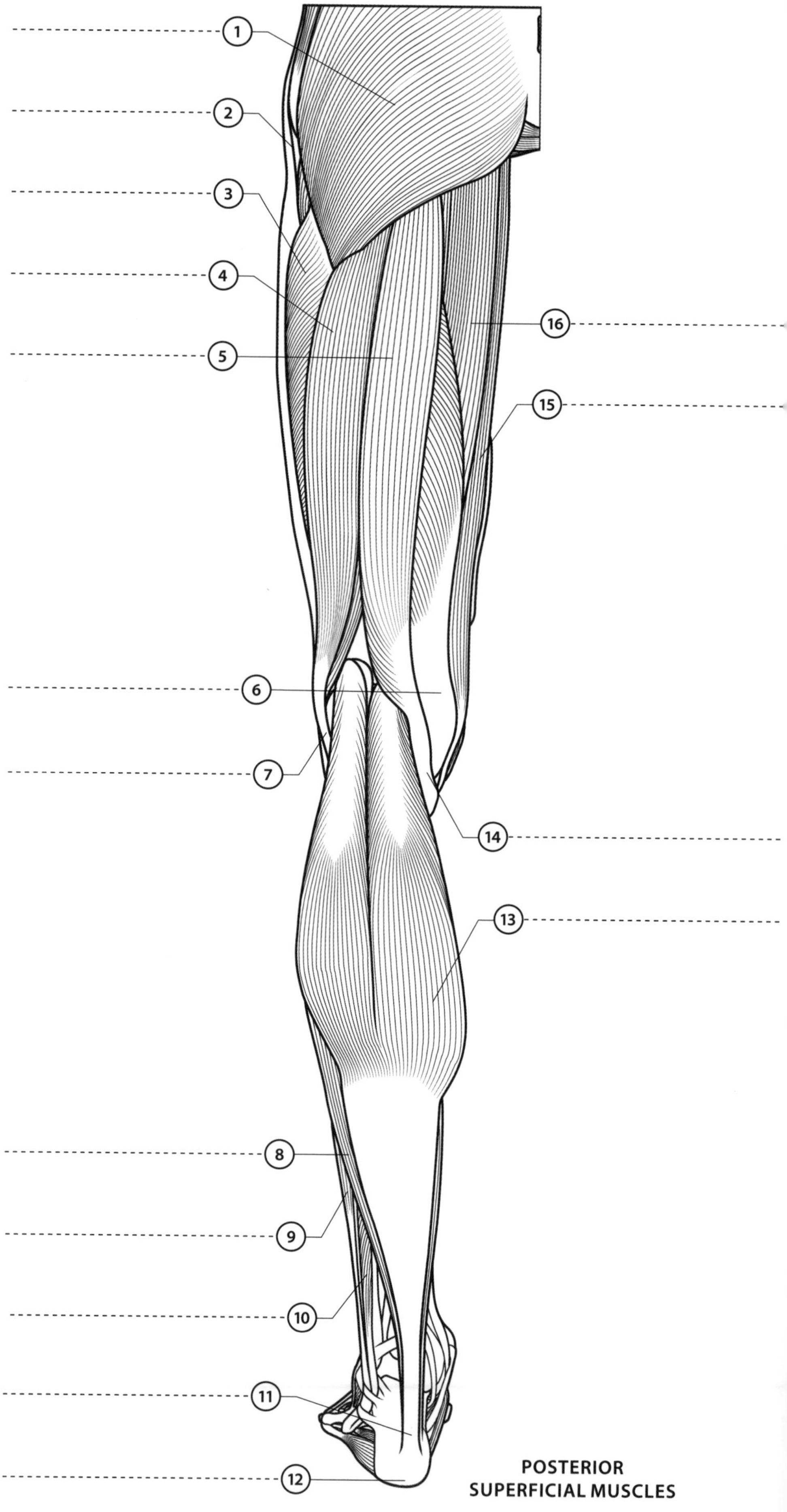

POSTERIOR SUPERFICIAL MUSCLES

POSTERIOR SUPERFICIAL MUSCLES

Color and/or label the structures indicated on the diagram using the key below.

1. Gluteus maximus
2. Iliotibial tract
3. Vastus lateralis
4. Biceps femoris
5. Semitendinosus
6. Semimembranosus
7. Biceps femoris tendon
8. Soleus
9. Fibularis (peroneus) longus
10. Fibularis (peroneus) brevis
11. Calcaneal (Achilles) tendon
12. Calcaneus
13. Gastrocnemius
14. Semitendinosus
15. Gracilis
16. Adductor magnus

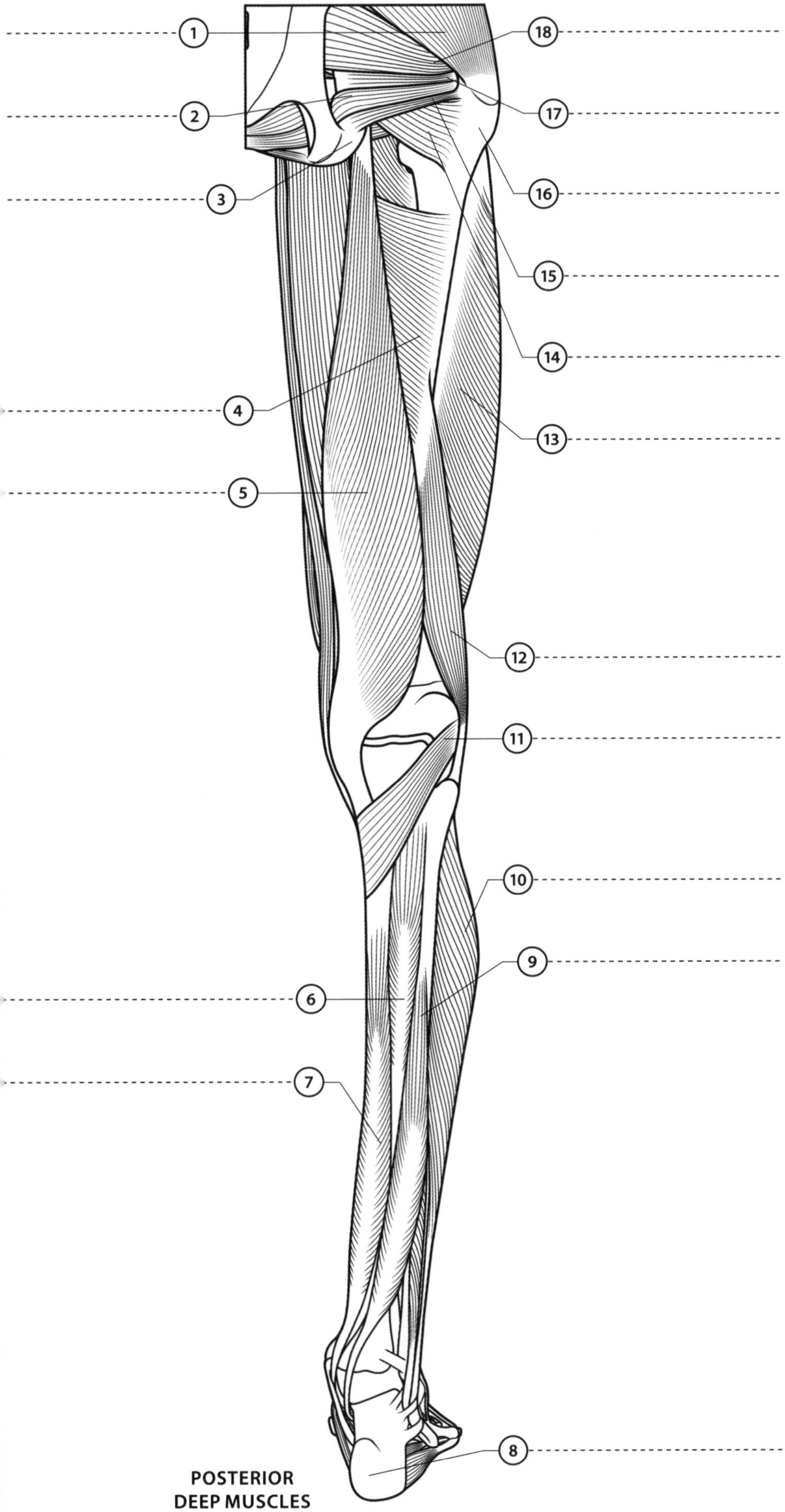

POSTERIOR
DEEP MUSCLES

POSTERIOR DEEP MUSCULATURE

On the back of the hip, with the gluteus maximus removed, the short muscles that rotate the hip are revealed. These include the piriformis, obturator internus, and quadratus femoris muscles. With the long head of the biceps femoris removed, the deeper, short head is visible, attaching to the linea aspera on the back of the femur. The semitendinosus has been cut away to reveal the semimembranosus underneath it. The popliteus muscle is also visible at the back of the knee joint. The deep muscles of the lower leg include flexor muscles and two fibular (peroneal) muscles. The flexor muscles move the foot downward at the ankle (plantarflexion) and flex or curl the toes. The fibular muscles move the foot outwards (eversion).

POSTERIOR DEEP MUSCLES

Color and/or label the structures indicated on the diagram using the key below.

1. Gluteus medius
2. Obturator internus
3. Ischial tuberosity
4. Adductor magnus
5. Semimembranosus
6. Tibialis posterior
7. Flexor digitorum longus
8. Calcaneus
9. Flexor hallucis longus
10. Fibular (peroneal) muscles
11. Popliteus
12. Short head of biceps femoris
13. Vastus lateralis
14. Quadratus femoris
15. Inferior gemellus
16. Greater trochanter of femur
17. Superior gemellus
18. Piriformis

See also pp. 46, 74, 104, 150, 152, 165 »

INTEGUMENTARY SYSTEM

THE MAIN COMPONENTS OF THE BODY'S INTEGUMENT ARE THE SKIN AND STRUCTURES SUCH AS NAILS.

SKIN STRUCTURE

The skin has two main structural layers: an outer epidermis and an underlying dermis. The epidermis is chiefly protective, whereas the dermis contains many different tissues with varied functions. The epidermis continually renews itself by cell division. The cells of its basal layer multiply rapidly and gradually move to the surface, slowly becoming keratinized. By the time they reach the surface, they are fully keratinized and have died; they then flake away with normal wear and tear. The dermis contains thousands of sensors that enable the sense of touch, as well as sweat glands, blood vessels, and hair follicles. Under the dermis is a layer of subcutaneous fat, which is sometimes regarded as part of the skin.

SKIN STRUCTURE

Color and/or label the structures indicated on the diagram using the key below.

1. Hair shaft
2. Basal epidermal layer
3. Touch sensor
4. Arrector pili muscle
5. Sebaceous gland
6. Hair bulb
7. Hair follicle
8. Sweat gland
9. Arteriole
10. Venule
11. Subcutaneous fat
12. Dermis
13. Epidermis
14. Capillary
15. Sweat duct

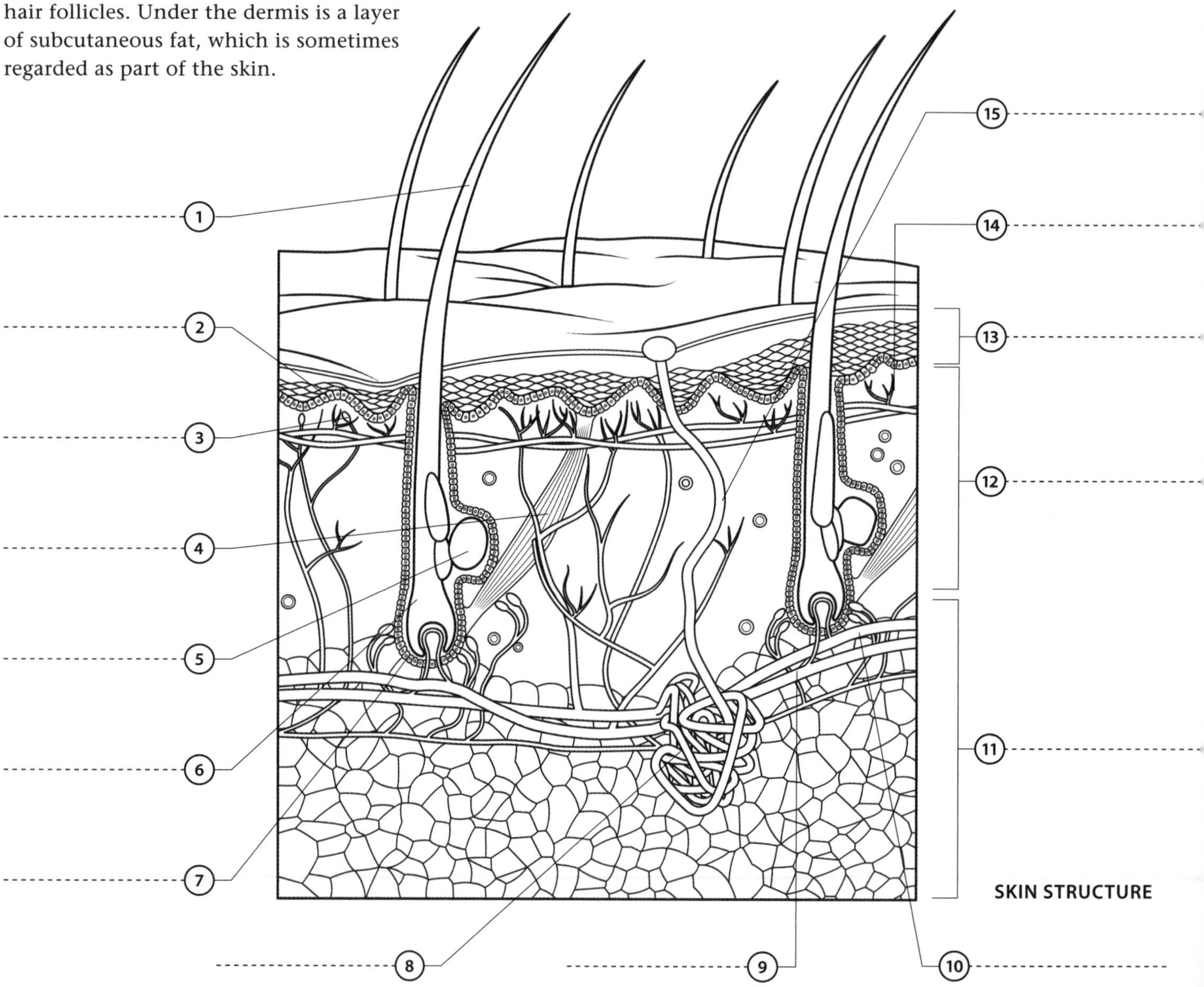

SKIN STRUCTURE

SKIN COLOR

Skin color depends on the amount and distribution of melanin in the skin. Melanin is made and packaged into melanosomes by melanocytes. Each melanocyte has branching dendrites that contact nearby keratinocytes, and through which melanosomes are released. Darker skin has larger (not more) melanocytes that produce more melanosomes, releasing melanin, which is distributed throughout the keratinocytes. Lighter skin has smaller melanocytes and little distribution of melanin.

SKIN COLOR

Color and/or label the structures indicated on the diagram using the key below.

1. Upper keratinocyte
2. Melanosome
3. Basal keratinocyte
4. Melanocyte
5. Basal keratinocyte
6. Upper layer of skin

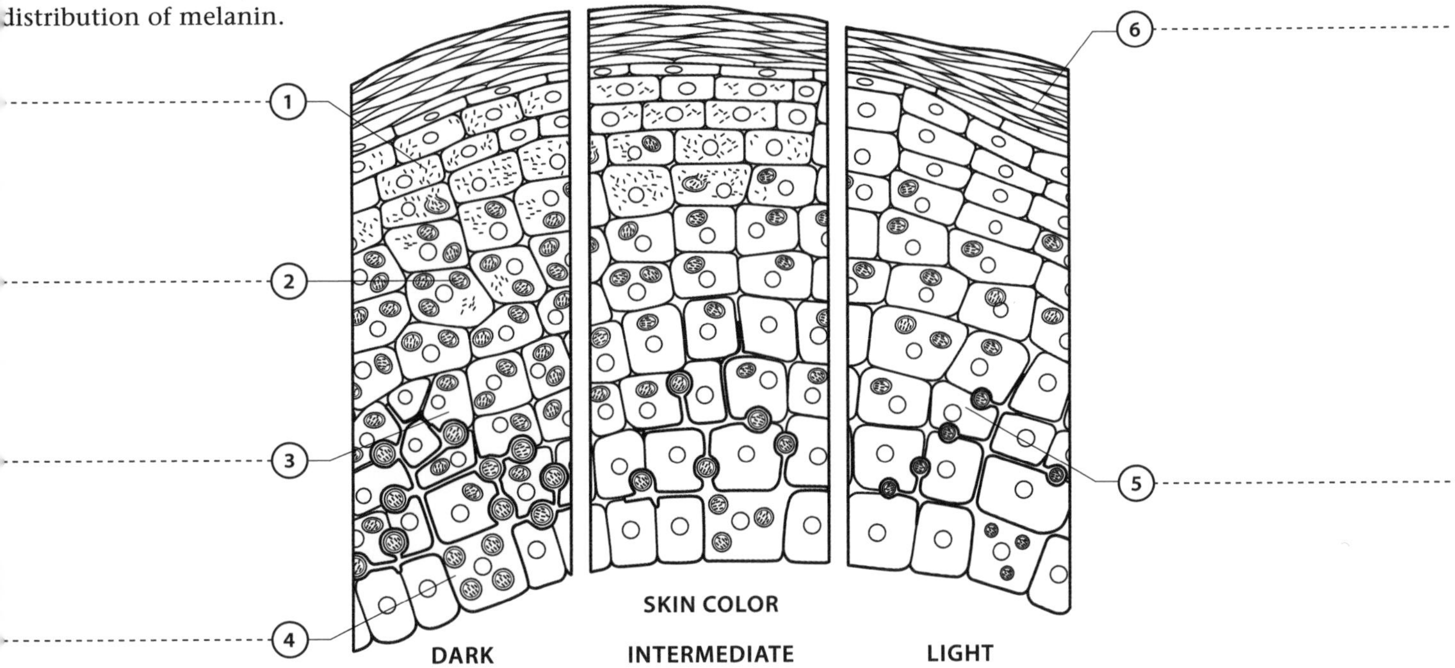

SKIN COLOR

NAILS

Nails are made of dead, flattened cells filled with the tough, structural protein keratin. Each nail has a root embedded in the skin, a body, and a free edge. Nail cells produced by the matrix push forward, becoming filled with keratin as the nail slides over the nail bed.

CROSS SECTION THROUGH NAIL AND FINGERTIP

Color and/or label the structures indicated on the diagram using the key below.

1. Nail plate
2. Nail bed
3. Free edge of nail
4. Bone (distal phalanx)
5. Fat
6. Matrix
7. Cuticle

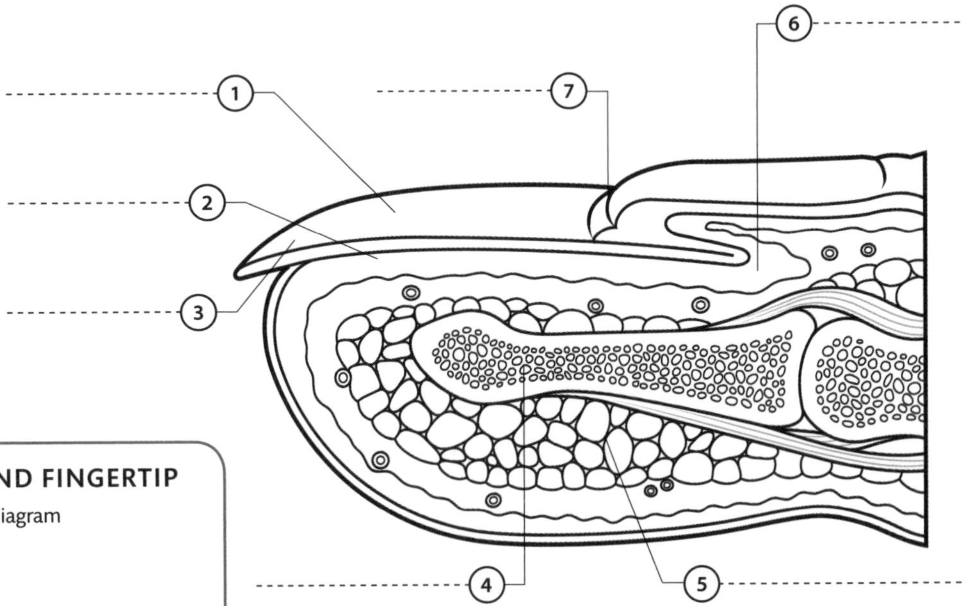

CROSS SECTION THROUGH NAIL AND FINGERTIP

See also pp. 14, 16, 112 »

NERVOUS SYSTEM

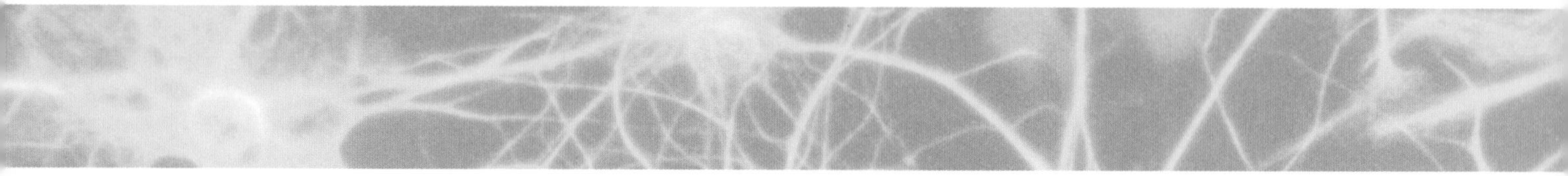

04

NERVOUS SYSTEM 1

THE BRAIN, SPINAL CORD, AND THE NERVES EXTENDING FROM THEM FORM THE NERVOUS SYSTEM. THIS COMPLEX ORGANIZATION MONITORS AND REGULATES ALMOST ALL BODILY PROCESSES AND ACTIVITIES, MANY OF WHICH OCCUR OUTSIDE CONSCIOUS CONTROL.

NERVOUS SYSTEM DIVISIONS

The nervous system comprises three main divisions defined by anatomy and function: central, peripheral, and autonomic. The central nervous system (CNS) is composed of the brain and the spinal cord. From the CNS, 43 pairs of nerves branch: 12 from the brain and 31 from the spinal cord. As these nerves divide and reach out to every organ and tissue in the body they form the network of the peripheral nervous system (PNS). The function of the CNS is to analyze sensory input and initiate responses; the PNS sends sensory data to the CNS and carries response signals to the body. The third component of the nervous system is the autonomic nervous system (ANS), which shares elements with both the CNS and the PNS; it also has its own chains of ganglia alongside the spinal cord and individual ganglia in the head. Autonomic nerves primarily control involuntary activities of body tissues including blood vessels, organs, and glands.

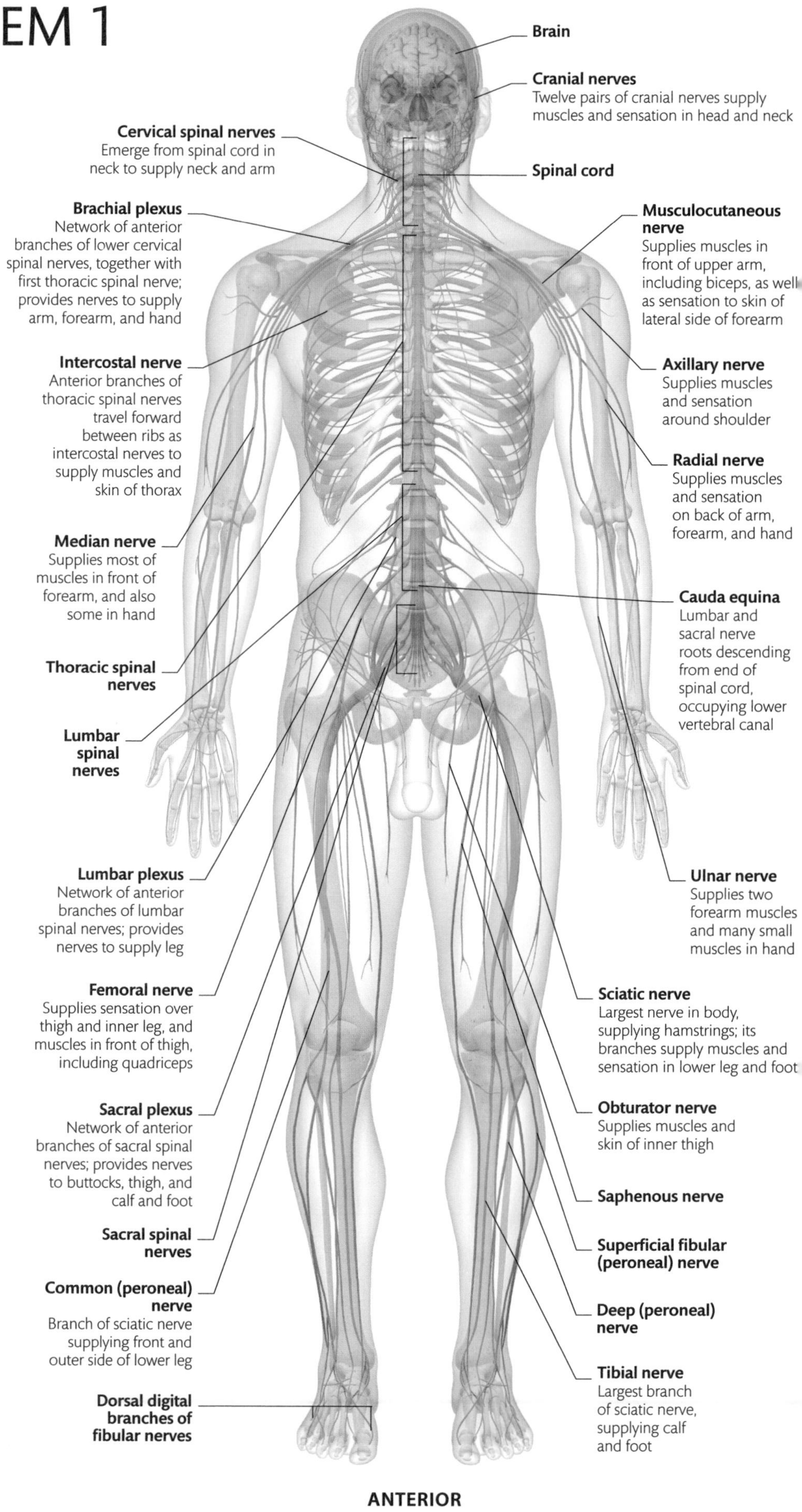
Brain
Cranial nerves
Twelve pairs of cranial nerves supply muscles and sensation in head and neck
Cervical spinal nerves
Emerge from spinal cord in neck to supply neck and arm
Spinal cord
Brachial plexus
Network of anterior branches of lower cervical spinal nerves, together with first thoracic spinal nerve; provides nerves to supply arm, forearm, and hand
Musculocutaneous nerve
Supplies muscles in front of upper arm, including biceps, as well as sensation to skin of lateral side of forearm
Intercostal nerve
Anterior branches of thoracic spinal nerves travel forward between ribs as intercostal nerves to supply muscles and skin of thorax
Axillary nerve
Supplies muscles and sensation around shoulder
Radial nerve
Supplies muscles and sensation on back of arm, forearm, and hand
Median nerve
Supplies most of muscles in front of forearm, and also some in hand
Cauda equina
Lumbar and sacral nerve roots descending from end of spinal cord, occupying lower vertebral canal
Thoracic spinal nerves
Lumbar spinal nerves
Lumbar plexus
Network of anterior branches of lumbar spinal nerves; provides nerves to supply leg
Ulnar nerve
Supplies two forearm muscles and many small muscles in hand
Femoral nerve
Supplies sensation over thigh and inner leg, and muscles in front of thigh, including quadriceps
Sciatic nerve
Largest nerve in body, supplying hamstrings; its branches supply muscles and sensation in lower leg and foot
Sacral plexus
Network of anterior branches of sacral spinal nerves; provides nerves to buttocks, thigh, and calf and foot
Obturator nerve
Supplies muscles and skin of inner thigh
Saphenous nerve
Sacral spinal nerves
Superficial fibular (peroneal) nerve
Common (peroneal) nerve
Branch of sciatic nerve supplying front and outer side of lower leg
Deep (peroneal) nerve
Tibial nerve
Largest branch of sciatic nerve, supplying calf and foot
Dorsal digital branches of fibular nerves
ANTERIOR

NEURAL NETWORK
The central nervous system (CNS) receives sensory information and sends out response signals via the bodywide neural network of the peripheral nervous system (PNS). Some responses are involuntary; others are dictated by conscious thought.

AUTONOMIC NERVOUS SYSTEM

The autonomic, or involuntary, nervous system (ANS) works largely independently of the conscious mind. Sensory nerve fibers send information about organs and internal activities, such as heart rate. This information is integrated in the hypothalamus, brain stem, or spinal cord. The ANS then sends commands, as motor nerve signals, to three main destinations: smooth muscle in viscera; cardiac muscle; and exocrine glands. The ANS is divided into sympathetic and parasympathetic systems, which usually produce contrasting responses. Sympathetic responses increase activity (the fight or flight response). Parasympathetic nerves maintain normal functions such as digestion and excretion. The enteric nervous system innervates most of the gastrointestinal tract and the abdominal organs.

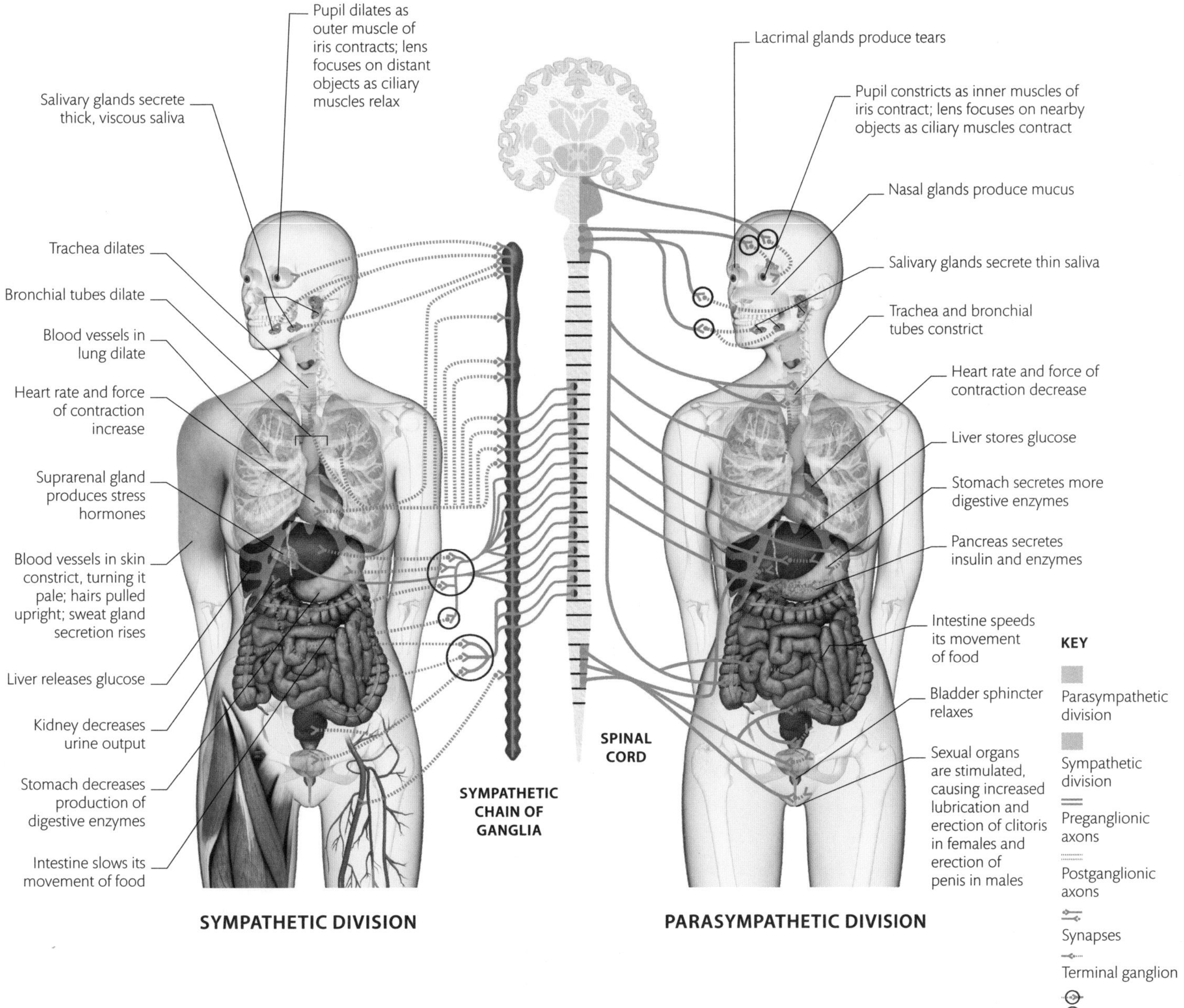

SYMPATHETIC DIVISION

- Pupil dilates as outer muscle of iris contracts; lens focuses on distant objects as ciliary muscles relax
- Salivary glands secrete thick, viscous saliva
- Trachea dilates
- Bronchial tubes dilate
- Blood vessels in lung dilate
- Heart rate and force of contraction increase
- Suprarenal gland produces stress hormones
- Blood vessels in skin constrict, turning it pale; hairs pulled upright; sweat gland secretion rises
- Liver releases glucose
- Kidney decreases urine output
- Stomach decreases production of digestive enzymes
- Intestine slows its movement of food

SYMPATHETIC CHAIN OF GANGLIA

SPINAL CORD

PARASYMPATHETIC DIVISION

- Lacrimal glands produce tears
- Pupil constricts as inner muscles of iris contract; lens focuses on nearby objects as ciliary muscles contract
- Nasal glands produce mucus
- Salivary glands secrete thin saliva
- Trachea and bronchial tubes constrict
- Heart rate and force of contraction decrease
- Liver stores glucose
- Stomach secretes more digestive enzymes
- Pancreas secretes insulin and enzymes
- Intestine speeds its movement of food
- Bladder sphincter relaxes
- Sexual organs are stimulated, causing increased lubrication and erection of clitoris in females and erection of penis in males

KEY

- Parasympathetic division
- Sympathetic division
- Preganglionic axons
- Postganglionic axons
- Synapses
- Terminal ganglion
- Collateral ganglia

DIVISIONS OF ANS

In the sympathetic nervous system, ganglia are arranged into two chains on either side of the vertebral column (only one illustrated here). The ganglia of the parasympathetic system are located in the head and in the pelvis.

See also pp. 84, 86, 88, 90

NERVOUS SYSTEM 2

THE NERVOUS SYSTEM CONTAINS BILLIONS OF NEURONS; THESE CELLS ARE OF SEVERAL TYPES, ALL HIGHLY SPECIALIZED IN STRUCTURE AND FUNCTION.

NEURON STRUCTURE

A typical neuron has a cell body (soma) and nucleus. Most neurons also have long, wirelike processes that reach out to transmit messages to other neurons at junctions (synapses). These processes are of two main kinds. Dendrites receive messages from other neurons, or from nervelike cells in sense organs, and conduct them toward the cell body of the neuron. Axons convey messages away from the cell body, to other neurons or to muscle or gland cells. Dendrites tend to be short and have many branches. Axons are usually longer and branch less along their length. While some axons within the brain are less than $^{1}/_{25}$ in (1 mm) in length, others, stretching from the spinal cord to muscles in the limbs, or from the skin of the limbs to the brain stem, can measure over 39 in (1 m) long. Neurons in the brain and spinal cord are protected and nourished by glial cells.

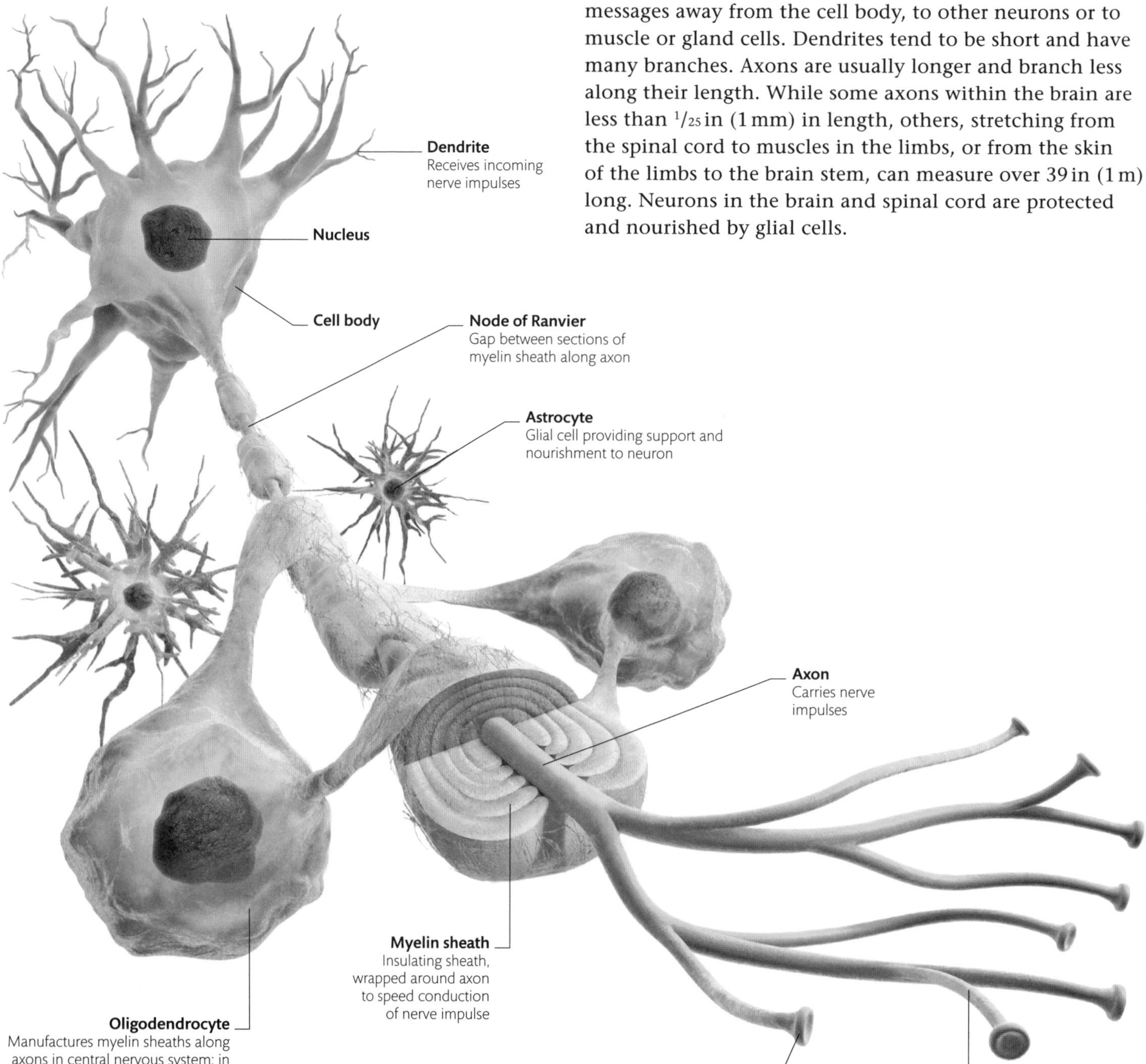
Dendrite
Receives incoming nerve impulses
Nucleus
Cell body
Node of Ranvier
Gap between sections of myelin sheath along axon
Astrocyte
Glial cell providing support and nourishment to neuron
Axon
Carries nerve impulses
Myelin sheath
Insulating sheath, wrapped around axon to speed conduction of nerve impulse
Oligodendrocyte
Manufactures myelin sheaths along axons in central nervous system; in peripheral nervous system, Schwann cells carry out this function
Synaptic knob
Transmits impulse via a synapse
Axon terminal

NEURON STRUCTURE
This image shows the structure of a neuron in the central nervous system. The dendrites and axons reaching out from the cell body can make contact with thousands of other neurons, creating a complex network of communication.

TYPES OF NEURON

Neurons are classified according to the number of processes that extend from the cell body. Most neurons in the brain and spinal cord are multipolar, with three or more processes. Unipolar neurons, present mainly in the sensory nerves of the peripheral nervous system, have a single process that splits into two processes. Bipolar neurons are found only in a few locations, such as the retina of the eye and the olfactory epithelium in the nose.

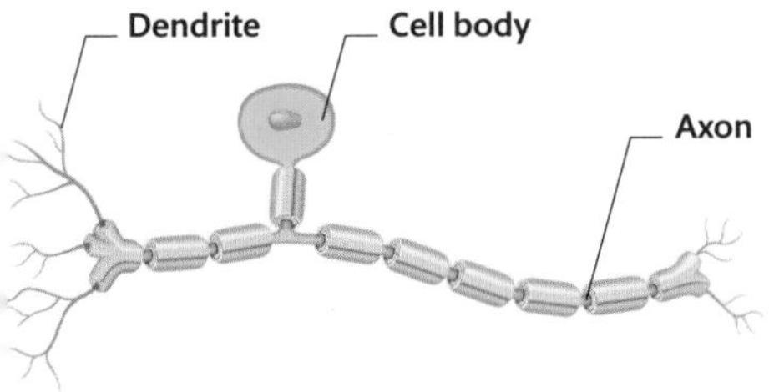
Dendrite
Cell body
Axon

UNIPOLAR NEURON
A single short process extends from the cell body and splits into two—an axon and a dendrite.

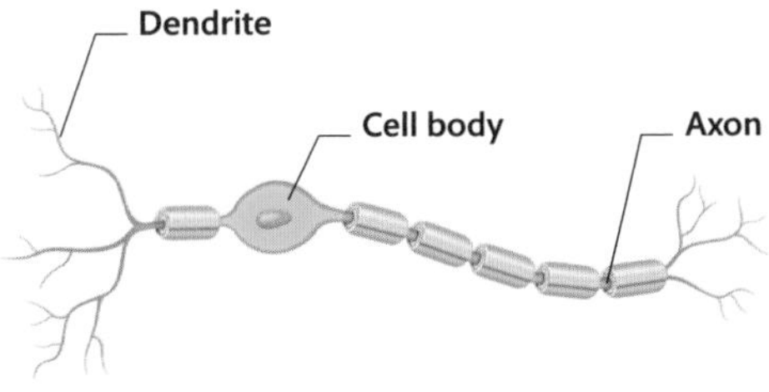
Dendrite
Cell body
Axon

BIPOLAR NEURON
The cell body is located between two processes—an axon and a dendrite.

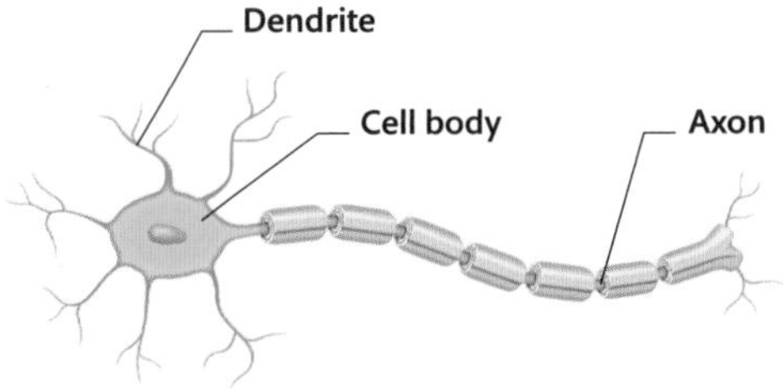
Dendrite
Cell body
Axon

MULTIPOLAR NEURON
These have three or more processes—several dendrites and one axon.

PERIPHERAL NERVE STRUCTURE

Peripheral nerves comprise hundreds of nerve fibers tightly grouped together in bundles. Axons are wrapped in a layer of connective tissue called endoneurium. Small bundles of these nerve fibers are bound together in a tough elastic tissue, the perineurium, to form fascicles. Several fascicles are then grouped in a protective outer covering, the epineurium, to form the nerve.

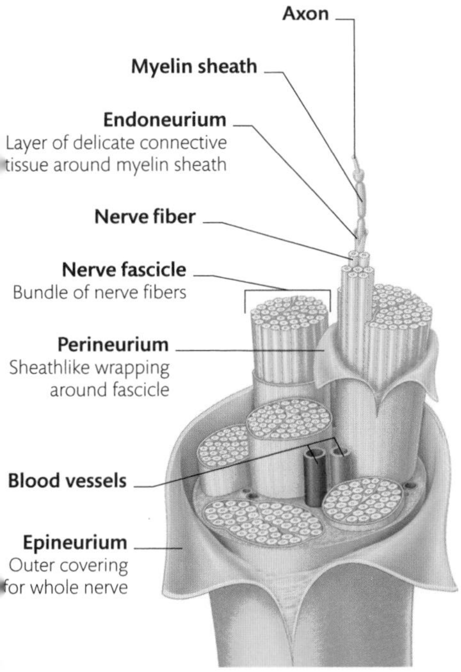
Axon
Myelin sheath
Endoneurium
Layer of delicate connective tissue around myelin sheath
Nerve fiber
Nerve fascicle
Bundle of nerve fibers
Perineurium
Sheathlike wrapping around fascicle
Blood vessels
Epineurium
Outer covering for whole nerve

PERIPHERAL NERVE
Peripheral nerve fibers are packaged in a series of bundles enclosed within an outer covering of epineurium.

SPINAL CORD STRUCTURE

The spinal cord extends from the brain stem down to the first lumbar vertebra. Shaped like a flattened cylinder, the spinal cord is slightly wider than a pencil for most of its length, tapering to a threadlike tail at the end. Inside, the cord has a core of gray matter made up of neuron cell bodies and axons, most of which are unmyelinated. Surrounding the core is an outer layer of white matter, composed mainly of myelinated nerve fiber tracts that carry nerve impulses up and down the spinal cord. Branching out from the spinal cord are 31 pairs of spinal nerves, which connect the cord to the skin, muscles, and other parts of the limbs, thorax, and abdomen.

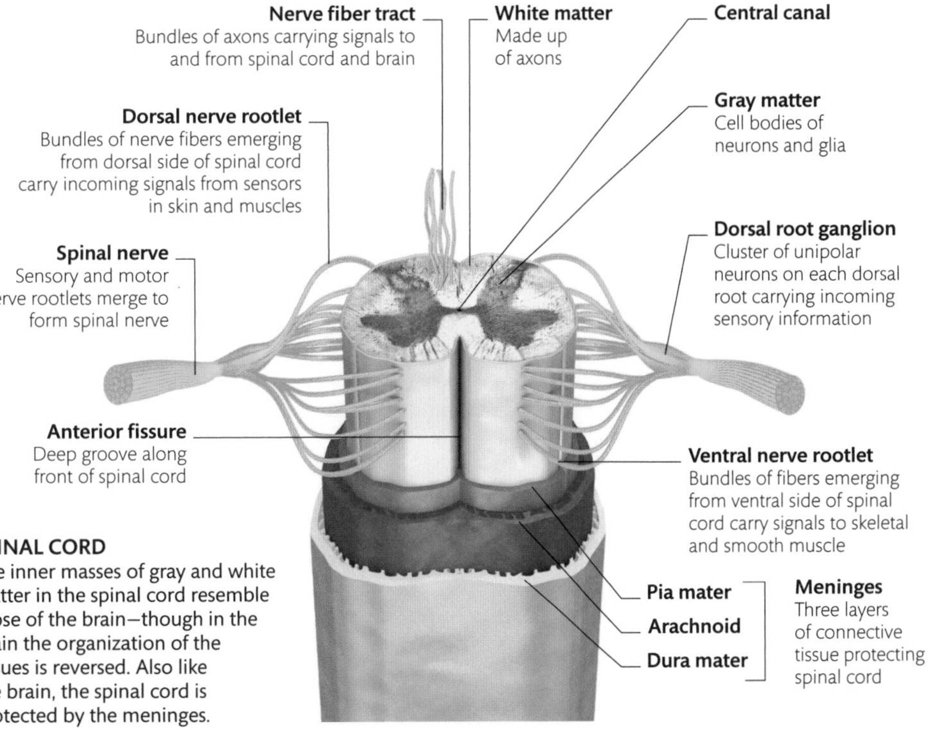
Nerve fiber tract
Bundles of axons carrying signals to and from spinal cord and brain
White matter
Made up of axons
Central canal
Gray matter
Cell bodies of neurons and glia
Dorsal nerve rootlet
Bundles of nerve fibers emerging from dorsal side of spinal cord carry incoming signals from sensors in skin and muscles
Dorsal root ganglion
Cluster of unipolar neurons on each dorsal root carrying incoming sensory information
Spinal nerve
Sensory and motor nerve rootlets merge to form spinal nerve
Anterior fissure
Deep groove along front of spinal cord
Ventral nerve rootlet
Bundles of fibers emerging from ventral side of spinal cord carry signals to skeletal and smooth muscle
Pia mater
Arachnoid
Dura mater
Meninges
Three layers of connective tissue protecting spinal cord

SPINAL CORD
The inner masses of gray and white matter in the spinal cord resemble those of the brain—though in the brain the organization of the tissues is reversed. Also like the brain, the spinal cord is protected by the meninges.

See also pp. 82, 86, 88, 90

NERVOUS SYSTEM 3

THE BRAIN CONTAINS MILLIONS OF NEURONS AND THEIR PROCESSES. DEEP WITHIN ITS INTERIOR, AGGREGATIONS OF NEURONS CALLED NUCLEI FORM THE MAIN CONTROL REGIONS OF THE CENTRAL NERVOUS SYSTEM.

EXTERNAL STRUCTURE OF THE BRAIN

The most obvious feature of the external brain is the large domed cerebrum. It is partly separated into two halves, the cerebral hemispheres, by a deep longitudinal fissure. The heavily folded surface of the cerebrum, the cerebral cortex, is traditionally divided by patterns of fissures into four major functional areas: the frontal, parietal, occipital, and temporal lobes. A smaller structure, the cerebellum, lies beneath the cerebrum. It is mainly concerned with the organization of motor information, including the control of balance and posture. The brain stem—consisting of the medulla oblongata, pons, and midbrain—regulates basic functions such as heartbeat, respiration, and digestion.

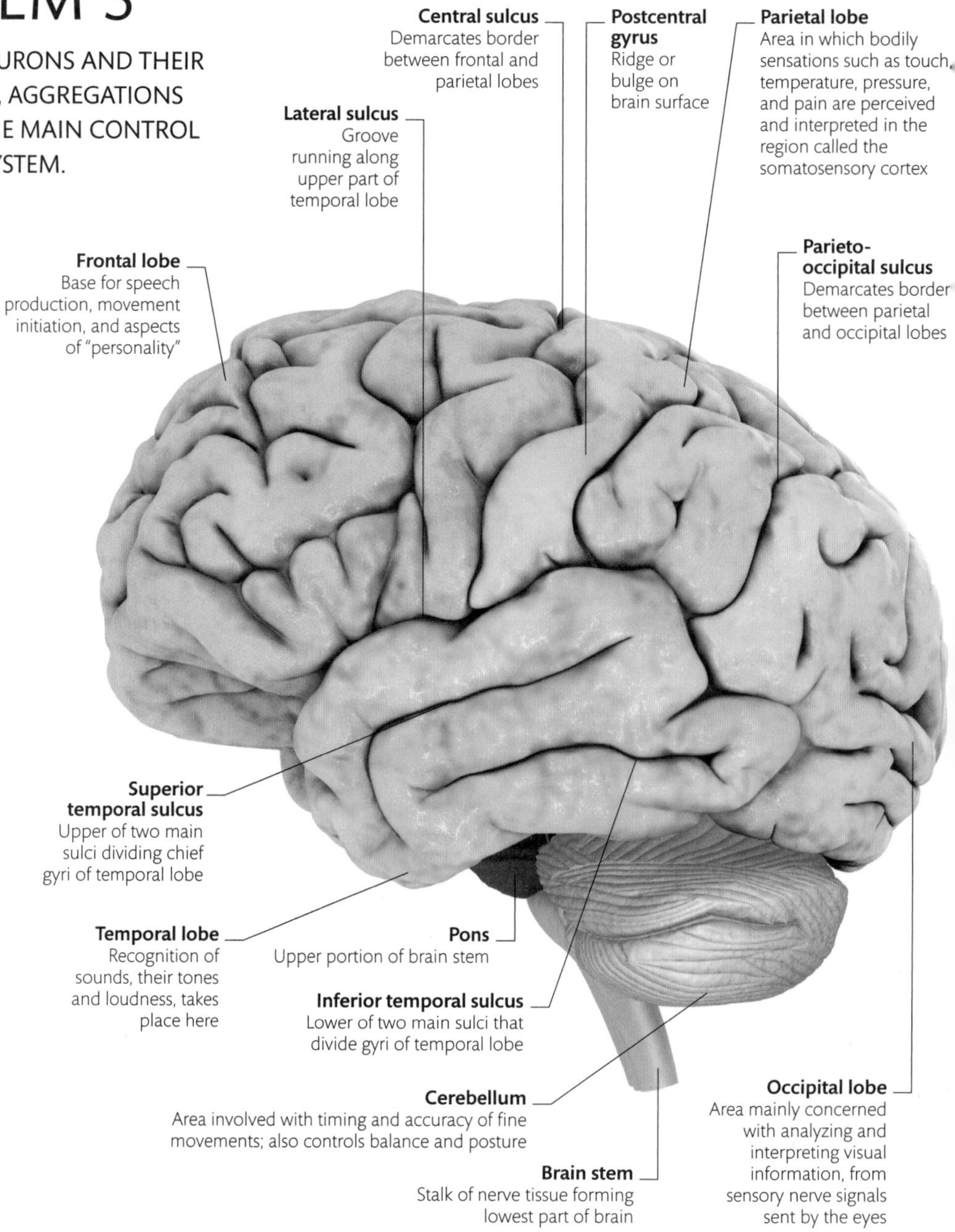

Central sulcus Demarcates border between frontal and parietal lobes
Postcentral gyrus Ridge or bulge on brain surface
Parietal lobe Area in which bodily sensations such as touch, temperature, pressure, and pain are perceived and interpreted in the region called the somatosensory cortex
Lateral sulcus Groove running along upper part of temporal lobe
Frontal lobe Base for speech production, movement initiation, and aspects of "personality"
Parieto-occipital sulcus Demarcates border between parietal and occipital lobes
Superior temporal sulcus Upper of two main sulci dividing chief gyri of temporal lobe
Temporal lobe Recognition of sounds, their tones and loudness, takes place here
Pons Upper portion of brain stem
Inferior temporal sulcus Lower of two main sulci that divide gyri of temporal lobe
Cerebellum Area involved with timing and accuracy of fine movements; also controls balance and posture
Brain stem Stalk of nerve tissue forming lowest part of brain
Occipital lobe Area mainly concerned with analyzing and interpreting visual information, from sensory nerve signals sent by the eyes

EXTERNAL BRAIN STRUCTURE
The surface of the cerebrum, the cerebral cortex, has a pattern of fissures (deep grooves), sulci (shallow grooves), and gyri (bulges) that is unique to each person. The fissures outline the lobes of the brain, four broad regions that are named after the overlying bones of the skull.

MENINGES

Apart from being encased in bone, the brain (together with the spinal cord) is also protected by a triple layer of membranes, the meninges. A tough, fibrous outmost layer, the dura mater, lines the inside of the skull. The middle membrane, the arachnoid, is elastic and weblike. The innermost membrane, the pia mater, lies closest to the surface of the brain. For additional protection, a space between the pia mater and the arachnoid—the subarachnoid space—contains cerebrospinal fluid, a watery liquid that acts as a shock-absorber.

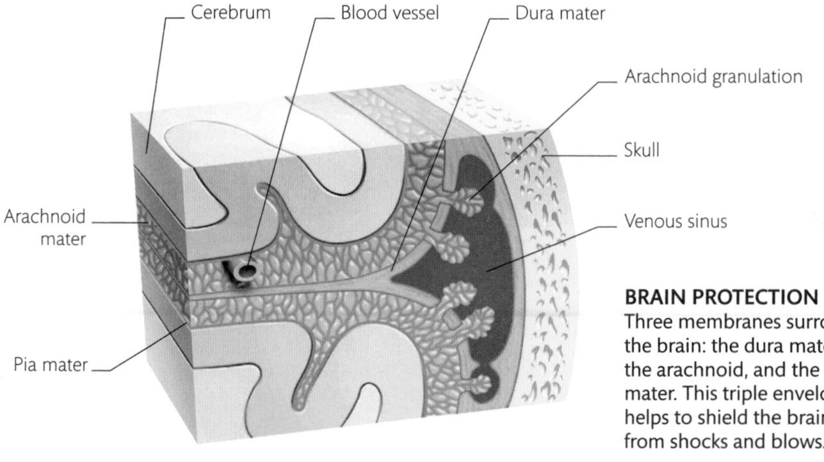

Cerebrum
Blood vessel
Dura mater
Arachnoid granulation
Skull
Venous sinus
Arachnoid mater
Pia mater

BRAIN PROTECTION
Three membranes surround the brain: the dura mater, the arachnoid, and the pia mater. This triple envelope helps to shield the brain from shocks and blows.

INNER BRAIN STRUCTURE

Below the cerebral cortex, which consists of gray matter, the interior of the cerebrum is largely made up of white matter with small islands of gray matter. These include the basal ganglia, thalamus and hypothalamus, and the limbic system. The thalamus monitors incoming sensory information and relays it to the cortex. The limbic system, which is concerned with emotion and memory, influences the autonomic nervous system and the closely linked hypothalamus, which plays a pivotal role in the endocrine system.

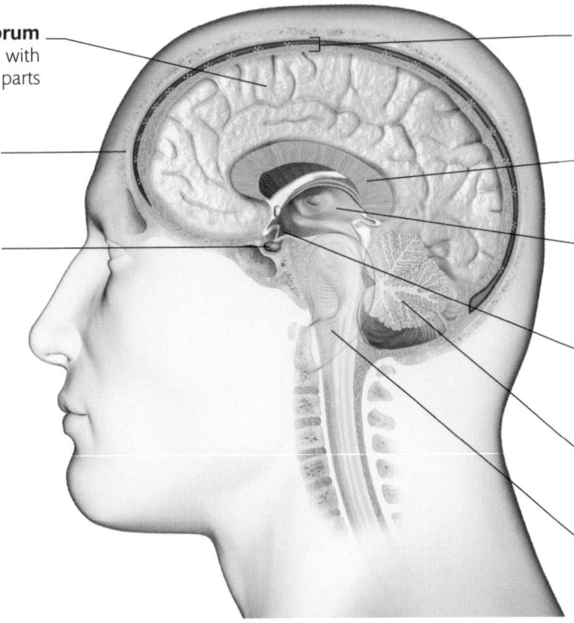

Cerebrum
Largest part of brain, with connections to all body parts
Skull
Pituitary gland
Most important endocrine gland; regulates activities of other glands
Meninges
Three membranes of connective tissue surrounding and protecting brain and spinal cord
Corpus callosum
Bundle of nerve fibers connecting two brain hemispheres
Thalamus
Area that relays nerve signals between cerebral cortex and rest of brain and spinal cord
Hypothalamus
Situated under thalamus; functions include regulation of body temperature and control of autonomic nervous system
Cerebellum
Responsible for control and coordination of movements
Brain stem
Regulates vital functions such as heartbeat and respiration

THE INNER BRAIN
The central structures of the brain include the thalamus, hypothalamus, and pituitary gland. A thick band of nerve fibers, the corpus callosum, links the two cerebral hemispheres.

CEREBROSPINAL FLUID

The soft tissue of the brain floats in cerebrospinal fluid (CSF) within the casing of the skull. CSF is a clear liquid containing proteins and glucose that provide energy for brain cell function as well as lymphocytes that guard against infection. It also protects and nourishes both the brain and the spinal cord as it flows around them. CSF is produced by the choroid plexuses in the two lateral ventricles (chambers) of the brain, which drain into the third ventricle. The fluid then flows into the fourth ventricle, located in front of the cerebellum.

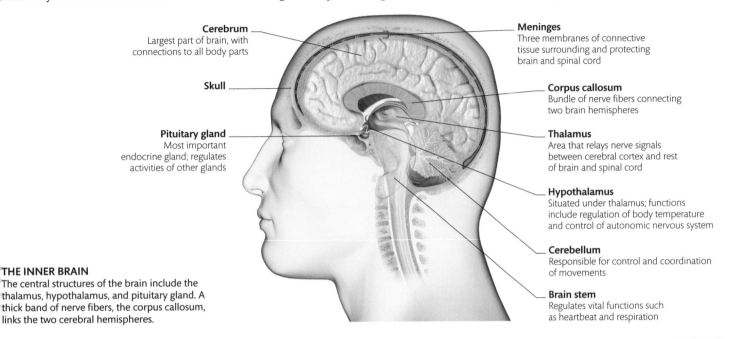

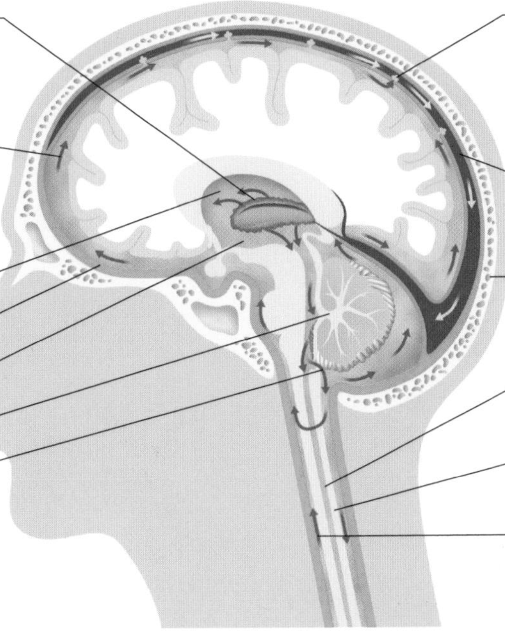

1 Site of fluid production (choroid plexuses)
CSF is produced in clusters of thin-walled capillaries, the choroid plexuses, lining the walls of the lateral ventricles
2 Direction of flow
Fluid moves from the lateral ventricles into the third and fourth ventricles. The CSF then flows up the back of the brain, down around the spinal cord, and up to the front of the brain (see arrows)
Lateral ventricle
Dura mater
Third ventricle
Cerebellum
Fourth ventricle
4 Site of reabsorption (arachnoid granulation)
After circulating around the brain, CSF is reabsorbed into the blood via arachnoid granulations, which are projections of the arachnoid layer into large venous sinuses lying between the dura mater and the skull
Sagittal sinus
Occipital bone
Central canal
Spinal cord
3 Circulation around spinal cord
Aided by vertebral movement, CSF flows downward around the spinal cord and in the central canal

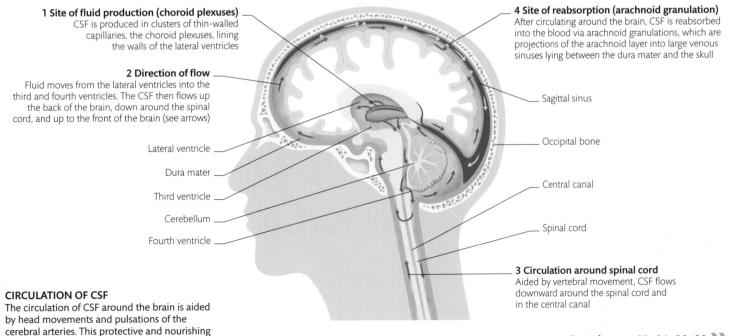

CIRCULATION OF CSF
The circulation of CSF around the brain is aided by head movements and pulsations of the cerebral arteries. This protective and nourishing fluid is renewed four to five times a day.

See also pp. 82, 84, 88, 90 »

NERVOUS SYSTEM 4

WHEN STIMULATED, NEURONS UNDERGO CHEMICAL CHANGES THAT PRODUCE TRAVELING WAVES OF ELECTRICITY—NERVE SIGNALS, OR IMPULSES. THESE IMPULSES PASS TO OTHER NEURONS, ELICITING SIMILAR RESPONSES FROM THEM.

NERVE IMPULSE

Information is conveyed through the nervous system as nerve impulses, or action potentials. Impulses are about 0.1 volts (100 millivolts) in strength and last just 0.001 s (1 millisecond). The information carried depends on the location of the impulses in the nervous system, and on their frequency—from one impulse every few seconds to several hundred per second. When impulses reach a junction (synapse), they trigger the release of chemicals (neurotransmitters). Molecules of the neurotransmitter cross the synapse and stimulate the receiving neuron to fire an impulse of its own, as wavelike movements of ions. Neurotransmitters may also inhibit a receiving neuron from firing.

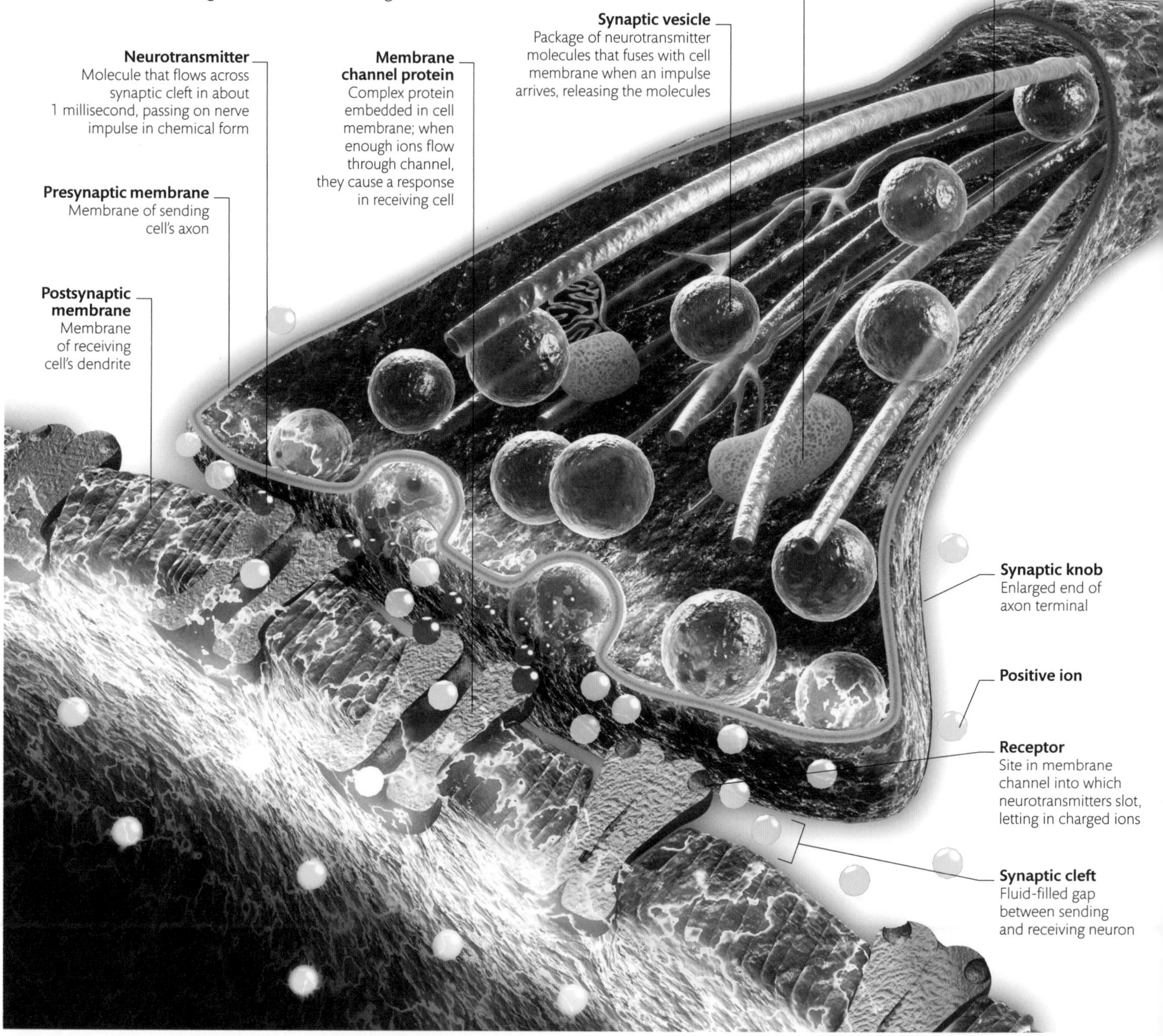

Neurotubule Specialized microtubule that conveys synaptic vesicles from cell body to axon terminal
Mitochondrion Standard cellular component that provides energy
Synaptic vesicle Package of neurotransmitter molecules that fuses with cell membrane when an impulse arrives, releasing the molecules
Neurotransmitter Molecule that flows across synaptic cleft in about 1 millisecond, passing on nerve impulse in chemical form
Membrane channel protein Complex protein embedded in cell membrane; when enough ions flow through channel, they cause a response in receiving cell
Presynaptic membrane Membrane of sending cell's axon
Postsynaptic membrane Membrane of receiving cell's dendrite
Synaptic knob Enlarged end of axon terminal
Positive ion
Receptor Site in membrane channel into which neurotransmitters slot, letting in charged ions
Synaptic cleft Fluid-filled gap between sending and receiving neuron

Dendrites
Projections of neuron; collect nerve impulses from other neurons or sensory nerve endings
Neuron cell body
Node of Ranvier
Unmyelinated portion of axon between two myelinating glial cells
Myelin sheath
Fatty sheath that helps to speed impulse, conduction, and prevent fading or leaking of current
Myelinating cell
Oligodendrocyte (CNS) or Schwann cell (PNS)
Axon
Conveys impulses away from cell body

IMPULSE MOVEMENT

An impulse within a neuron is based chiefly on movement of positively charged sodium and potassium ions through the cell membrane. Impulses travel at speeds of between 1 and 3–400 ft/s (120 m/s), depending on the type of nerve. Movement is many times faster in myelinated axons.

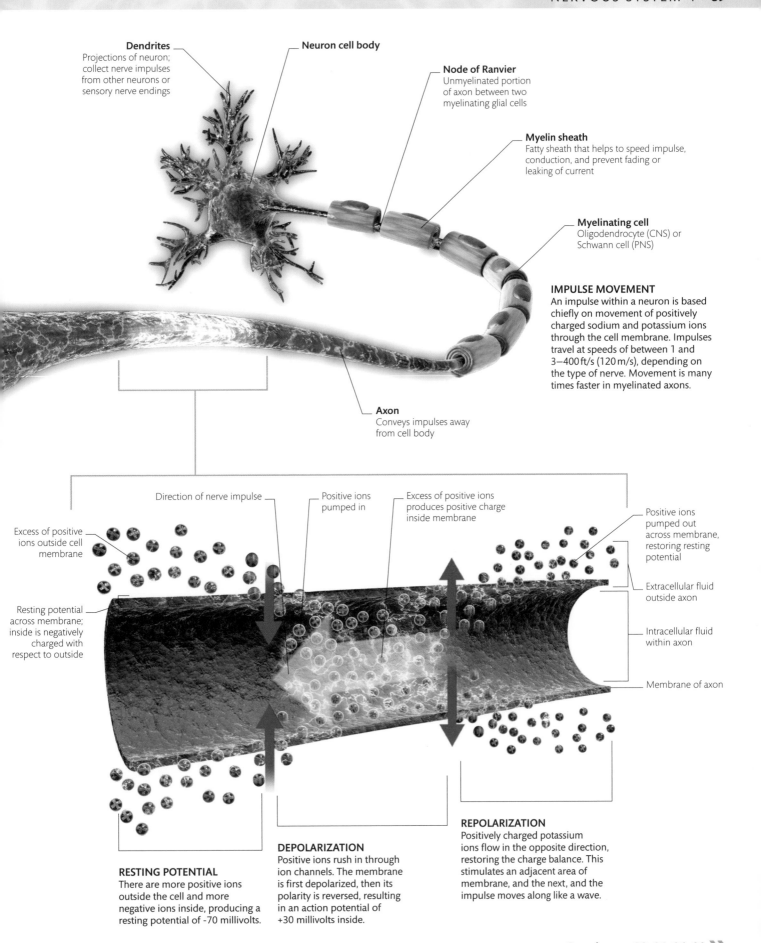

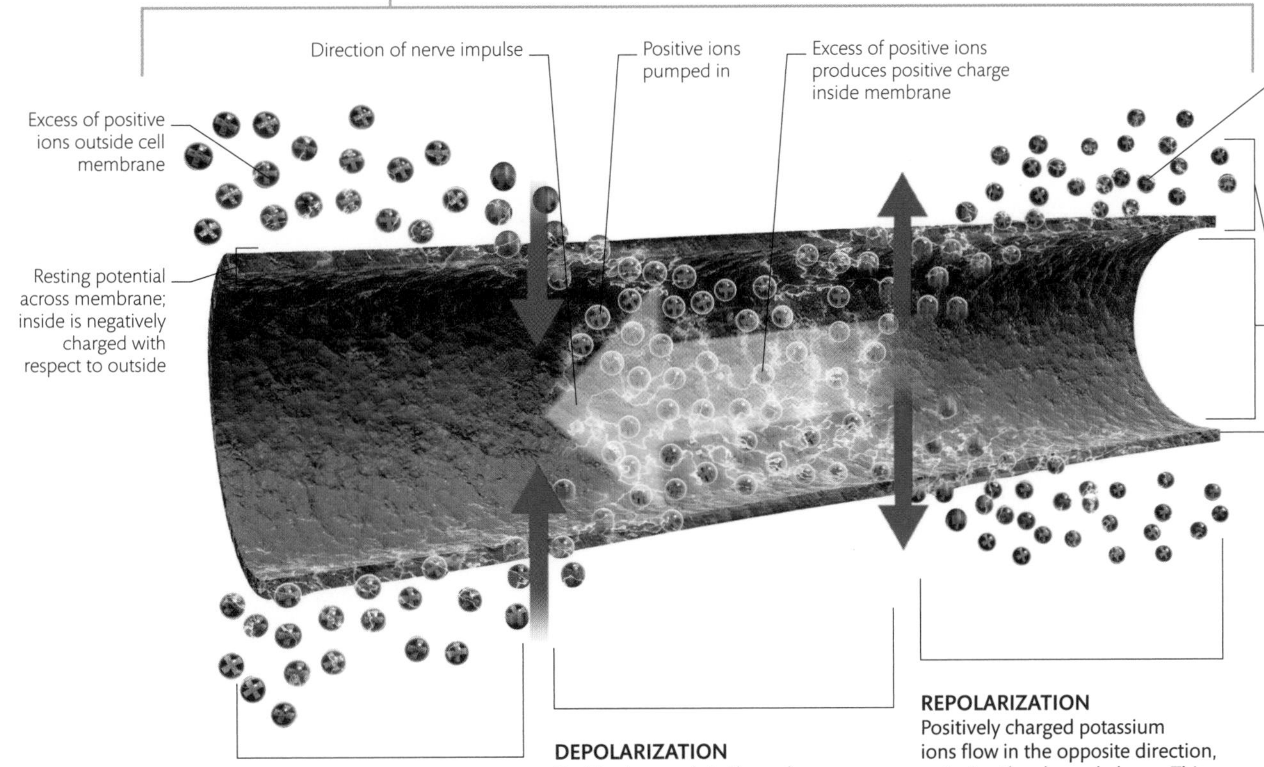
Direction of nerve impulse
Positive ions pumped in
Excess of positive ions produces positive charge inside membrane
Positive ions pumped out across membrane, restoring resting potential
Excess of positive ions outside cell membrane
Extracellular fluid outside axon
Resting potential across membrane; inside is negatively charged with respect to outside
Intracellular fluid within axon
Membrane of axon

RESTING POTENTIAL

There are more positive ions outside the cell and more negative ions inside, producing a resting potential of -70 millivolts.

DEPOLARIZATION

Positive ions rush in through ion channels. The membrane is first depolarized, then its polarity is reversed, resulting in an action potential of +30 millivolts inside.

REPOLARIZATION

Positively charged potassium ions flow in the opposite direction, restoring the charge balance. This stimulates an adjacent area of membrane, and the next, and the impulse moves along like a wave.

See also pp. 82, 84, 86, 90

NERVOUS SYSTEM 5

THE CENTRAL NERVOUS SYSTEM WORKS CEASELESSLY TO PROCESS IMPULSES RECEIVED FROM THE PERIPHERAL NERVOUS SYSTEM AND TO SEND RESPONSES BACK DOWN THE PERIPHERAL NERVES. SOME RESPONSES OCCUR UNDER CONSCIOUS CONTROL; OTHERS DO NOT INVOLVE AWARENESS.

THE CNS IN ACTION

The components of the central nervous system (CNS) interpret incoming signals from the peripheral nervous system (PNS) and send responses back out. The process begins when external stimuli are converted into nerve impulses by specialized receptor cells. These impulses travel through the sensory nerves of the PNS and on to the higher centers of the brain; the route to the cerebral cortex may involve a series of up to 10 neurons linked by synapses. At each relay station, additional signals are sent out along other pathways. When the CNS receives a nerve signal, it analyzes the information and initiates a response, such as the contraction of a muscle. The nerve tracts that convey impulses between the CNS and PNS cross over from the left side of the body to the right side, and vice versa. So, for example, sensory information from the body's left side is received in the right hemisphere of the brain, and motor instructions from the left hemisphere control muscles on the right side of the body.

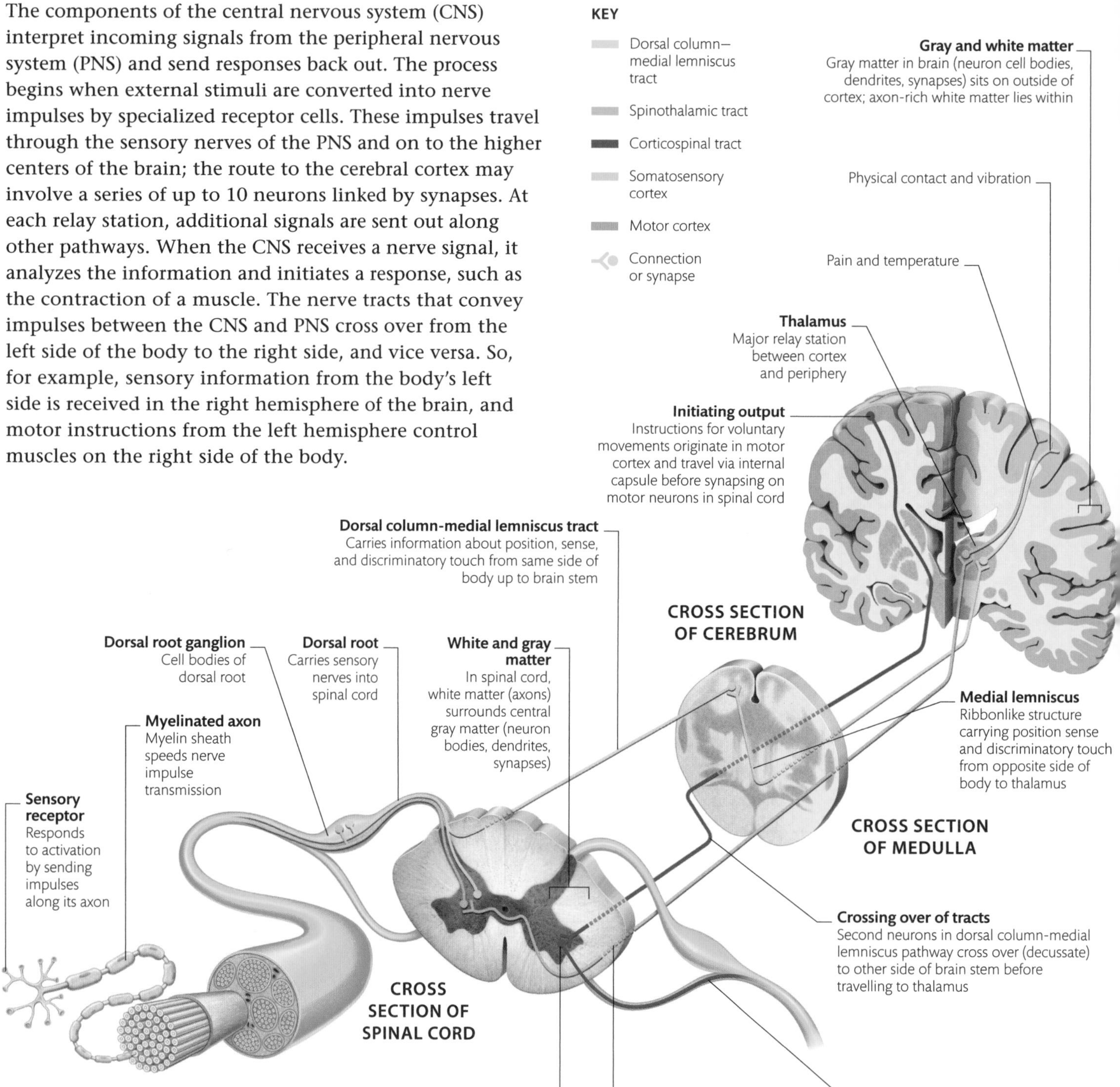
KEY
Dorsal column–medial lemniscus tract
Spinothalamic tract
Corticospinal tract
Somatosensory cortex
Motor cortex
Connection or synapse
Gray and white matter
Gray matter in brain (neuron cell bodies, dendrites, synapses) sits on outside of cortex; axon-rich white matter lies within
Physical contact and vibration
Pain and temperature
Thalamus
Major relay station between cortex and periphery
Initiating output
Instructions for voluntary movements originate in motor cortex and travel via internal capsule before synapsing on motor neurons in spinal cord
Dorsal column-medial lemniscus tract
Carries information about position, sense, and discriminatory touch from same side of body up to brain stem
CROSS SECTION OF CEREBRUM
Dorsal root ganglion
Cell bodies of dorsal root
Dorsal root
Carries sensory nerves into spinal cord
White and gray matter
In spinal cord, white matter (axons) surrounds central gray matter (neuron bodies, dendrites, synapses)
Myelinated axon
Myelin sheath speeds nerve impulse transmission
Sensory receptor
Responds to activation by sending impulses along its axon
Medial lemniscus
Ribbonlike structure carrying position sense and discriminatory touch from opposite side of body to thalamus
CROSS SECTION OF MEDULLA
Crossing over of tracts
Second neurons in dorsal column-medial lemniscus pathway cross over (decussate) to other side of brain stem before travelling to thalamus
CROSS SECTION OF SPINAL CORD
Motor messages
Motor nerve impulses descend corticospinal tract and relay along motor axons to arm and hand muscles
Spinothalamic tract
Carries information about pain, temperature, and light touch from opposite side of body to thalamus
Ventral root
Motor axons leave spinal cord here, carrying instructions to muscles

CNS IN ACTION
Information from any of the body's sensory receptors travels to the brain up one of two pathways: the spinothalamic or the dorsal column-medial lemniscus tracts. Motor instructions travel along several tracts, the fastest being the corticospinal tract.

NVOLUNTARY RESPONSES

ome responses do not involve the higher regions of the rain associated with awareness. Many of these involuntary esponses are controlled by the autonomous nervous ystem (ANS). One category of involuntary response, reflex ction, mainly affects muscles normally under voluntary control. In a reflex, sensory nerve fibers feed information to the spinal cord and then synapse directly, or via an intermediate neuron, on to motor neurons. The other main type of involuntary response includes autonomic motor actions, which are initiated in the lower brain.

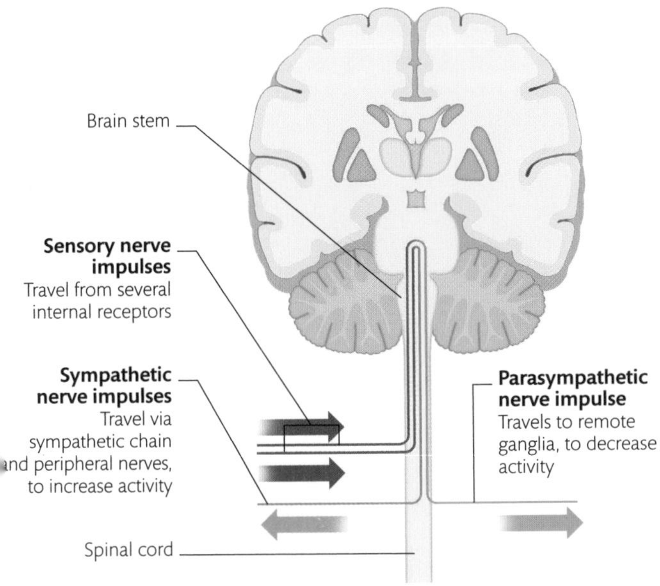
Brain stem
Sensory nerve impulses
Travel from several internal receptors
Sympathetic nerve impulses
Travel via sympathetic chain and peripheral nerves, to increase activity
Parasympathetic nerve impulse
Travels to remote ganglia, to decrease activity
Spinal cord

AUTONOMIC RESPONSES
Nerve signals from viscera travel through the ANS and spinal cord tracts to nuclei in the brain stem, where they initiate appropriate motor responses.

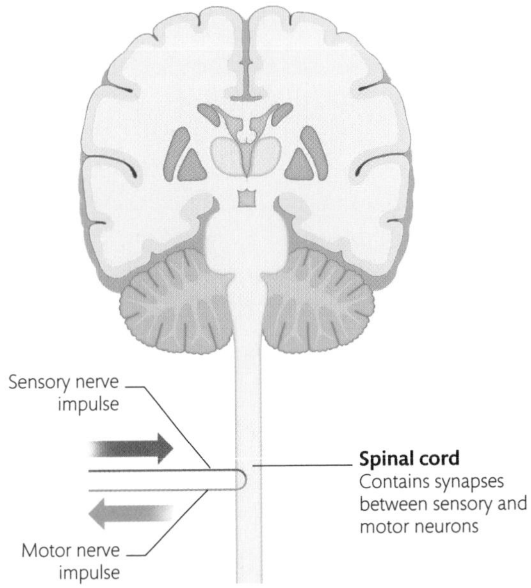
Sensory nerve impulse
Spinal cord
Contains synapses between sensory and motor neurons
Motor nerve impulse

SIMPLE REFLEXES
Sensory signals arrive, and motor signals depart, wholly within the spinal cord, and without brain involvement—although the brain becomes aware soon afterwards.

RESPONSES UNDER VOLUNTARY CONTROL

Actions under voluntary control are initiated by conscious thought, or by incoming sensory stimuli. When the cerebral cortex receives sensory data from the cerebellum and from the basal ganglia, it formulates a central motor plan for a particular movement and sends out instructions as motor nerve signals to voluntary muscles. As the movement progresses, it is monitored by sensory endings in the muscles, tendons, and joints. The sensory endings update the brain, so that the cerebral cortex can send corrective nerve signals back to the muscles.

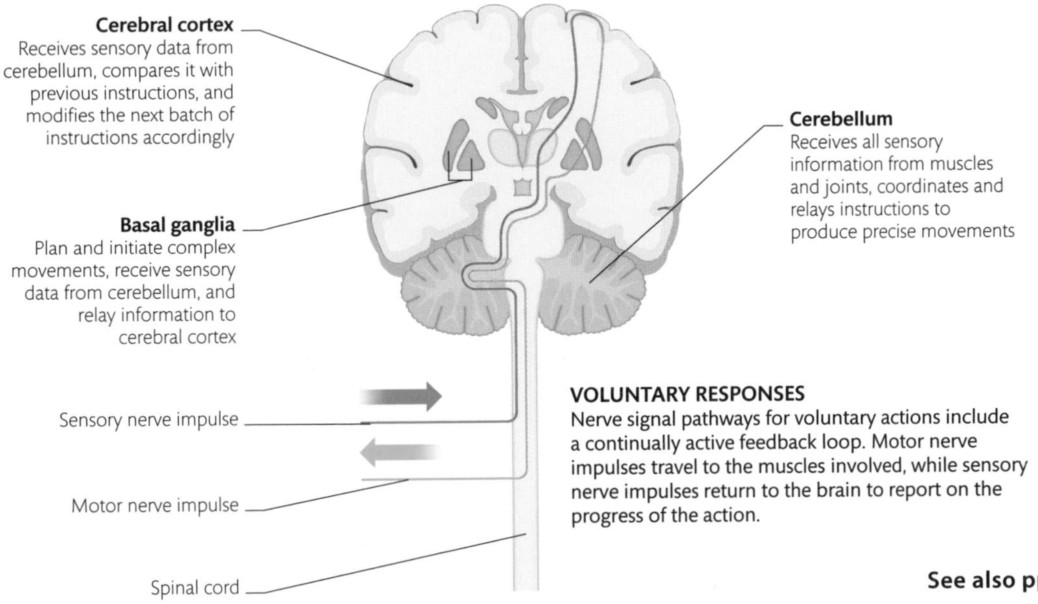
Cerebral cortex
Receives sensory data from cerebellum, compares it with previous instructions, and modifies the next batch of instructions accordingly
Basal ganglia
Plan and initiate complex movements, receive sensory data from cerebellum, and relay information to cerebral cortex
Cerebellum
Receives all sensory information from muscles and joints, coordinates and relays instructions to produce precise movements
Sensory nerve impulse
Motor nerve impulse
Spinal cord

VOLUNTARY RESPONSES
Nerve signal pathways for voluntary actions include a continually active feedback loop. Motor nerve impulses travel to the muscles involved, while sensory nerve impulses return to the brain to report on the progress of the action.

See also pp. 82, 84, 86, 88 »

HEAD AND NECK 1

THE LARGEST PART OF THE BRAIN—THE CEREBRUM—IS ALMOST COMPLETELY DIVIDED INTO TWO HEMISPHERES, A DIVISION THAT CAN BE CLEARLY SEEN FROM THE ANTERIOR AND POSTERIOR VIEWS SHOWN HERE.

ANTERIOR VIEW

Color and/or label the structures indicated on the diagram using the key below.

1. Longitudinal (cerebral) fissure
2. Frontal lobe
3. Frontal pole
4. Lateral sulcus
5. Temporal lobe
6. Optic nerve
7. Optic chiasma
8. Cerebellar hemisphere
9. Pons
10. Medulla oblongata
11. Spinal cord
12. Pituitary gland
13. Horizontal fissure of cerebellum
14. Temporal pole
15. Olfactory tract
16. Olfactory bulb
17. Corpus callosum

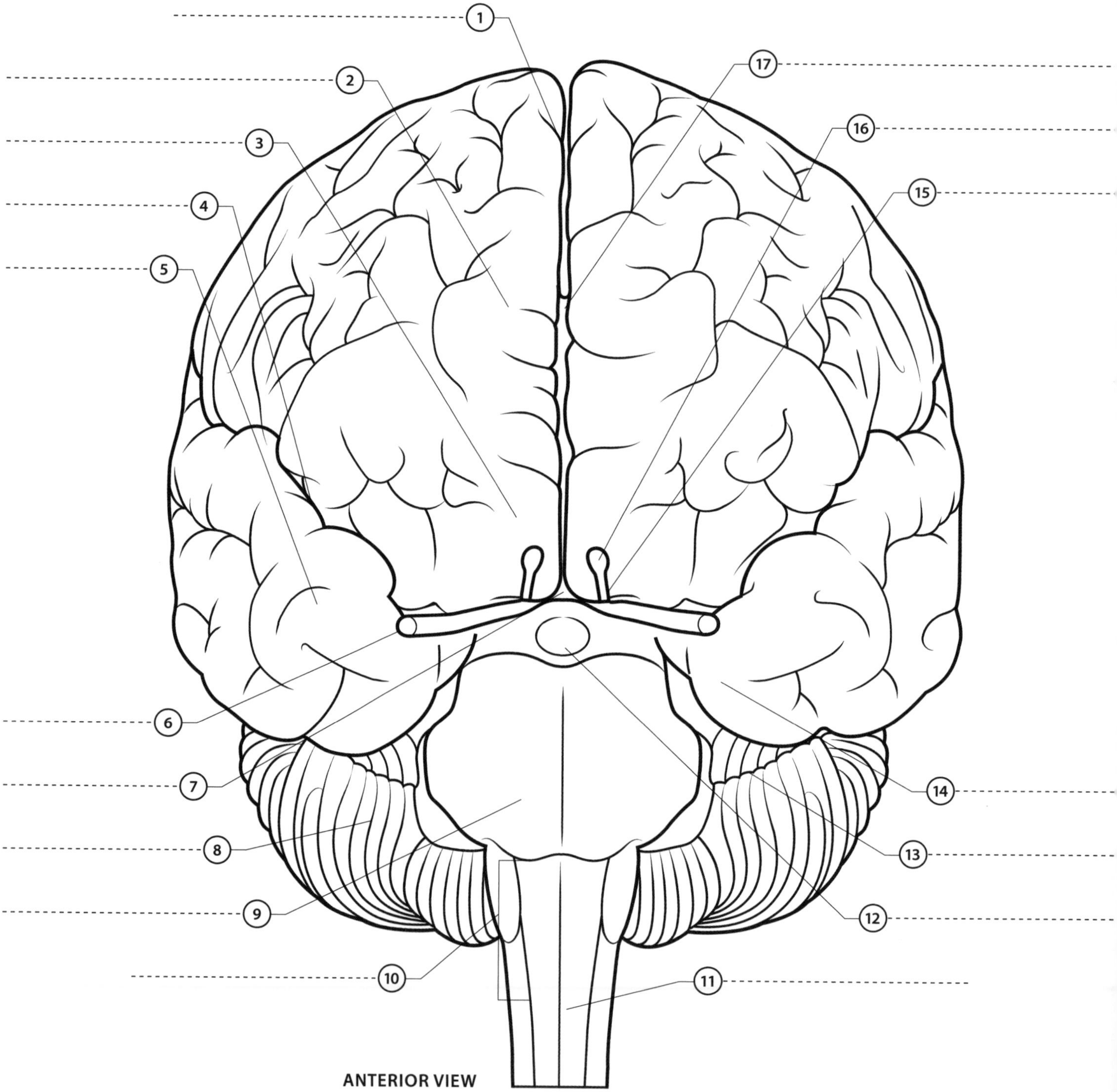

ANTERIOR VIEW

ɪNTERIOR AND POSTERIOR ANATOMY

'he cerebral hemispheres are divided by the deep ɔngitudinal (cerebral) fissure. At the bottom of this fissure es the corpus callosum, forming a bridge of nerve fibers etween the two sides of the brain. Areas of the brain that eceive and process different types of information or hat control movements can be very widely separated. or example, the visual pathways from the two eyes end n the cortex of the occipital lobe at the back of the brain, vhere most visual information is processed. However, he nerve impulses that eventually reach the extraocular nuscles that move the eyes begin in neurons in the ortex of the frontal lobe.

POSTERIOR VIEW

Color and/or label the structures indicated on the diagram using the key below.

1. Longitudinal (cerebral) fissure
2. Occipital pole
3. Cerebellar vermis
4. Medulla oblongata
5. Spinal cord
6. Cerebellar hemisphere
7. Folia
8. Fissures
9. Corpus callosum

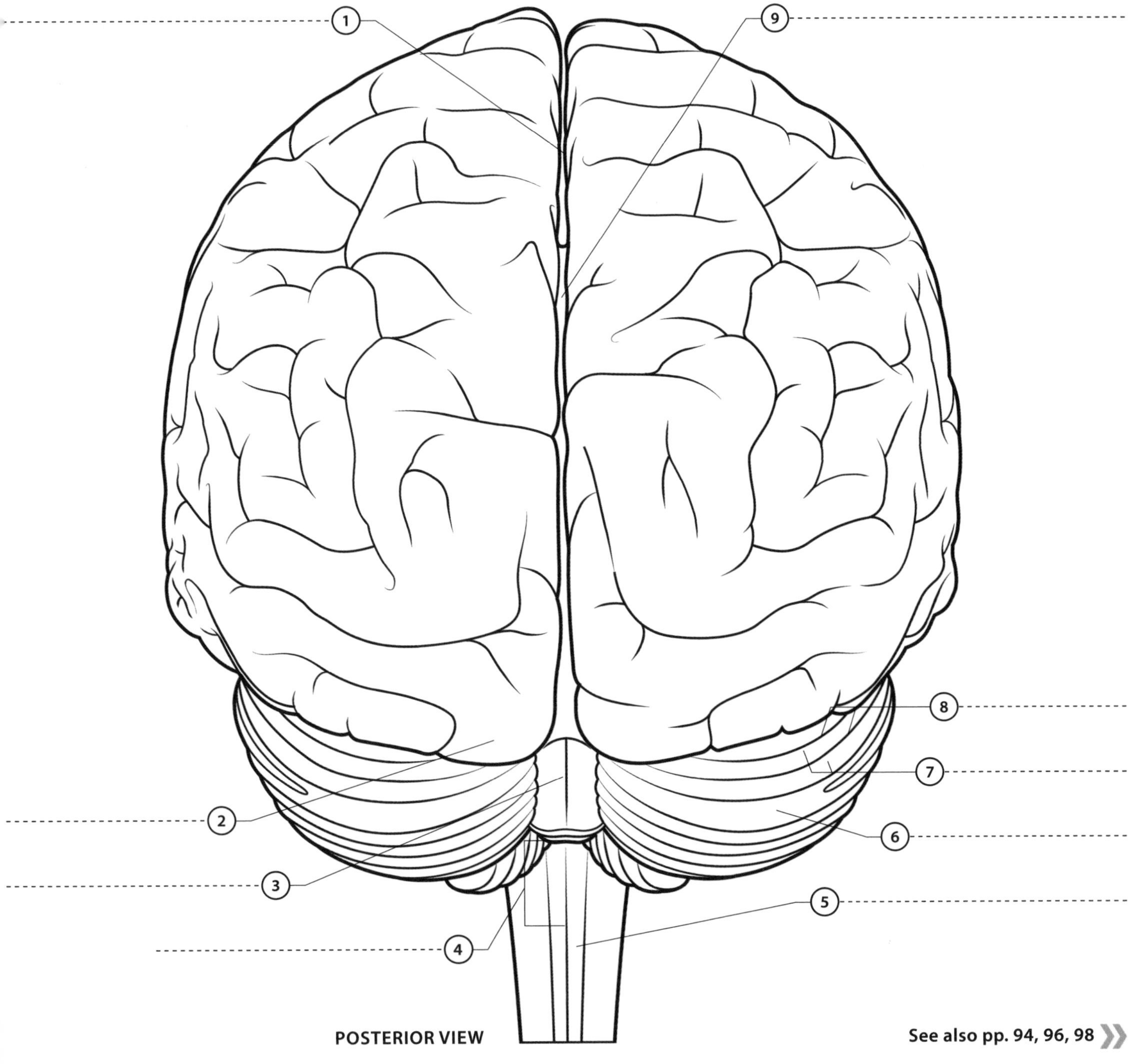

POSTERIOR VIEW

See also pp. 94, 96, 98

HEAD AND NECK 2

LIKE THE ANTERIOR AND POSTERIOR VIEWS (SEE PREVIOUS PAGES), THE SUPERFICIAL VIEW OF THE BRAIN SHOWS ITS TWO HEMISPHERES, WHEREAS THE INFERIOR VIEW REVEALS ALL BUT ONE PAIR OF CRANIAL NERVES.

EXTERNAL ANATOMY

As well as showing the two cerebral hemispheres—separated by the longitudinal (cerebral) fissure—the superficial view also clearly reveals that the outer layer of gray matter—the cortex—is highly folded, with clefts (sulci) separating the raised areas (gyri). The names of areas of the brain correspond to the bones of the skull that cover them; for example, the anterior extremity of the brain is the frontal pole, and its posterior extremity is the occipital pole.

SUPERFICIAL VIEW OF BRAIN

Color and/or label the structures indicated on the diagram using the key below.

1. Longitudinal (cerebral) fissure
2. Middle frontal gyrus
3. Inferior frontal gyrus
4. Precentral sulcus
5. Precentral gyrus
6. Postcentral gyrus
7. Central sulcus
8. Supramarginal gyrus
9. Postcentral sulcus
10. Occipital pole
11. Parieto-occipital sulcus
12. Inferior parietal lobule
13. Angular gyrus
14. Intraparietal sulcus
15. Superior parietal lobule
16. Superior frontal gyrus
17. Inferior frontal sulcus
18. Superior frontal sulcus
19. Frontal pole

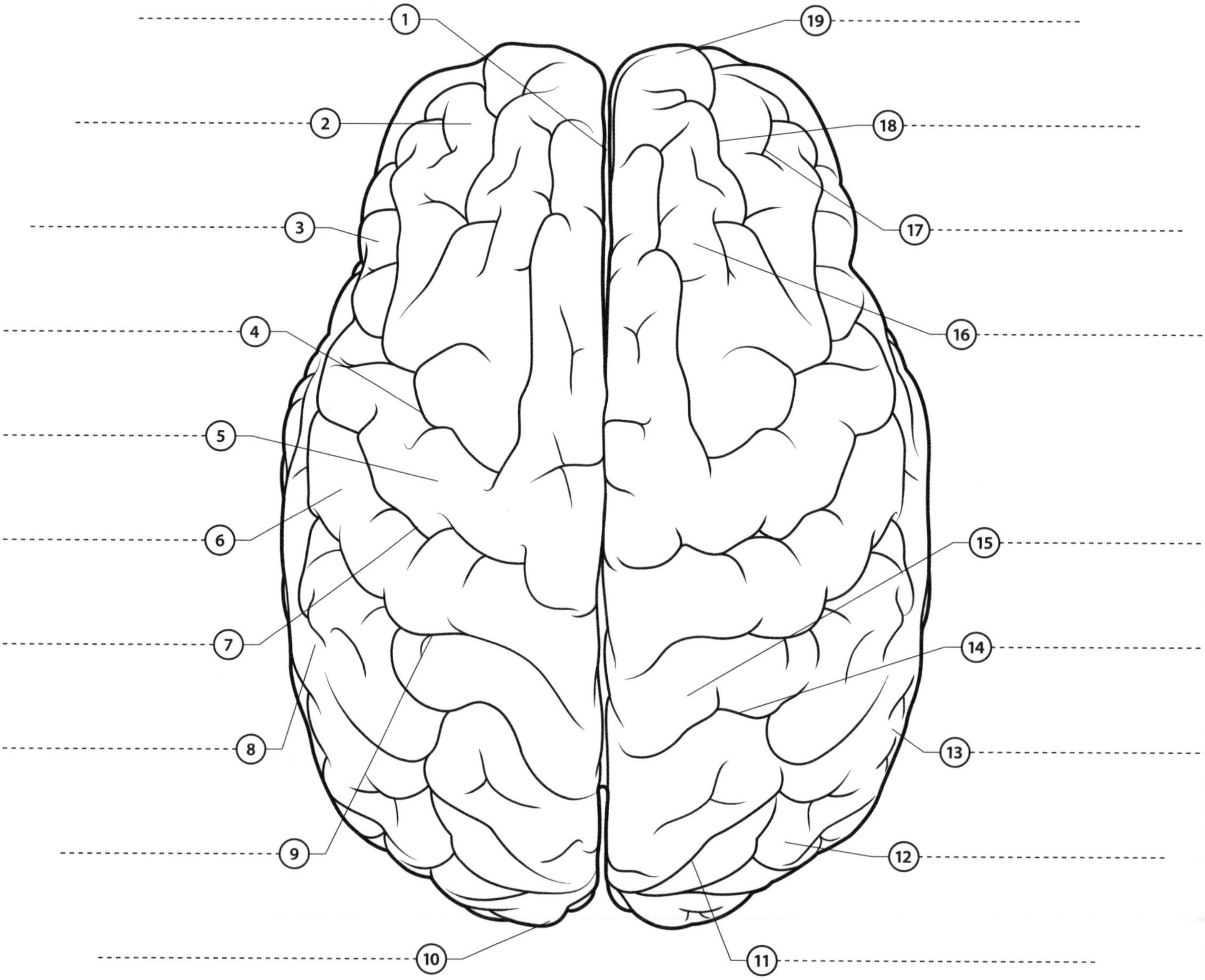

EXTERNAL SURFACE OF BRAIN FROM ABOVE

CRANIAL NERVES

Twelve pairs of cranial nerves emerge from the brain and brain stem, and leave the skull via foramina in the skull. The olfactory and optic nerves emerge from the brain itself; the other cranial nerves emerge from the brain stem. The olfactory nerve (I) carries olfactory information from the specialized epithelium in the roof of the nasal cavity to the limbic system. The optic nerve (II) carries visual information from the retina to the visual cortex. The oculomotor (III), trochlear (IV), and abducens (VI) nerves innervate the extraocular muscles; the oculomotor also innervates smooth muscles that control the size of the pupil and shape of the lens. The trigeminal nerve (V) carries sensory information from the eyes, face, and teeth, and innervates the muscles used for chewing. The facial nerve (VII) innervates the muscles of facial expression, carries sensory information from the taste buds, and also carries autonomic fibers controlling secretions of the salivary and lacrimal glands. The vestibulocochlear nerve (VIII) carries information that is essential for maintaining balance and posture from the inner ear to the cerebellum, and auditory information from the middle ear to the auditory cortex. The glossopharyngeal nerve (IX) carries sensory information about taste, touch, and temperature from the tongue and oropharynx. The vagus nerve (X) has sensory, motor, and autonomic fibers and is involved in many vital functions, including breathing and controlling the heartbeat. The accessory nerve (XI) supplies motor fibers to the head, neck, and shoulders, and to the pharynx and larynx. The hypoglossal nerve (XII) innervates most of the muscles of the tongue.

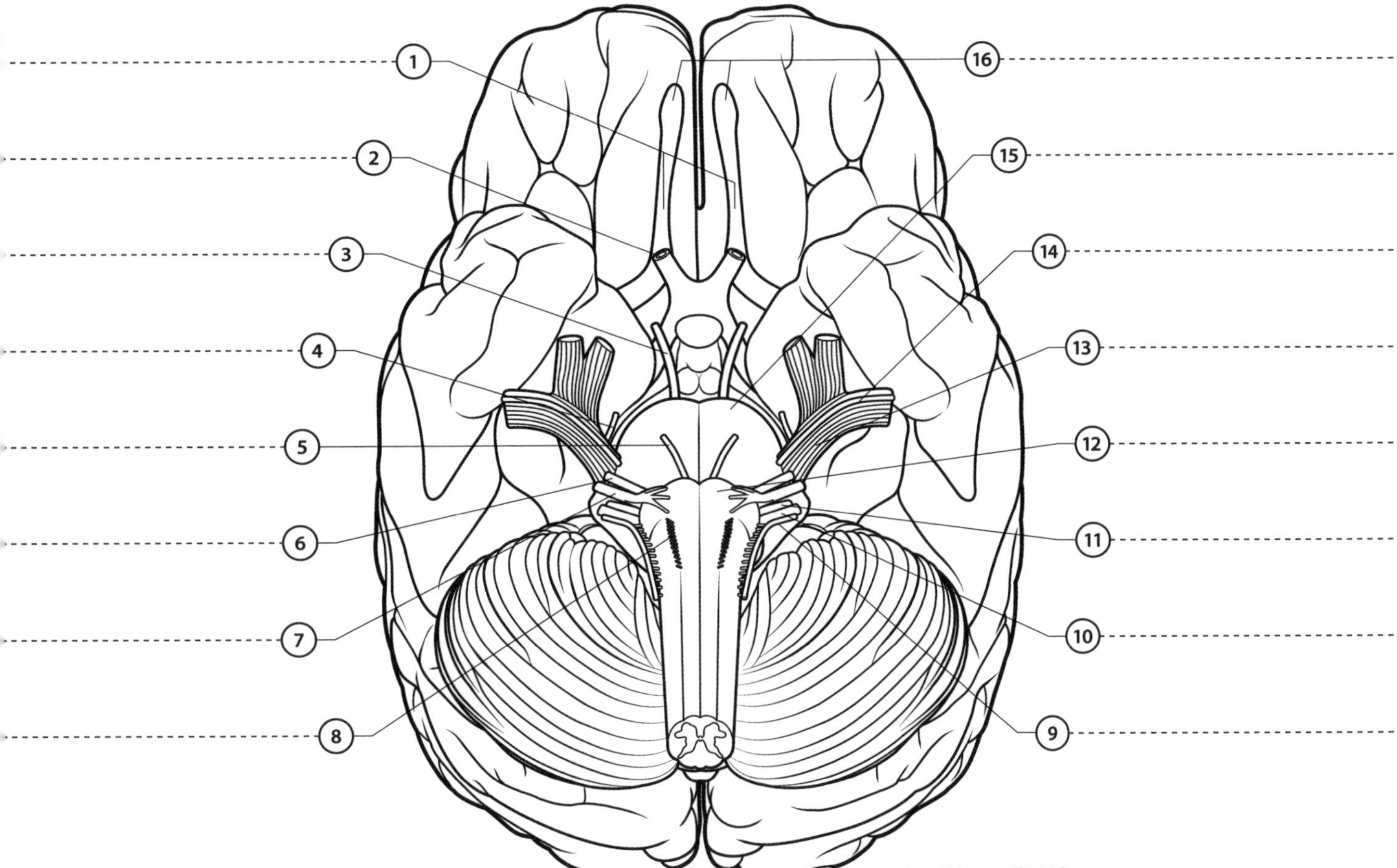

CRANIAL NERVES

CRANIAL NERVES

Color and/or label the structures indicated on the diagram using the key below.

1. Olfactory tracts
2. Optic nerve (II)
3. Oculomotor nerve (III)
4. Trochlear nerve (IV)
5. Abducens nerve (VI)
6. Facial nerve (VII)
7. Vestibulocochlear nerve (VIII)
8. Hypoglossal nerve (XII)
9. Accessory nerve (XI)
10. Vagus nerve (X)
11. Glossopharyngeal nerve (IX)
12. Pyramid
13. Sensory root of trigeminal nerve (V)
14. Motor root of trigeminal nerve (V)
15. Pons
16. Olfactory bulbs

See also pp. 32, 92, 96, 98, 136

HEAD AND NECK 3

SECTIONS THROUGH THE BRAIN REVEAL ITS INTERNAL STRUCTURE. THE OUTER CORTICAL LAYER IS RELATIVELY THIN AND SURROUNDS WHITE MATTER, WHICH CONTAINS ISLANDS OF GRAY MATTER. THE VENTRICLES ARE FILLED WITH CEREBROSPINAL FLUID.

CORONAL SECTION THROUGH THE BRAIN

Gray matter contains neuronal cell bodies; white matter contains bundles of axons (tracts), some of which are myelinated (hence the name "white matter"). In this section through the cerebral hemispheres, a number of nuclei can be seen, including the caudate and lentiform nuclei, which are components of the basal ganglia and play an important role in selecting appropriate actions to perform. The fornix and mammillary bodies are parts of the limbic system (see p. 99). The thalamus is a major relay station for both sensory and motor fibers passing between the periphery and the cortex.

CORONAL SECTION OF BRAIN

Color and/or label the structures indicated on the diagram using the key below.

1. Body of corpus callosum
2. Anterior horn of lateral ventricle
3. Fornix
4. Third ventricle
5. Mammillary body
6. Hypothalamus
7. Thalamus
8. Lentiform nucleus
9. Caudate nucleus
10. Septum pellucidum

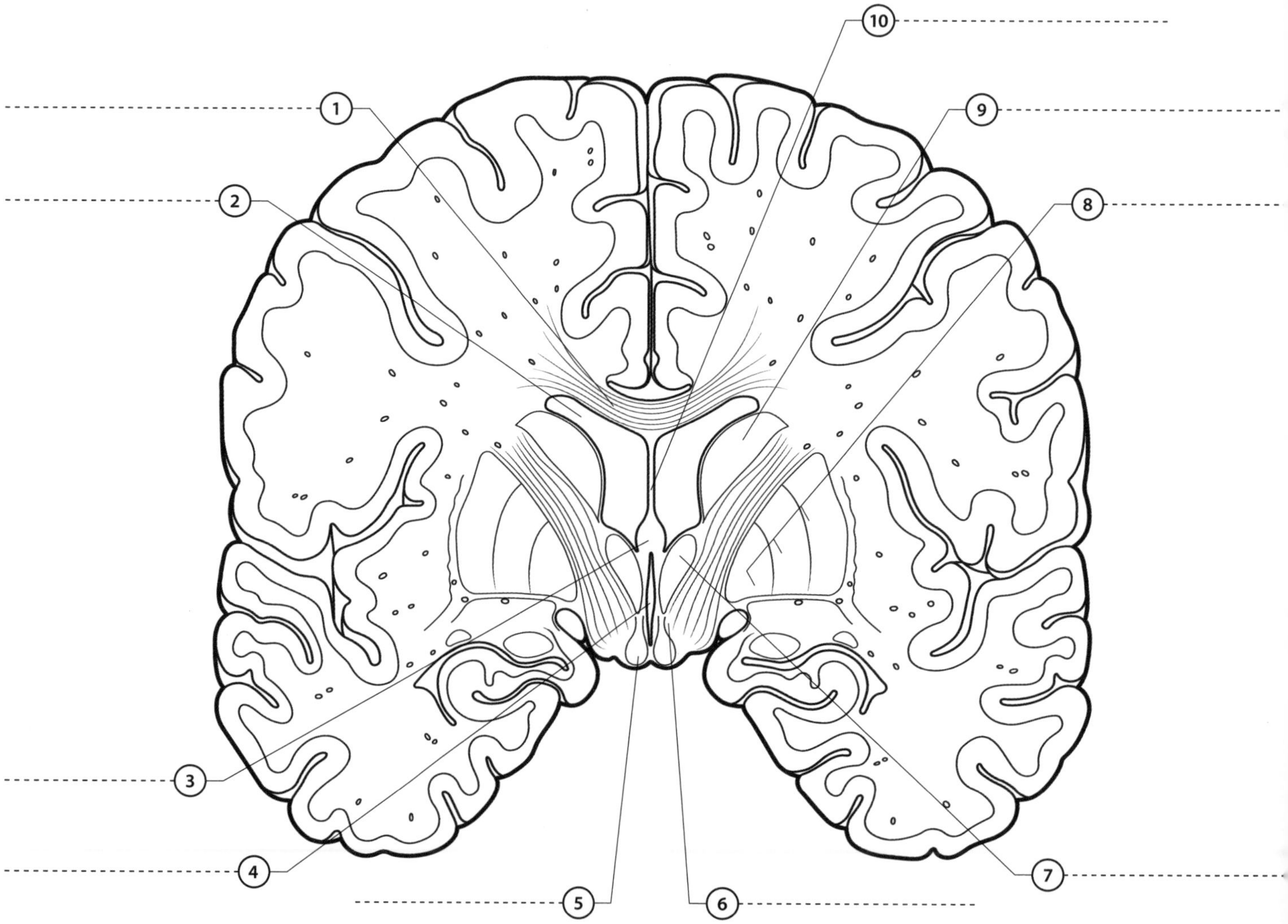

CORONAL SECTION OF BRAIN

TRANSVERSE SECTION THROUGH THE BRAIN

The internal capsule is a major pathway in the brain: its fibers convey sensory information from the thalamus to the sensory cortex, and motor information heading for the brain stem and spinal cord from the motor areas of the cortex. This region is particularly vulnerable to strokes. Cerebrospinal fluid is formed in the choroid plexuses, in the lateral ventricles.

TRANSVERSE SECTION OF BRAIN

Color and/or label the structures indicated on the diagram using the key below.

1. Septum pellucidum
2. Anterior horn of lateral ventricle
3. Caudate nucleus
4. Fornix
5. Inferior horn of lateral ventricle
6. Splenium of corpus callosum
7. Thalamus
8. Optic radiation
9. Lentiform nucleus
10. Internal capsule
11. Genu of corpus callosum

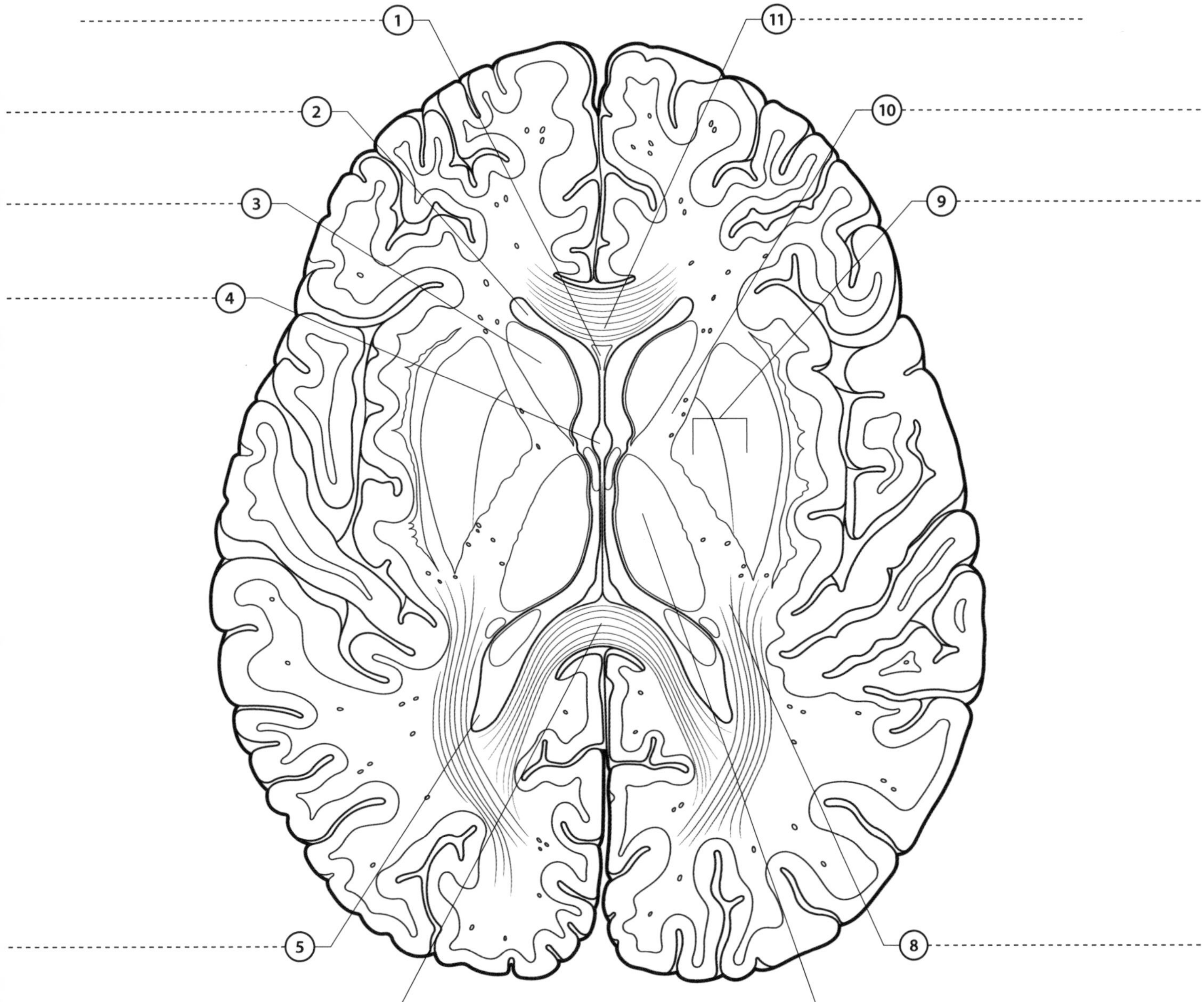

TRANSVERSE SECTION OF BRAIN

See also pp. 92, 94, 98

HEAD AND NECK 4

ALTHOUGH TO THE NAKED EYE THE BRAIN APPEARS LARGELY UNDIFFERENTIATED, DIFFERENT FUNCTIONAL AREAS CAN BE MAPPED OUT USING SPECIALIZED TECHNIQUES.

FUNCTIONAL AREAS OF THE BRAIN

Each part of the cerebral cortex has a designated number known as a Brodmann number (which run from 1 to 52), devised by the German neurologist Korbinian Brodmann (1868—1918) and based on features of microscopic anatomy. Distinct from these numbers, but partly overlapping with them, are cortical areas dealing with certain functions, such as the visual cortex for input from the retinas, and Broca's and Wernicke's areas for language.

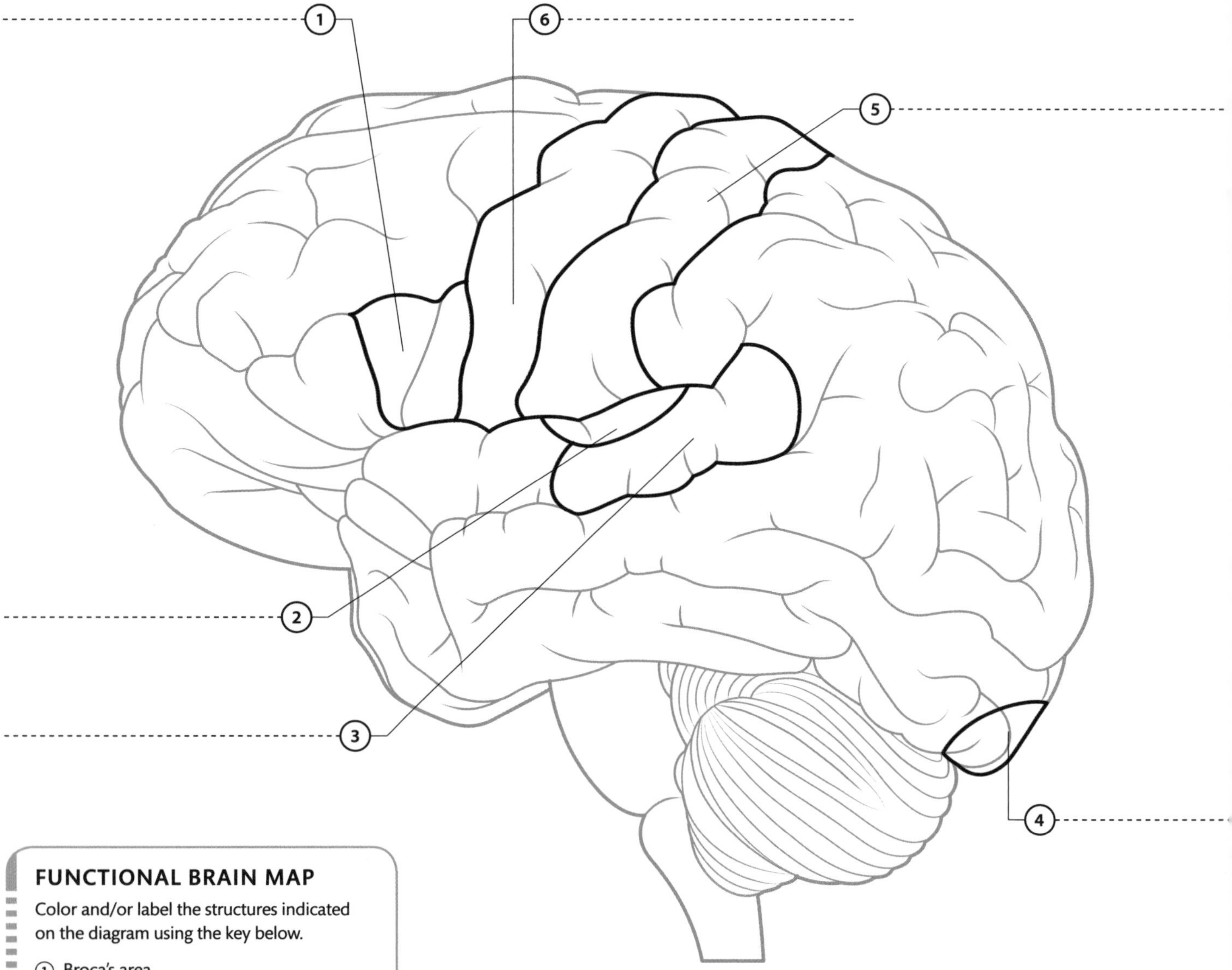

FUNCTIONAL BRAIN MAP

FUNCTIONAL BRAIN MAP

Color and/or label the structures indicated on the diagram using the key below.

1. Broca's area
2. Auditory cortex
3. Wernicke's area
4. Visual cortex
5. Somatosensory cortex
6. Motor cortex

.IMBIC SYSTEM

'he limbic system is a complex network of areas, nuclei, .nd pathways involved in instinctive behavior and ·motions. It is also involved in memory formation .nd forms a link between areas of higher consciousness n the cerebral cortex and the brain stem. The limbic ystem includes areas of the cortex and adjacent regions uch as the cingulate gyrus and parahippocampal gyrus, .s well as the amygdala, hippocampus, hypothalamus, halamus, fornix, and mammillary bodies.

LIMBIC SYSTEM

Color and/or label the structures indicated on the diagram using the key below.

1. Caudate nucleus
2. Cingulate gyrus
3. Hypothalamus
4. Olfactory tract
5. Pituitary gland
6. Mammillary body
7. Amygdala
8. Pons
9. Cerebellum
10. Hippocampus
11. Thalamus
12. Fornix

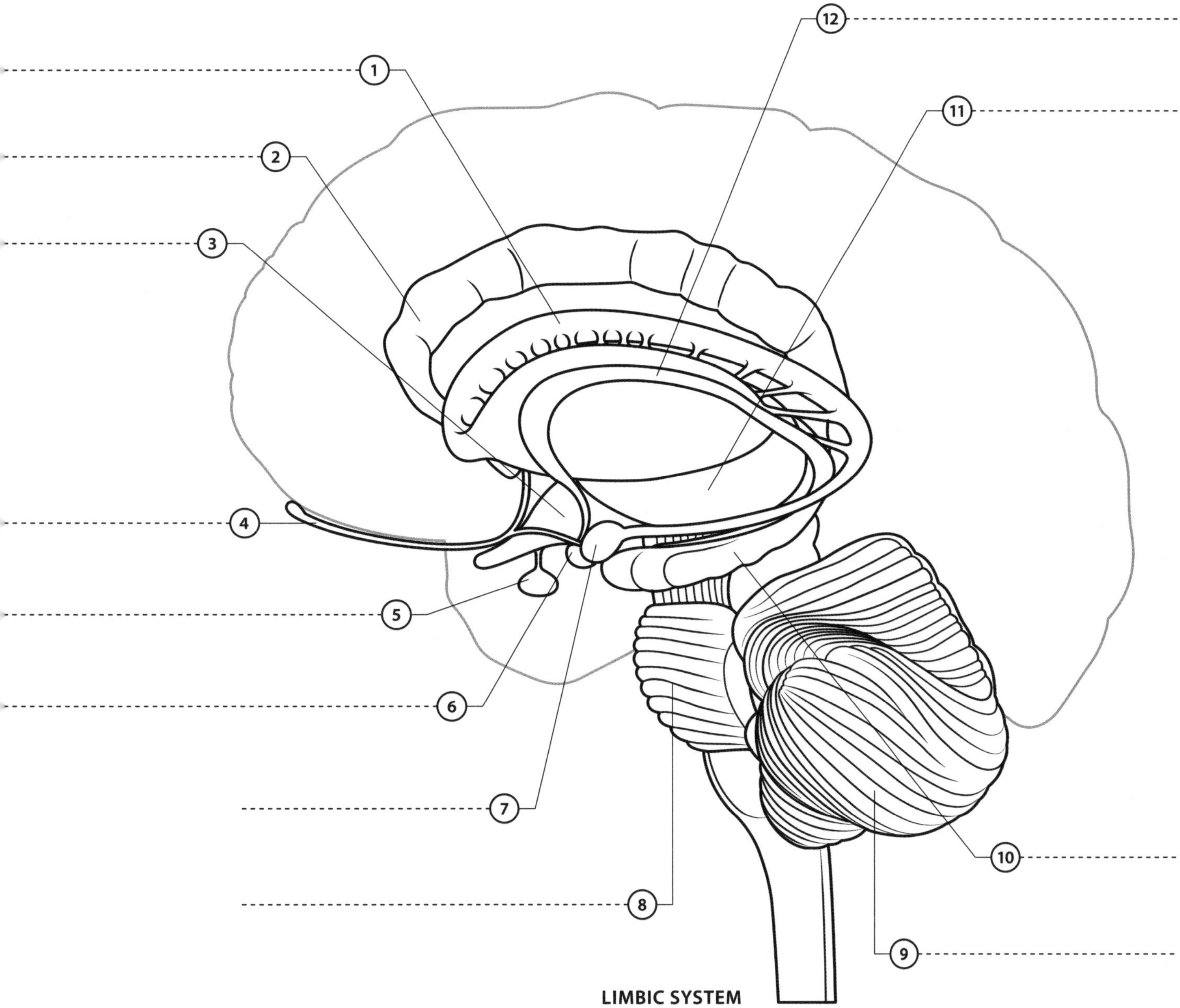

LIMBIC SYSTEM

See also pp. 30, 92, 94, 96, 134, 136, 160

THORAX, ABDOMEN, AND PELVIS

THE THORAX CONTAINS INTERCOSTAL NERVES, THE SYMPATHETIC TRUNKS, AND THE VAGUS AND PHRENIC NERVES. THE ABDOMEN AND PELVIS CONTAIN THE LUMBAR AND SACRAL PLEXUSES OF NERVES.

THORAX

Pairs of spinal nerves emerge via the intervertebral foramina. Each nerve splits into an anterior and a posterior branch. The posterior branches supply the muscles and skin of the back. The anterior branches of the upper 11 thoracic spinal nerves run, one under each rib, as intercostal nerves, supplying the intercostal muscles and overlying skin. The anterior branch of the last thoracic spinal nerve runs under the twelfth rib as the subcostal nerve. As well as motor and sensory fibers, thoracic spinal nerves contain sympathetic nerve fibers that are linked by small connecting branches to the sympathetic trunk.

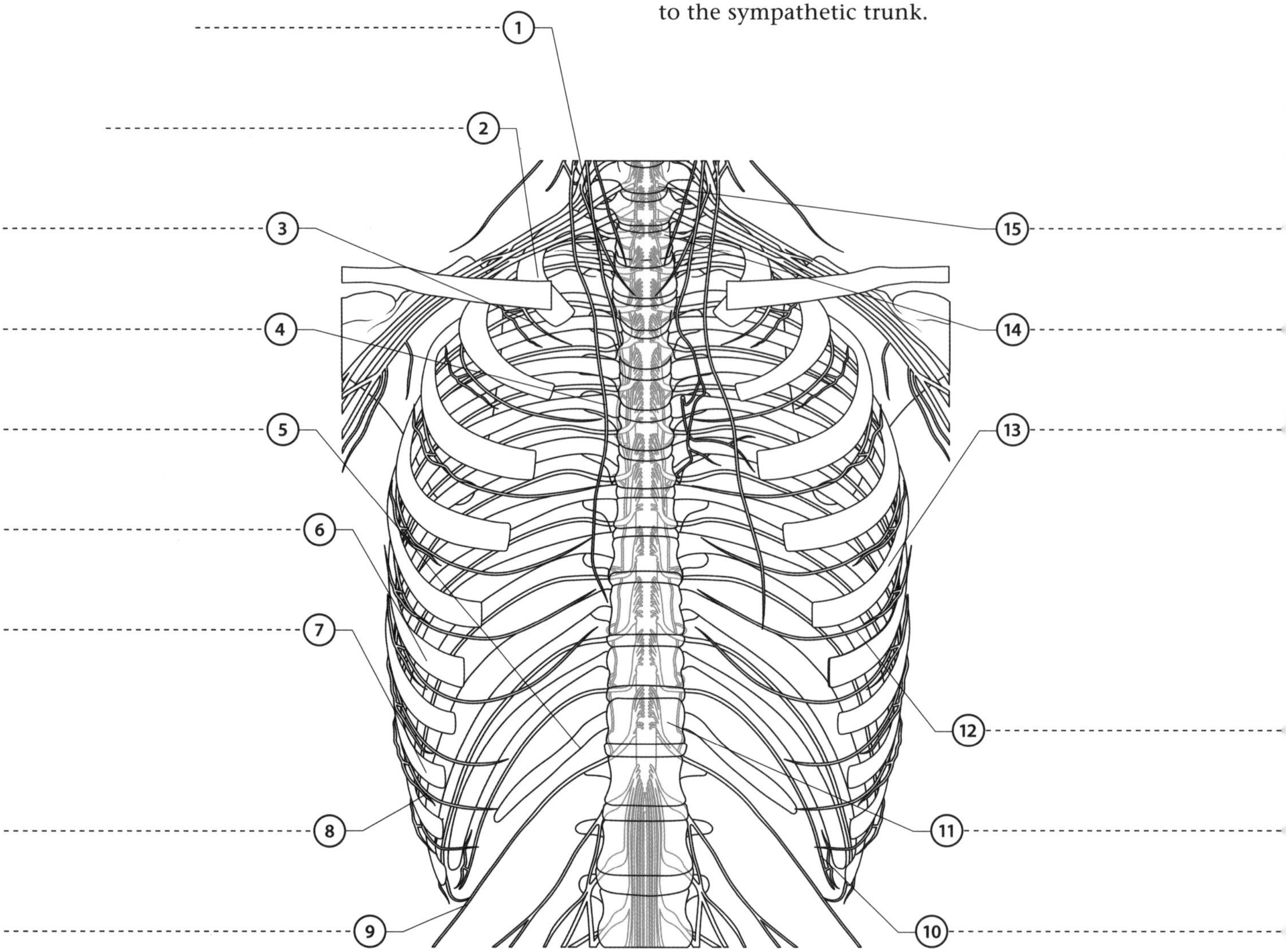

THORAX (ANTERIOR VIEW)

THORAX (ANTERIOR VIEW)

Color and/or label the structures indicated on the diagram using the key below.

1. Vagus nerve
2. First rib
3. First intercostal nerve
4. Phrenic nerve
5. Twelfth rib
6. Sixth rib
7. Eighth rib
8. Eighth intercostal nerve
9. Subcostal nerve
10. Eleventh intercostal nerve
11. T12 vertebra
12. Fifth intercostal nerve
13. Fifth rib
14. T1 spinal nerve
15. T1 vertebra

ABDOMEN AND PELVIS

The lower intercostal nerves continue past the lower edges of the rib cage at the front to supply the muscles and skin of the abdominal wall. The lower parts of the abdomen are supplied by the subcostal and iliohypogastric nerves. The abdominal portion of the sympathetic trunk receives nerves from the thoracic and first two lumbar spinal nerves, and sends nerves back to all the spinal nerves. The lumbar spinal nerves emerge from the vertebral column and run into the psoas major muscle at the back of the abdomen. Inside the muscle, the nerves join up to form a plexus. Branches of this lumbar plexus emerge around and through the psoas muscle, and run to the thigh. Lower down, branches of the sacral plexus supply pelvic organs and some enter the buttock. One of these branches is the sciatic nerve—the largest nerve in the body—which runs down the leg.

ABDOMEN AND PELVIS (ANTERIOR VIEW)

Color and/or label the structures indicated on the diagram using the key below.

1. T12 vertebra
2. Twelfth rib
3. Genitofemoral nerve
4. Iliohypogastric nerve
5. Ilioinguinal nerve
6. Femoral nerve
7. Sacral plexus
8. Lateral cutaneous nerve of the thigh
9. Obturator nerve
10. Sciatic nerve
11. Superior gluteal nerve
12. Anterior sacral foramen
13. Iliac crest
14. Lumbosacral trunk
15. Subcostal nerve
16. Intercostal nerve
17. Lumbar plexus

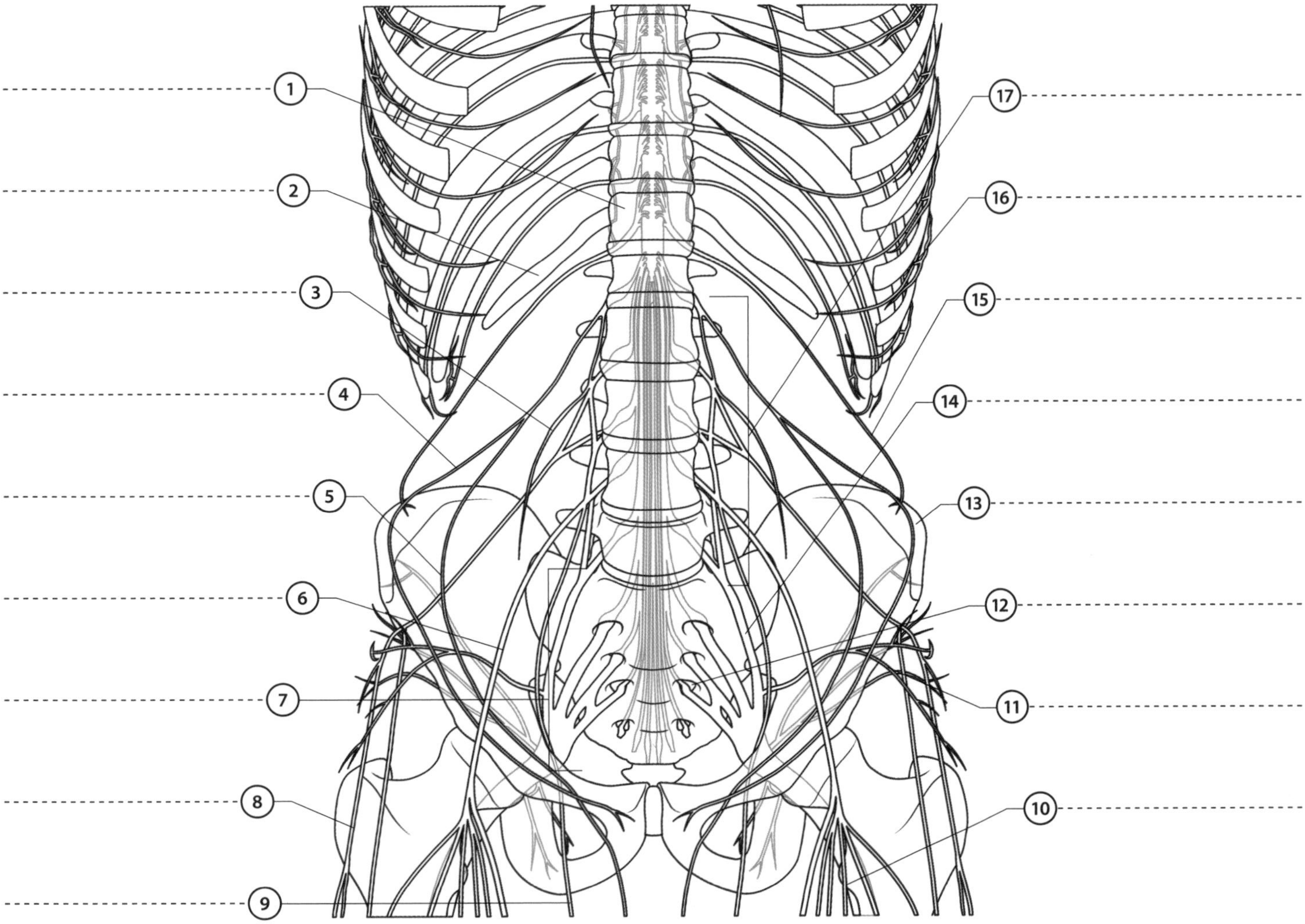

ABDOMEN AND PELVIS (ANTERIOR VIEW)

See also pp. 36, 38, 62, 66, 120, 138, 144, 162, 178, 198, 204, 218, 224 »

SHOULDER, ARM, WRIST, AND HAND

THE UPPER LIMB IS SUPPLIED BY FIVE NERVE ROOTS, WHICH FORM THE BRACHIAL PLEXUS. MAJOR NERVES (MUSCULOCUTANEOUS, MEDIAN, ULNAR, AXILLARY, AND RADIAL) PROVIDE SENSATION TO THE SKIN AND JOINTS OF THE UPPER LIMB AND SUPPLY ITS MUSCLES.

ANTERIOR ANATOMY

The musculocutaneous nerve supplies the muscles in the front of the arm (the biceps, brachialis, and coracobrachialis) and sensation to the lateral side of the forearm. The median nerve runs down the middle of the forearm, supplying most of the flexor muscles. It then travels over the wrist and into the hand to supply some of the thumb muscles, as well as sensation to the palm, thumb, and some fingers. The ulnar nerve runs down the inner side of the forearm, where it supplies two muscles. It continues on to supply most of the small muscles in the hand and provide sensation to the little finger and also to the inner side of the ring finger.

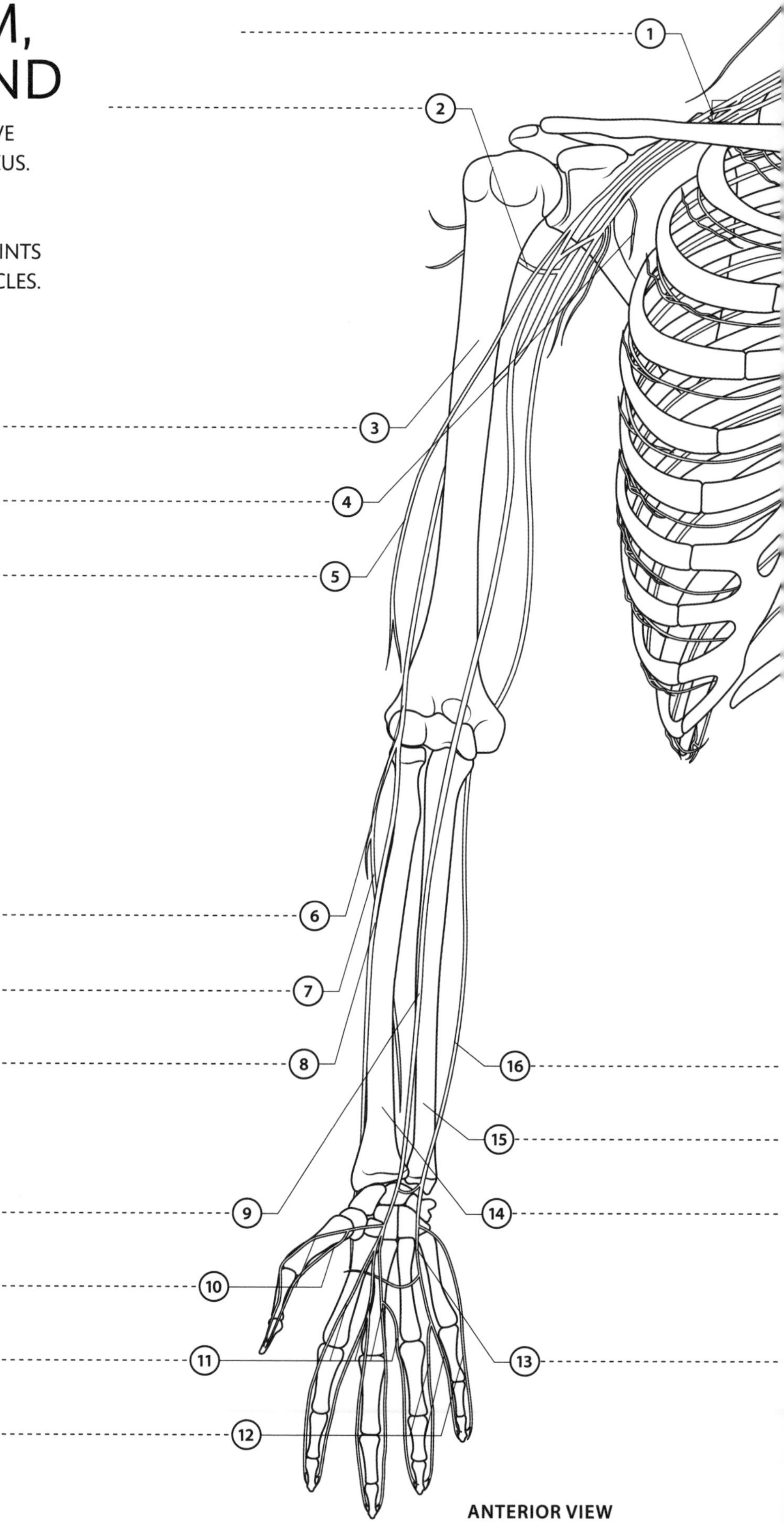

ANTERIOR VIEW

ANTERIOR VIEW

Color and/or label the structures indicated on the diagram using the key below.

1. Divisions of the brachial plexus
2. Axillary nerve
3. Humerus
4. Medial pectoral nerve
5. Musculocutaneous nerve
6. Radial nerve
7. Posterior interosseus nerve
8. Superficial radial nerve
9. Median nerve
10. Palmar digital branches of median nerve
11. Palmar digital branches of median nerve
12. Palmar digital branches of ulnar nerve
13. Palmar branch of ulnar nerve
14. Radius
15. Ulna
16. Ulnar nerve

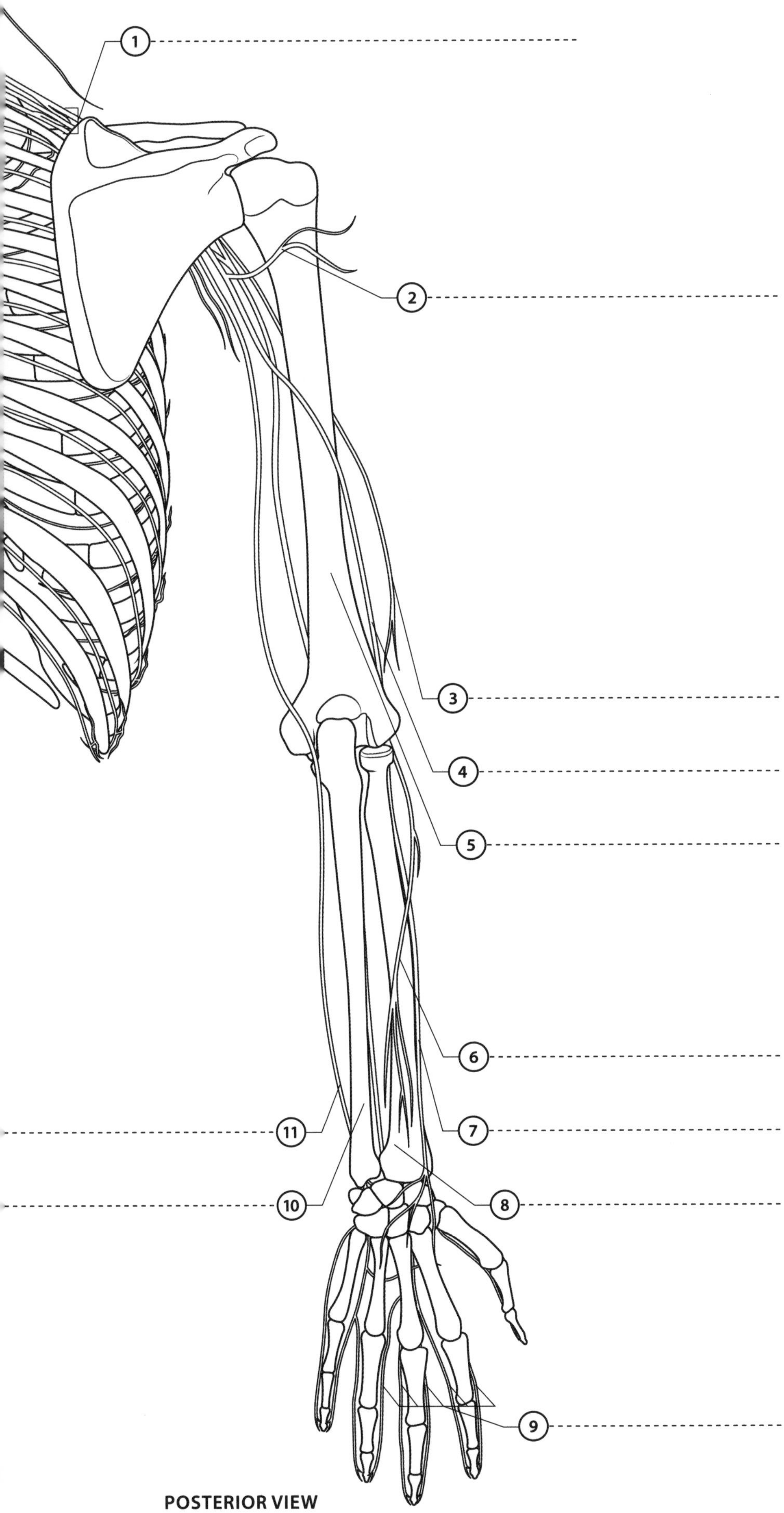

POSTERIOR VIEW

POSTERIOR ANATOMY

The axillary and radial nerves run behind the humerus. The axillary nerve wraps around the neck of the humerus, just under the shoulder joint, and supplies the deltoid muscle. The radial nerve supplies all the extensor muscles in the upper arm and forearm. It spirals around the back of the humerus and sends branches to the heads of the triceps muscle. The radial nerve then continues in its spiral, running forward to just in front of the lateral epicondyle of the humerus at the elbow. On the back of the forearm, the radial nerve and its branches supply all the extensor muscles. Branches of the radial nerve also fan out over the back of the hand, where they provide sensation.

POSTERIOR VIEW

Color and/or label the structures indicated on the diagram using the key below.

1. Divisions of the brachial plexus
2. Axillary nerve
3. Musculocutaneous nerve
4. Radial nerve
5. Humerus
6. Posterior interosseus nerve
7. Superficial branch of radial nerve
8. Radius
9. Dorsal digital branches of median nerve
10. Ulna
11. Ulnar nerve

See also pp. 40, 42, 44, 70, 72

HIP, LEG, ANKLE, AND FOOT

THE LOWER LIMB RECEIVES NERVES FROM THE LUMBAR AND SACRAL PLEXUSES.

ANTERIOR ANATOMY

Three main nerves supply the thigh muscles: the femoral, obturator, and sciatic nerves. The femoral nerve runs over the pubic bone to supply the quadriceps and sartorius muscles. The saphenous nerve (a branch of the femoral nerve) continues past the knee and supplies skin on the inside of the lower leg and inner side of the foot. The obturator nerve passes through the obturator foramen in the pelvic bone to supply the adductor muscles of the inner thigh and provide sensation to the skin there. The common fibular nerve runs past the knee and wraps around the neck of the fibula. It then splits into the deep and superficial fibular nerves. The deep fibular supplies the extensor muscles of the shin, then fans out to provide sensation to the skin at the back of the foot. The superficial fibular stays on the side of the leg and supplies the fibular muscles.

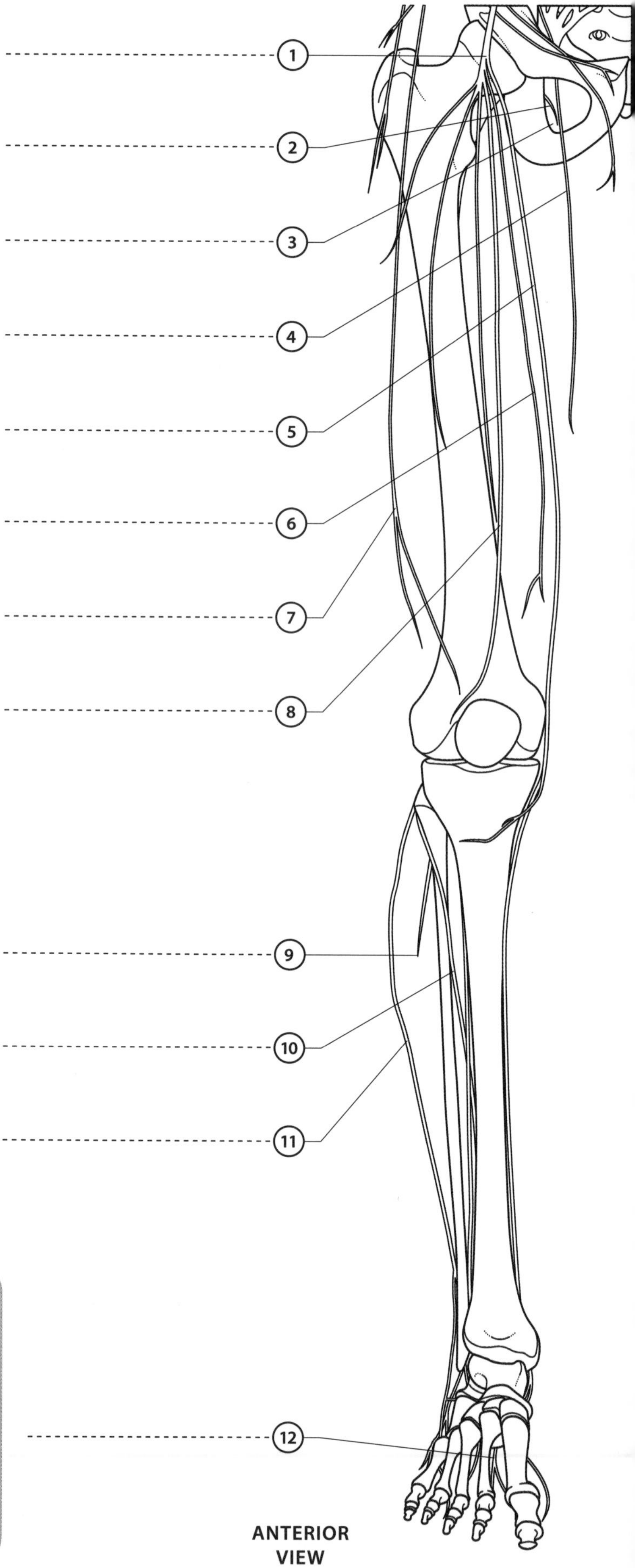

ANTERIOR VIEW

ANTERIOR VIEW

Color and/or label the structures indicated on the diagram using the key below.

1. Femoral nerve
2. Pudendal nerve
3. Obturator foramen
4. Obturator nerve
5. Saphenous nerve
6. Medial femoral cutaneous nerve
7. Lateral femoral cutaneous nerve
8. Intermediate femoral cutaneous nerve
9. Lateral sural cutaneous nerve (shown cut)
10. Deep fibular nerve
11. Superficial fibular nerve
12. Deep fibular nerve

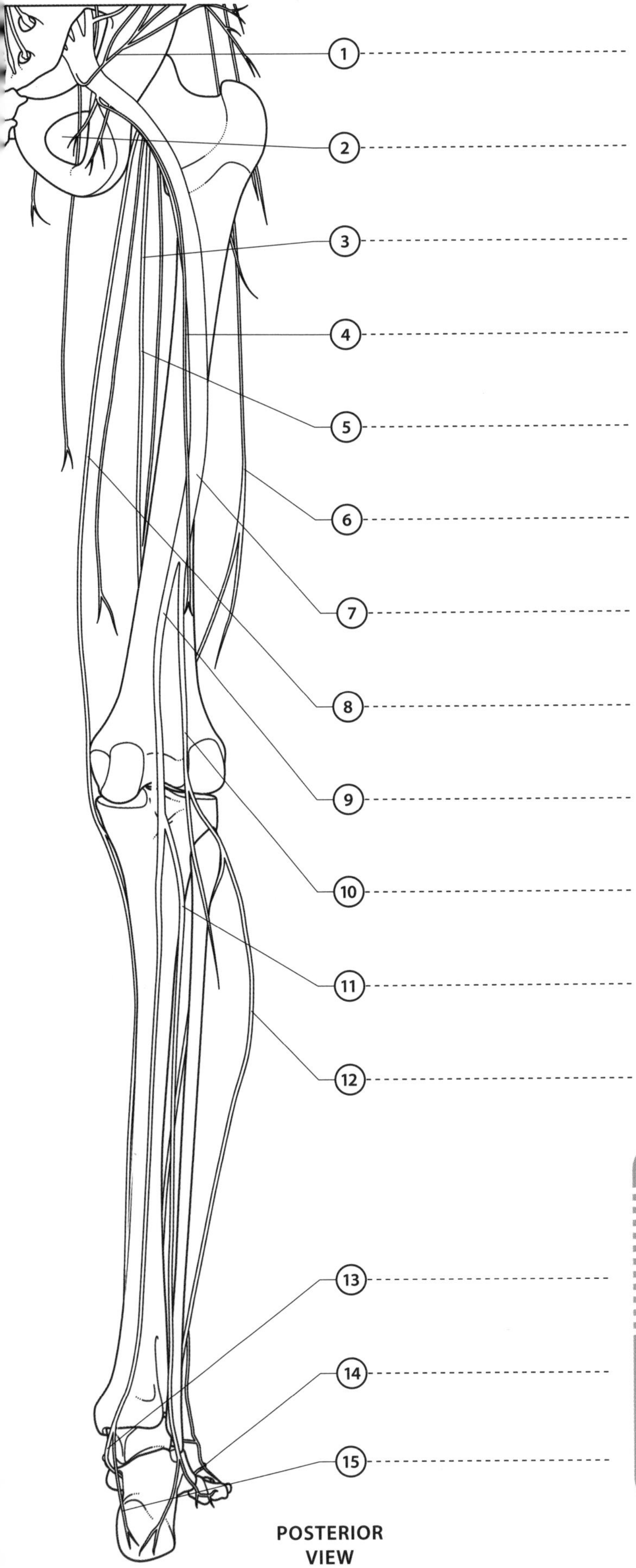

POSTERIOR VIEW

POSTERIOR ANATOMY

Gluteal nerves from the sacral plexus emerge via the greater sciatic foramen to supply the muscles and skin of the buttock. The sciatic nerve also emerges through the greater sciatic foramen into the buttock and runs down the back of the thigh, supplying the hamstrings. In most people, the sciatic nerve runs halfway down the thigh, then splits into two branches, the tibial and common fibular nerves, which continue into the popliteal fossa (back of the knee). From the popliteal fossa, the tibial nerve runs under the soleus muscle, and between the deep and superficial calf muscles, which it supplies. It continues behind the medial malleolus and under the foot, then splits into two plantar nerves that supply the small muscles of the foot and skin of the sole.

POSTERIOR VIEW

Color and/or label the structures indicated on the diagram using the key below.

1. Superior gluteal nerve
2. Obturator foramen
3. Medial femoral cutaneous nerve
4. Posterior cutaneous nerve of thigh
5. Intermediate femoral cutaneous nerve
6. Lateral femoral cutaneous nerve
7. Sciatic nerve
8. Saphenous nerve
9. Tibial nerve
10. Common fibular nerve
11. Sural nerve
12. Superficial fibular nerve
13. Medial plantar nerve
14. Dorsal digital nerves
15. Calcaneal branch of tibial nerve

See also pp. 46, 74, 76, 150, 152, 165 »

VISION

VISION IS THE MOST IMPORTANT SENSE FOR MOST PEOPLE. THE ROUGHLY 125 MILLION PHOTORECEPTORS IN THE RETINA SEND A VAST AMOUNT OF VISUAL INFORMATION TO THE BRAIN FOR PROCESSING INTO IMAGES.

HORIZONTAL SECTION THROUGH EYE

Color and/or label the structures indicated on the diagram using the key below.

1. Sclera
2. Conjunctiva
3. Iris
4. Cornea
5. Pupil
6. Aqueous humor
7. Lens
8. Suspensory ligament
9. Ciliary body
10. Extraocular muscle
11. Retina
12. Blind spot
13. Optic nerves
14. Choroid
15. Vitreous humor
16. Extraocular muscle

EYE ANATOMY

Each eyeball is about 1 in (2.5 cm) in diameter and sits within the bony orbit of the skull. The eye has three main layers: the sclera, choroid, and retina. The sclera can be seen as the white of the eye, which becomes the transparent cornea. Behind the cornea is the anterior chamber, filled with aqueous humor; at the rear of this chamber is the iris, with its central pupil. Behind the iris, and in contact with it, is the lens, which is suspended by fibers from the ciliary body. The posterior chamber of the eye is filled with vitreous humor, a transparent jellylike substance that helps to maintain the eye's shape. On the inside of the back of the eye is the retina, which contains millions of light-sensitive rods and cones. Immediately under the retina is the choroid, which is packed with blood vessels. At the front, the choroid is continuous with the ciliary body and iris.

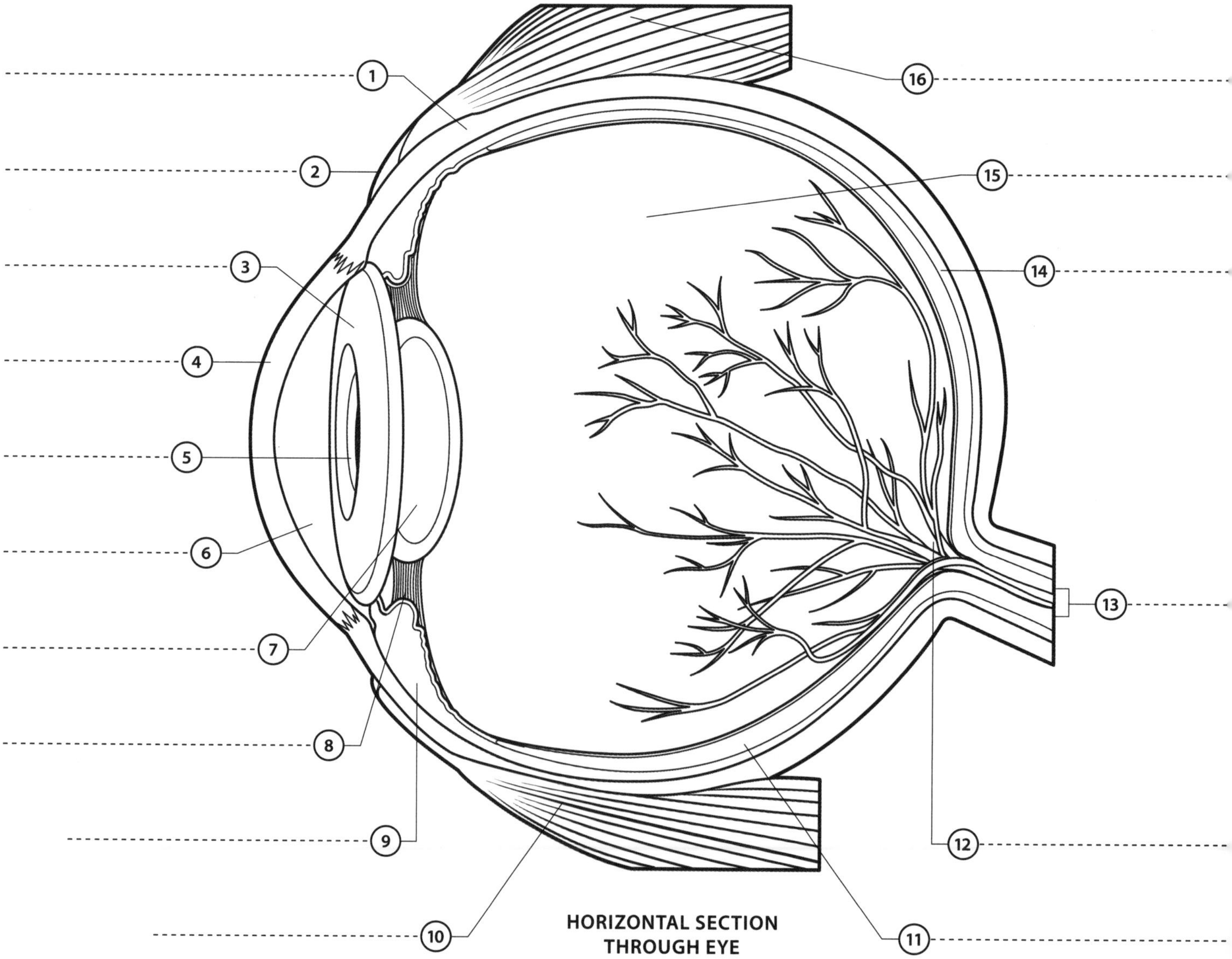

HORIZONTAL SECTION THROUGH EYE

RETINA

The retina contains three layers of cells, each one communicating with the next by synapses. The first two layers (containing ganglion, amacrine, and bipolar cells) send signals to the visual cortex of the brain via the optic nerve but do not respond directly to light. The third layer, at the back of the retina, contains the photoreceptors—rods and cones. The five million cones are concentrated at the fovea. There are three types of cone cell—red, green, and blue—which allow us to see in color. Each type responds to a certain color (wavelength) of light; their combined signals are analyzed by the brain to produce the millions of colors we can perceive. Cones need more light than rods do to respond. In dim light, the cones work less well and the 120 million or so rods provide most of the visual information. However, rods are not color-sensitive, so vision in dim light is in black and white.

LAYERS OF RETINAL CELLS

Color and/or label the structures indicated on the diagram using the key below.

1. Bipolar cell
2. Amacrine cell
3. Ganglion cell
4. Inner surface of retina
5. Blood vessel
6. Axons extending from ganglion cells
7. Horizontal cell
8. Cone cell
9. Rod cell

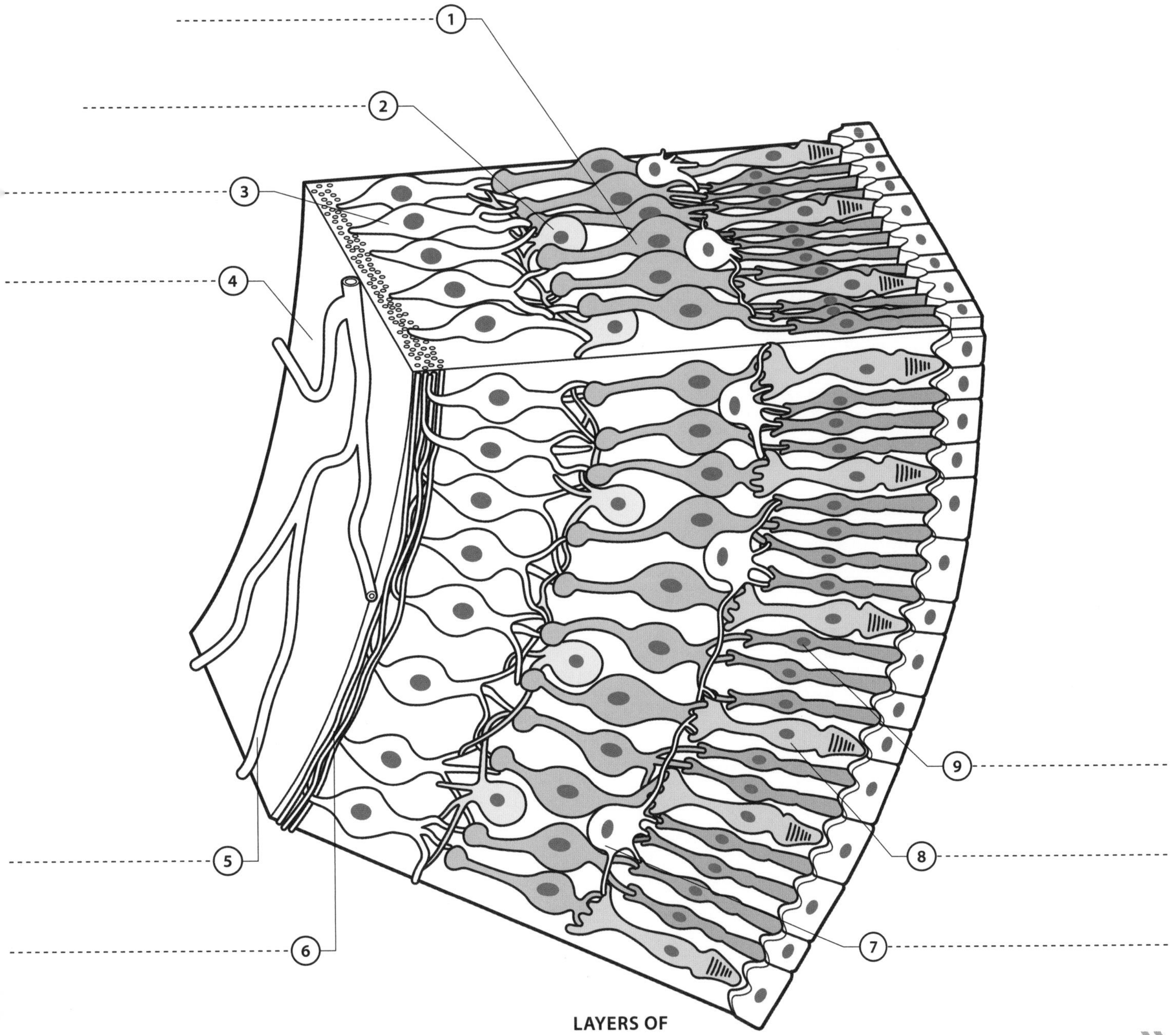

LAYERS OF RETINAL CELLS

See also pp. 59, 95

HEARING AND BALANCE

THE EARS PROVIDE THE SENSE OF HEARING, AND THEY ALSO DETECT HEAD POSITION AND MOTION, WHICH ARE ESSENTIAL FOR THE SENSE OF BALANCE.

EAR AND HEARING

The ear can be divided into external, middle, and inner parts. The external ear includes the auricle on the outside of the head and the external acoustic meatus. The middle ear contains the ossicles and is linked to the pharynx by the Eustachian (pharyngotympanic) tube. The inner ear comprises the cochlea, which detects sounds, and the vestibular apparatus (comprising the semicircular canals, utricle, and saccule), which detect motion and position of the head. When sound waves hit the tympanic membrane, they cause it to vibrate. These vibrations are transmitted along a chain of three bony ossicles (malleus, incus, and stapes). The stapes butts against the oval window, which is set into the fluid-filled cochlea. Vibrations of the stapes are changed into pressure waves in the fluid in the cochlea. Within the cochlea lies the organ of Corti, containing a fine membrane in which hair cells are embedded. The pressure waves in the fluid distort these hairs, causing them to produce nerve signals, which pass along the cochlear nerve to the auditory cortex in the brain.

EXTERNAL AND MIDDLE EAR

Color and/or label the structures indicated on the diagram using the key below.

1. Temporal bone
2. Cartilage in auricle
3. Auricle
4. External acoustic meatus
5. Tympanic membrane
6. Eustachian tube (pharyngotympanic tube)
7. Round window
8. Oval window
9. Vestibulocochlear nerve
10. Cochlea (cut)
11. Stapes
12. Semicircular canal
13. Incus
14. Malleus

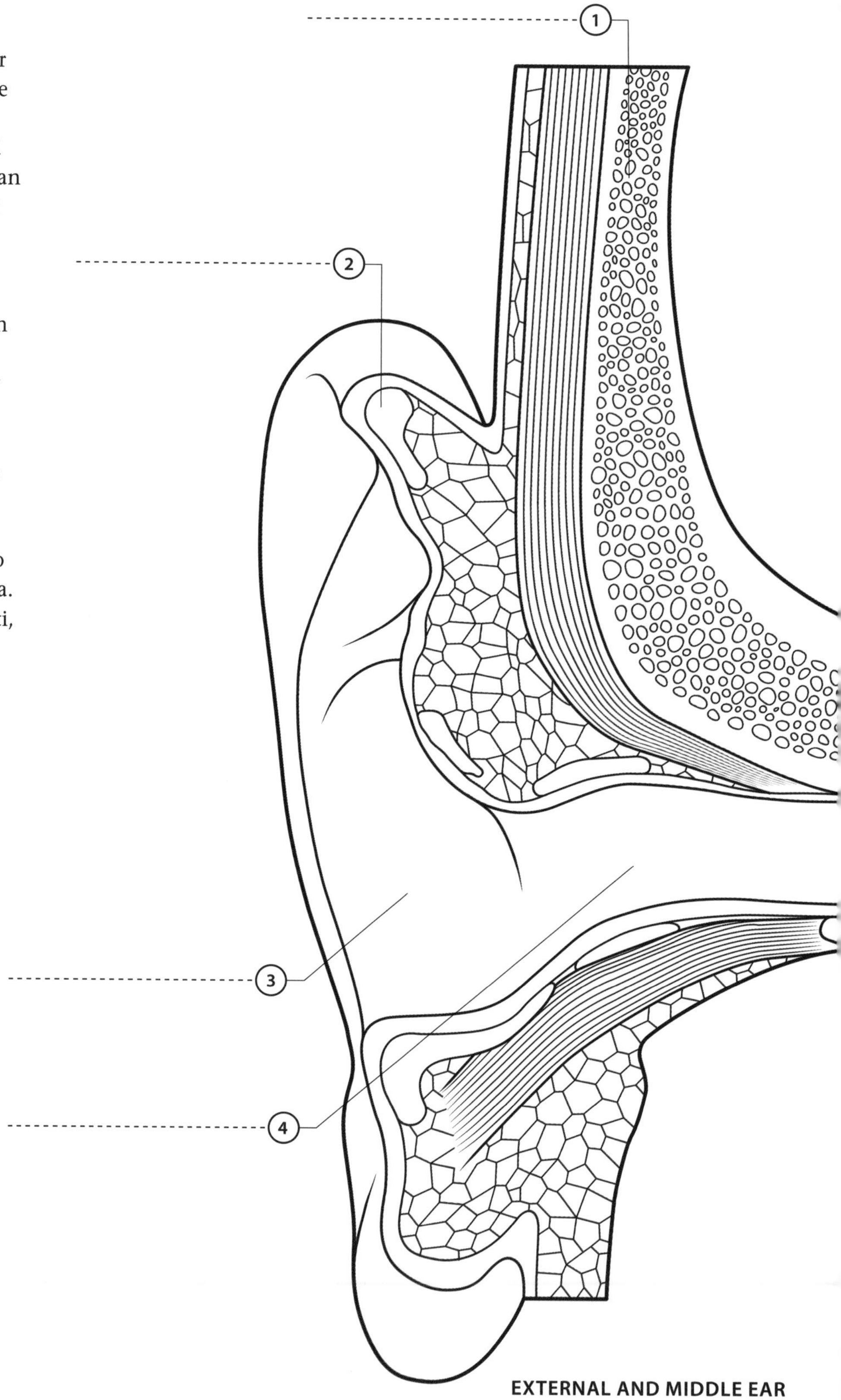

EXTERNAL AND MIDDLE EAR

BALANCE

Maintaining balance involves making constant motor changes in response to sensory information from many inputs (eyes, muscles, joints, and skin pressure sensors). The vestibular apparatus in the inner ear plays a key role. Specialized cells in three pairs of semicircular canals, each set at right angles to each other, and in the macula of the utricle and saccule, detect the velocity and direction of head movements and convert this information into nerve signals that pass along the vestibular nerve to appropriate brain areas, particularly the cerebellum, where they are processed with other sensory inputs to stimulate muscles to maintain balance.

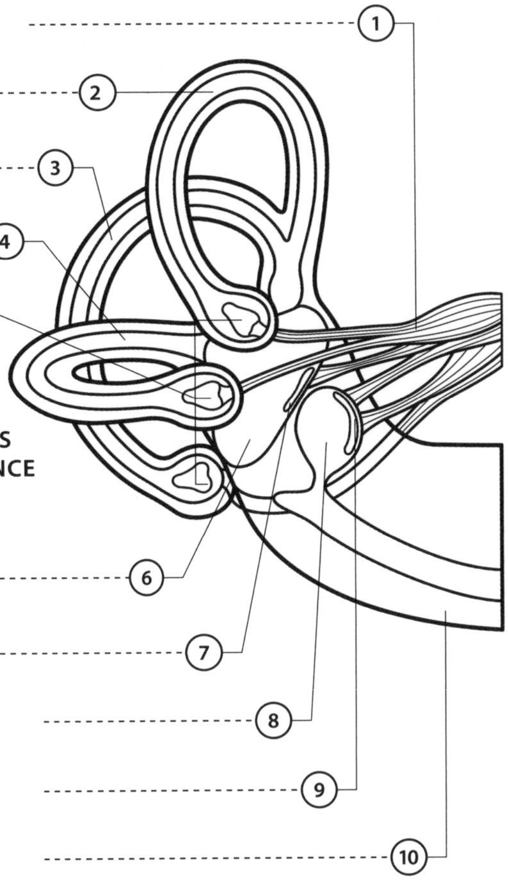

ORGANS OF BALANCE

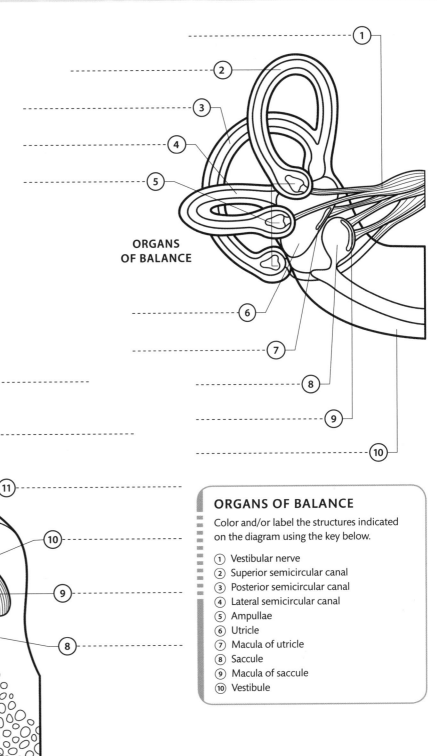

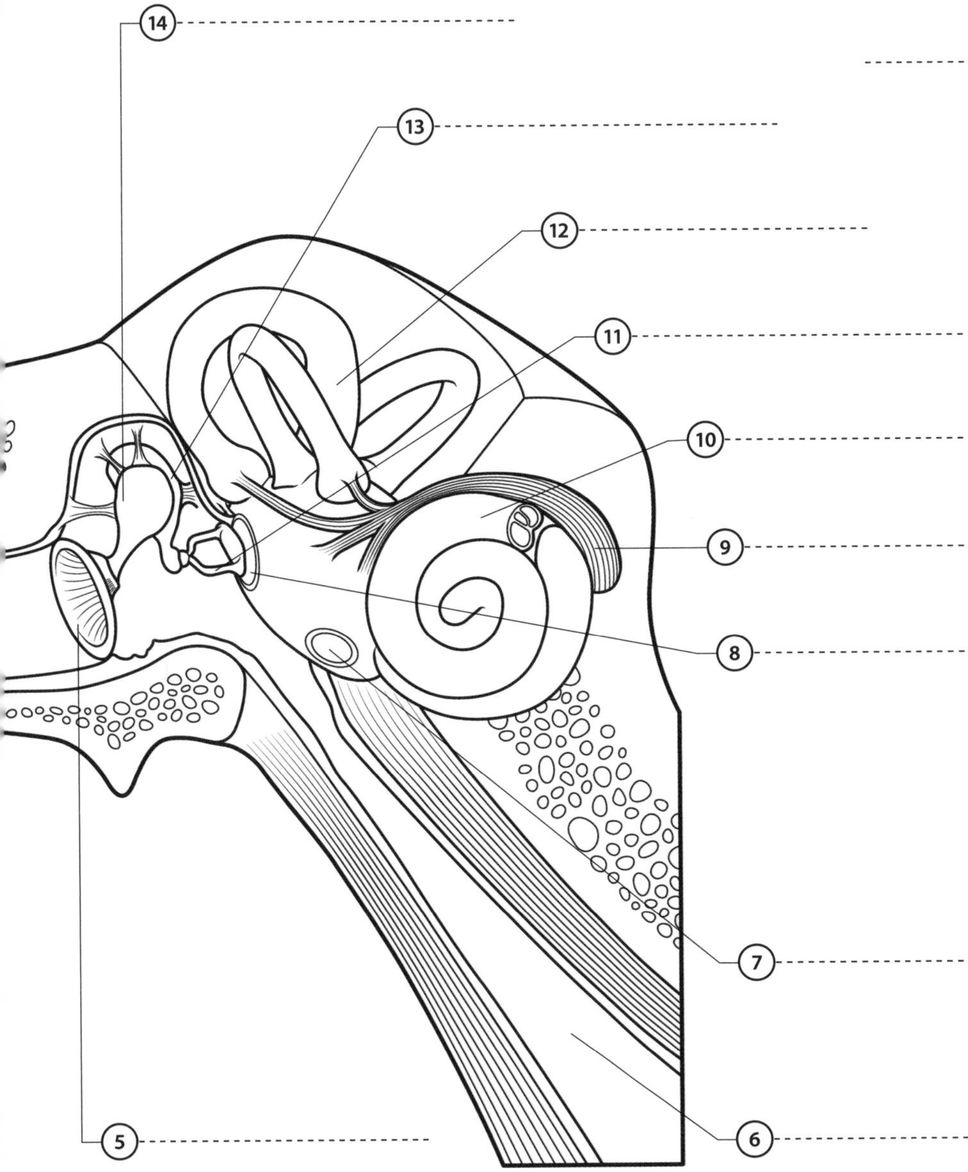

ORGANS OF BALANCE

Color and/or label the structures indicated on the diagram using the key below.

1. Vestibular nerve
2. Superior semicircular canal
3. Posterior semicircular canal
4. Lateral semicircular canal
5. Ampullae
6. Utricle
7. Macula of utricle
8. Saccule
9. Macula of saccule
10. Vestibule

See also pp. 94, 96, 98 »

TASTE AND OLFACTION

THE SENSES OF TASTE AND OLFACTION BOTH DETECT CHEMICALS AND WORK IN SIMILAR WAYS.

TASTE

The main organ of taste is the tongue, which has several thousand tiny cell clusters (taste buds) distributed mainly on its tip and along its upper sides and rear. The taste buds detect different combinations of the five main tastes: sweet, salty, savory (umami), sour, and bitter. Most of these are detected equally in all parts of the tongue that have taste buds. It is thought that, like olfactory receptors, the taste buds work using a "lock and key" system. Receptor sites for different taste molecules are located on the hairlike processes of gustatory receptor cells in each taste bud. When a suitably shaped taste molecule (the "key") contacts a receptor site (the "lock"), it fits into the site and causes the receptor cell to fire a nerve signal. The signals pass along branches of the trigeminal and glossopharyngeal nerves to the medulla, and continue to the thalamus, then to the primary gustatory areas of the cerebral cortex.

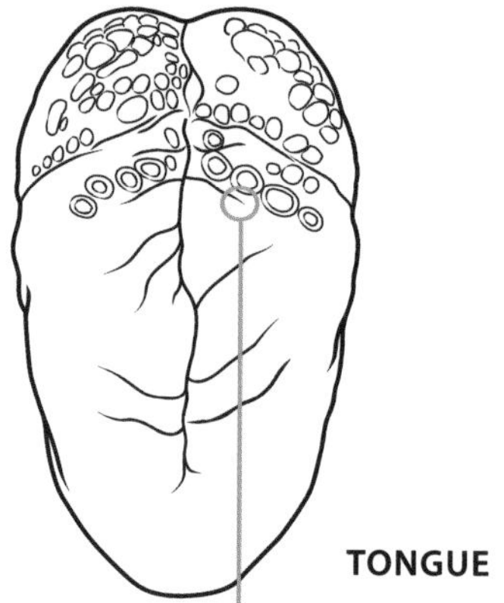

TONGUE

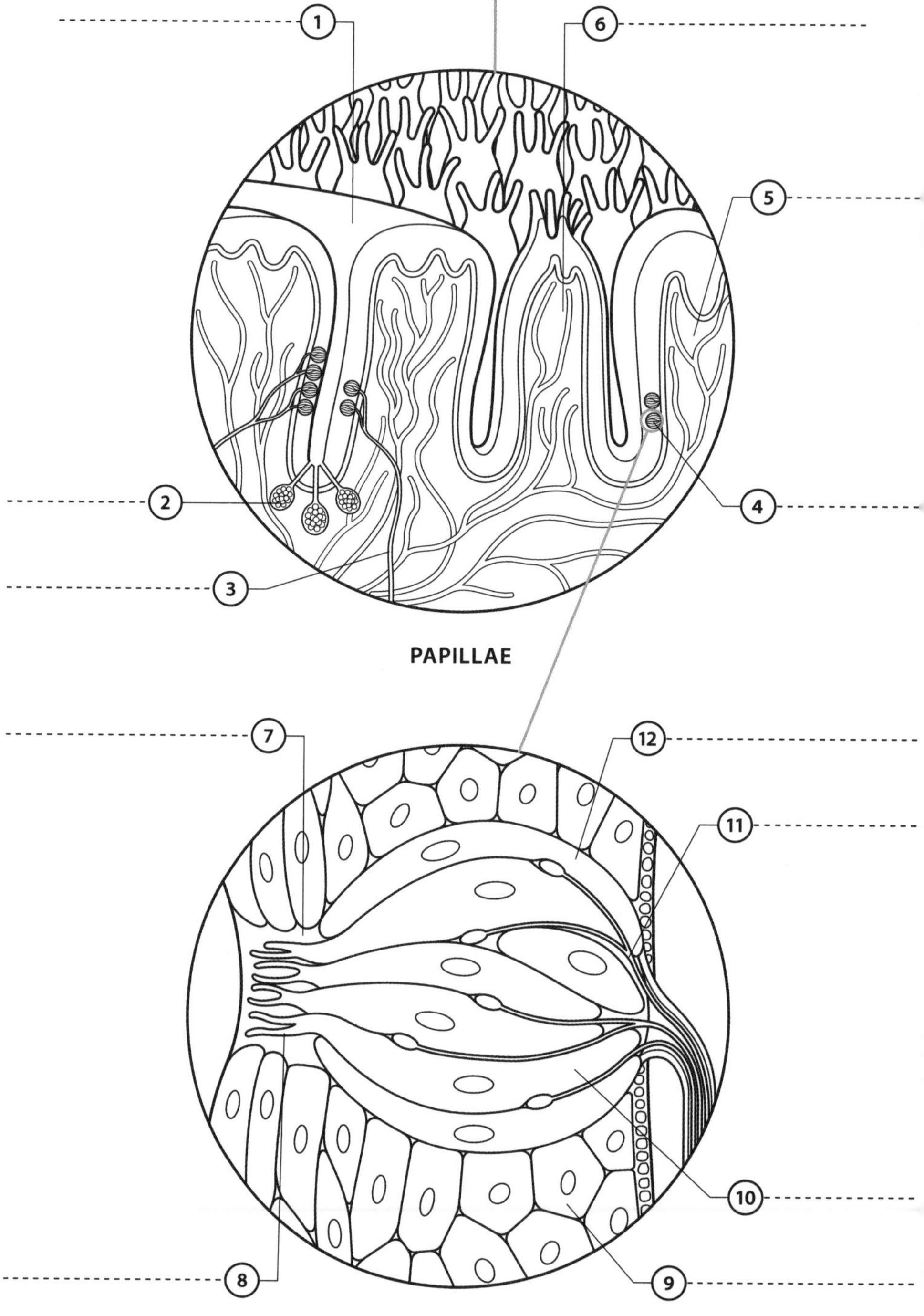

PAPILLAE

TASTE BUD

PAPILLAE AND TASTE BUD

Color and/or label the structures indicated on the diagram using the key below.

1. Vallate papillae
2. Mucus-secreting gland
3. Nerve fiber
4. Taste bud
5. Fungiform papilla
6. Filiform papilla
7. Taste pore
8. Taste hair
9. Epithelium of tongue
10. Gustatory receptor cell
11. Nerve fiber
12. Supporting cell

OLFACTION

Odorant molecules are detected by the olfactory epithelium located in the roof of the nasal cavity. This epithelium contains several million specialized olfactory receptor cells, whose lower ends project into the mucus lining the nasal cavity and bear cilia on which are located receptor sites. As with taste, these sites are thought to work using a "lock and key" system, responding only to suitably shaped molecules. However, there is also a "fuzzy coding" component that is less well understood, where each odor produces a variable pattern or signature of impulses. Olfactory information passes via the olfactory nerve to the olfactory bulb, then along the olfactory tract to the olfactory cortex in the brain. The olfactory cortex has close links with limbic areas that presumably mediate emotional responses to smells.

LATERAL SECTION THROUGH HEAD

Color and/or label the structures indicated on the diagram using the key below.

1. Olfactory tract
2. Olfactory bulb
3. Olfactory epithelium
4. Odor in inhaled air
5. Orthonasal flow
6. Airborne odor molecule
7. Retronasal flow
8. Odor molecules from food

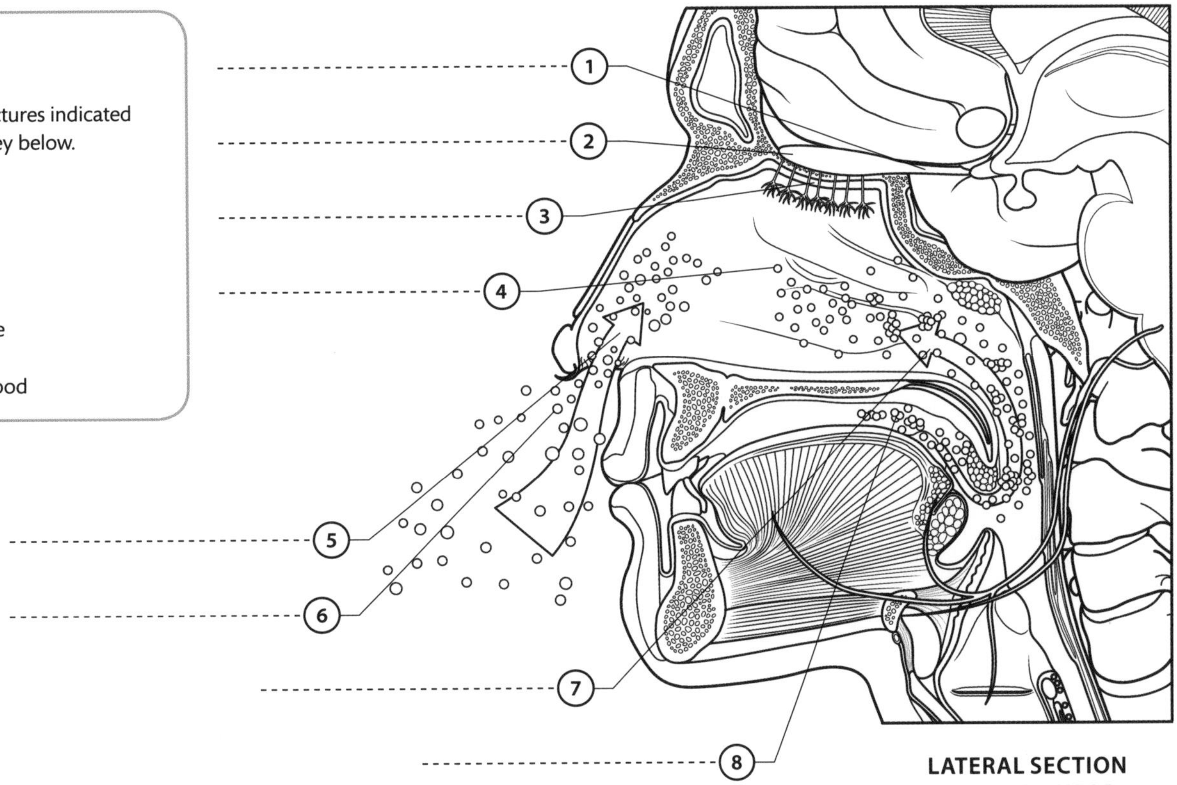

LATERAL SECTION THROUGH HEAD

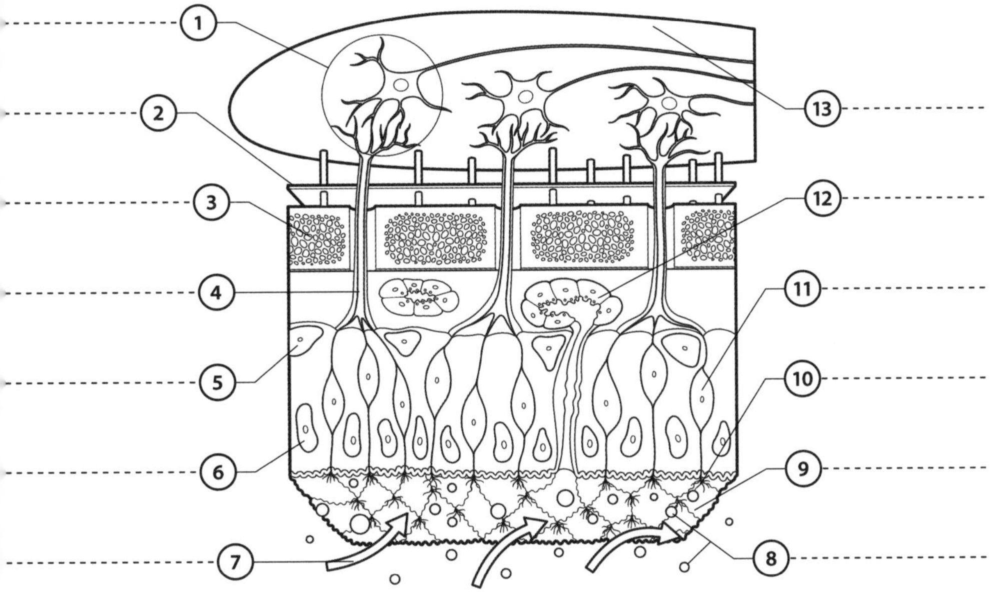

OLFACTORY EPITHELIUM

OLFACTORY EPITHELIUM

Color and/or label the structures indicated on the diagram using the key below.

1. Glomerulus
2. Dura mater
3. Ethmoid bone (cribriform plate)
4. Nerve fiber (axon)
5. Basal cell
6. Supporting cell
7. Air flow
8. Odor molecules
9. Mucus
10. Cilia
11. Receptor cell
12. Mucus-secreting gland
13. Olfactory bulb

See also pp. 95, 118, 174

TOUCH

THERE ARE MANY KINDS OF TOUCH SENSATION, INCLUDING LIGHT TOUCH, PRESSURE, AND VIBRATION. THE SKIN IS THE MAIN TOUCH SENSE ORGAN.

TOUCH RECEPTORS

Discriminative or analytical touch is the perception of pressure, vibration, and texture, and is well localized. It is mediated by four different receptors in the skin: Meissner's corpuscles, located in dermal papillae, close to the dermo-epidermal junction and especially abundant on the palms, soles of the feet, eyelids, genitalia and nipples; Pacinian corpuscles, located deep in the dermis or in the hypodermis, particularly of the digits; Merkel's disks; and Ruffini endings. Deep pressure receptors like Pacinian corpuscles consist of nerve endings encapsulated by specialized connective tissues. Light touch is not well localized and is detected by free nerve endings in the epidermis and by the root hair plexus, which responds when a hair moves. The distribution of touch receptors varies: the fingertips have many more touch receptors than the skin of the back.

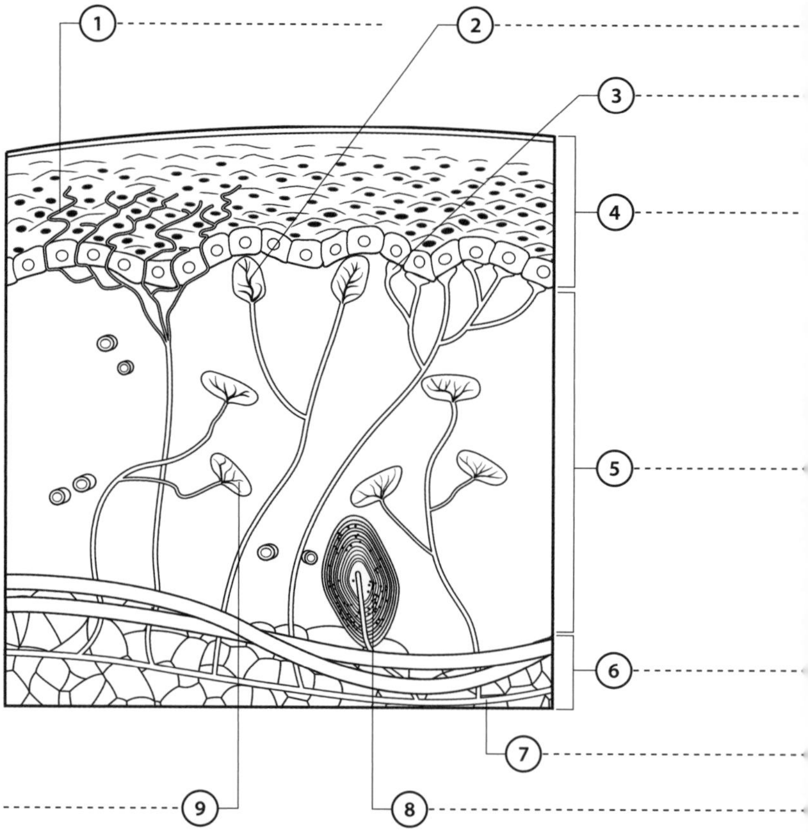

TOUCH RECEPTORS IN SKIN

SOMATOSENSORY CORTEX

Color and/or label the areas indicated on the diagram using the key below.

1. Head and neck
2. Arm
3. Hand
4. Fingers and thumb
5. Eye
6. Face
7. Lips
8. Tongue
9. Genitals
10. Toes
11. Foot
12. Leg
13. Trunk

TOUCH RECEPTORS IN SKIN

Color and/or label the structures indicated on the diagram using the key below.

1. Free nerve ending
2. Meissner's corpuscle
3. Merkel's disc
4. Epidermis
5. Dermis
6. Subcutaneous fat
7. Axon
8. Pacinian corpuscle
9. Ruffini corpuscle

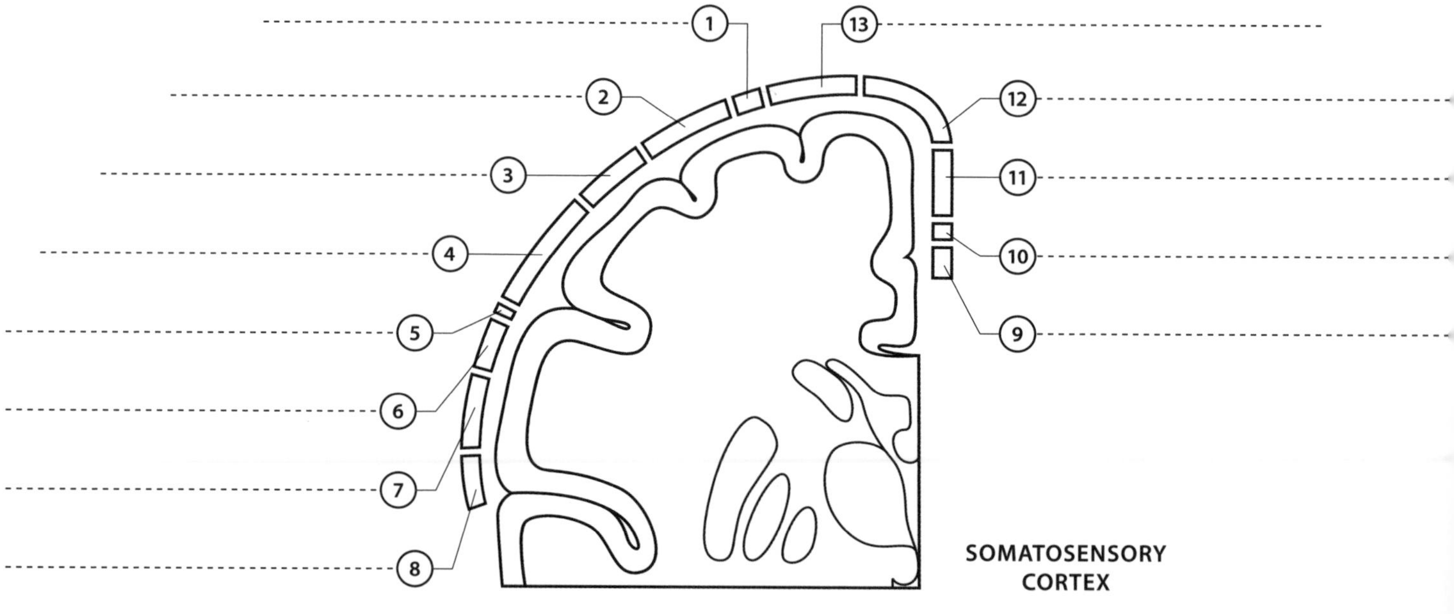

SOMATOSENSORY CORTEX

SPINAL NERVES AND DERMATOMES

There are 31 pairs of spinal nerves. They branch out from the spinal cord and divide into peripheral nerves carrying sensory and/or motor information. A dermatome is an area of skin supplied by all the cutaneous branches of a spinal or cranial nerve. Dermatomes innervated by sensory nerves derived from adjacent segments in the spinal cord usually overlap, often quite extensively, and there is considerable individual variation. Dermatome maps, such as those shown here, help clinicians to localize pathological changes in patients with peripheral sensory deficits.

ANTERIOR DERMATOMES

Color and/or label the structures indicated on the diagram using the key below.

1. C2
2. S2
3. S3
4. S1
5. L4
6. L5
7. L3
8. L2
9. L1
10. C7–C8
11. T2–T12
12. T1–T2
13. C4–C6
14. C3
15. C2
16. V_{i}
17. V_{ii}
18. V_{iii}

POSTERIOR DERMATOMES

Color and/or label the structures indicated on the diagram using the key below.

1. C2
2. C3
3. C4
4. C5
5. S1
6. S2
7. L5
8. S2
9. S1
10. L3–L4
11. L2–L3
12. S3–S5
13. L1
14. C7–T2
15. T2–T12
16. C6

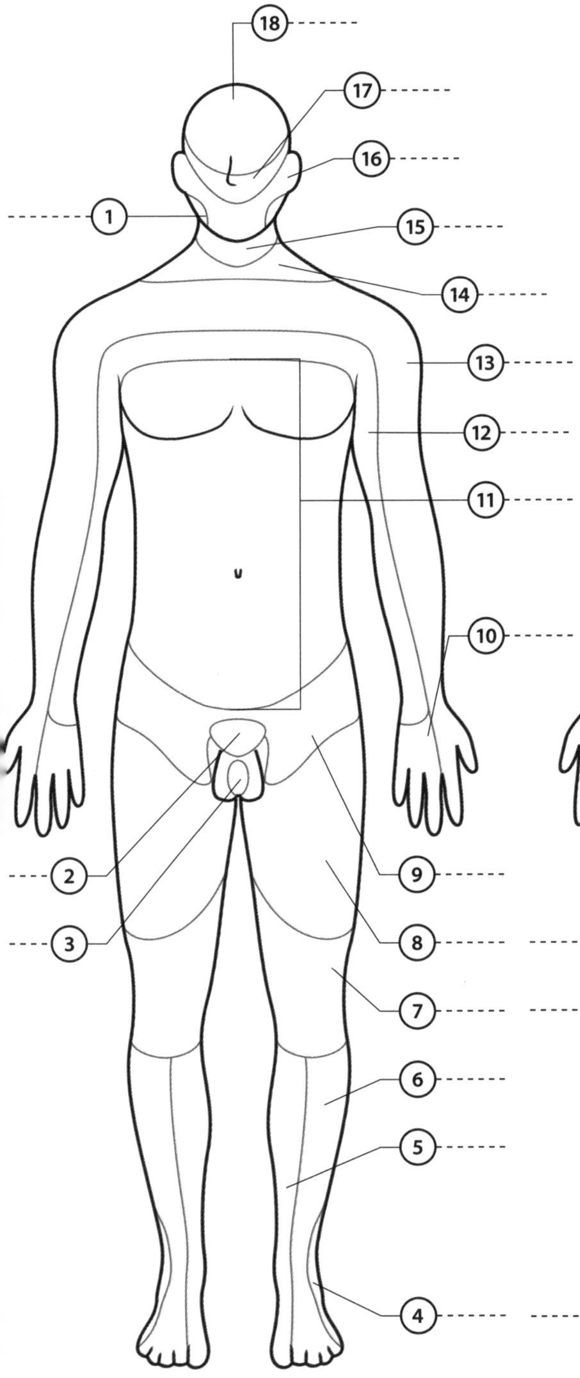

ANTERIOR DERMATOMES

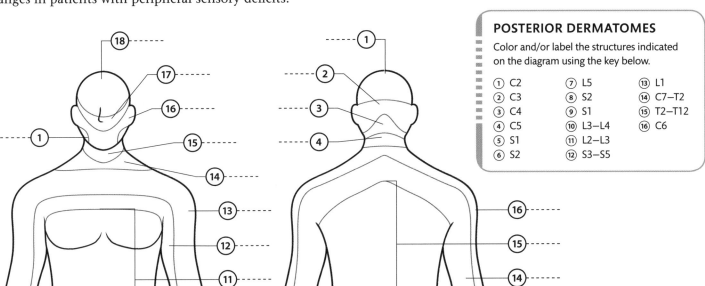

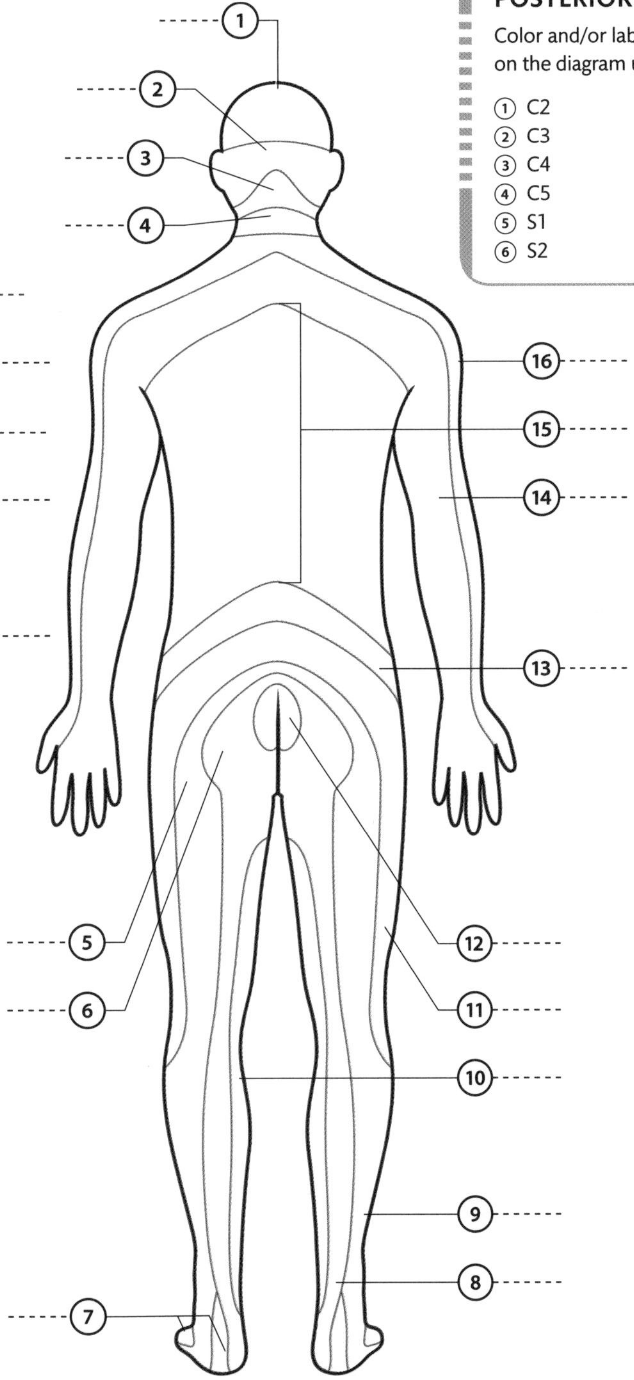

POSTERIOR DERMATOMES

See also pp. 82, 95, 97, 100, 102, 104 »

RESPIRATORY SYSTEM

05

RESPIRATORY SYSTEM

THE RESPIRATORY SYSTEM, IN CONJUNCTION WITH THE CIRCULATORY SYSTEM, IS RESPONSIBLE FOR SUPPLYING ALL BODY CELLS WITH OXYGEN AND REMOVING CARBON DIOXIDE FROM THE BODY.

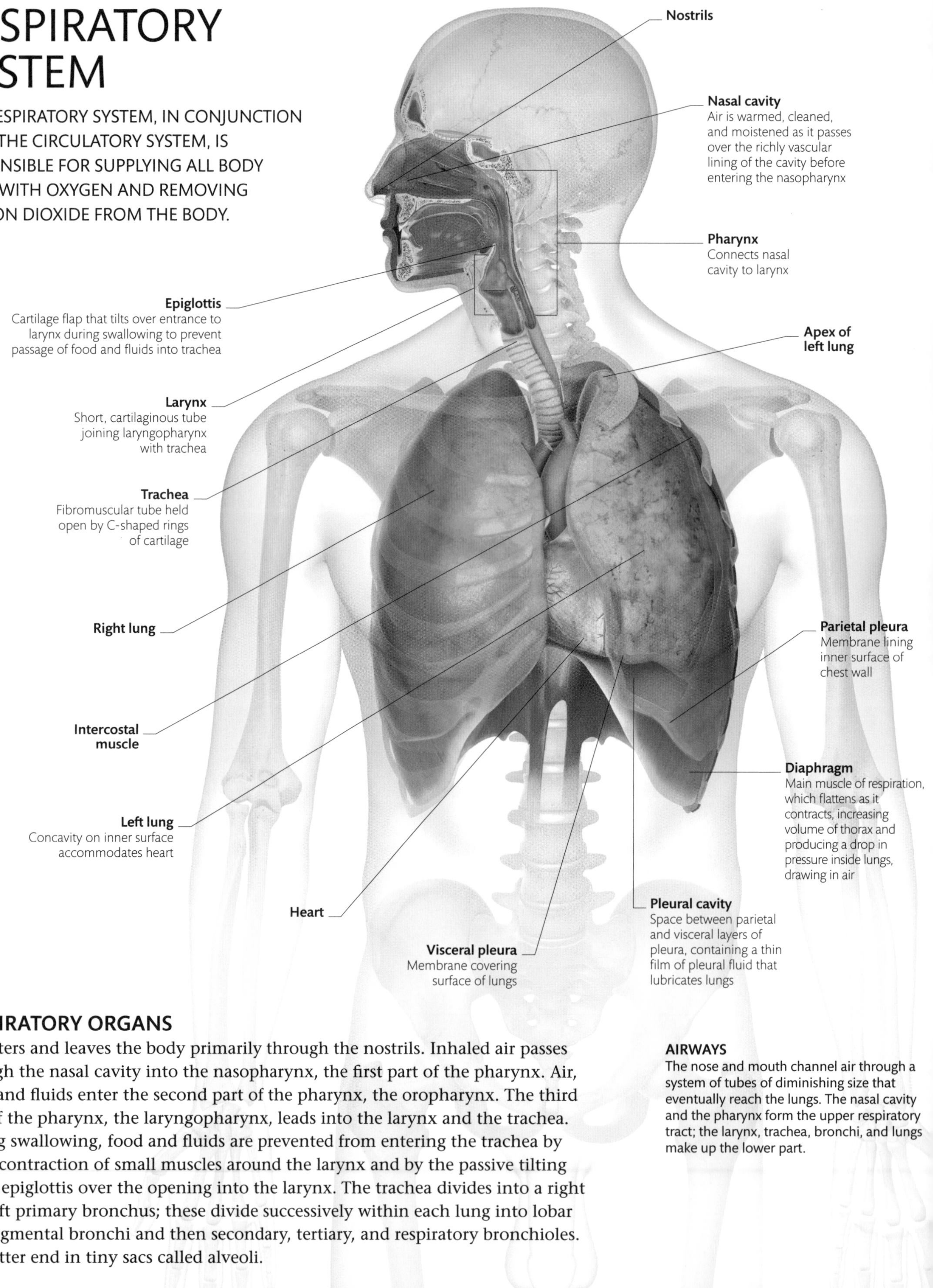
Nostrils
Nasal cavity Air is warmed, cleaned, and moistened as it passes over the richly vascular lining of the cavity before entering the nasopharynx
Pharynx Connects nasal cavity to larynx
Apex of left lung
Epiglottis Cartilage flap that tilts over entrance to larynx during swallowing to prevent passage of food and fluids into trachea
Larynx Short, cartilaginous tube joining laryngopharynx with trachea
Trachea Fibromuscular tube held open by C-shaped rings of cartilage
Right lung
Parietal pleura Membrane lining inner surface of chest wall
Intercostal muscle
Diaphragm Main muscle of respiration, which flattens as it contracts, increasing volume of thorax and producing a drop in pressure inside lungs, drawing in air
Left lung Concavity on inner surface accommodates heart
Heart
Pleural cavity Space between parietal and visceral layers of pleura, containing a thin film of pleural fluid that lubricates lungs
Visceral pleura Membrane covering surface of lungs

RESPIRATORY ORGANS

Air enters and leaves the body primarily through the nostrils. Inhaled air passes through the nasal cavity into the nasopharynx, the first part of the pharynx. Air, food, and fluids enter the second part of the pharynx, the oropharynx. The third part of the pharynx, the laryngopharynx, leads into the larynx and the trachea. During swallowing, food and fluids are prevented from entering the trachea by reflex contraction of small muscles around the larynx and by the passive tilting of the epiglottis over the opening into the larynx. The trachea divides into a right and left primary bronchus; these divide successively within each lung into lobar and segmental bronchi and then secondary, tertiary, and respiratory bronchioles. The latter end in tiny sacs called alveoli.

AIRWAYS

The nose and mouth channel air through a system of tubes of diminishing size that eventually reach the lungs. The nasal cavity and the pharynx form the upper respiratory tract; the larynx, trachea, bronchi, and lungs make up the lower part.

GAS EXCHANGE

All cells in the body require a continual supply of oxygen, which they combine with glucose to produce energy. Carbon dioxide is generated as a waste product of this process and is exchanged for oxygen in the alveoli. The transfer of oxygen from the outside air to the blood flowing through the lungs, and of carbon dioxide from the lungs to the outside air requires three processes: ventilation (movement of air in and out of the lungs); diffusion (movement of gases across the alveolar walls into the pulmonary capillary bed); and perfusion (movement of blood through the pulmonary vessels).

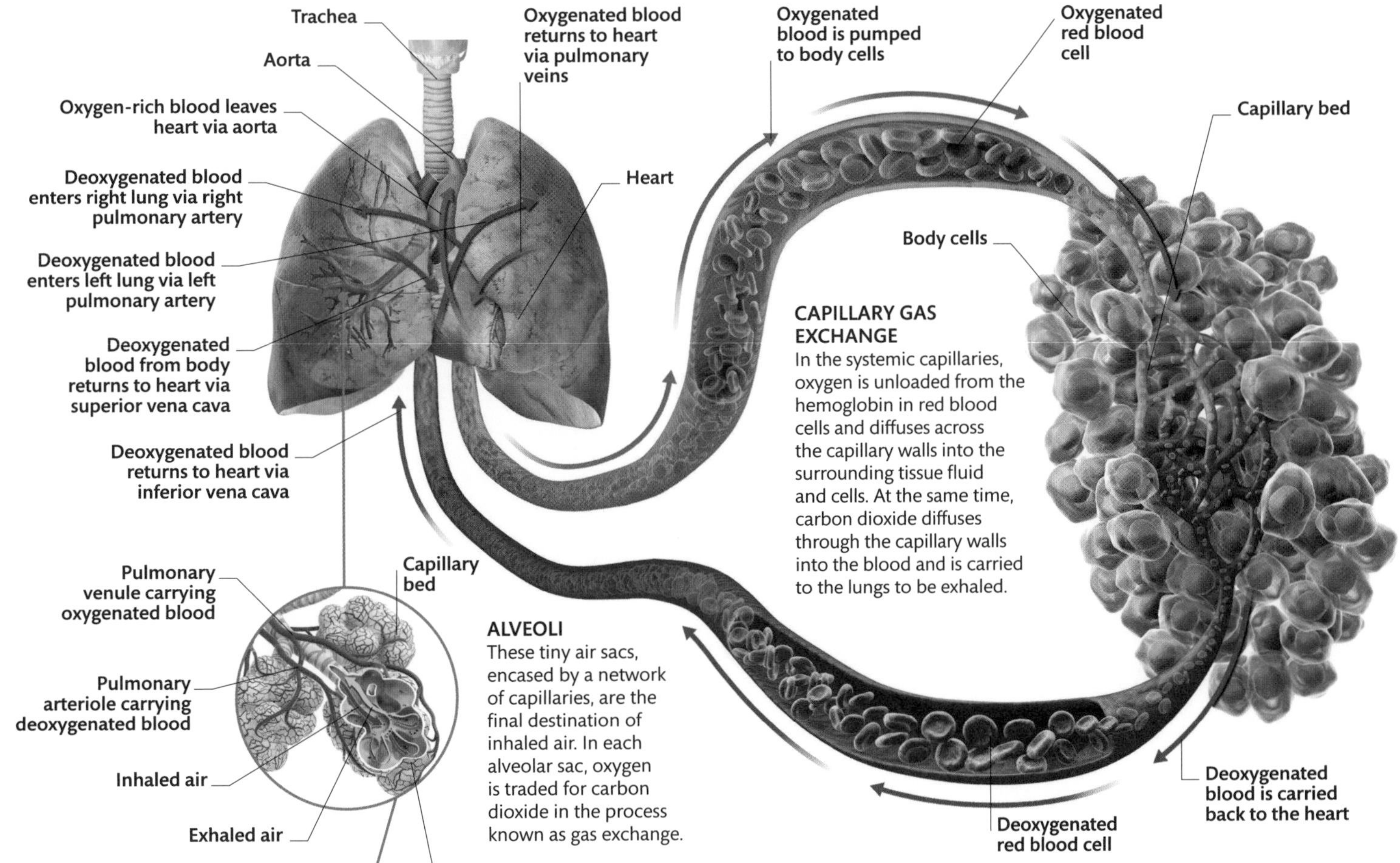
Trachea
Aorta
Oxygen-rich blood leaves heart via aorta
Deoxygenated blood enters right lung via right pulmonary artery
Deoxygenated blood enters left lung via left pulmonary artery
Deoxygenated blood from body returns to heart via superior vena cava
Deoxygenated blood returns to heart via inferior vena cava
Oxygenated blood returns to heart via pulmonary veins
Heart
Oxygenated blood is pumped to body cells
Oxygenated red blood cell
Capillary bed
Body cells
Pulmonary venule carrying oxygenated blood
Pulmonary arteriole carrying deoxygenated blood
Inhaled air
Exhaled air
Capillary bed
Alveolar sac
Deoxygenated red blood cell
Deoxygenated blood is carried back to the heart

CAPILLARY GAS EXCHANGE
In the systemic capillaries, oxygen is unloaded from the hemoglobin in red blood cells and diffuses across the capillary walls into the surrounding tissue fluid and cells. At the same time, carbon dioxide diffuses through the capillary walls into the blood and is carried to the lungs to be exhaled.

ALVEOLI
These tiny air sacs, encased by a network of capillaries, are the final destination of inhaled air. In each alveolar sac, oxygen is traded for carbon dioxide in the process known as gas exchange.

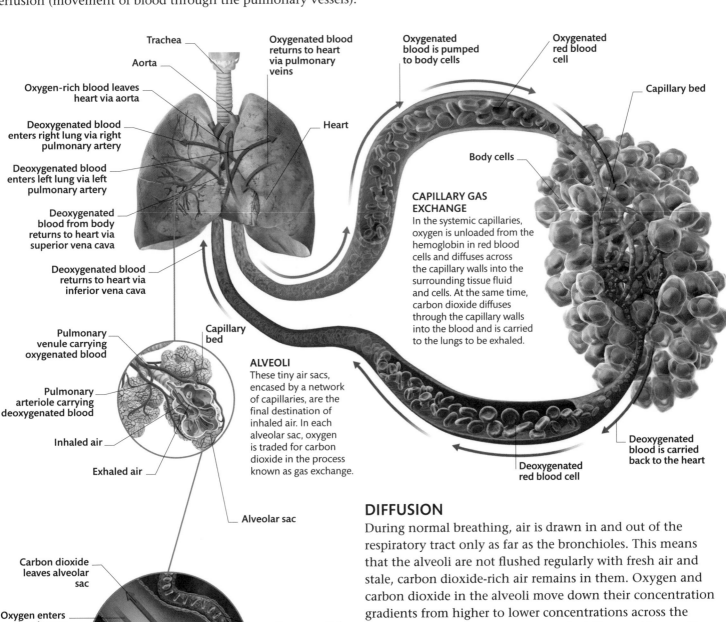

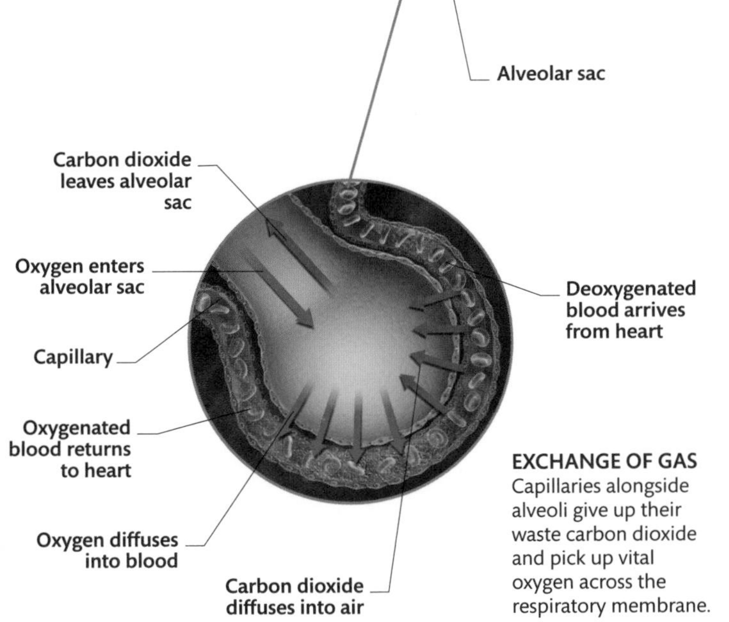
Carbon dioxide leaves alveolar sac
Oxygen enters alveolar sac
Capillary
Oxygenated blood returns to heart
Oxygen diffuses into blood
Carbon dioxide diffuses into air
Deoxygenated blood arrives from heart

EXCHANGE OF GAS
Capillaries alongside alveoli give up their waste carbon dioxide and pick up vital oxygen across the respiratory membrane.

DIFFUSION

During normal breathing, air is drawn in and out of the respiratory tract only as far as the bronchioles. This means that the alveoli are not flushed regularly with fresh air and stale, carbon dioxide-rich air remains in them. Oxygen and carbon dioxide in the alveoli move down their concentration gradients from higher to lower concentrations across the alveolar walls, a process called simple diffusion. Almost all the oxygen (about 98.5 percent) that enters the blood binds to the pigment hemoglobin in the red blood cells. The human lung contains over 300 million alveoli, each one covered in a dense network of capillaries. Diffusion is very effective in the lungs because the surface area of the pulmonary capillary bed is enormous and the barrier between blood and alveolar air is extremely thin (less than 1 micron).

See also pp. 118, 120, 122, 124 »

HEAD AND NECK

DURING INSPIRATION, AIR IS PULLED IN THROUGH THE NOSTRILS, INTO THE NASAL CAVITIES, WHERE IT IS WARMED, CLEANED, AND MOISTENED BEFORE PASSING INTO THE LOWER RESPIRATORY TRACT.

NASAL CONCHAE

The nasal conchae are three shelf-like projections in the nasal cavity that provide a natural obstruction to inhaled air, forcing it to spread out as it passes over their surfaces. This fulfils several roles. The moist, mucus-lined conchae humidify the passing air and trap inhaled particles, while their many capillary networks warm the air to body temperature before it reaches the lower parts of the respiratory tract. Nerves within the conchae sense the condition of the air and, if necessary, cause them to enlarge. If the air is cold, for example, a larger surface area helps to warm it more effectively. This enlargement is what gives the feeling of nasal congestion.

NASAL CONCHAE

Color and/or label the structures indicated on the diagram using the key below.

1. Sphenoidal sinus
2. Superior nasal concha
3. Middle nasal concha
4. Air warmed as it passes conchae
5. Inhaled air
6. Inferior nasal concha
7. Hard palate

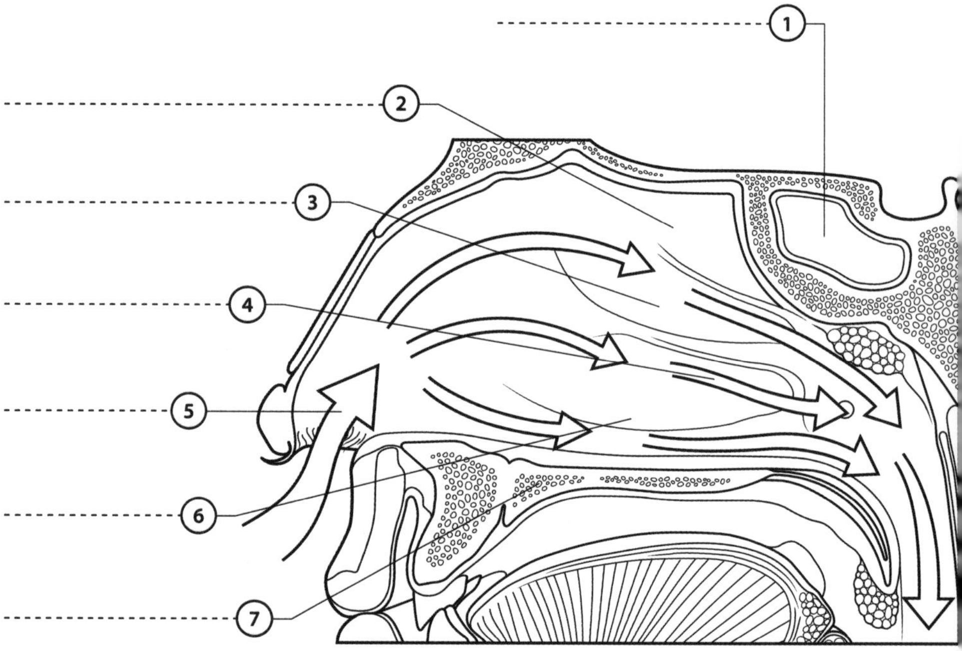

NASAL CONCHAE

PARANASAL SINUSES

The paranasal sinuses are four pairs of air-filled cavities that sit within the bones of the skull and drain into the nasal cavity. They are lined with cells that produce mucus, which flows into the nasal passageways through very small openings. The sinuses affect the resonance of the voice. This is obvious during a cold, when the nose becomes blocked, giving a nasal quality to the voice.

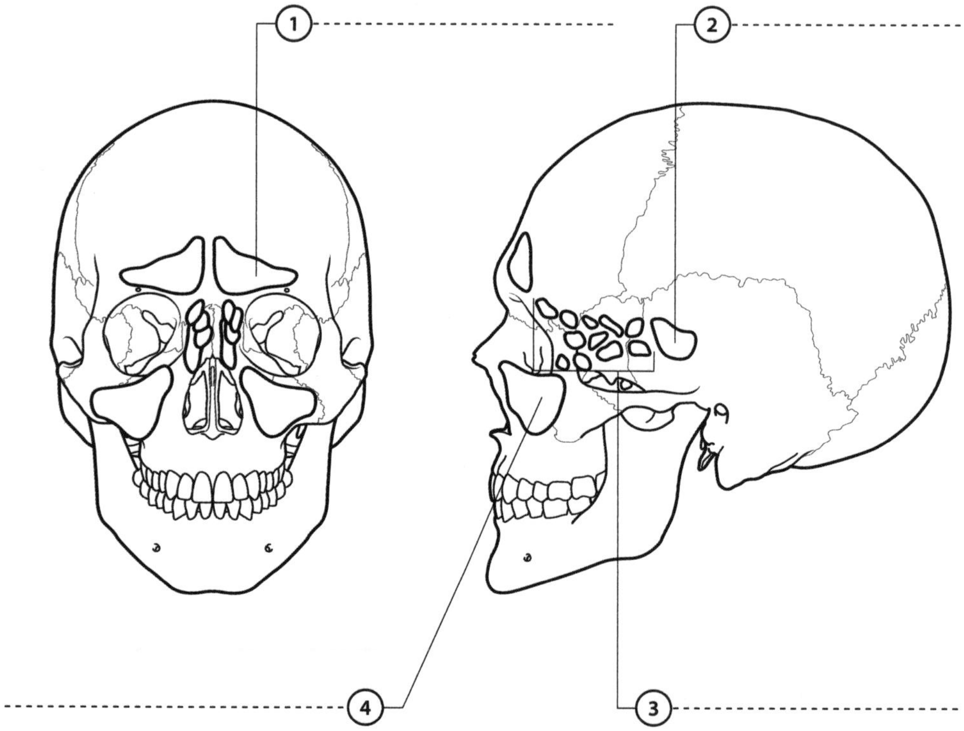

PARANASAL SINUSES

PARANASAL SINUSES

Color and/or label the structures indicated on the diagram using the key below.

1. Frontal sinus
2. Sphenoidal sinus
3. Ethmoid sinuses
4. Maxillary sinus

ANATOMY

The nasal cavities are divided by the thin partition of the nasal septum, composed of plates of cartilage and bone. The lateral walls of the nasal cavity are more elaborate, with bony curls called conchae (see left). The nasal cavity is lined with mucosa, which produces mucus that traps particles in the incoming air and also moistens it. The paranasal sinuses (see left), also lined with mucosa, open via tiny orifices into the nasal cavity. Below and in front of the pharynx is the larynx. The way in which exhaled air passes through the larynx is modulated to produce sounds.

SAGITTAL SECTION

Color and/or label the structures indicated on the diagram using the key below.

1. Superior meatus
2. Frontal sinus
3. Inferior meatus
4. Atrium
5. Vestibule
6. Nostril
7. Hard palate
8. Thyroid cartilage
9. Vocal cord
10. Trachea
11. Cricoid cartilage
12. Epiglottis
13. Pharyngeal opening of Eustachian tube
14. Cut edge of inferior concha
15. Sphenoidal sinus
16. Cut edge of middle concha
17. Middle meatus
18. Cut edge of superior concha

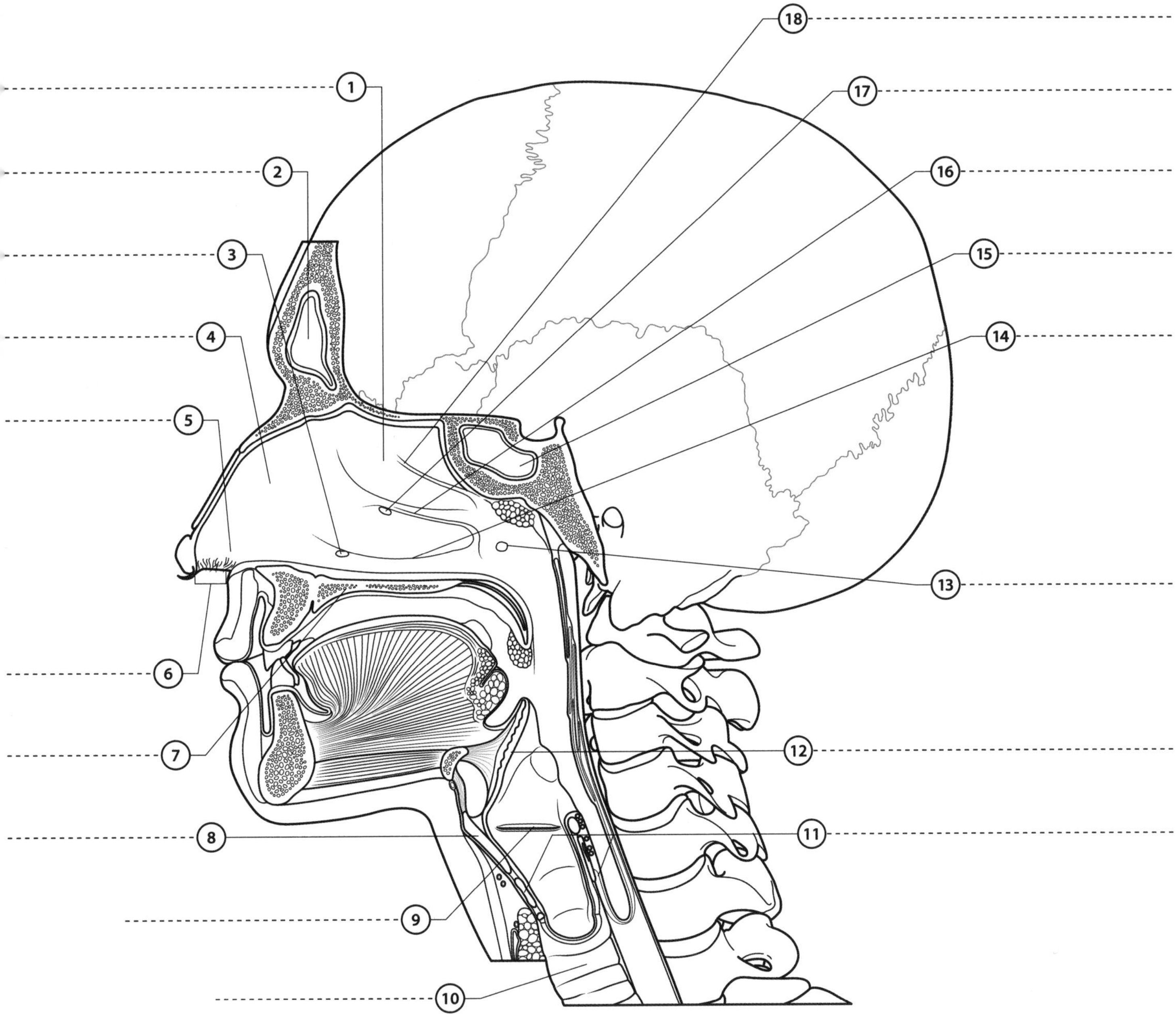

SAGITTAL SECTION

See also pp. 134, 160, 174 »

THORAX 1

THE PARTS OF THE RESPIRATORY SYSTEM IN THE THORAX ARE THE LUNGS, BRONCHI, AND THE LOWER PART OF THE TRACHEA, WHICH PASSES INTO THE THORAX FROM THE NECK. THE HEART, THE OTHER MAJOR ORGAN IN THE THORAX, IS COVERED IN THE CARDIOVASCULAR SYSTEM.

ANATOMY

In the thorax, the trachea divides into two main bronchi, each supplying one lung. The trachea is supported and held open by 15–20 C-shaped pieces of cartilage, and there is smooth muscle in its wall that can alter its width. The cartilages in the walls of the bronchi prevent them from collapsing when air enters the lungs under low pressure. Inside the lungs, the bronchi branch repeatedly, forming smaller airways (bronchioles), which are just muscular tubes with no cartilage. The smallest bronchioles end in a cluster of alveoli. These are air sacs surrounded by capillaries, where oxygen passes into the blood and carbon dioxide passes in the opposite direction into the airway.

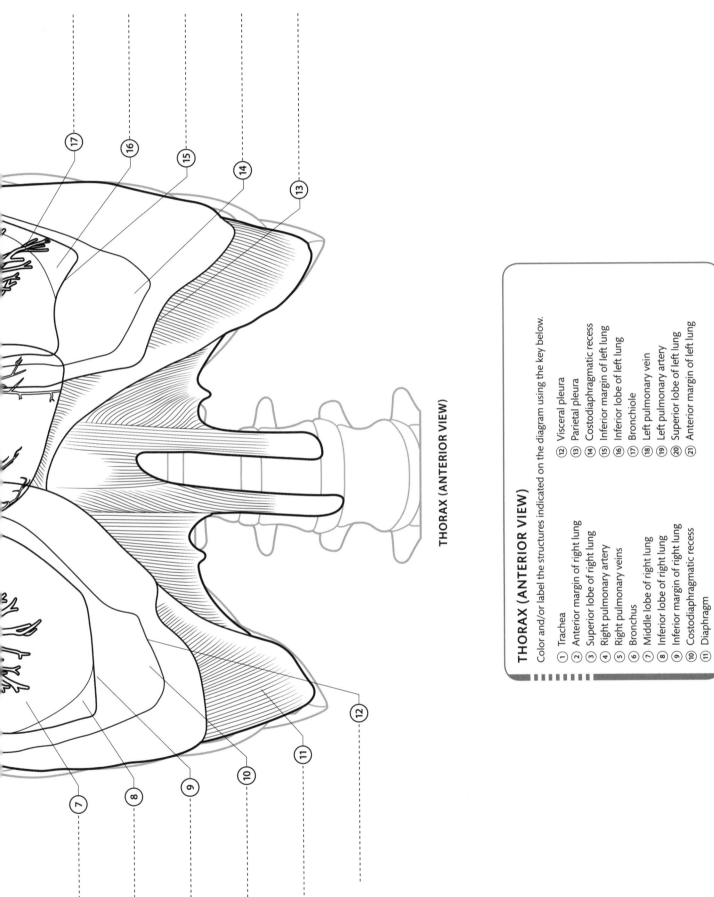

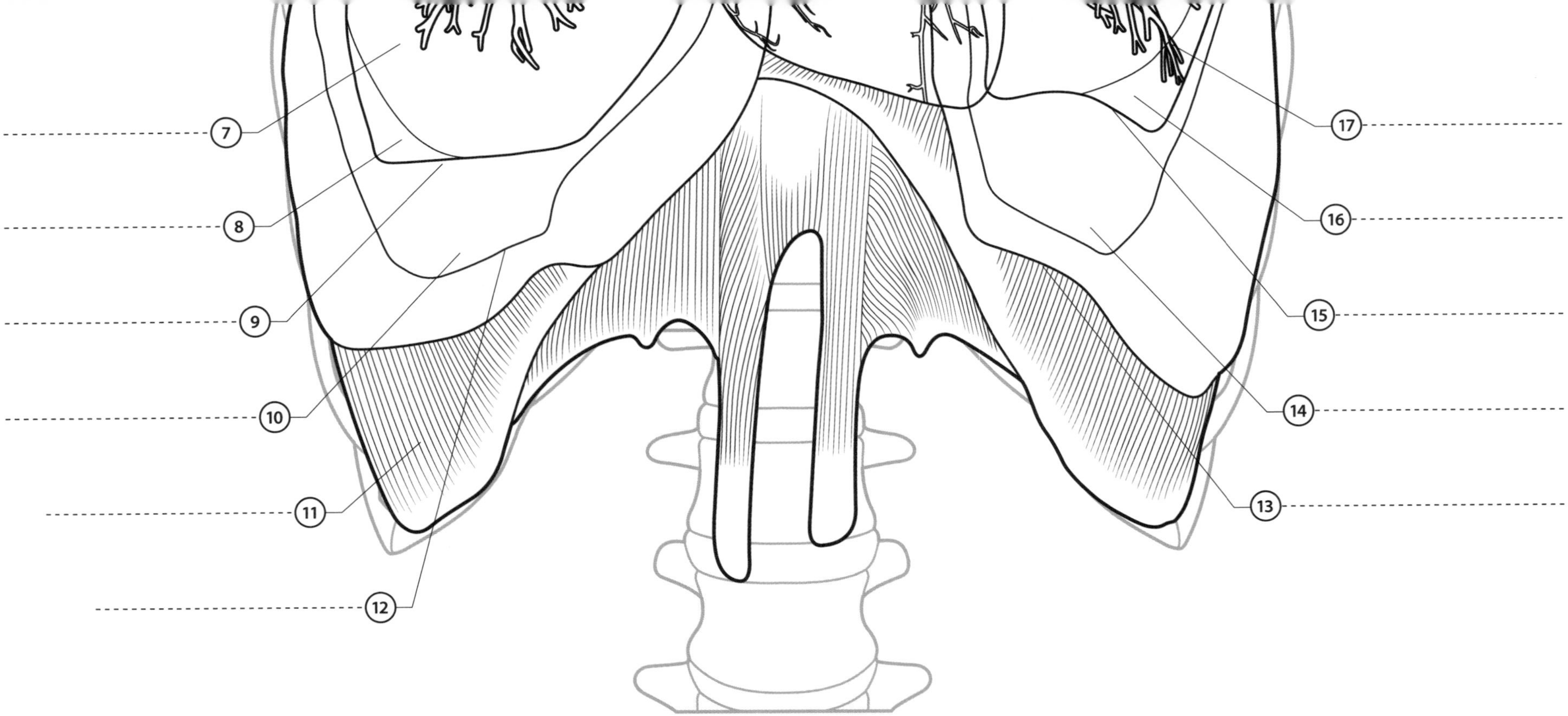

THORAX (ANTERIOR VIEW)

THORAX (ANTERIOR VIEW)

Color and/or label the structures indicated on the diagram using the key below.

1. Trachea
2. Anterior margin of right lung
3. Superior lobe of right lung
4. Right pulmonary artery
5. Right pulmonary veins
6. Bronchus
7. Middle lobe of right lung
8. Inferior lobe of right lung
9. Inferior margin of right lung
10. Costodiaphragmatic recess
11. Diaphragm
12. Visceral pleura
13. Parietal pleura
14. Costodiaphragmatic recess
15. Inferior margin of left lung
16. Inferior lobe of left lung
17. Bronchiole
18. Left pulmonary vein
19. Left pulmonary artery
20. Superior lobe of left lung
21. Anterior margin of left lung

See also pp. 36, 62, 64, 100, 116, 122, 124, 138, 162 »

THORAX 2

EACH LUNG FITS SNUGLY INSIDE ITS HALF OF THE THORACIC CAVITY. THE TWO LUNGS APPEAR SIMILAR AT FIRST GLANCE, BUT THERE IS ACTUALLY SOME ASYMMETRY.

LUNGS

The surface of each lung is covered with a thin pleural membrane (visceral pleura), and the inside of the chest wall is also lined with pleura (parietal pleura). Between the two pleural layers lies a thin film of lubricating liquid that allows the lungs to slide against the chest wall during breathing movements, and that also creates a fluid seal, effectively sticking the lungs to the ribs and diaphragm. The bronchi and blood vessels enter each lung at the hilum on its medial surface. Although the two lungs may appear to be the same at first glance, there are differences between them. The left lung is concave to fit around the heart and has two lobes, whereas the right lung has three lobes, marked out by two deep fissures.

LEFT LUNG (MEDIAL VIEW)

Color and/or label the structures indicated on the diagram using the key below.

1. Superior lobe
2. Left pulmonary artery
3. Left main bronchus
4. Pulmonary ligament
5. Inferior lobe
6. Inferior margin
7. Diaphragmatic surface of lung
8. Oblique fissure
9. Lingula
10. Anterior margin
11. Left inferior pulmonary vein
12. Left superior pulmonary vein

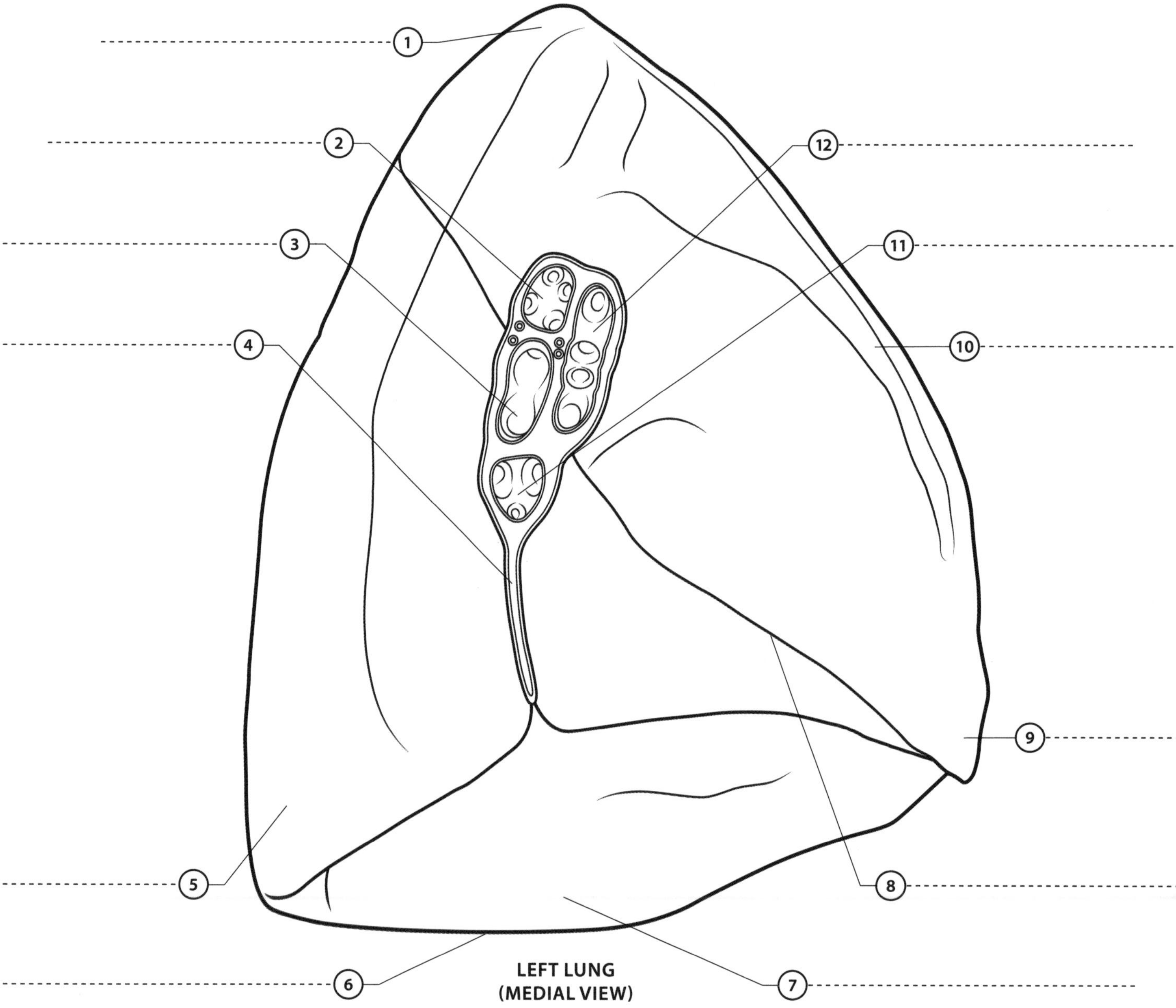

LEFT LUNG
(MEDIAL VIEW)

RIGHT LUNG (MEDIAL VIEW)

Color and/or label the structures indicated on the diagram using the key below.

1. Groove for esophagus
2. Branches of right pulmonary artery
3. Superior lobe
4. Branches of right superior pulmonary vein
5. Anterior margin
6. Horizontal fissure
7. Middle lobe
8. Diaphragmatic surface
9. Inferior margin
10. Inferior lobe
11. Pulmonary ligament
12. Pleura
13. Right inferior pulmonary veins
14. Right main bronchus
15. Superior lobar bronchus (eparterial)

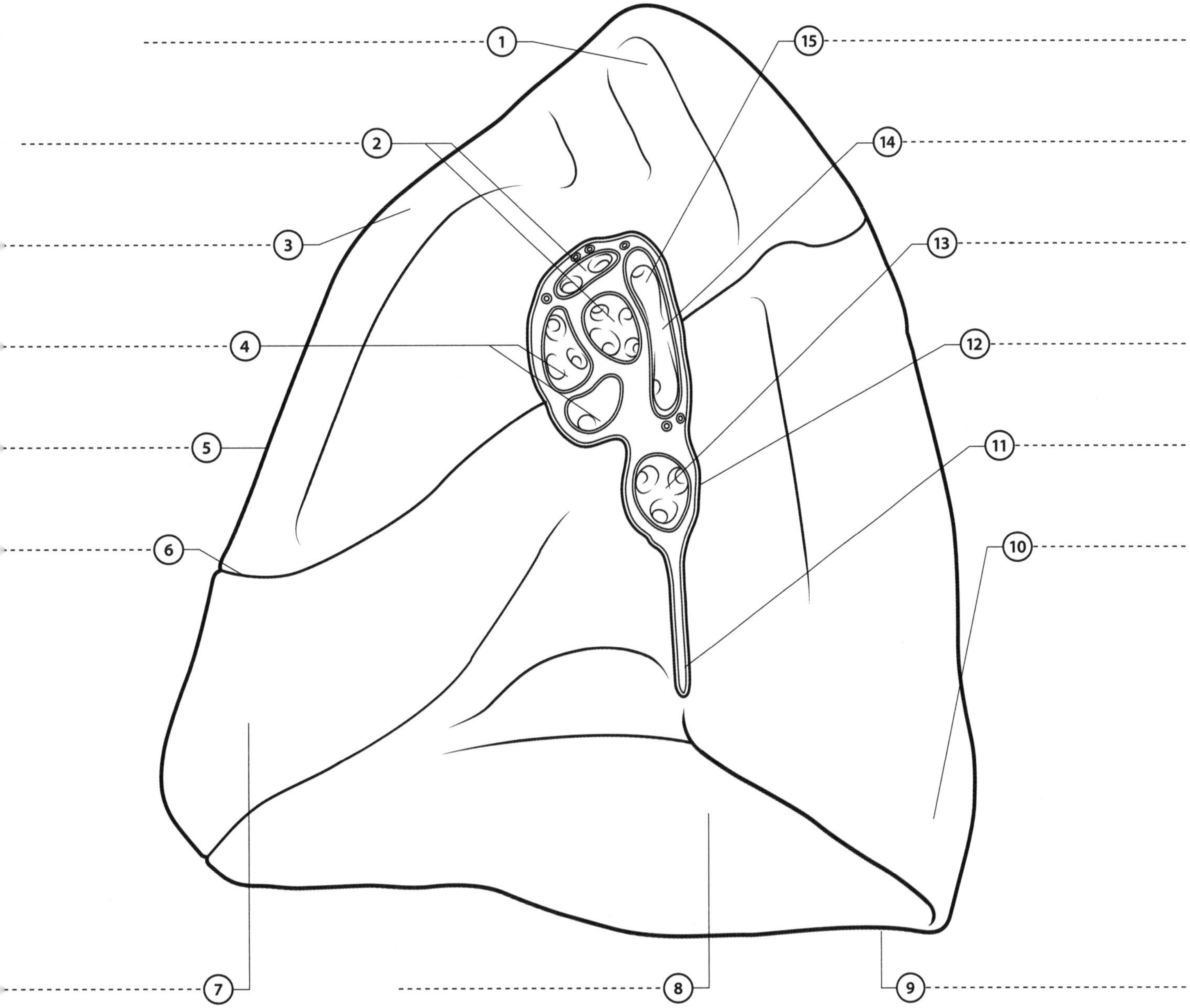

RIGHT LUNG
(MEDIAL VIEW)

See also pp. 36, 62, 64, 100, 116, 120, 124, 138, 162

BREATHING

THE MOVEMENTS OF BREATHING—INHALATION AND EXHALATION—BRING FRESH, OXYGEN-RICH AIR INTO THE LUNGS AND REMOVE STALE, CARBON DIOXIDE-RICH AIR FROM THE LUNGS.

INHALATION AND EXHALATION

The movement of air into and out of the lungs is generated by differences in air pressure within the lungs, compared to the surrounding atmospheric pressure. The pressure differences are created by expanding the chest and lungs by muscular action, and then passively allowing them to return to their former size. The chief muscles used in inhalation at rest are the diaphragm at the base of the thorax and the external intercostals between the ribs. For forceful inhalation, additional muscles assist in moving the ribs and sternum. Exhalation is largely passive and is due to elastic recoil of the enlarged lungs and abdominal pressure forcing the relaxed diaphragm upward. Forced expiration brings into play additional muscles to actively compress the lungs beyond their usual resting volume.

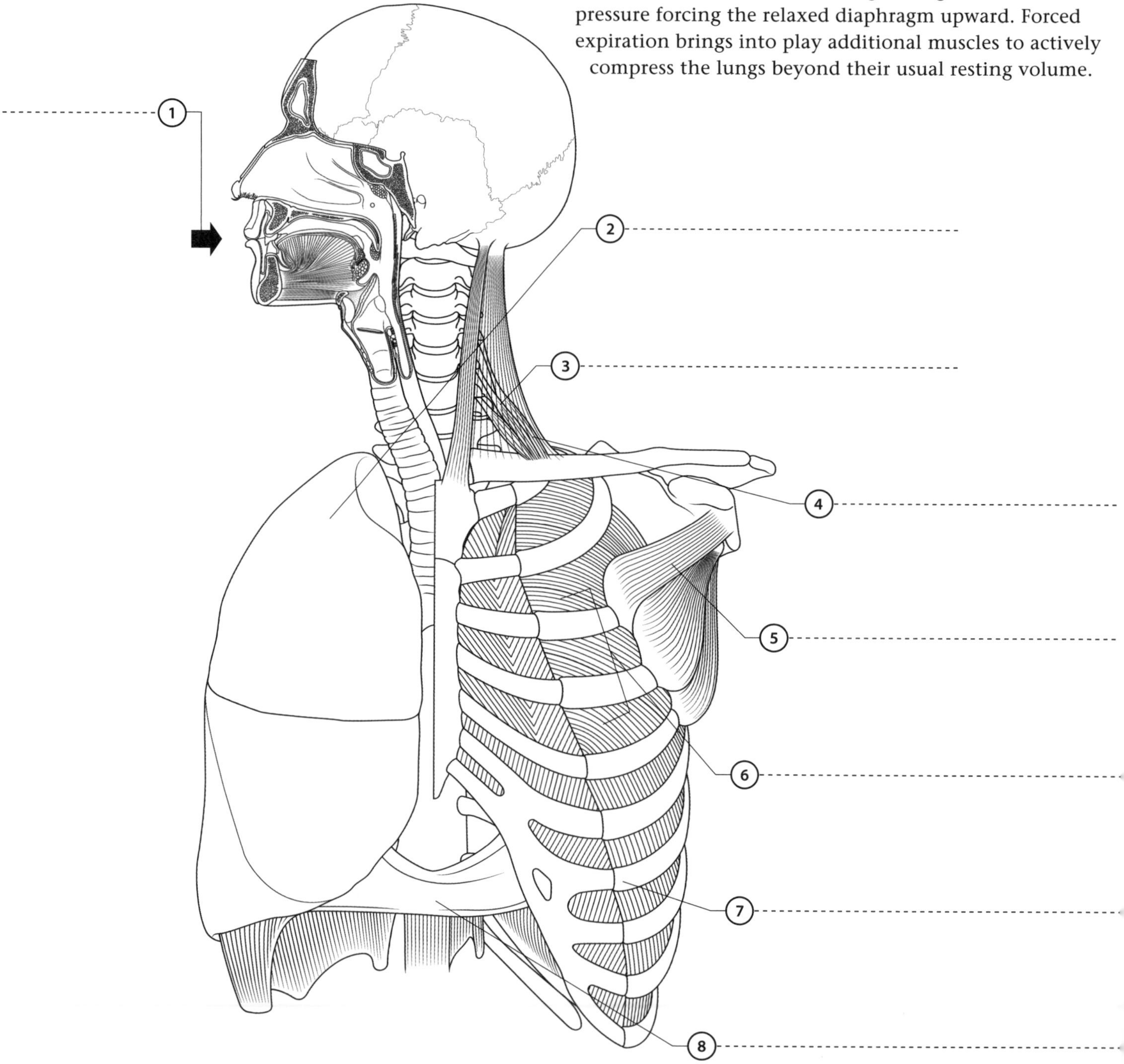

INHALATION

INHALATION

Color and/or label the structures indicated on the diagram using the key below.

1. Air inhaled
2. Lungs expand
3. Sternocleidomastoid pulls up clavicle and sternum
4. Scalenes help raise upper two ribs
5. Pectoralis minor pulls up on 3rd, 4th, and 5th ribs
6. External intercostals pull ribs together
7. Ribs tilt up and out
8. Diaphragm contracts and moves down

EXHALATION

Color and/or label the structures indicated on the diagram using the key below.

1. Air exhaled
2. Lungs shrink
3. Trachea
4. Sternum moves down and in
5. Internal intercostals pull ribs down for forced exhalation
6. Diaphragm relaxes and moves up
7. Rectus abdominis pulls in 5th and 7th ribs and sternum
8. Ribs drawn in and down

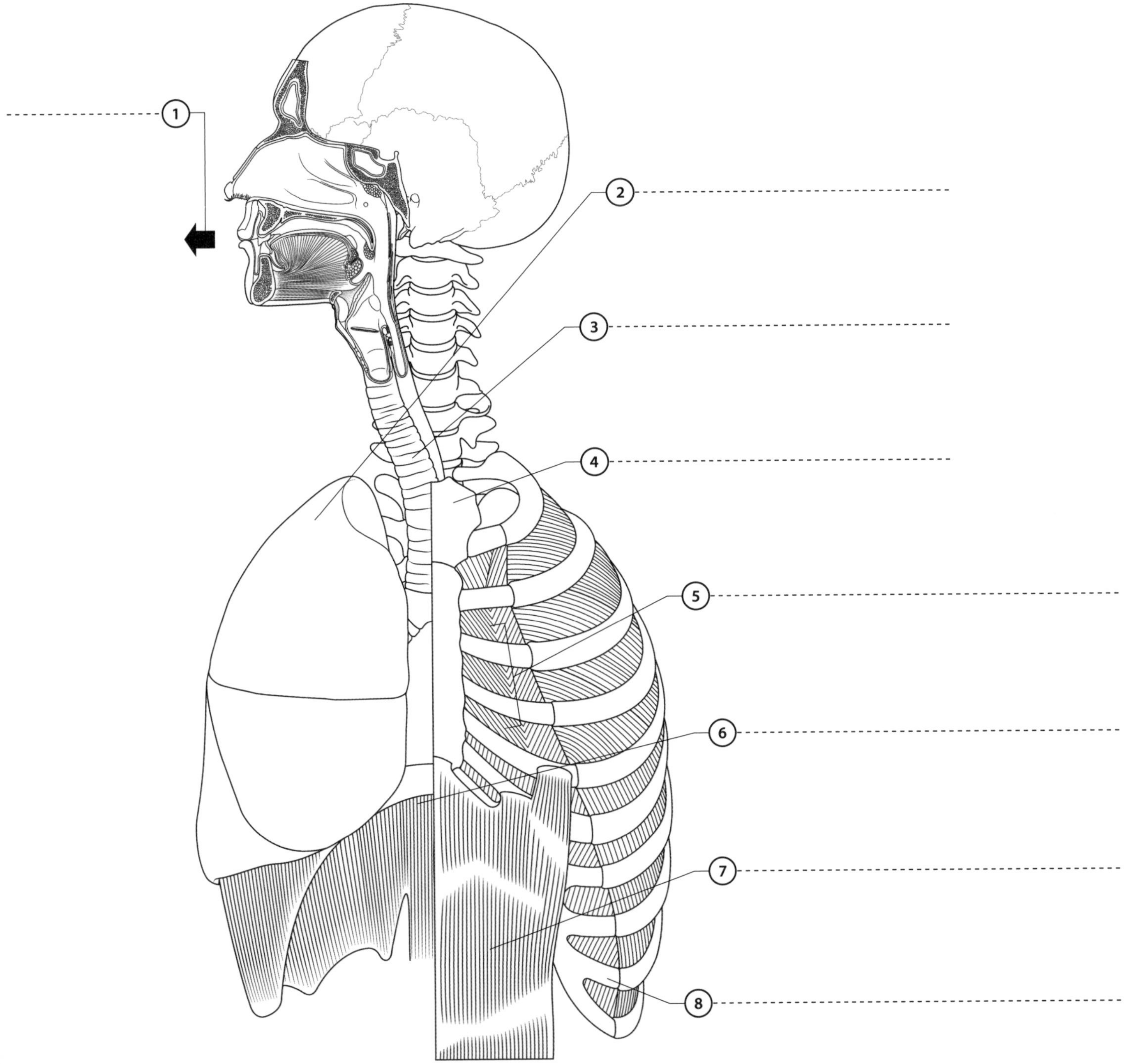

EXHALATION

See also pp. 62, 120, 122

CARDIOVASCULAR SYSTEM

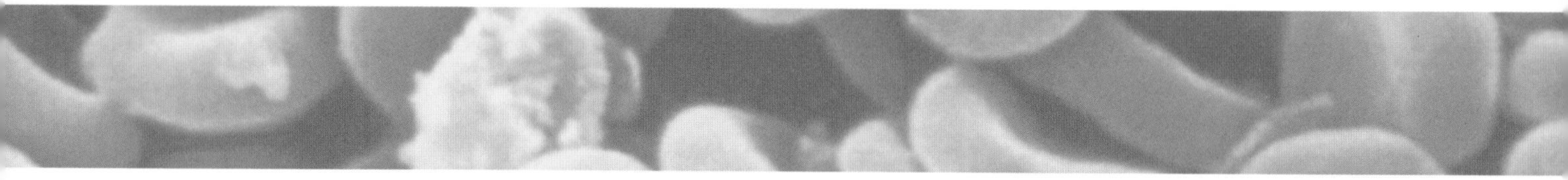

06

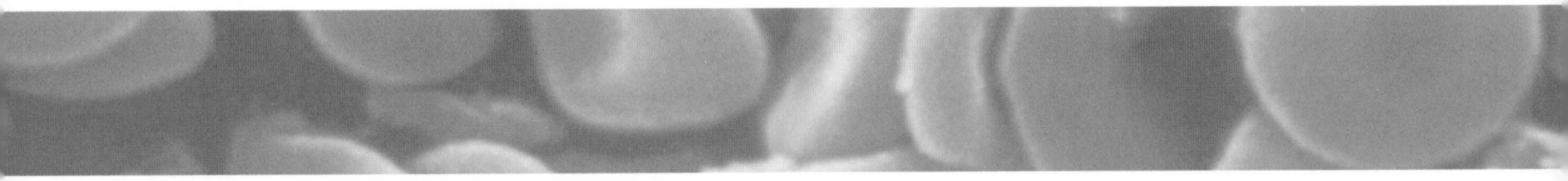

CARDIOVASCULAR SYSTEM 1

THE CARDIOVASCULAR, OR CIRCULATORY, SYSTEM TRANSPORTS OXYGEN, NUTRIENTS, AND OTHER USEFUL SUBSTANCES, INCLUDING HORMONES AND IMMUNE CELLS, AROUND THE BODY. CIRCULATING BLOOD ALSO REMOVES WASTES, SUCH AS CARBON DIOXIDE, PRODUCED BY CELL METABOLISM.

CARDIOVASCULAR COMPONENTS

The components of the cardiovascular system are the heart, blood vessels, and blood. The heart is a powerful muscular pump, beating over three million times in an average human lifetime to send blood to every cell in the body. Each heartbeat forces blood through a network of vessels. Arteries—large vessels with thick, elastic walls—carry blood rich in oxygen and nutrients away from the heart; they branch into progressively smaller vessels that eventually lead to networks of thin-walled capillaries. The smallest of the blood vessels, capillaries, join up to form tiny veins, which merge into larger veins as they return deoxygenated blood to the heart.

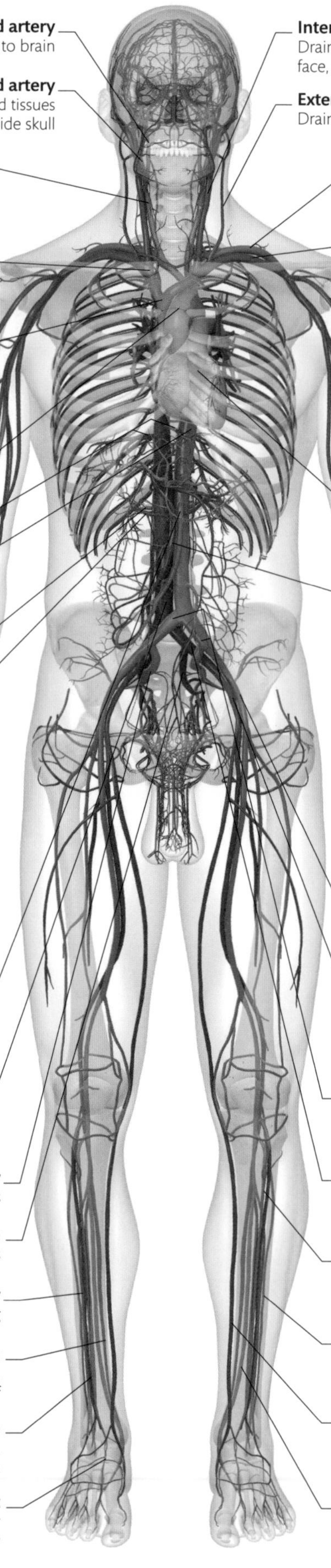

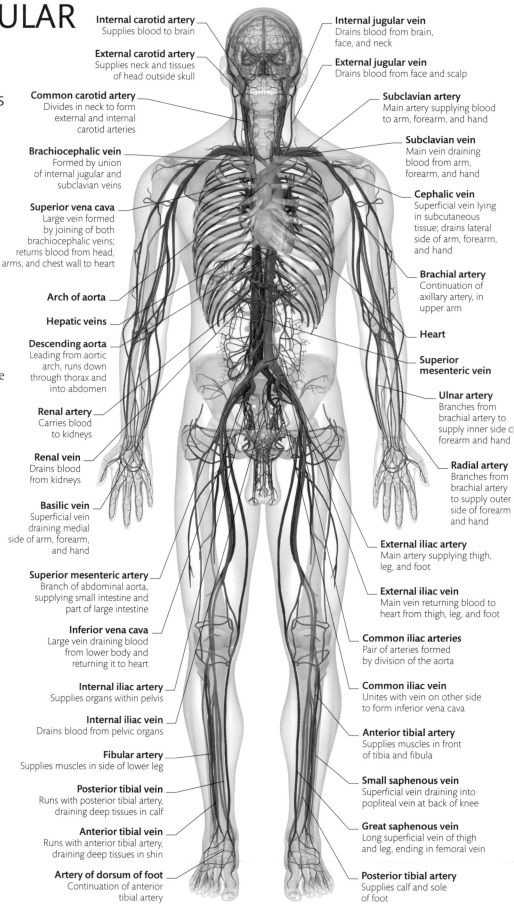

ANTERIOR VIEW

ANTERIOR VIEW OF CARDIOVASCULAR SYSTEM
The complex pathways of the cardiovascular system extend throughout the entire body. Vessels containing oxygenated blood (usually arteries) are shown in red, while those carrying deoxygenated blood (usually veins) are blue.

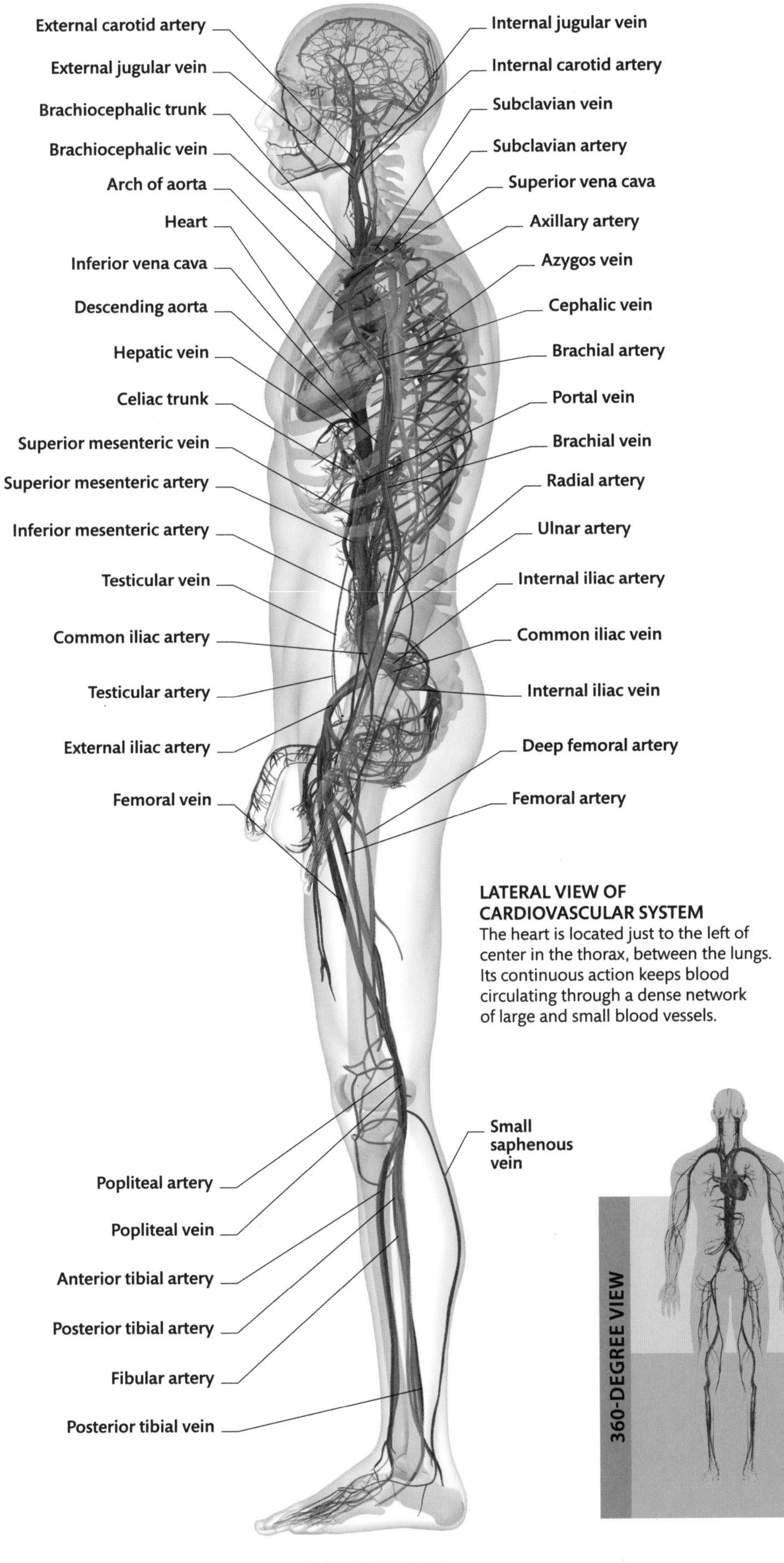

LATERAL VIEW

LATERAL VIEW OF CARDIOVASCULAR SYSTEM
The heart is located just to the left of center in the thorax, between the lungs. Its continuous action keeps blood circulating through a dense network of large and small blood vessels.

BLOOD SUPPLY ROUTES

The main artery in the body is the aorta, which supplies blood to the head via the carotid arteries and to the lower body via the descending aorta, which branches in the abdomen into the common iliac arteries. Continuing from the external iliac arteries, the femoral arteries run down the legs to the back of the knees. From here, the popliteal arteries and then the tibial arteries supply the lower legs. The subclavian arteries, which branch sideways at the collarbone, transport blood to the upper limbs.

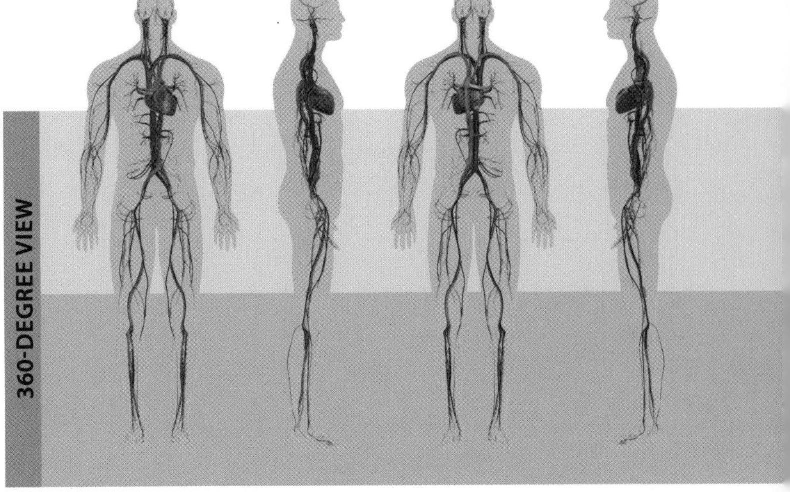

See also pp. 130, 132, 156

CARDIOVASCULAR SYSTEM 2

THE HEART IS A TWO-SIDED PUMP THAT KEEPS BLOOD IN CONTINUOUS CIRCULATION THROUGH THE LUNGS AND AROUND THE BODY. IN RESPONSE TO ELECTRICAL IMPULSES, THE MUSCULAR WALLS OF THE HEART CHAMBERS PULSE IN A REGULAR SEQUENCE OF CONTRACTIONS THAT IS REPEATED WITH EACH HEARTBEAT.

BLOOD CIRCULATION

Blood is pumped around the body from the heart in two linked circuits: the pulmonary circulation and the systemic circulation. In the pulmonary circulation, the right side of the heart receives deoxygenated blood from the body and pumps it to the lungs, at relatively low pressure. From the lungs, the oxygen-replenished blood returns to the left side of the heart. The left side of the heart pumps the oxygenated blood to the rest of body via the systemic circulation, at a much higher pressure.

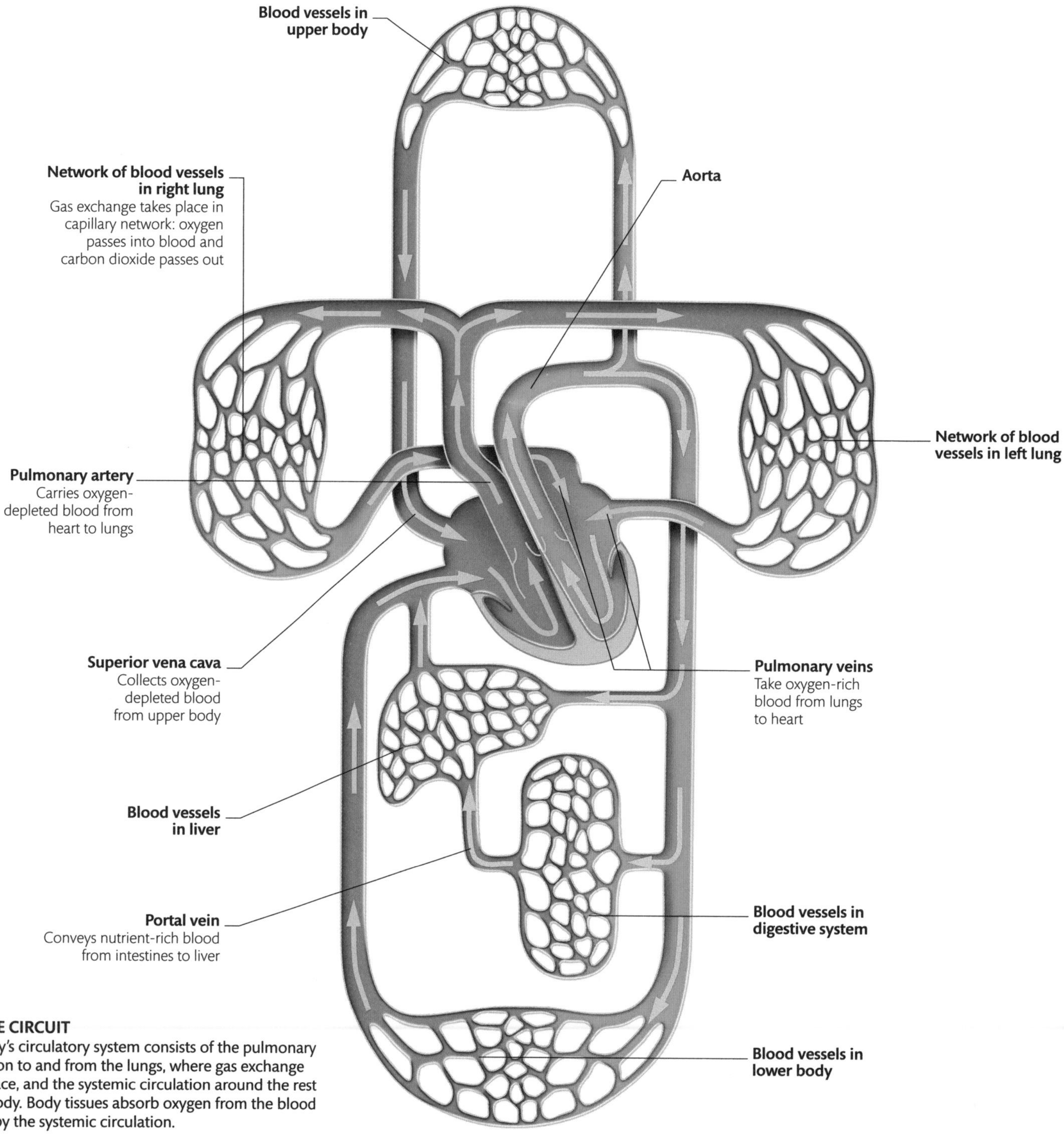

DOUBLE CIRCUIT

The body's circulatory system consists of the pulmonary circulation to and from the lungs, where gas exchange takes place, and the systemic circulation around the rest of the body. Body tissues absorb oxygen from the blood carried by the systemic circulation.

CARDIAC CYCLE

Circulating blood is pumped through the heart by the rhythmic action of cardiac muscle. Every heartbeat involves the coordinated contraction (systole) and relaxation (diastole) of the heart's four chambers. These regulated muscular pulses transfer blood from the upper two chambers (atria) into the lower two chambers (ventricles) via two valves (mitral and tricuspid). From the ventricles, blood is ejected out of the heart through the aorta and the pulmonary artery. This process, the cardiac cycle, involves several distinct stages; on average a complete cycle takes less than a second. When at rest, the heart beats on average 100,000 times per day.

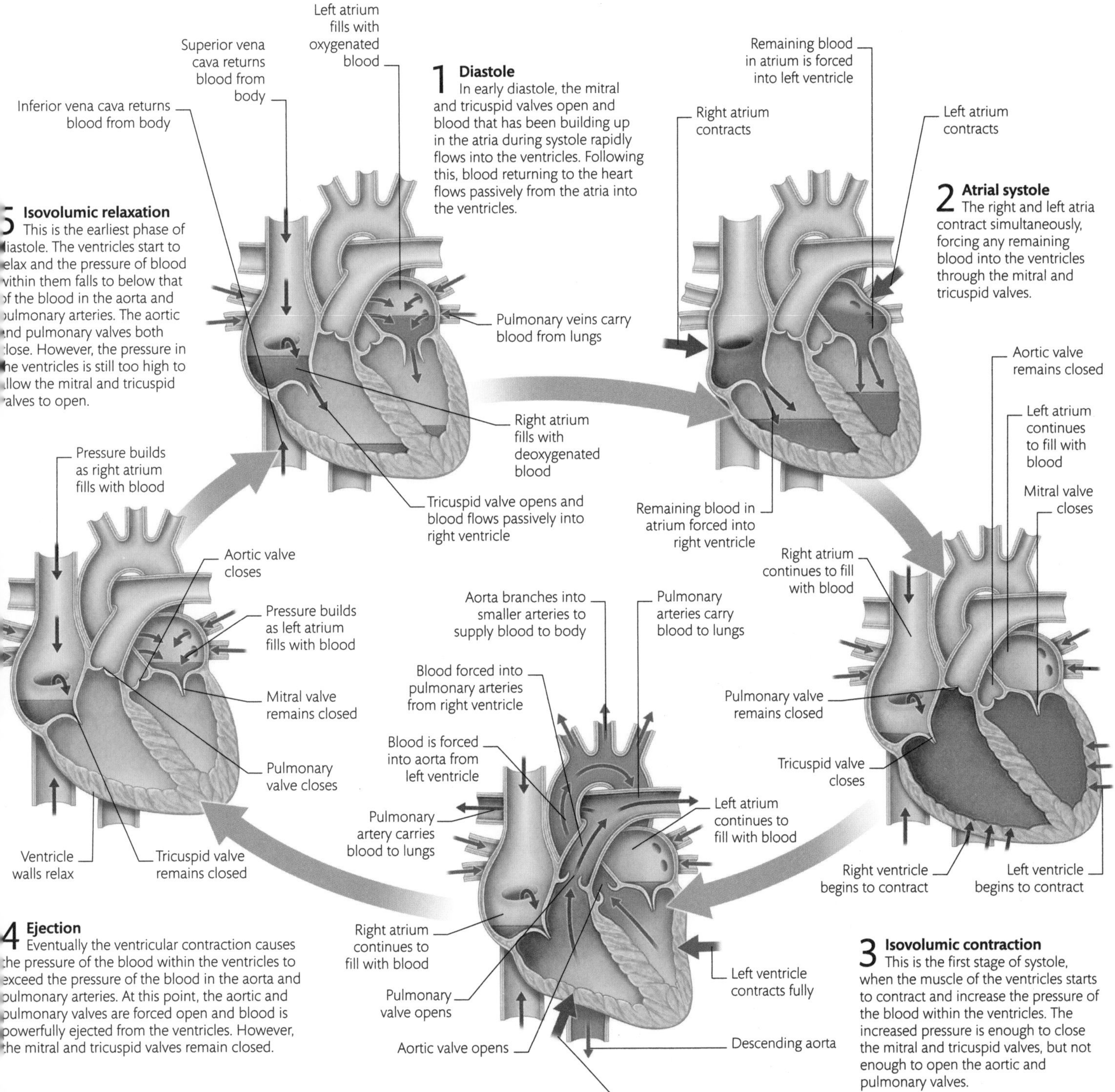
Left atrium fills with oxygenated blood
Superior vena cava returns blood from body
Inferior vena cava returns blood from body
Pulmonary veins carry blood from lungs
Right atrium fills with deoxygenated blood
Tricuspid valve opens and blood flows passively into right ventricle
Remaining blood in atrium is forced into left ventricle
Right atrium contracts
Left atrium contracts
Aortic valve remains closed
Left atrium continues to fill with blood
Mitral valve closes
Remaining blood in atrium forced into right ventricle
Right atrium continues to fill with blood
Pulmonary valve remains closed
Tricuspid valve closes
Right ventricle begins to contract
Left ventricle begins to contract
Pressure builds as right atrium fills with blood
Aortic valve closes
Pressure builds as left atrium fills with blood
Mitral valve remains closed
Pulmonary valve closes
Ventricle walls relax
Tricuspid valve remains closed
Aorta branches into smaller arteries to supply blood to body
Pulmonary arteries carry blood to lungs
Blood forced into pulmonary arteries from right ventricle
Blood is forced into aorta from left ventricle
Pulmonary artery carries blood to lungs
Left atrium continues to fill with blood
Right atrium continues to fill with blood
Left ventricle contracts fully
Pulmonary valve opens
Aortic valve opens
Descending aorta
Right ventricle contracts fully

1 Diastole
In early diastole, the mitral and tricuspid valves open and blood that has been building up in the atria during systole rapidly flows into the ventricles. Following this, blood returning to the heart flows passively from the atria into the ventricles.

2 Atrial systole
The right and left atria contract simultaneously, forcing any remaining blood into the ventricles through the mitral and tricuspid valves.

3 Isovolumic contraction
This is the first stage of systole, when the muscle of the ventricles starts to contract and increase the pressure of the blood within the ventricles. The increased pressure is enough to close the mitral and tricuspid valves, but not enough to open the aortic and pulmonary valves.

4 Ejection
Eventually the ventricular contraction causes the pressure of the blood within the ventricles to exceed the pressure of the blood in the aorta and pulmonary arteries. At this point, the aortic and pulmonary valves are forced open and blood is powerfully ejected from the ventricles. However, the mitral and tricuspid valves remain closed.

5 Isovolumic relaxation
This is the earliest phase of diastole. The ventricles start to relax and the pressure of blood within them falls to below that of the blood in the aorta and pulmonary arteries. The aortic and pulmonary valves both close. However, the pressure in the ventricles is still too high to allow the mitral and tricuspid valves to open.

HEARTBEAT

Contractions of the heart occur in response to the electrical activity of specialized cells that carry electrical impulses through the cardiac muscle. Under normal circumstances, this activity follows a regular pattern, with contractions of the heart chambers following suit. The heart can adapt to the demands of the body by altering the rate and force of its contractions.

See also pp. 128, 132 »

CARDIOVASCULAR SYSTEM 3

BRANCHING BLOOD VESSELS OF VARYING SIZE AND STRUCTURE JOIN TOGETHER IN A VAST NETWORK TO FORM PART OF THE CIRCULATORY SYSTEM. THE MAIN COMPONENT OF THE BLOOD TRANSPORTED BY THESE VESSELS IS WATER, IN WHICH A COLLECTION OF SPECIALIZED CELLS IS SUSPENDED.

BLOOD VESSELS

The three main types of blood vessel are arteries, veins, and capillaries. Both arteries and veins have walls made up of multiple layers of tissue: an innermost lining, or tunica intima; an internal elastic lamina; a muscular middle layer, the tunica media; and a tough outer wrapping, the tunica adventitia. In capillaries, the smallest blood vessels, the walls are just one cell thick.

Arteries are the primary suppliers of blood to the body—taking blood away from the heart, collecting oxygen from the lungs, and transporting oxygenated blood to different body areas. These are large vessels with thick, elastic walls strong enough to withstand the waves of blood pumped at high pressure with each contraction of the heart. Veins are responsible for taking deoxygenated blood from the body back to the heart. Blood in veins flows at relatively low pressure, so these vessels have much thinner, less elastic walls than arteries. Some of the larger veins, such as the jugular and the major veins in the legs, contain one-way valves to prevent backflow of blood.

Capillaries form fine networks, or capillary beds, that carry blood between the smallest arteries (arterioles) and the smallest veins (venules). The thin walls of capillaries, comprising a single layer of endothelial cells, are permeable and allow the exchange of oxygen, nutrients, and wastes.

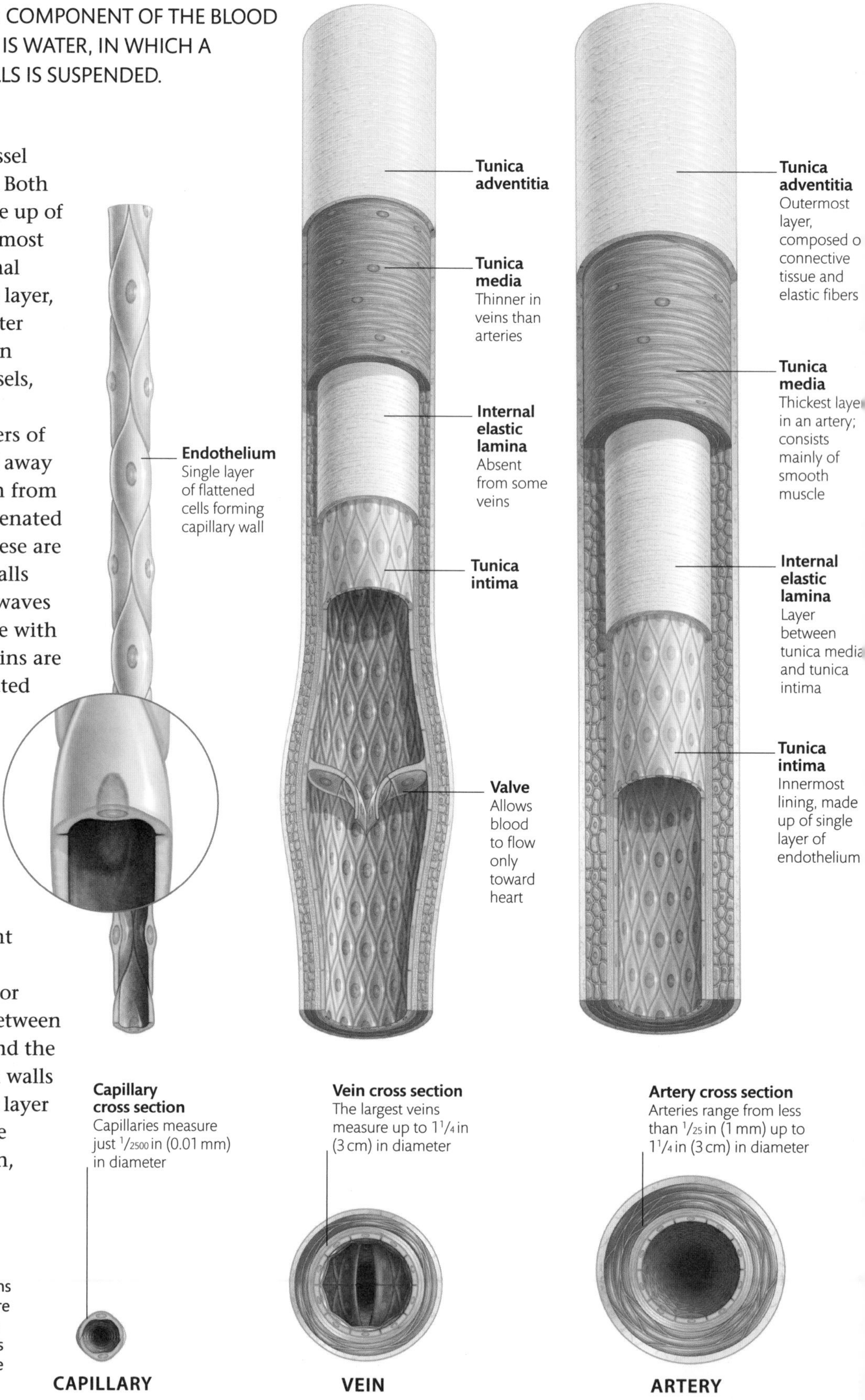

STRUCTURE OF BLOOD VESSELS
The major arteries are muscular and highly elastic to withstand high blood pressure. Veins contain proportionately less muscle and more connective and elastic tissue than arteries. In capillaries, the delicate walls of flattened cells are thin enough to allow substance exchange between blood and tissues.

BLOOD COMPOSITION

The liquid component of blood, plasma, is 92 percent water, but also contains glucose, minerals, enzymes, hormones, and waste products. Some of these substances are dissolved in the plasma; others are attached to specialized plasma transport proteins. Plasma also contains antibodies. The remainder of blood comprises red blood cells (erythrocytes), white blood cells (leucocytes), and platelets (thrombocytes).

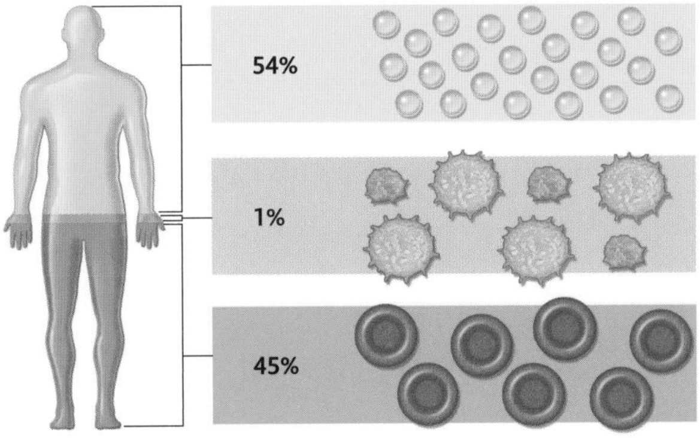

MAINLY WATER
Blood is 54 percent liquid plasma. The remaining 46 percent of blood is made up of solids, in the form of blood cells, suspended in the fluid.

BLOOD CELLS

There are three major types of blood cell: red blood cells, or erythrocytes; white blood cells, or leucocytes; and platelets, or thrombocytes. They are formed in bone marrow, from where they are released into the bloodstream. Erythrocytes, the most numerous blood cells, are oxygen-carriers and have a lifespan of about 120 days. Their flexibility allows them to squeeze through narrow blood vessels. Leucocytes, of which there are various types, are larger than erythrocytes but far less numerous. All leucocytes play important roles in the immune system. Thrombocytes, the smallest blood cells, are fragments of large cells called megakaryocytes and are involved in blood clotting.

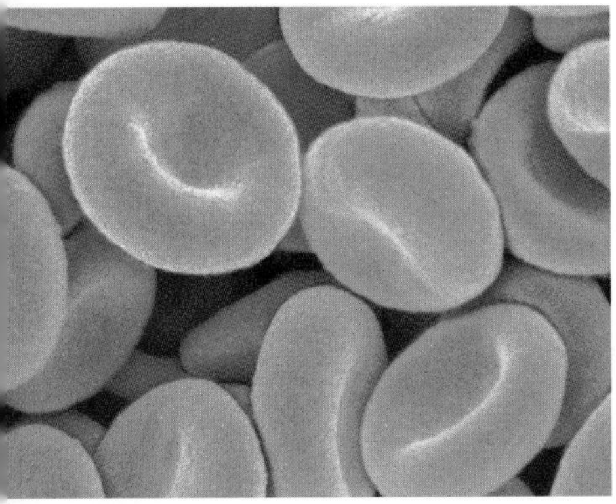

ERYTHROCYTE
Erythrocytes contain hemoglobin, which binds to oxygen and also creates the cells' red pigmentation. The biconcave disk shape of these cells increases their surface area and flexibility.

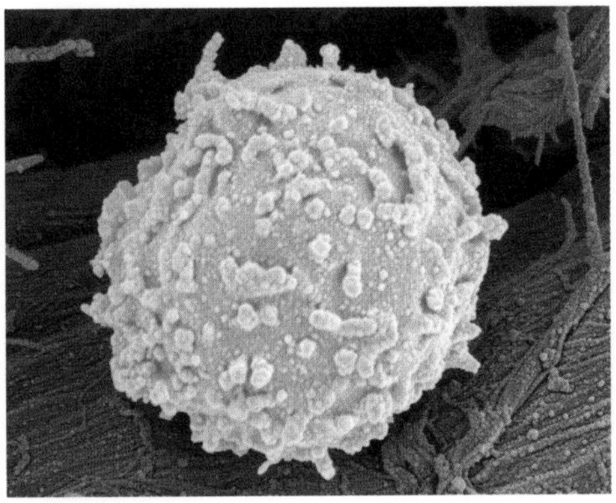

LEUCOCYTE
The different types of leucocyte have specific roles, which include fighting infections, triggering allergic reactions, and destroying foreign bodies. Some leucocytes may have a lifespan of up to ten years.

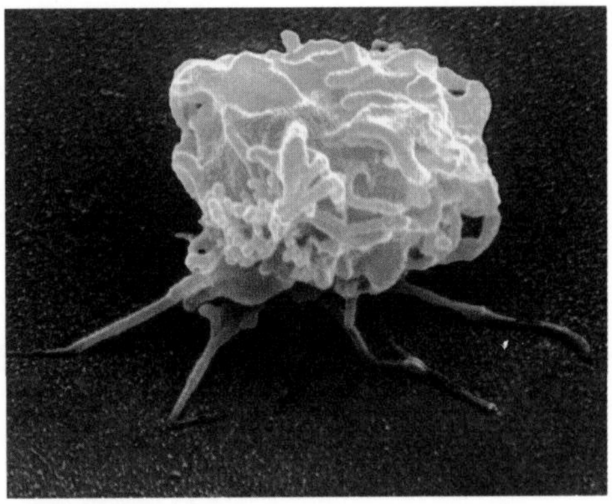

THROMBOCYTE
These cell fragments are important in the blood clotting process, clumping together at injury sites when activated. Thrombocytes lack a nucleus and are short-lived, lasting 8–12 days.

BLOOD TYPES

Blood type is hereditary and is determined by antigens on the surface of red blood cells (erythrocytes). The main antigens are called A and B, and cells can display A antigens (blood group A), B antigens (group B), both together (AB), or none (group O). Antigens are triggers for the immune system. An individual's immune system ignores antigens on his or her own cells but produces antibodies to recognize and help destroy foreign cells that display new antigens.

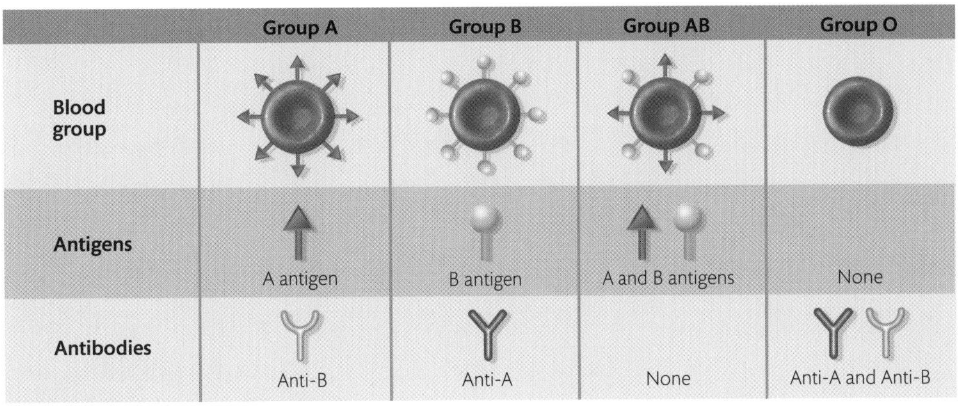

ANTIGENS
There are 30 different antigens that blood cells can display. The ABO antigens, illustrated here, are the most well known.

See also pp. 128, 130 »

HEAD AND NECK 1

THE MAIN VESSELS SUPPLYING OXYGENATED BLOOD TO THE HEAD AND NECK ARE THE COMMON CAROTID AND VERTEBRAL ARTERIES. THE INTERNAL JUGULAR VEIN IS THE MAIN VESSEL THAT DRAINS BLOOD FROM THE HEAD AND NECK.

EXTERNAL ARTERIES OF THE HEAD

Color and/or label the structures indicated on the diagram using the key below. Conventionally, arteries are colored red.

1. Maxillary artery
2. Buccal artery
3. Infraorbital artery
4. Angular artery
5. Superior labial artery
6. Inferior labial artery
7. Mental artery
8. Submental artery
9. Facial artery
10. Superior thyroid artery
11. Common carotid artery
12. Vertebral artery
13. Internal carotid artery
14. External carotid artery
15. Occipital artery
16. Posterior auricular artery
17. Superficial temporal artery

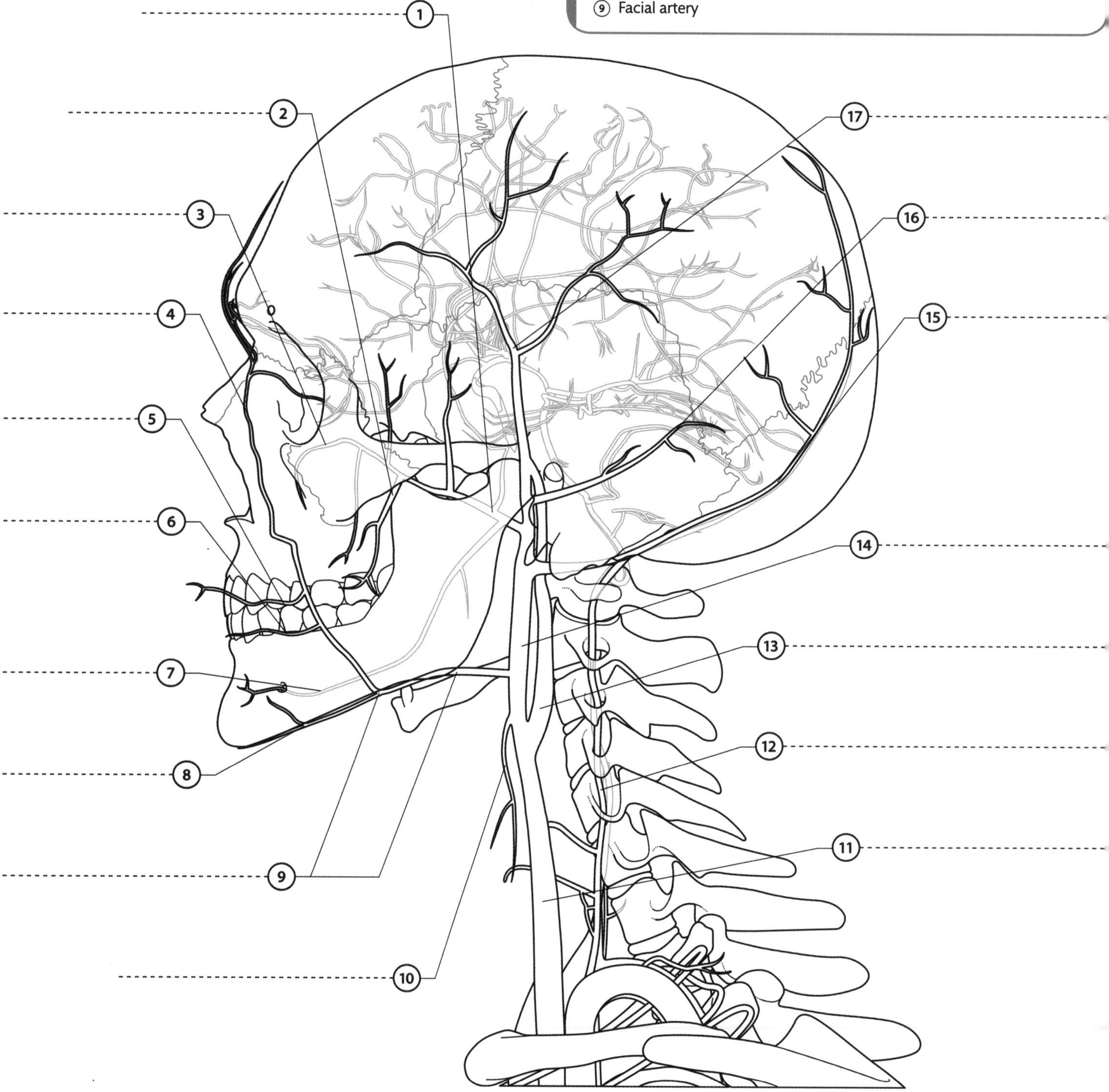

EXTERNAL ARTERIES OF THE HEAD

EXTERNAL VESSELS OF THE HEAD

The common carotid artery runs up into the neck and divides into two—the internal carotid artery supplies the brain, and the external carotid artery gives rise to a profusion of branches, some of which supply the thyroid gland, mouth, tongue, and nasal cavity. The vertebral artery runs up through holes in the cervical vertebrae and eventually enters the skull through the foramen magnum. The veins of the head and neck come together like river tributaries, draining into the large internal jugular vein, behind the sternocleidomastoid muscle, and into the subclavian vein, low in the neck.

EXTERNAL VEINS OF THE HEAD

Color and/or label the structures indicated on the diagram using the key below. Conventionally, veins are colored blue.

1. Superficial temporal vein
2. Infraorbital vein
3. Angular vein
4. Inferior labial vein
5. Superior labial vein
6. Mental vein
7. Submental vein
8. Facial vein
9. Superior thyroid vein
10. Internal jugular vein
11. External jugular vein
12. Retromandibular vein
13. Occipital vein
14. Posterior auricular vein
15. Maxillary vein

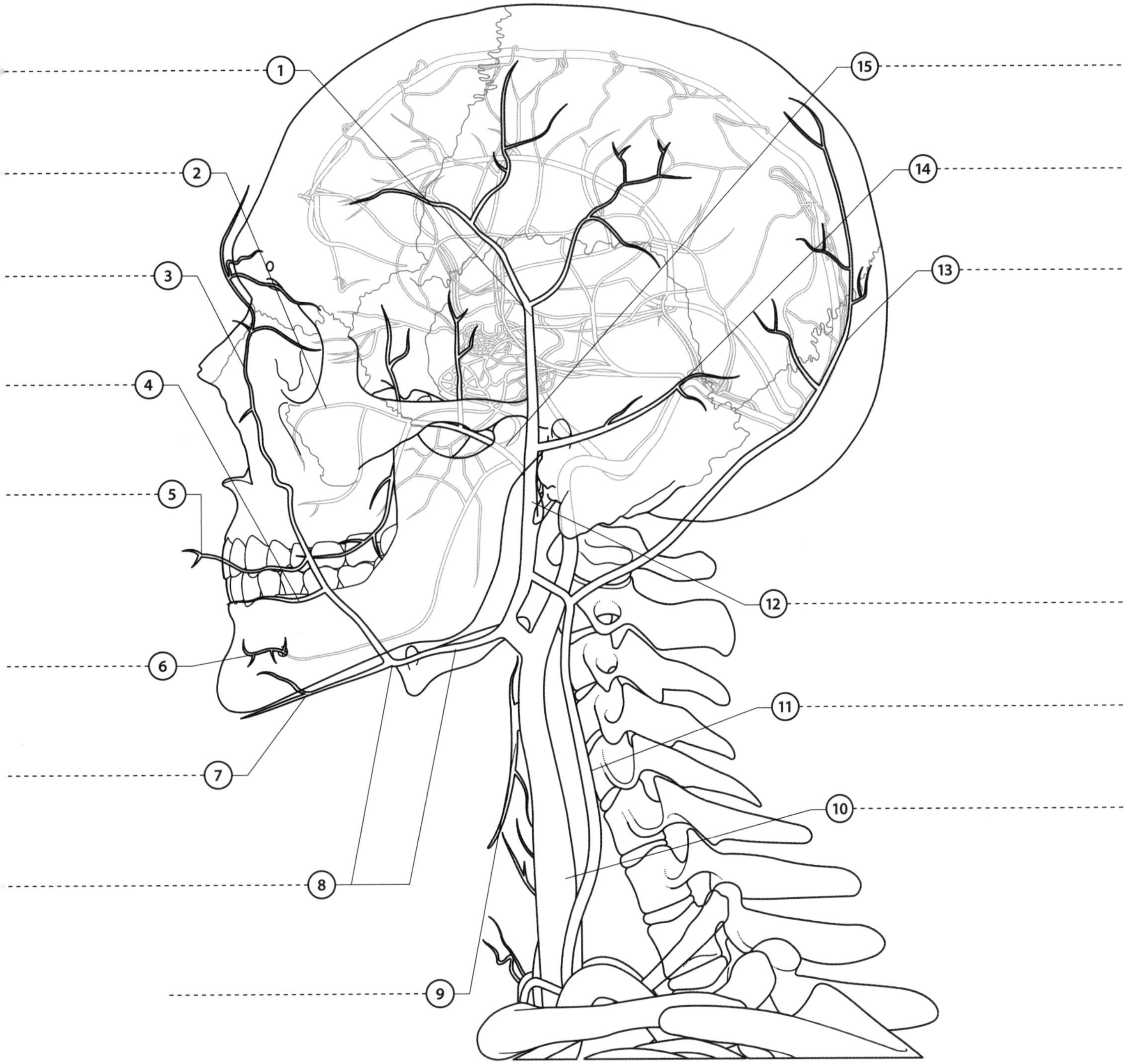

EXTERNAL VEINS OF THE HEAD

See also pp. 28, 136, 160

HEAD AND NECK 2

THE BRAIN HAS A RICH BLOOD SUPPLY, ARRIVING VIA THE VERTEBRAL AND INTERNAL CAROTID ARTERIES.

VESSELS AROUND THE BRAIN

The vertebral arteries join together to form the basilar artery. The internal carotid arteries and basilar artery join up on the undersurface of the brain to form the circle of Willis. From there, three pairs of cerebral arteries make their way into the brain. The veins of the brain and skull drain into the dural venous sinuses.

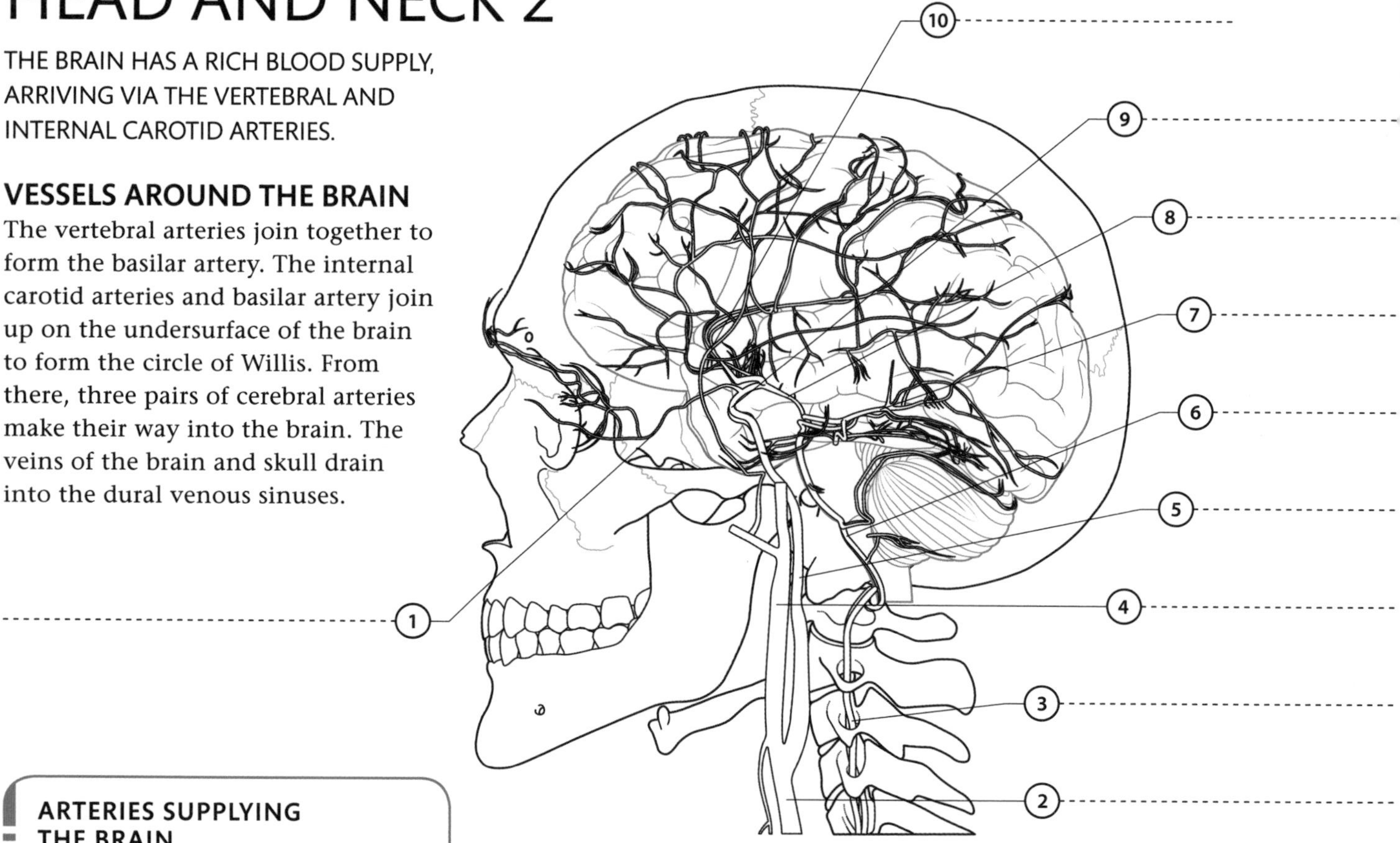

ARTERIES SUPPLYING THE BRAIN

ARTERIES SUPPLYING THE BRAIN

Color and/or label the structures indicated on the diagram using the key below.

1. Ophthalmic artery
2. Common carotid artery
3. Vertebral artery
4. External carotid artery
5. Internal carotid artery
6. Basilar artery
7. Posterior cerebral artery
8. Posterior communicating artery
9. Cavernous part of internal carotid artery
10. Anterior cerebral artery

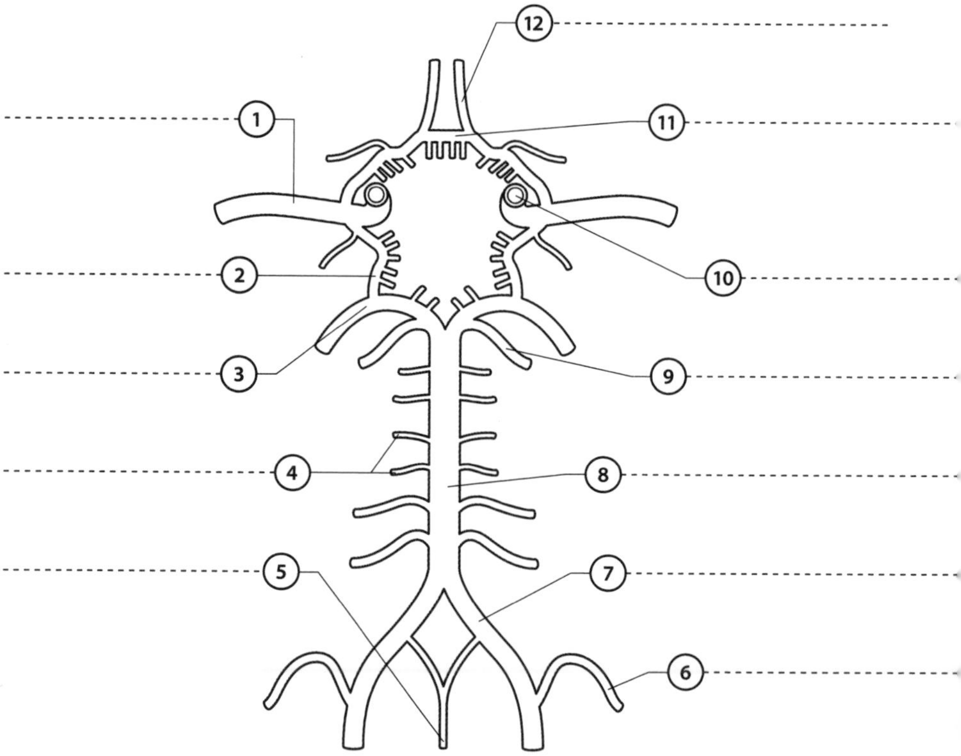

CIRCLE OF WILLIS

CIRCLE OF WILLIS

Color and/or label the structures indicated on the diagram using the key below.

1. Middle cerebral artery
2. Posterior communicating artery
3. Posterior cerebral artery
4. Pontine arteries
5. Anterior spinal artery
6. Posterior inferior cerebellar artery
7. Vertebral artery
8. Basilar artery
9. Superior cerebellar artery
10. Internal carotid artery
11. Anterior communicating artery
12. Anterior cerebral artery

VEINS DRAINING THE BRAIN

Color and/or label the structures indicated on the diagram using the key below.

1. Inferior ophthalmic vein
2. Pterygoid venous plexus
3. Internal jugular vein
4. Sigmoid sinus
5. Straight sinus
6. Great cerebral vein
7. Inferior sagittal sinus
8. Superior sagittal sinus
9. Cavernous sinus
10. Superior ophthalmic vein

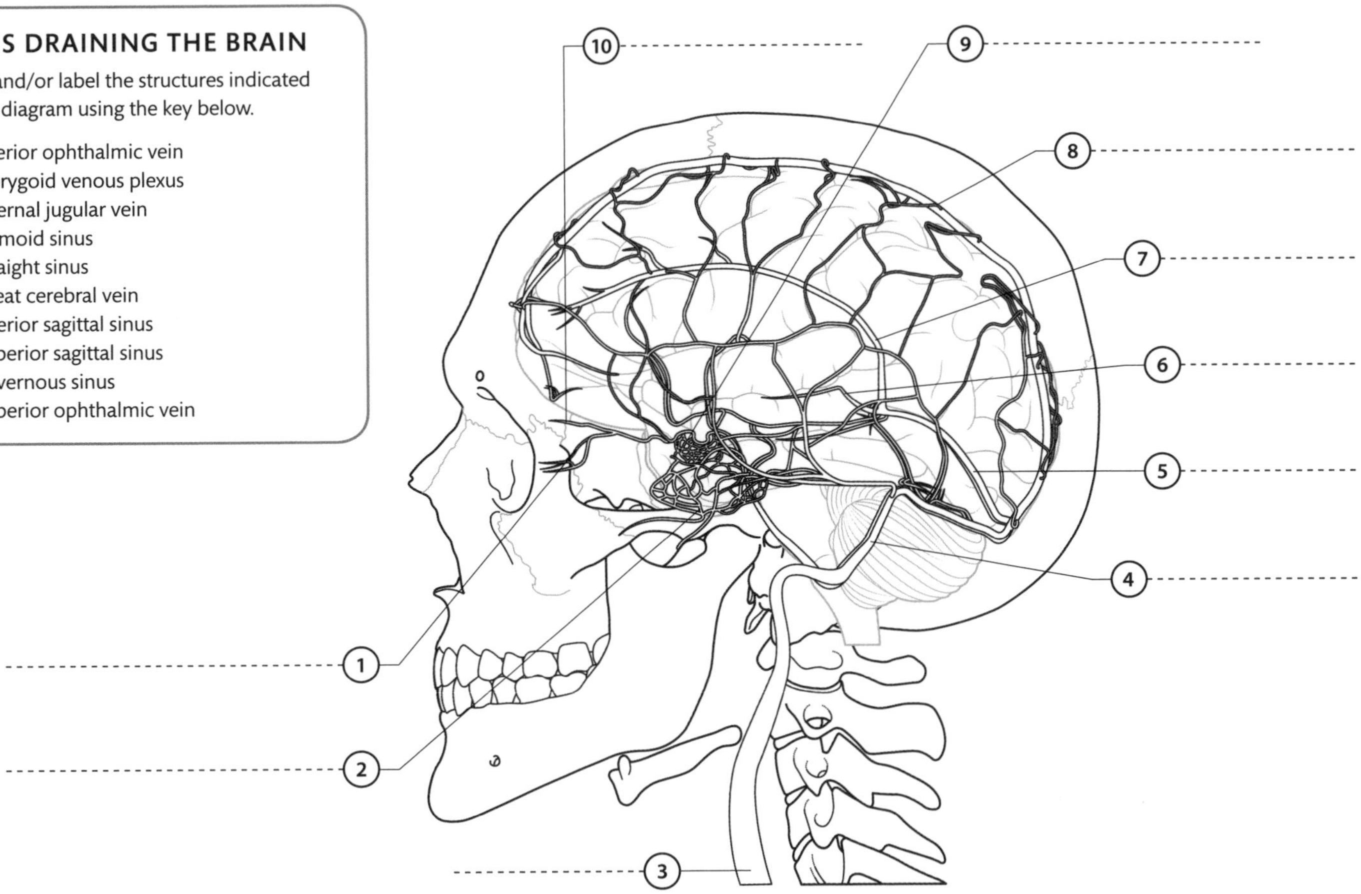

VEINS DRAINING THE BRAIN

DURAL VENOUS SINUSES

There are eight dural venous sinuses, which are enclosed within the dura mater (outermost layer of the meninges), lying in grooves on the inner surface of the skull. The sinuses join up and eventually drain out of the base of the skull, mostly into the internal jugular vein.

DURAL VENOUS SINUSES

Color and/or label the structures indicated on the diagram using the key below.

1. Marginal sinus
2. Sphenoparietal sinus
3. Cavernous sinus
4. Inferior petrosal sinus
5. Superior petrosal sinus
6. Sigmoid sinus
7. Transverse sinus
8. Superior sagittal sinus

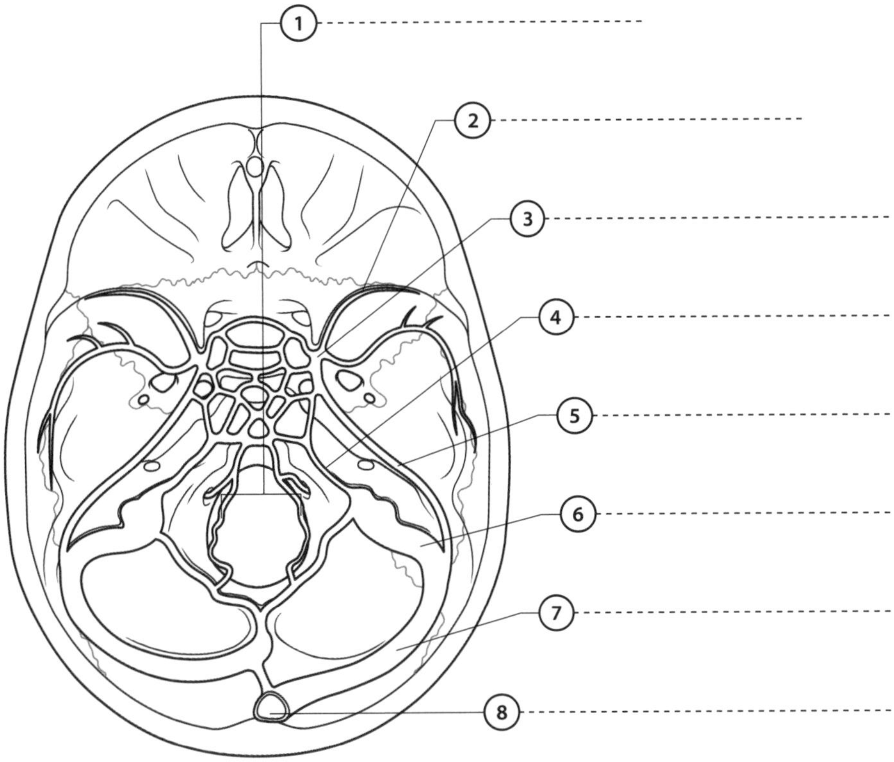

DURAL VENOUS SINUSES

See also pp. 28, 98, 134, 160

THORAX

THE THORAX CONTAINS THE MOST IMPORTANT STRUCTURE IN THE CARDIOVASCULAR SYSTEM, THE HEART, AS WELL AS MAJOR BLOOD VESSELS SUCH AS THE VENAE CAVAE AND AORTA.

ANATOMY

The heart is located centrally in the thorax, but skewed and twisted to the left, so that the frontal view of the heart is formed mainly by the right ventricle, and the apex of the heart usually lies along a line dropped down from the midpoint of the left clavicle. The chest walls, including the skin on the chest, are supplied with blood vessels—intercostal arteries and veins—that run with the nerves in the spaces between the ribs. Intercostal arteries arise from the aorta at the back and from the two internal thoracic arteries at the front (which lie vertically along either side of the sternum, behind the ribs). Intercostal veins drain into similar veins alongside the sternum at the front, and into the large azygos vein at the back, on the right side, and into the hemiazygos vein at the left side.

THORACIC CAVITY (HEART AND LUNGS REMOVED)

Color and/or label the structures indicated on the diagram using the key below.

1. Right common carotid artery
2. Right subclavian artery
3. Trachea
4. Brachiocephalic trunk
5. Right main bronchus
6. Arch of aorta
7. Azygos vein
8. Intercostal neurovascular bundle
9. Descending thoracic aorta
10. Diaphragm
11. Esophagus
12. Sympathetic chain
13. Left subclavian artery
14. Left common carotid artery

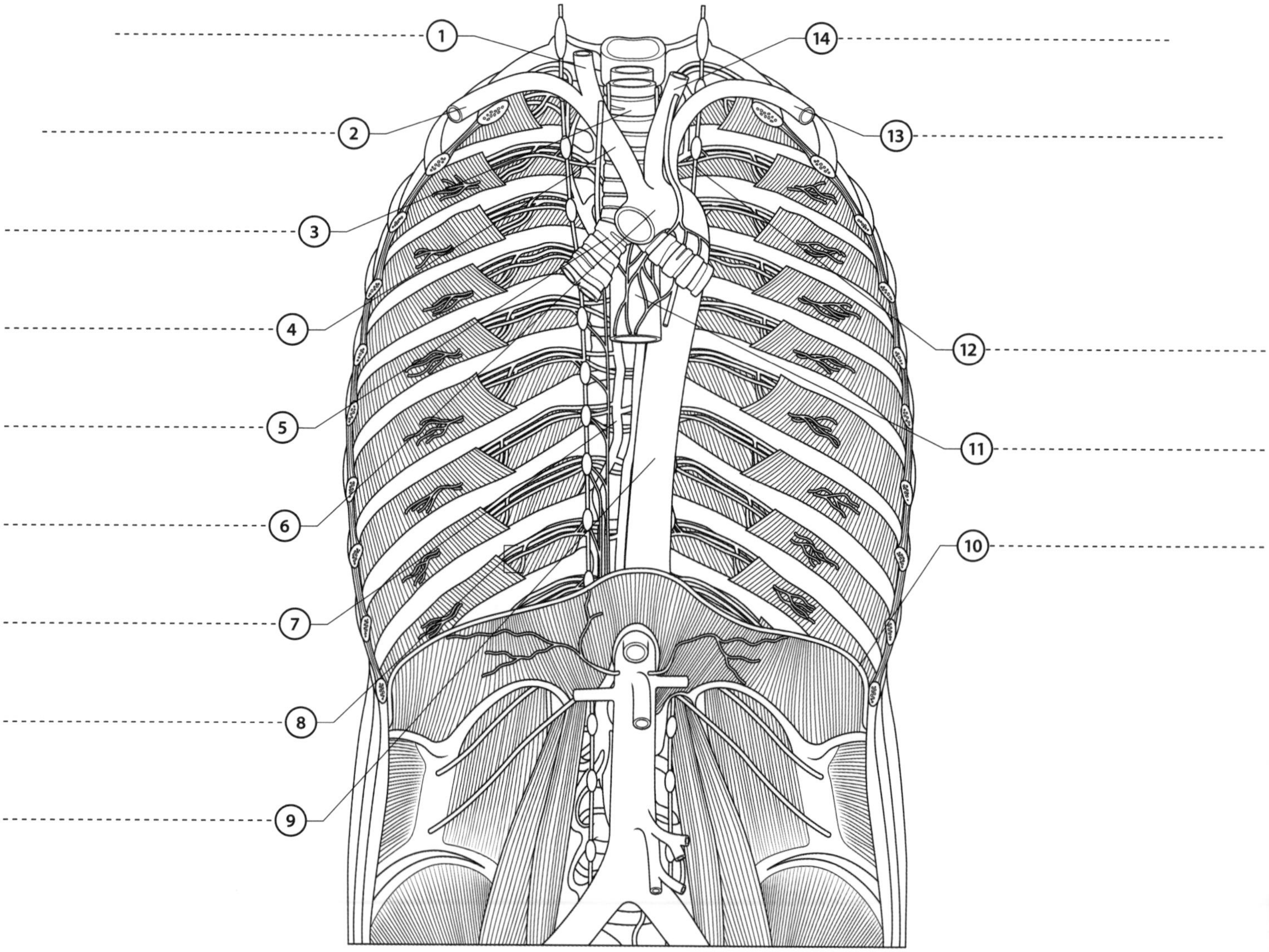

THORACIC CAVITY
(HEART AND LUNGS REMOVED)

THORAX (ANTERIOR VIEW)

Color and/or label the structures indicated on the diagram using the key below.

1. Right common carotid artery
2. Right internal jugular vein
3. Brachiocephalic trunk
4. Right subclavian artery
5. Right subclavian vein
6. Right brachiocephalic vein
7. Superior vena cava
8. Right pulmonary artery
9. Intercostal neurovascular bundle
10. Right atrium
11. Right ventricle
12. Inferior vena cava
13. Pulmonary trunk
14. Ascending aorta
15. Arch of aorta
16. Left pulmonary artery
17. Left brachiocephalic vein
18. Left subclavian vein
19. Left subclavian artery
20. Left internal jugular vein
21. Left common carotid artery

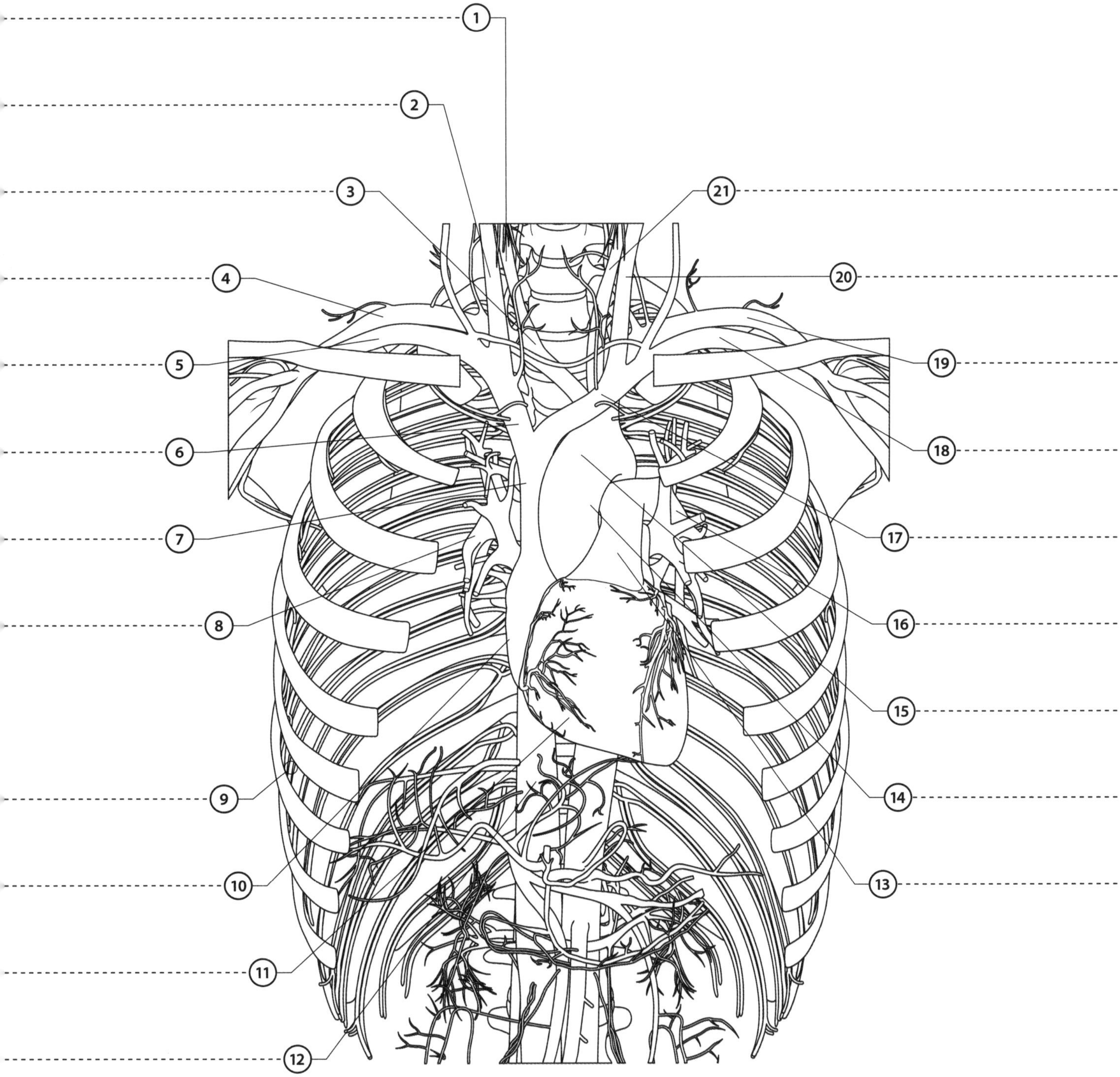

THORAX (ANTERIOR VIEW)

See also pp. 36, 62, 64, 100, 120, 162, 179

HEART 1

LOCATED BEHIND THE STERNUM IN THE THORAX, THE HEART IS A FOUR-CHAMBERED MUSCULAR PUMP ABOUT THE SIZE OF A CLENCHED FIST.

EXTERNAL ANTERIOR ANATOMY

The heart is encased in a two-layered membrane called the pericardium. Running over the heart's surface are nerves and the coronary arteries and their branches, which send smaller blood vessels into the heart muscle.

ANTERIOR EXTERNAL VIEW

Color and/or label the structures indicated on the diagram using the key below.

1. Right vagus nerve
2. Right phrenic nerve
3. Superior vena cava
4. Right pulmonary artery
5. Right auricle
6. Right coronary artery
7. Muscular wall of ventricles
8. Marginal artery
9. Pericardium (cut)
10. Anterior interventricular artery
11. Left auricle
12. Pulmonary trunk
13. Left pulmonary artery
14. Cut edge of pericardium
15. Left recurrent laryngeal nerve
16. Arch of aorta
17. Left phrenic nerve
18. Left vagus nerve

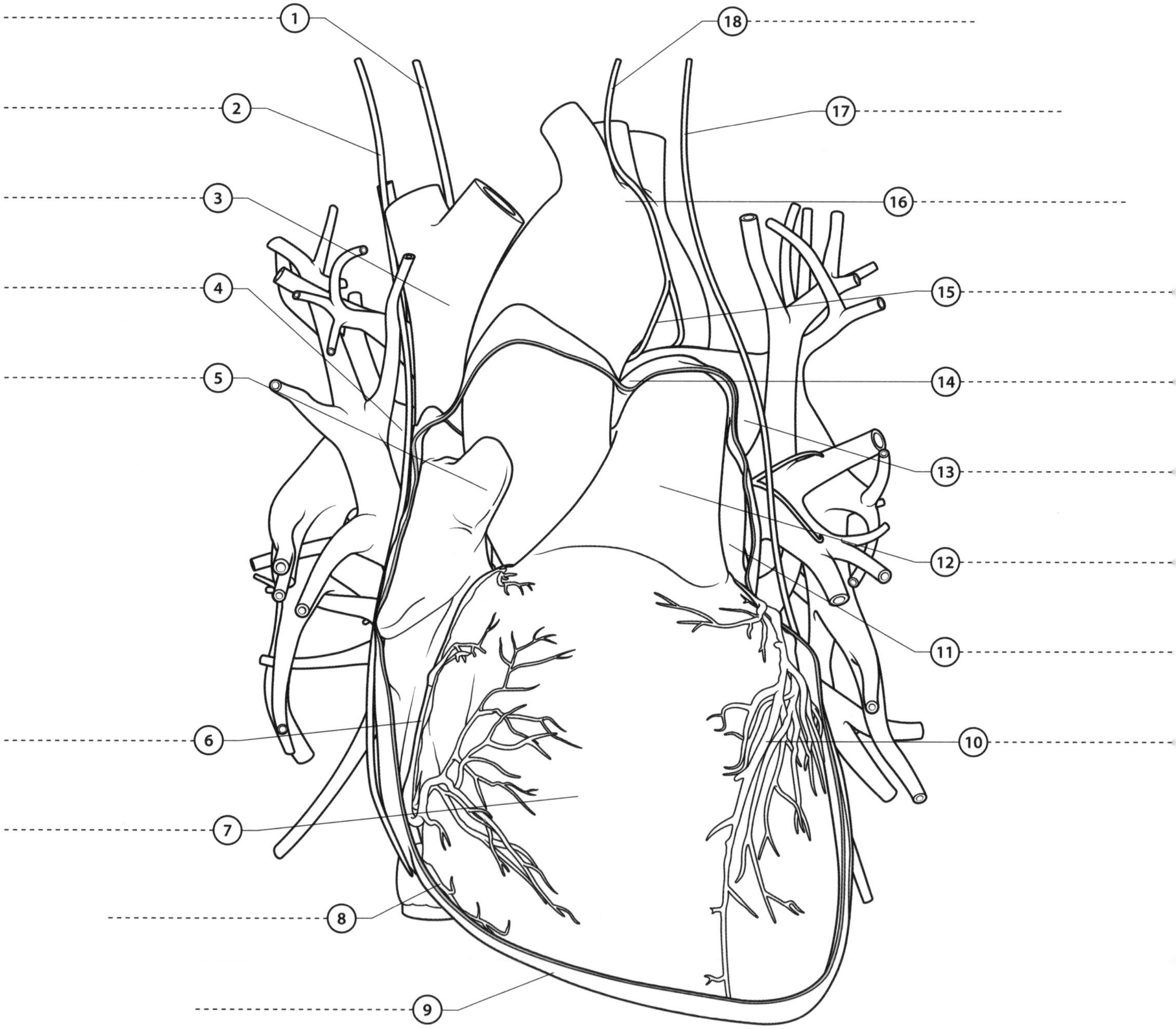

ANTERIOR EXTERNAL VIEW

INTERNAL ANATOMY

The heart is divided into two halves by a septum, each half comprising an upper chamber (atrium) and a lower chamber (ventricle). The right atrium receives deoxygenated blood from the body via the inferior and superior venae cavae, and pumps the blood through the tricuspid valve into the right ventricle. This then pumps the blood through the pulmonary valve into the pulmonary trunk, then into the lungs. Oxygenated blood from the lungs passes via the pulmonary veins into the left atrium, then through the mitral valve into the left ventricle to be pumped out through the aortic valve into the aorta and then on to the body.

ANTERIOR INTERNAL VIEW

Color and/or label the structures indicated on the diagram using the key below.

1. Superior vena cava
2. Pulmonary trunk
3. Pulmonary valve
4. Right pulmonary veins
5. Right atrium
6. Tricuspid valve
7. Chordae tendineae
8. Right ventricle
9. Inferior vena cava
10. Descending aorta
11. Pericardium
12. Left ventricle
13. Papillary muscle
14. Septum
15. Mitral valve
16. Aortic valve
17. Left pulmonary veins
18. Left atrium
19. Arch of aorta

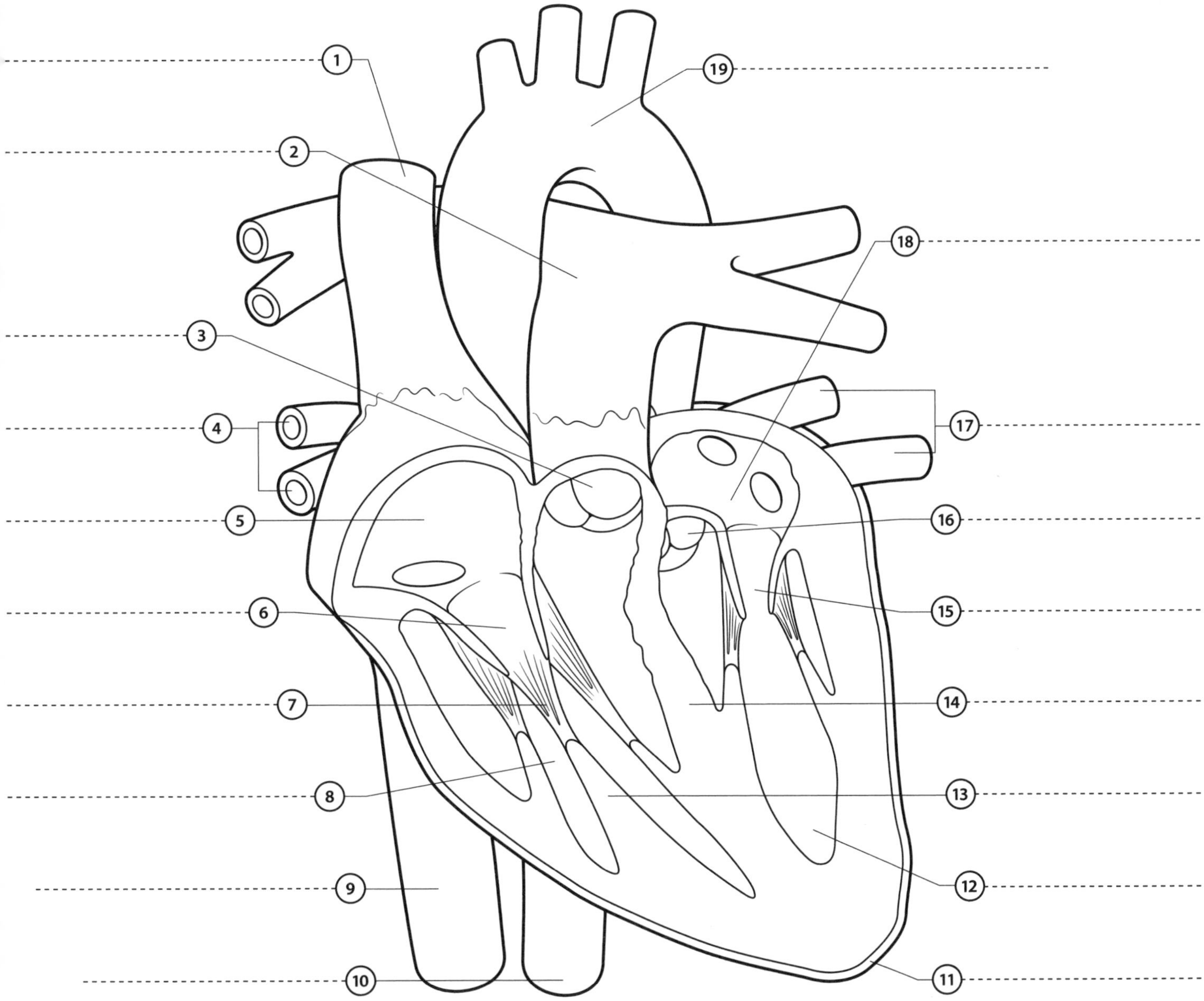

ANTERIOR INTERNAL VIEW

See also pp. 138, 142

HEART 2

THE HEART IS USUALLY SHOWN FROM THE ANTERIOR (SEE PREVIOUS PAGE), BUT THE POSTERIOR AND LATERAL VIEWS SHOWN HERE REVEAL THE FULL COMPLEXITY OF THE VESSELS AROUND IT.

EXTERNAL POSTERIOR AND LATERAL ANATOMY

The position of the pulmonary arteries and veins is particularly clearly seen in the posterior and lateral views of the heart. The posterior view also shows the coronary sinus, a large vein that receives many of the cardiac veins and empties into the right atrium. The right lateral view reveals the position of the superior and inferior venae cavae in relation to the right atrium.

EXTERNAL POSTERIOR VIEW

Color and/or label the structures indicated on the diagram using the key below.

1. Left subclavian artery
2. Left pulmonary artery
3. Left pulmonary veins
4. Left atrium
5. Coronary sinus
6. Left ventricle
7. Posterior interventricular artery
8. Right ventricle
9. Inferior vena cava
10. Middle cardiac vein
11. Right coronary artery
12. Right atrium
13. Right pulmonary veins
14. Right pulmonary arteries
15. Superior vena cava
16. Arch of aorta
17. Brachiocephalic trunk
18. Left common carotid artery

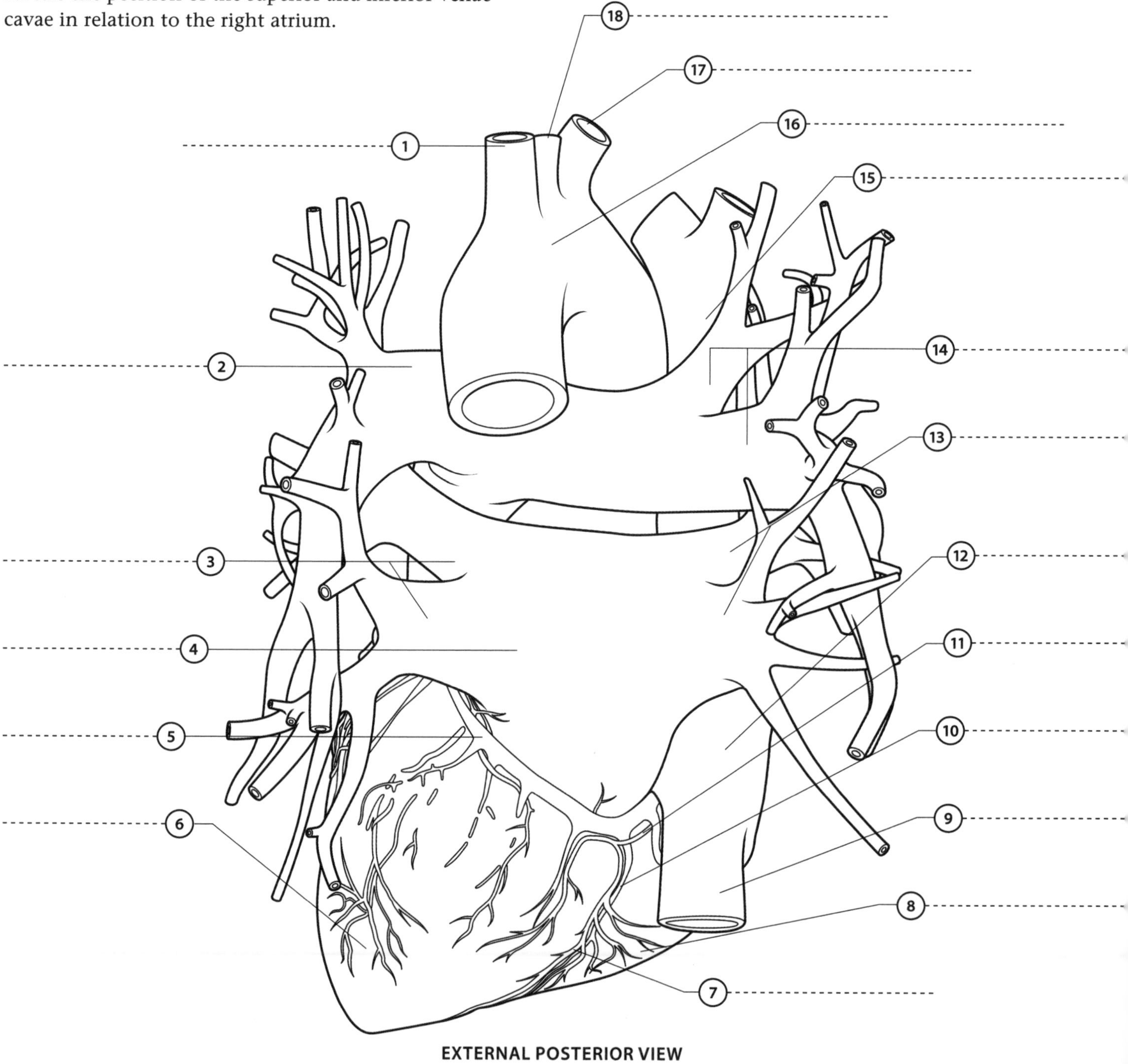

EXTERNAL POSTERIOR VIEW

EXTERNAL LEFT LATERAL VIEW

Color and/or label the structures indicated on the diagram using the key below.

1. Left auricle
2. Anterior interventricular artery
3. Left ventricle
4. Apex
5. Coronary sinus
6. Left atrium
7. Circumflex artery
8. Left pulmonary veins
9. Left pulmonary artery
10. Arch of aorta

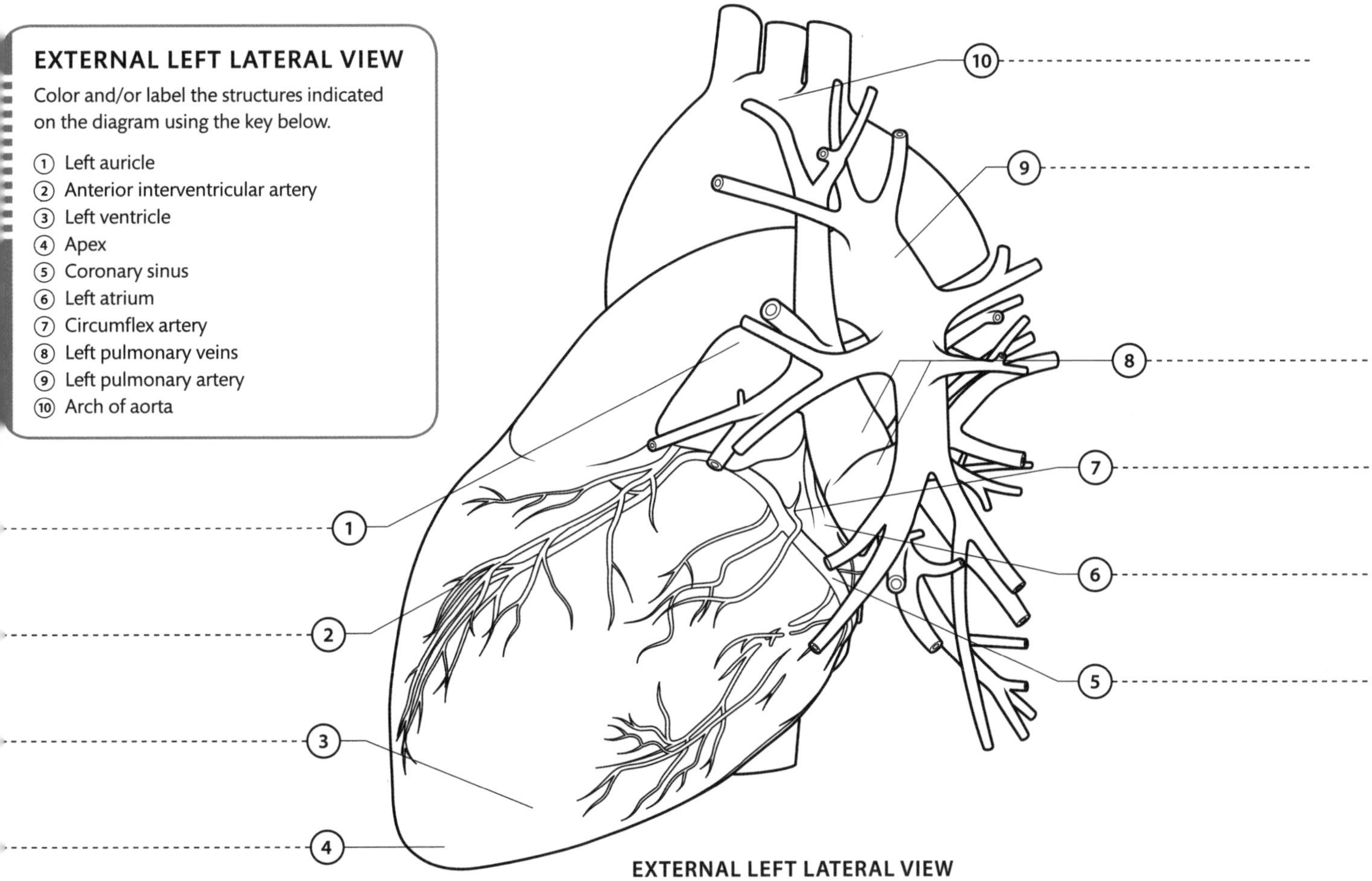

EXTERNAL LEFT LATERAL VIEW

EXTERNAL RIGHT LATERAL VIEW

Color and/or label the structures indicated on the diagram using the key below.

1. Superior vena cava
2. Arch of aorta
3. Right pulmonary veins
4. Coronary sinus
5. Inferior vena cava
6. Small cardiac vein
7. Right ventricle
8. Right coronary artery
9. Right atrium

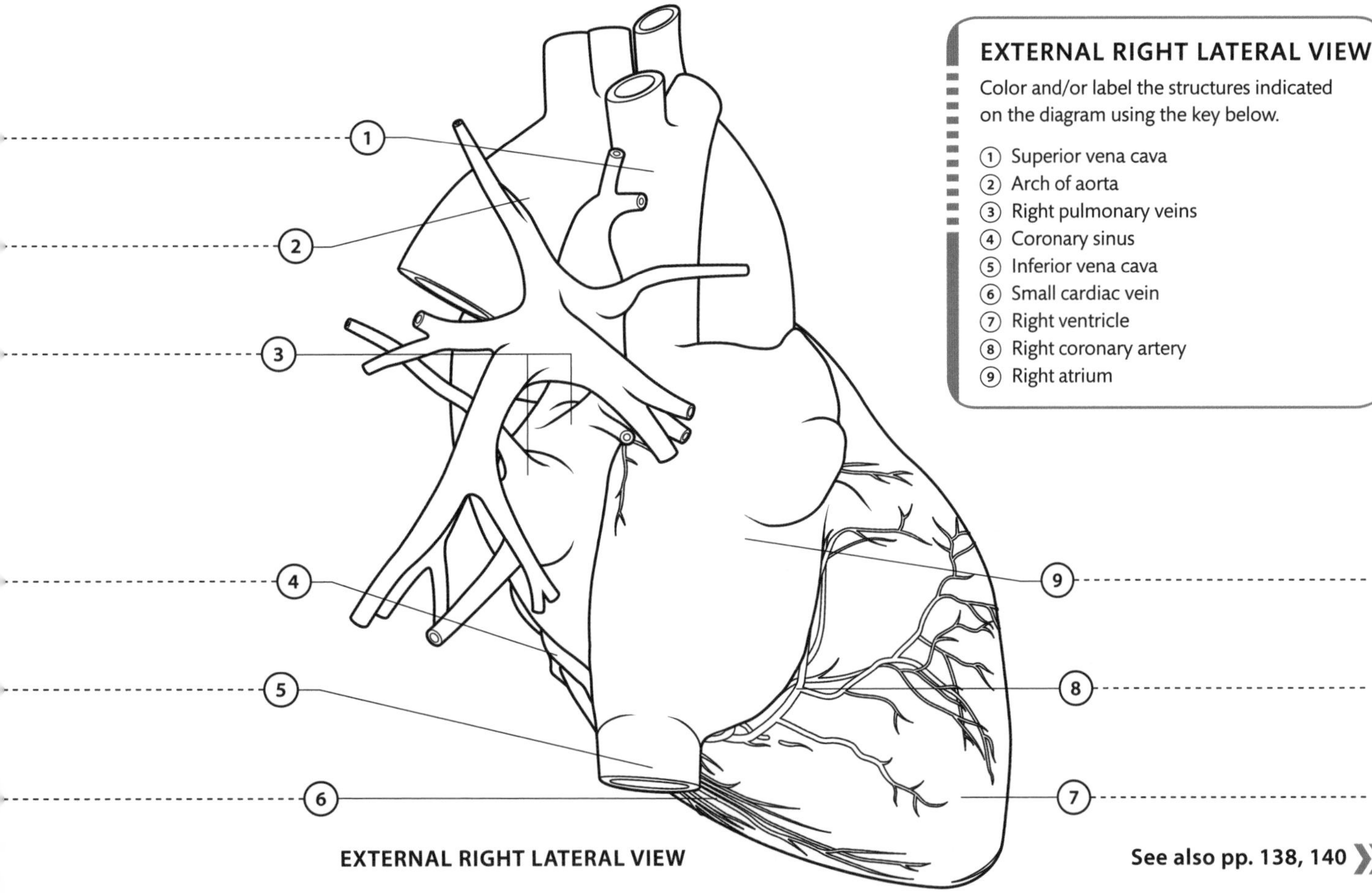

EXTERNAL RIGHT LATERAL VIEW

See also pp. 138, 140

ABDOMEN AND PELVIS

THE BODY'S TWO MAIN BLOOD VESSELS—THE AORTA AND INFERIOR VENA CAVA—RUN THROUGH THE CENTER OF THE ABDOMEN, SENDING BRANCHES TO THE ORGANS AND DIVIDING IN THE PELVIS TO FORM THE MAIN BLOOD VESSELS OF THE LEGS.

ANATOMY

Pairs of arteries branch from the sides of the aorta to supply the walls of the abdomen, kidneys, suprarenal glands, and the testes or ovaries. Branches from the front of the aorta supply the abdominal organs: the celiac trunk gives branches to the liver, stomach, pancreas, and spleen, and the mesenteric arteries supply the gut. The abdominal aorta ends by splitting in two, forming the common iliac arteries. Each of these then divides, forming an internal iliac artery (which supplies the pelvic organs) and an external iliac artery (which continues into the thigh, becoming the femoral artery). To the right of the aorta is the major abdominal vein: the inferior vena cava, which receives blood from parts of the body below the diaphragm.

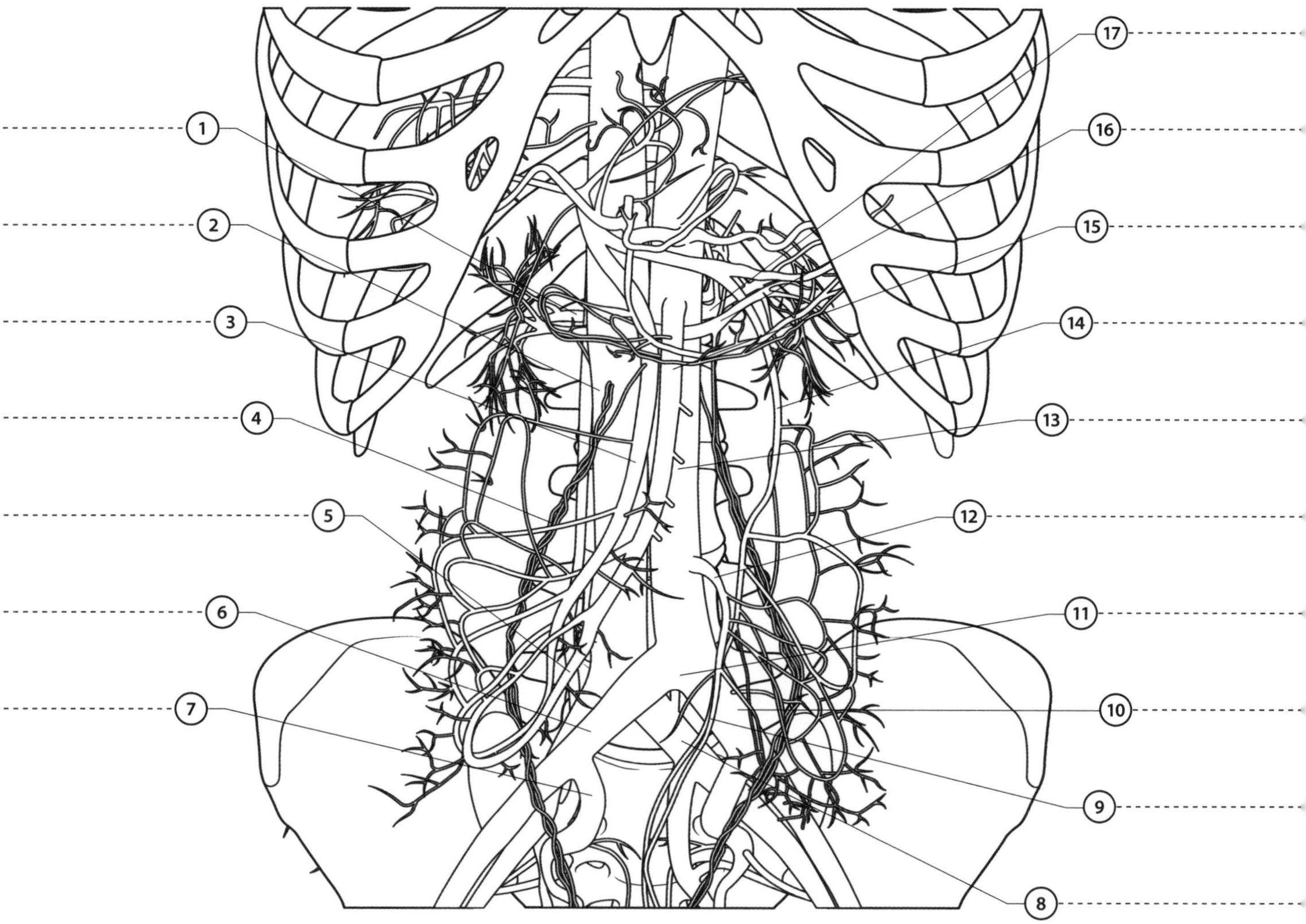

ANTERIOR VIEW

ANTERIOR VIEW

Color and/or label the structures indicated on the diagram using the key below.

1. Right renal artery
2. Inferior vena cava
3. Superior mesenteric vein
4. Right gonadal vessels
5. Iliocolic artery
6. Right common iliac artery
7. Right internal iliac artery
8. Left common iliac vein
9. Superior rectal artery
10. Left common iliac artery
11. Bifurcation of aorta
12. Inferior mesenteric artery
13. Abdominal aorta
14. Inferior mesenteric vein
15. Superior mesenteric artery
16. Left renal vein
17. Left renal artery

PELVIC BLOOD VESSELS

Branches of the internal iliac artery supply pelvic bones, muscles, and viscera. Gluteal and internal pudendal arteries leave the pelvis through the greater sciatic foramen. The internal pudendal artery re-enters the pelvis via the lesser sciatic foramen, and supplies the rectum, scrotum (or labia), perineum, bulb of the penis (or clitoris), and urethra before dividing into the deep and dorsal arteries of the penis (or clitoris). Other branches of the internal iliac arteries supply the bladder, uterus or vas deferens, vagina, and rectum. The internal iliac vein lies behind its artery; its tributaries generally correspond with the branches of the artery.

ANTEROLATERAL VIEW OF PELVIC VESSELS

Color and/or label the structures indicated on the diagram using the key below.

1. Internal iliac vein
2. Internal iliac artery
3. External iliac artery
4. External iliac vein
5. Superior gluteal artery
6. Prevesical veins
7. Obturator artery
8. Vesical artery
9. Middle rectal artery
10. Umbilical artery
11. Femoral artery
12. Femoral vein
13. Pudendal artery
14. Inferior gluteal artery
15. Presacral veins
16. Lateral sacral artery
17. Superior rectal artery
18. Iliolumbar artery
19. Inferior mesenteric artery

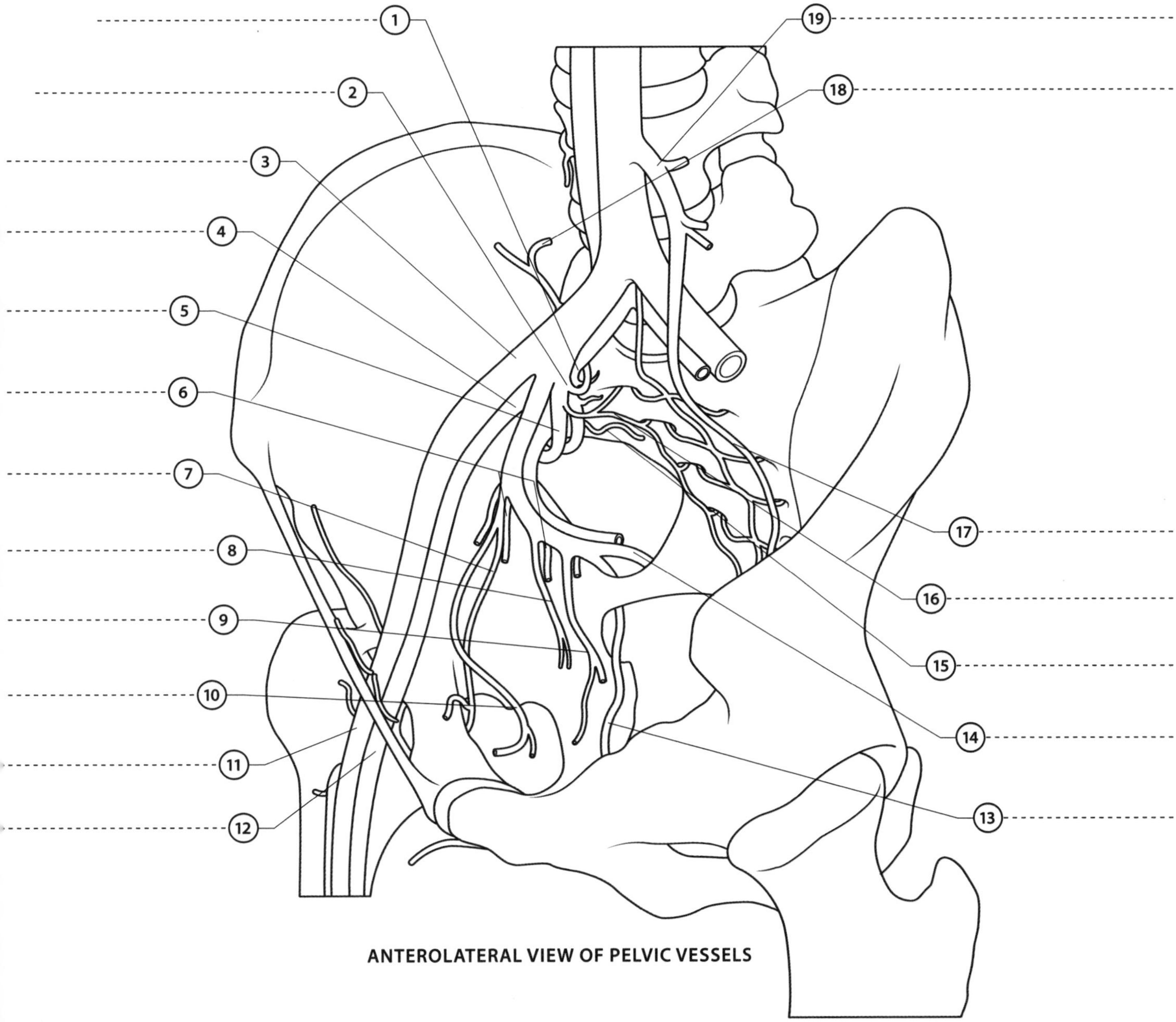

ANTEROLATERAL VIEW OF PELVIC VESSELS

See also pp. 38, 66, 100, 162, 178, 204, 218, 224 »

SHOULDER AND UPPER ARM

THE MAIN BLOOD VESSEL SUPPLYING THE UPPER LIMB IS THE SUBCLAVIAN ARTERY. THE MAIN VESSEL DRAINING THE UPPER LIMB IS THE AXILLARY VEIN.

ANTERIOR

The subclavian artery passes under the clavicle and into the axilla, where it becomes the axillary artery. Several branches spring off in this region, running backward toward the scapula, up to the shoulder, and around the humerus. Leaving the axilla, the axillary artery changes its name to the brachial artery, which runs down the front of the arm, usually accompanied by a pair of companion veins. Two superficial veins that drain blood from the back of the hand end in the arm by draining into deep veins: the basilic vein drains into the brachial veins; the cephalic vein runs up to the shoulder, then plunges deeper to join the axillary vein.

ANTERIOR VIEW

Color and/or label the structures indicated on the diagram using the key below.

1. Subclavian artery
2. Axillary vein
3. Anterior circumflex humeral artery
4. Subscapular artery
5. Cephalic vein
6. Basilic vein
7. Brachial artery
8. Deep brachial artery
9. Radial recurrent artery
10. Median cubital vein
11. Inferior ulnar collateral artery
12. Brachial veins
13. Superior ulnar collateral artery

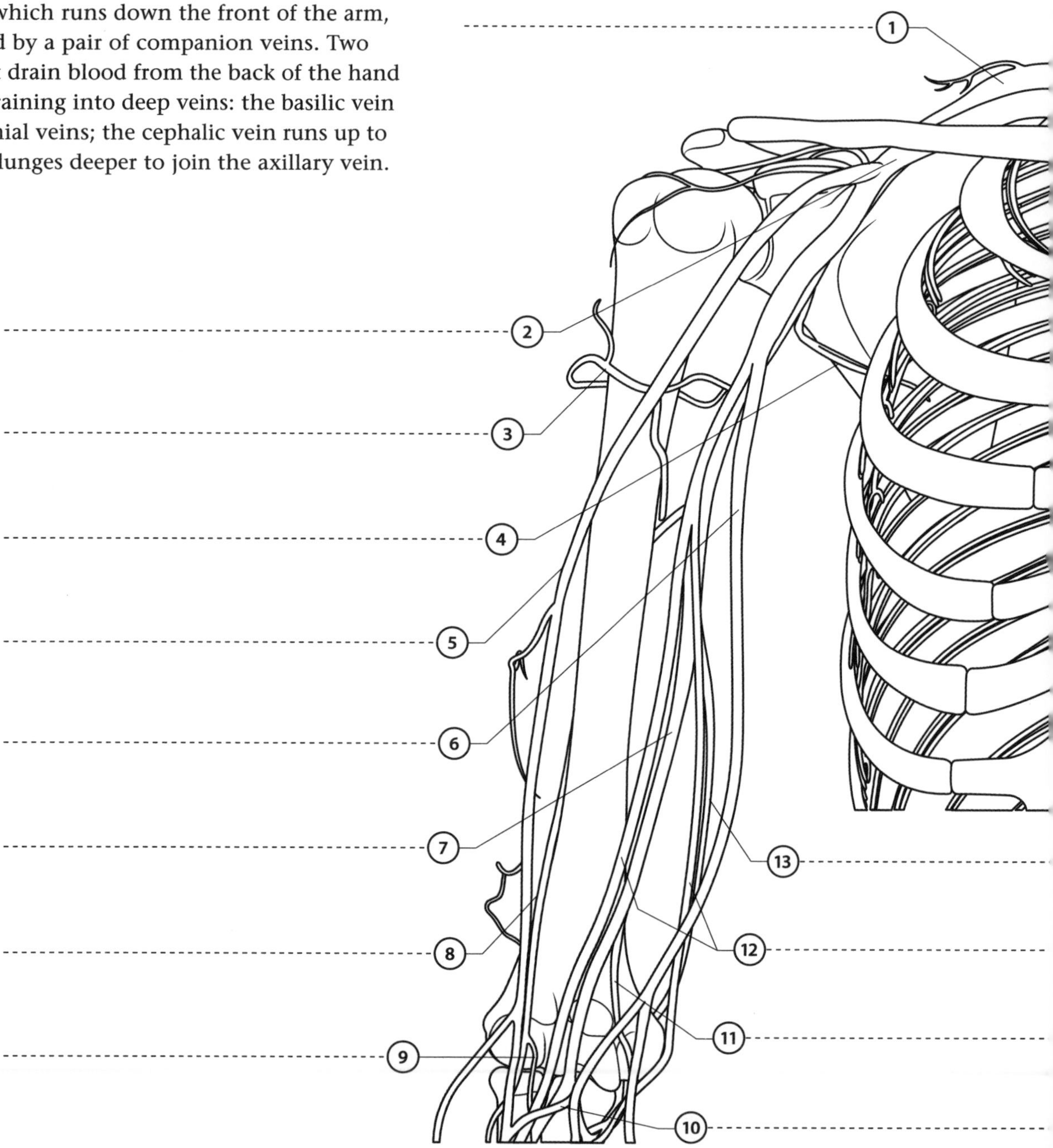

ANTERIOR VIEW

'OSTERIOR

'arious branches from the axillary and brachial arteries upply the back of the shoulder and upper arm. The osterior circumflex humeral artery, which runs with the xillary nerve, curls around the upper end of the humerus. 'he deep brachial artery runs with the radial nerve, spiraling round the back of the bone. From this artery, and from the rachial artery itself, collateral branches run down the arm nd anastomose with recurrent branches running back up rom the radial and ulnar arteries of the forearm around he elbow. Branches of the subclavian and axillary arteries nastomose around the shoulder.

POSTERIOR VIEW

Color and/or label the structures indicated on the diagram using the key below.

1. Basilic vein
2. Superior ulnar collateral artery
3. Ulnar recurrent artery
4. Radial collateral artery
5. Brachial veins
6. Deep brachial artery
7. Cephalic vein
8. Posterior circumflex humeral artery
9. Brachial artery
10. Subscapular artery
11. Axillary artery

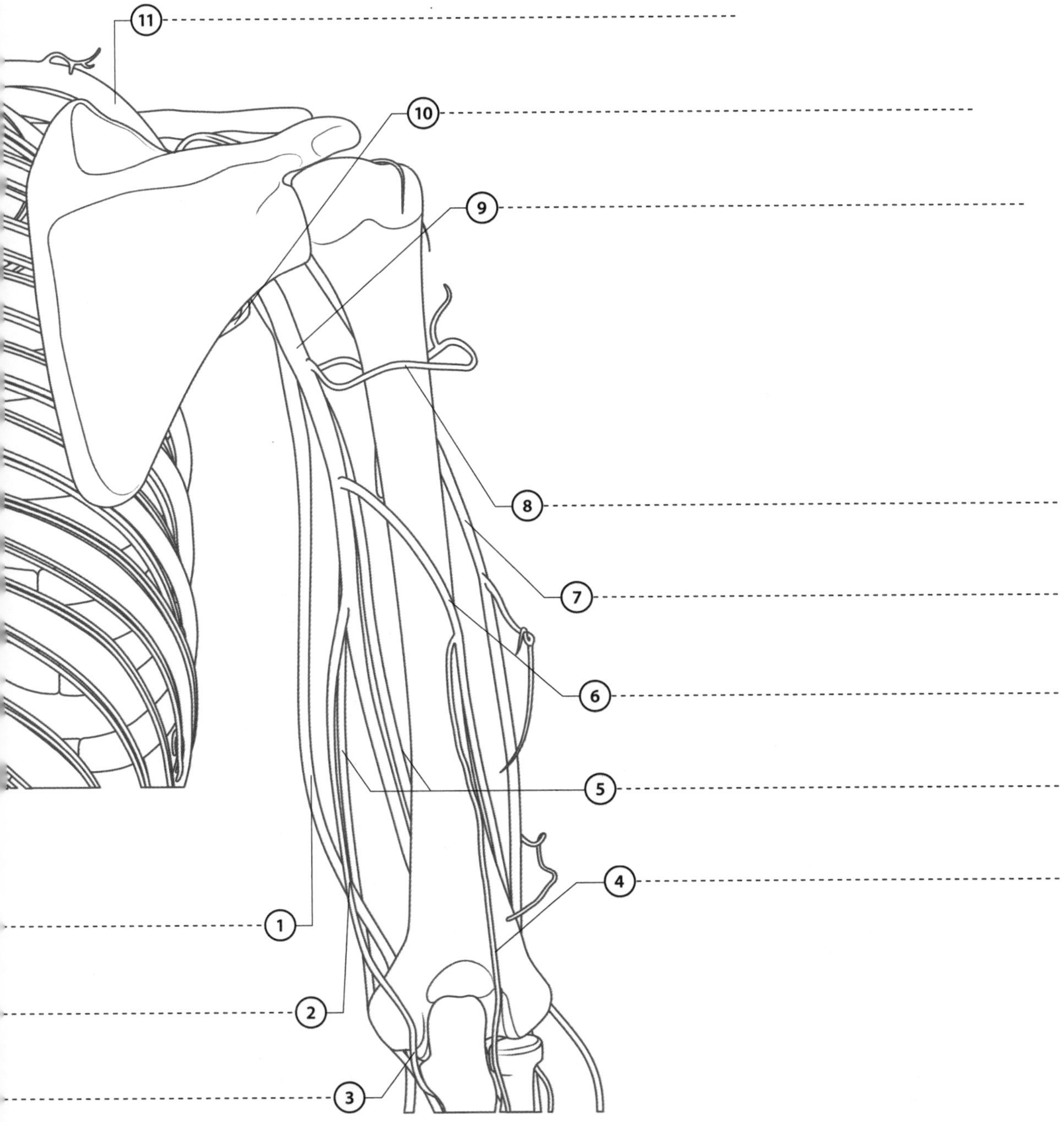

POSTERIOR VIEW

See also pp. 40, 70, 72, 102, 128, 148, 164 »

LOWER ARM AND HAND

THE MAIN VESSELS SUPPLYING BLOOD TO THE LOWER ARM ARE THE ULNAR AND RADIAL ARTERIES. THE MAIN VESSELS DRAINING BLOOD FROM THE LOWER ARM ARE THE CEPHALIC AND BASILIC VEINS.

ANTERIOR

The brachial artery divides into two arteries that take their names from the bones of the forearm: the radial and ulnar arteries. The radial artery can be felt at the wrist, and this is a common place for taking the pulse as the pulsations are easy to feel when the artery is pressed against the bone beneath it. The radial and ulnar arteries end by joining up to form arterial arches in the wrist and palm. Digital arteries, destined for the fingers, spring off from the palmar arches.

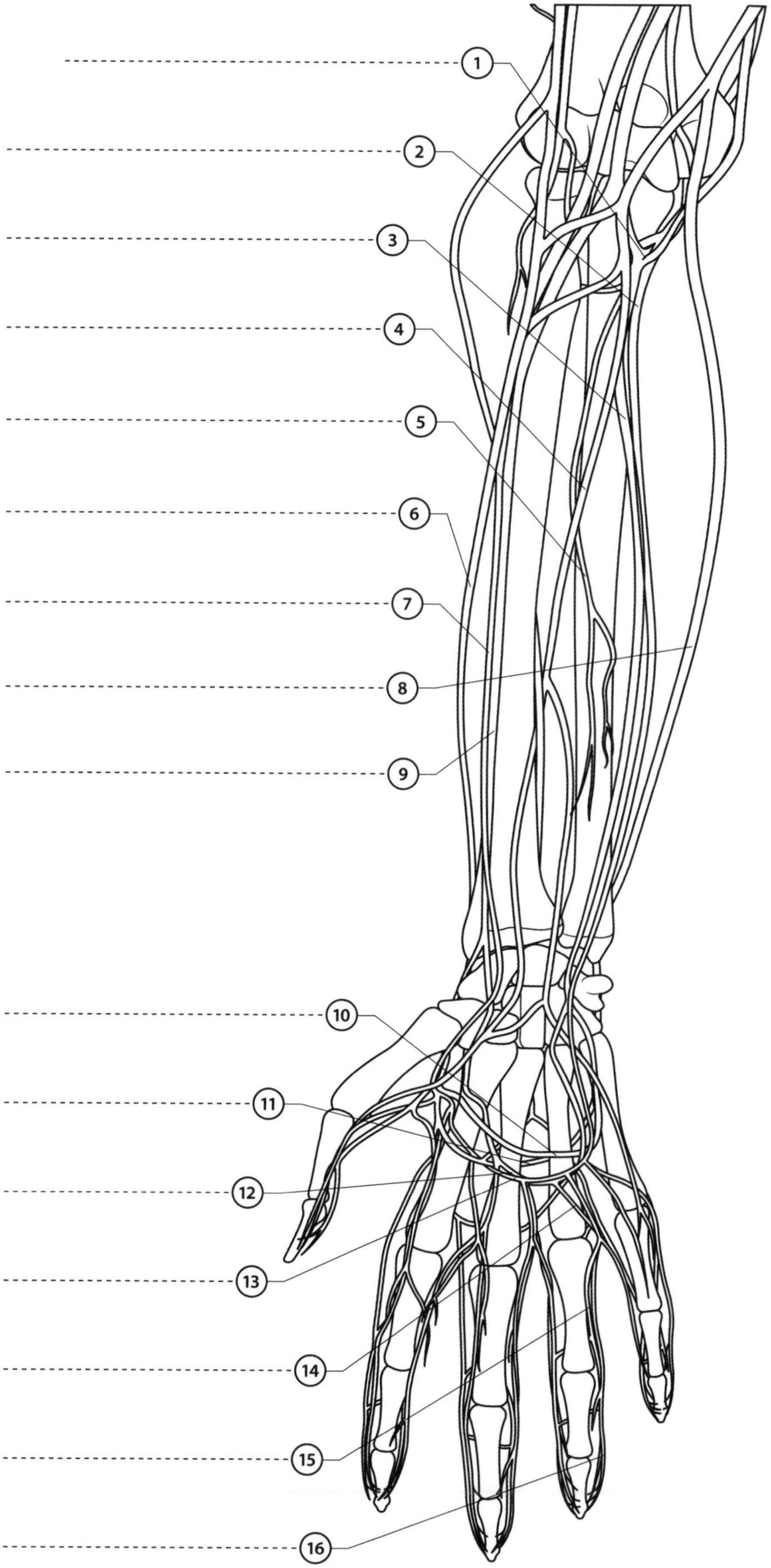

ANTERIOR VIEW

ANTERIOR VIEW

Color and/or label the structures indicated on the diagram using the key below.

1. Ulnar recurrent artery
2. Ulnar vein
3. Ulnar artery
4. Median vein of the forearm
5. Interosseus artery
6. Cephalic vein
7. Radial vein
8. Basilic vein
9. Radial artery
10. Deep palmar arch
11. Deep palmar venous arch
12. Superficial palmar arch
13. Superficial palmar venous arch
14. Common palmar digital artery
15. Palmar digital vein
16. Palmar digital artery

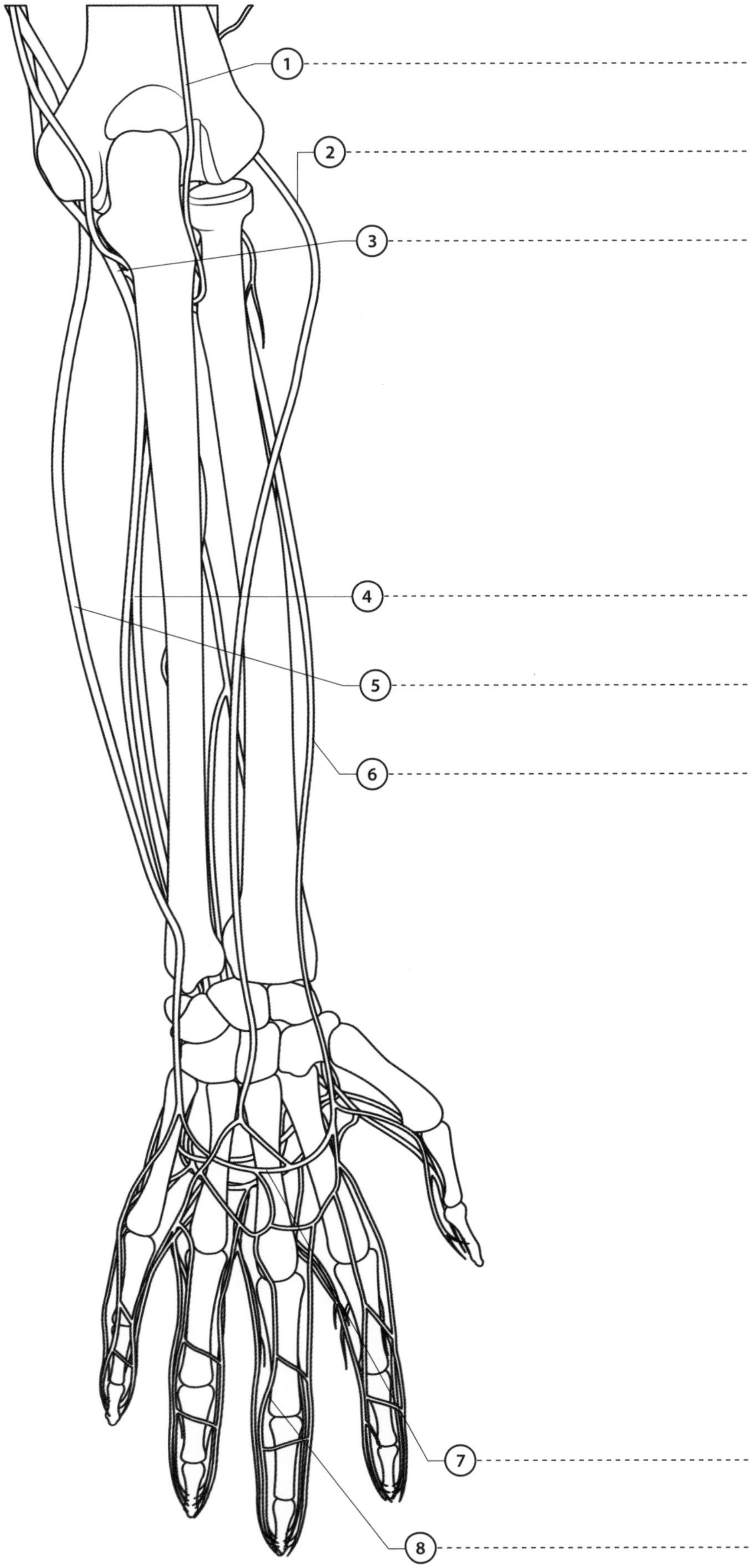

POSTERIOR VIEW

POSTERIOR

Concentrated on the back of the hand is a network of superficial veins. They are situated here, rather than on the palm, because otherwise they would be compressed every time something was gripped. The dorsal venous network of the hand drains into two main vessels: the basilic and cephalic veins.

POSTERIOR VIEW

Color and/or label the structures indicated on the diagram using the key below.

1. Middle collateral artery
2. Accessory cephalic vein
3. Ulnar vein
4. Ulnar artery
5. Basilic vein
6. Cephalic vein
7. Dorsal venous network
8. Dorsal digital vein

See also pp. 40, 70, 72, 102, 128, 164 »

HIP AND THIGH

THE MAIN BLOOD VESSEL SUPPLYING THE LOWER LIMB IS THE FEMORAL ARTERY. THE MAIN VESSEL DRAINING THE LOWER LIMB IS THE FEMORAL VEIN.

ANTERIOR

As the external iliac artery runs over the pubic bone and underneath the inguinal ligament, it changes its name to the femoral artery—the main vessel carrying blood to the lower limb. The femoral artery gives off a large branch (the deep femoral artery) that supplies the muscles of the thigh. The femoral artery then runs toward the inner thigh, passing through a hole in the adductor magnus tendon and changing its name to the popliteal artery. Deep veins run with the arteries but—just as in the arm—there are also superficial veins. The great (or long) saphenous vein drains up the inner side of the leg and thigh, and ends by joining the femoral vein near the hip.

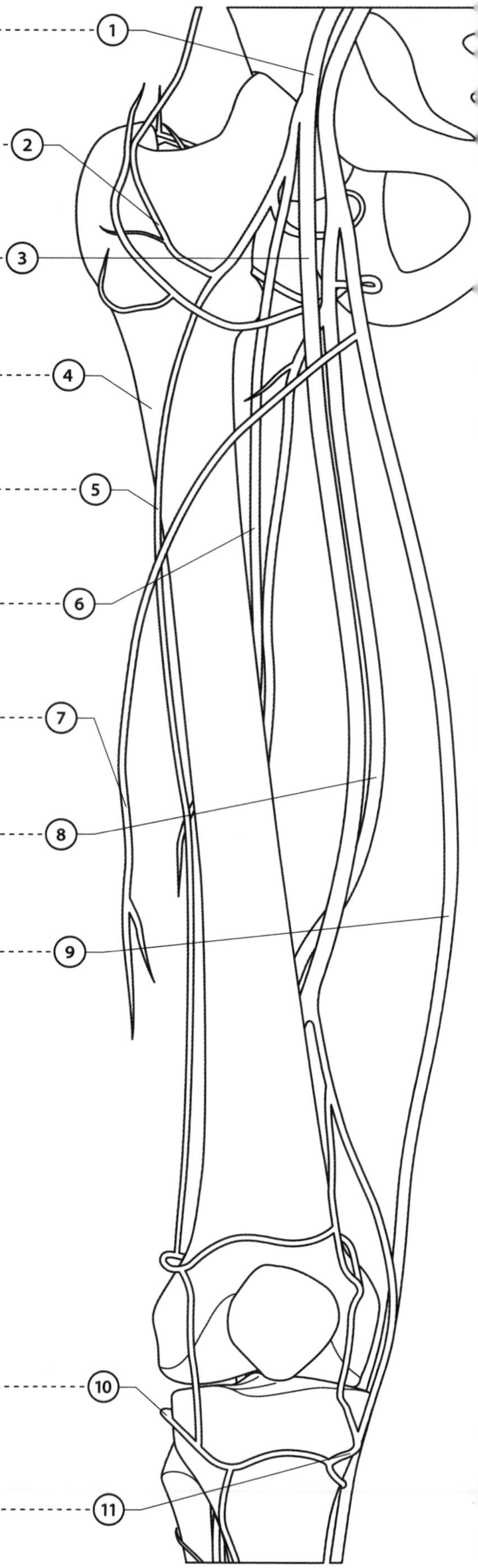

ANTERIOR VIEW

ANTERIOR VIEW

Color and/or label the structures indicated on the diagram using the key below.

1. External iliac artery
2. Ascending branch of lateral circumflex femoral artery
3. Femoral artery
4. Femur
5. Descending branch of the lateral circumflex femoral artery
6. Deep femoral artery
7. Accessory saphenous vein
8. Femoral vein
9. Great (long) saphenous vein
10. Lateral inferior genicular artery
11. Medial inferior genicular artery

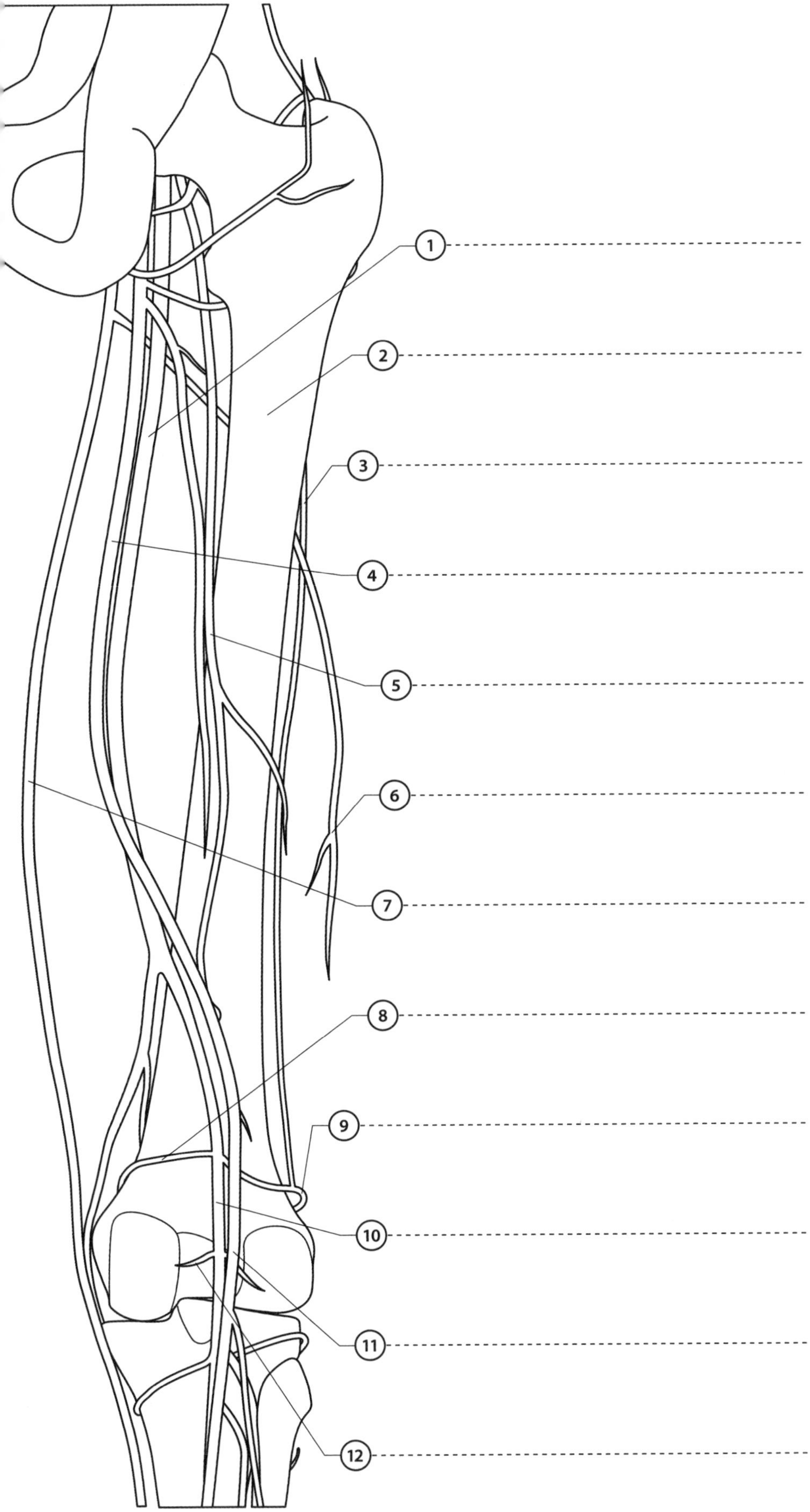

POSTERIOR VIEW

POSTERIOR

Gluteal branches of the internal iliac artery emerge through the greater sciatic foramen to supply the buttock. The muscles and skin of the inner part of the back of the thigh are supplied by branches of the deep femoral artery. These are the perforating arteries, so called because they pierce through the adductor magnus muscle. Higher up, the circumflex femoral arteries encircle the femur. The popliteal artery, formed after the femoral artery passes through a hiatus in the adductor magnus, lies on the back of the femur, deep to the popliteal vein.

POSTERIOR VIEW

Color and/or label the structures indicated on the diagram using the key below.

1. Femoral artery
2. Femur
3. Descending branch of lateral circumflex femoral artery
4. Femoral vein
5. Deep femoral artery
6. Accessory saphenous vein
7. Great (long) saphenous vein
8. Medial superior genicular artery
9. Lateral superior genicular artery
10. Popliteal artery
11. Popliteal vein
12. Sural artery

See also pp. 46, 74, 76, 104, 128, 152, 164 »

LOWER LEG AND FOOT

THE MAIN BLOOD VESSEL SUPPLYING THE LOWER LIMB IS THE POPLITEAL ARTERY. THE MAIN VESSELS DRAINING THE LOWER LIMB ARE THE GREAT (LONG) SAPHENOUS VEIN AND THE POPLITEAL VEIN.

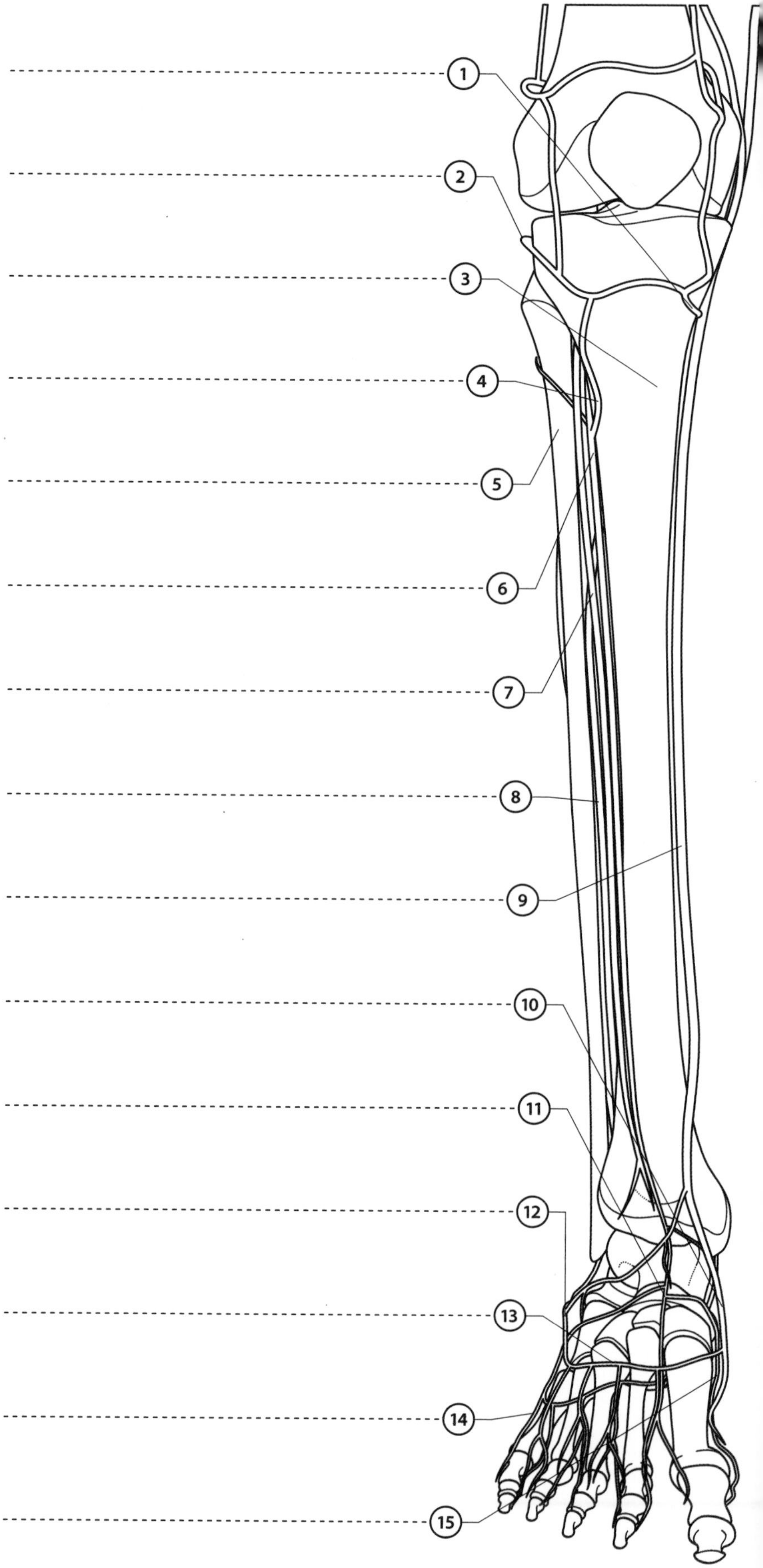

ANTERIOR VIEW

ANTERIOR VIEW

Color and/or label the diagram using the key below.

1. Medial inferior genicular artery
2. Lateral inferior genicular artery
3. Tibia
4. Anterior tibial recurrent artery
5. Fibula
6. Anterior tibial artery
7. Anterior tibial vein
8. Fibular artery
9. Great (long) saphenous vein
10. Medial marginal vein
11. Dorsalis pedis artery
12. Lateral marginal vein
13. Dorsal venous arch of the foot
14. Lateral plantar artery
15. Medial plantar artery

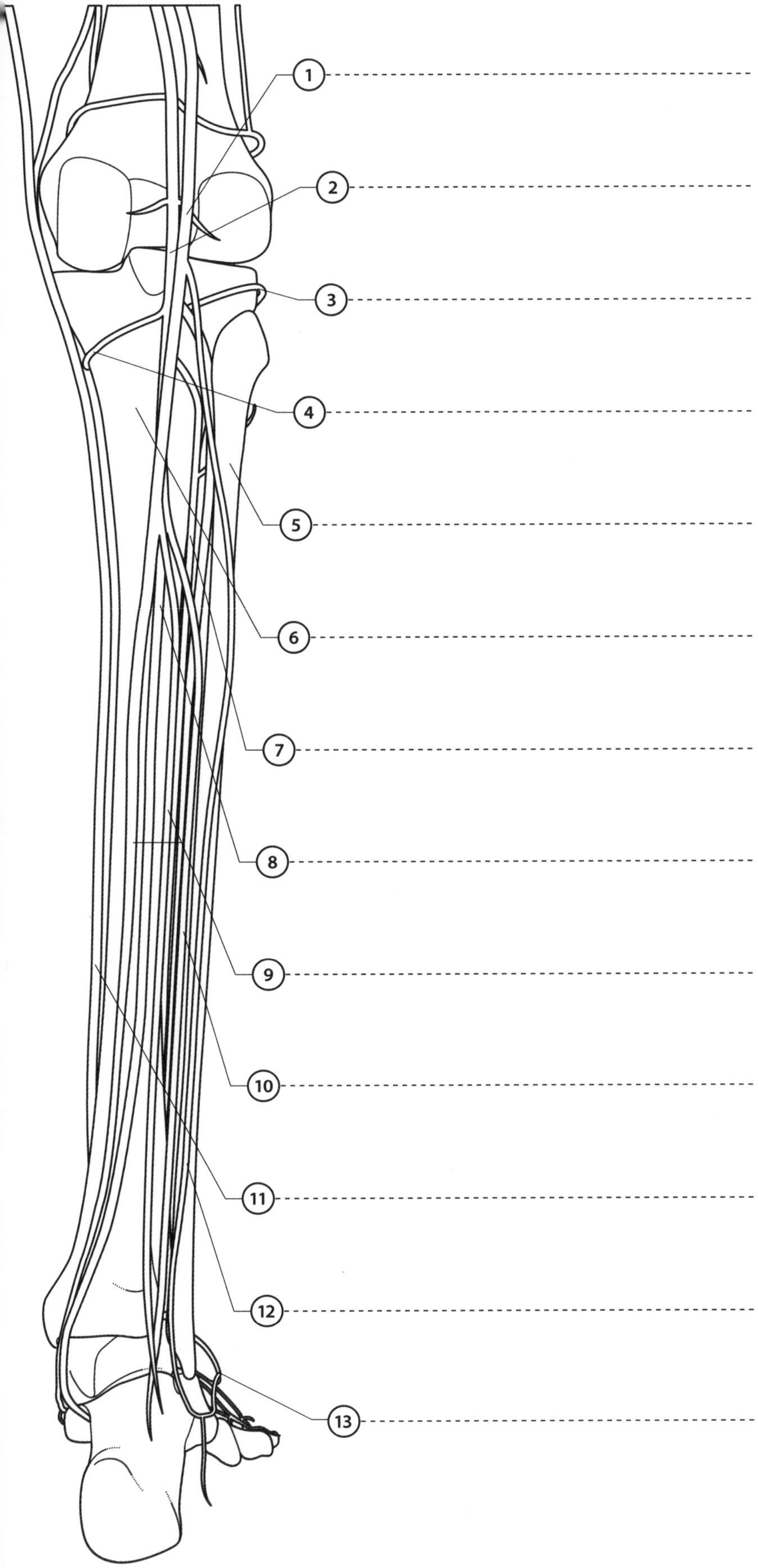

POSTERIOR VIEW

CIRCULATION

The popliteal artery runs deep across the back of the knee, dividing into two branches: the anterior and posterior tibial arteries. The anterior tibial artery runs forward, piercing the interosseus membrane between the tibia and fibula, to supply the extensor muscles of the shin. It runs down past the ankle, onto the top of the foot, as the dorsalis pedis artery. The posterior tibial artery gives off a fibular branch, supplying the muscles and skin on the outer side of the leg. The posterior tibial artery itself continues in the calf, running with the tibial nerves, and divides into plantar branches to supply the foot. A network of superficial veins on the dorsum of the foot is drained by the saphenous veins.

POSTERIOR VIEW

Color and/or label the diagram using the key below.

1. Popliteal vein
2. Popliteal artery
3. Lateral inferior genicular artery
4. Medial inferior genicular artery
5. Fibula
6. Tibia
7. Anterior tibial artery
8. Posterior tibial artery
9. Posterior tibial veins
10. Fibular artery
11. Great (long) saphenous vein
12. Small (short) saphenous vein
13. Lateral marginal vein

See also pp. 46, 74, 76, 104, 128, 150, 165 »

LYMPHATIC AND IMMUNE SYSTEMS

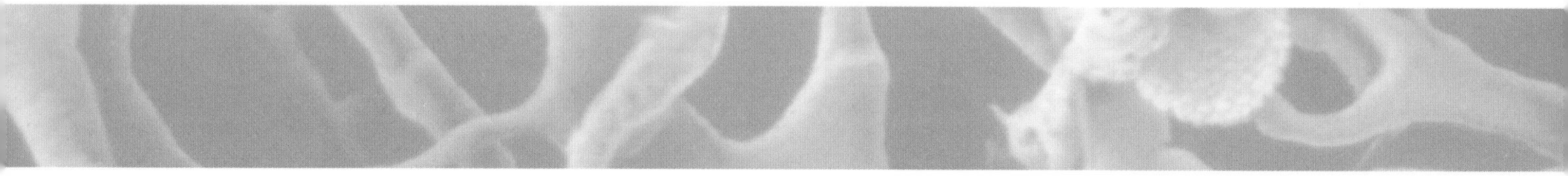

07

LYMPHATIC AND IMMUNE SYSTEMS 1

THE LYMPHATIC SYSTEM, WHICH RUNS ALMOST PARALLEL TO THE CIRCULATORY SYSTEM, COLLECTS EXCESS FLUID (LYMPH) FROM TISSUES AND RETURNS IT TO THE BLOOD. THIS SYSTEM IS AN INTEGRAL PART OF THE BODY'S IMMUNE SYSTEM.

LYMPHATIC SYSTEM

A network of vessels, tissues, and nodes makes up the lymphatic system. Lymph begins as the interstitial fluid that collects between cells. This fluid drains into capillaries that unite to form larger vessels, the lymphatics. As lymph circulates round the body, it passes through lymph nodes, within which lymph is cleansed of pathogens before entering the bloodstream via the right and left subclavian veins. By draining interstitial fluid, the lymphatic system helps maintain the body's fluid balance and prevent fluid build-up in tissues. Lymph also assists in the absorption of fats from the intestine.

LYMPH NODES

A lymph node is a fibrous capsule containing a network of meshlike tissue, which is packed with lymphocytes. As circulating lymph filters through channels in the node, cell debris and pathogens become trapped and are destroyed by the lymphocytes. Numerous lymph nodes occur throughout the lymphatic system, some grouped in clusters, such as those in the axilla and groin, and behind the knee.

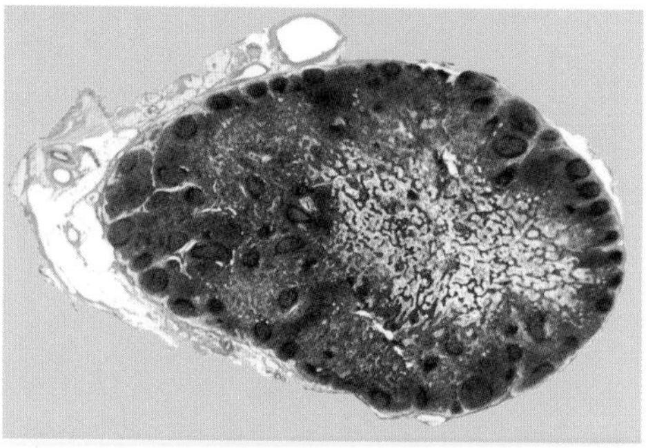

INSIDE A LYMPH NODE
This cross section of a lymph node shows the outer capsule (stained pink), an outer cortex packed full of immune cells (dark purple), and an inner medulla made up of lymphatic channels (blue).

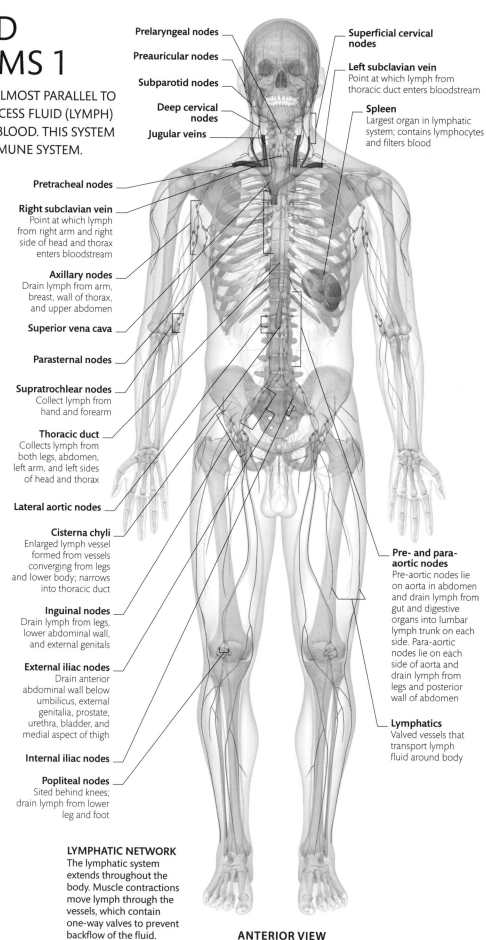

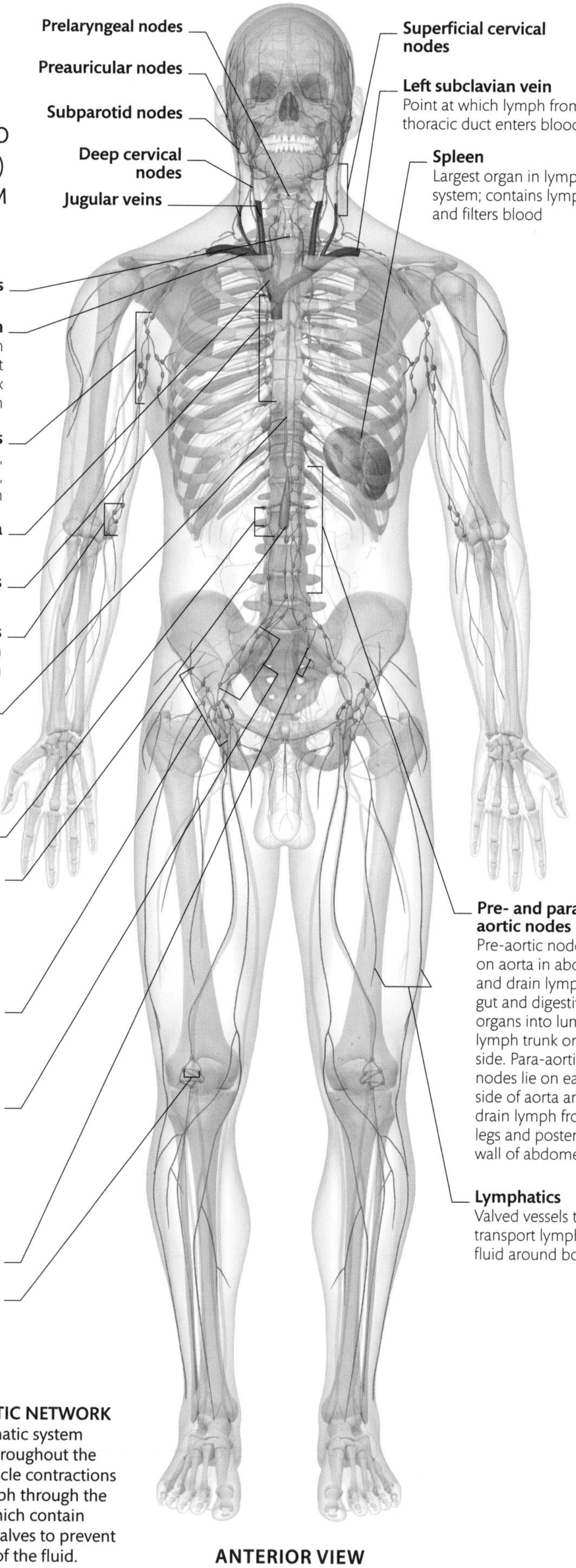
Prelaryngeal nodes
Preauricular nodes
Subparotid nodes
Deep cervical nodes
Jugular veins
Pretracheal nodes
Right subclavian vein
Point at which lymph from right arm and right side of head and thorax enters bloodstream
Axillary nodes
Drain lymph from arm, breast, wall of thorax, and upper abdomen
Superior vena cava
Parasternal nodes
Supratrochlear nodes
Collect lymph from hand and forearm
Thoracic duct
Collects lymph from both legs, abdomen, left arm, and left sides of head and thorax
Lateral aortic nodes
Cisterna chyli
Enlarged lymph vessel formed from vessels converging from legs and lower body; narrows into thoracic duct
Inguinal nodes
Drain lymph from legs, lower abdominal wall, and external genitals
External iliac nodes
Drain anterior abdominal wall below umbilicus, external genitalia, prostate, urethra, bladder, and medial aspect of thigh
Internal iliac nodes
Popliteal nodes
Sited behind knees; drain lymph from lower leg and foot
Superficial cervical nodes
Left subclavian vein
Point at which lymph from thoracic duct enters bloodstream
Spleen
Largest organ in lymphatic system; contains lymphocytes and filters blood
Pre- and para-aortic nodes
Pre-aortic nodes lie on aorta in abdomen and drain lymph from gut and digestive organs into lumbar lymph trunk on each side. Para-aortic nodes lie on each side of aorta and drain lymph from legs and posterior wall of abdomen
Lymphatics
Valved vessels that transport lymph fluid around body
ANTERIOR VIEW

LYMPHATIC NETWORK
The lymphatic system extends throughout the body. Muscle contractions move lymph through the vessels, which contain one-way valves to prevent backflow of the fluid.

IMMUNE SYSTEM

The immune system is the body's defense mechanism against disease and infection. Its main components are the lymphatic system and a range of specialized immune molecules, such as antibodies, and white blood cells, including lymphocytes. Lymphoid tissues and organs important to the immune system occur at key points throughout the body. The primary lymphoid tissues are bone marrow and (in children) the thymus, both sites of immune cell generation and maturation. Secondary lymphoid tissues are the lymph nodes, spleen, adenoid, tonsils, and gut-associated lymphoid tissue (GALT).

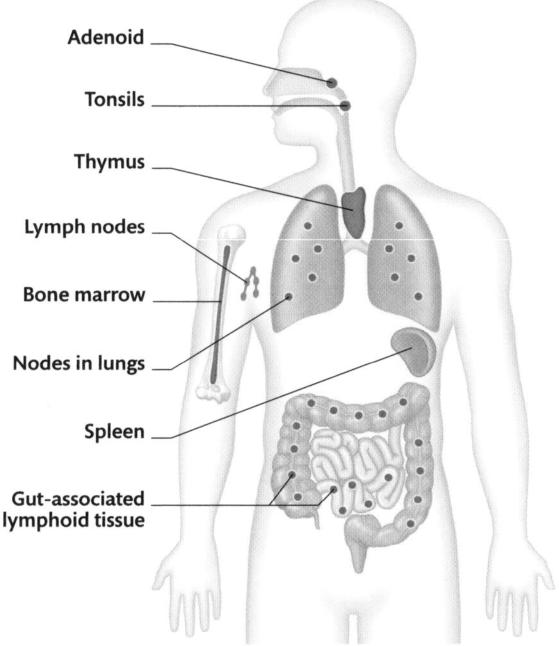

Adenoid
Tonsils
Thymus
Lymph nodes
Bone marrow
Nodes in lungs
Spleen
Gut-associated lymphoid tissue

LYMPHOID TISSUE
The main locations of lymphoid structures associated with the immune system are closely related to entry points for infection, such as the lungs and the gut. All types of immune cell are produced in bone marrow.

KEY
- Primary lymphoid tissues
- Lymph nodes and spleen
- Mucosa-associated lymphoid tissue (MALT)

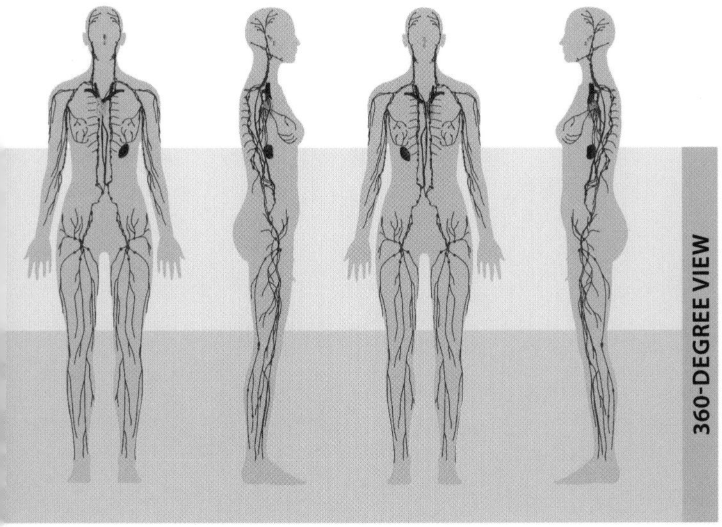

360-DEGREE VIEW

BARRIER IMMUNITY

As an initial strategy, the immune system uses barrier methods to defend the body against pathogens. Barrier, or passive, immunity provides protection via various physical and chemical barriers. These include both external surfaces of the body, such as the skin and microscopic hairs, as well as mucus-lined internal surfaces—for example, the airways and intestines. An initial physical barrier to infection is then supplemented by the secretion of antimicrobial substances, such as gastric enzymes, which break down bacteria. The body has additional mechanisms to expel or flush out unwanted microorganisms: for example, coughing, sweating, and urination.

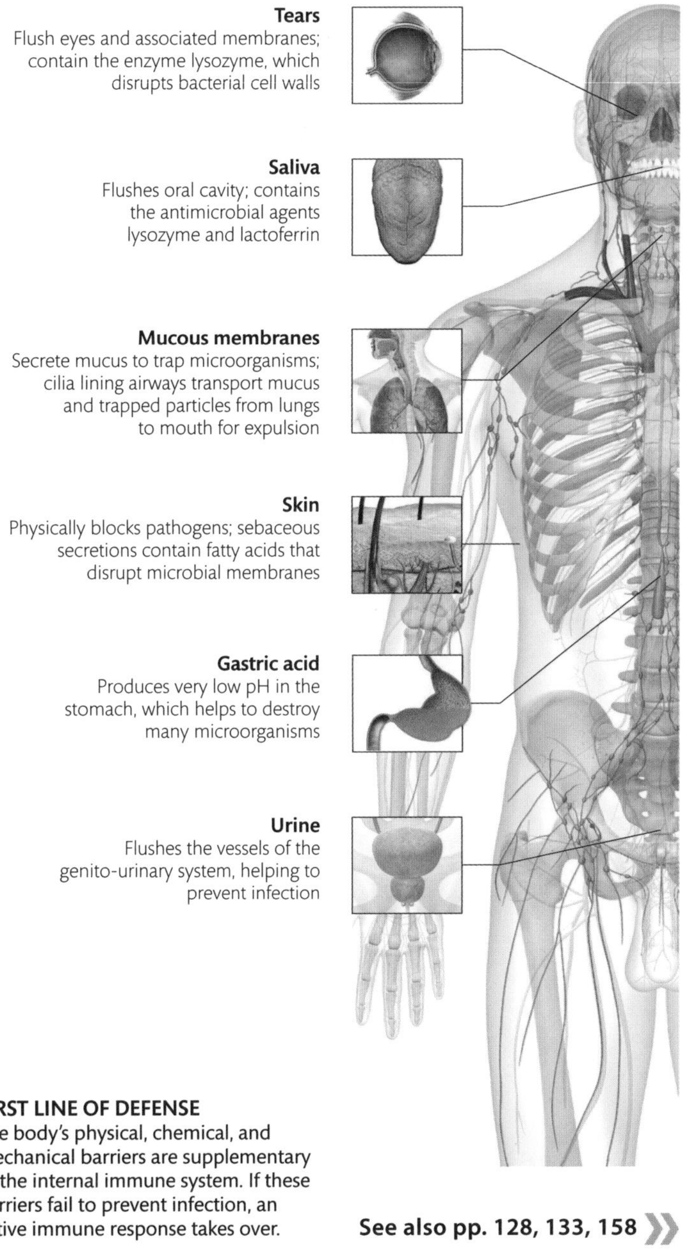

Tears
Flush eyes and associated membranes; contain the enzyme lysozyme, which disrupts bacterial cell walls
Saliva
Flushes oral cavity; contains the antimicrobial agents lysozyme and lactoferrin
Mucous membranes
Secrete mucus to trap microorganisms; cilia lining airways transport mucus and trapped particles from lungs to mouth for expulsion
Skin
Physically blocks pathogens; sebaceous secretions contain fatty acids that disrupt microbial membranes
Gastric acid
Produces very low pH in the stomach, which helps to destroy many microorganisms
Urine
Flushes the vessels of the genito-urinary system, helping to prevent infection

FIRST LINE OF DEFENSE
The body's physical, chemical, and mechanical barriers are supplementary to the internal immune system. If these barriers fail to prevent infection, an active immune response takes over.

See also pp. 128, 133, 158 »

LYMPHATIC AND IMMUNE SYSTEMS 2

ONCE EXTERNAL DEFENSE BARRIERS ARE BREACHED, THE IMMUNE SYSTEM RESPONDS RAPIDLY WITH VARIOUS MECHANISMS. INNATE IMMUNITY INVOLVES THE RECOGNITION OF GENERALIZED PATHOGENS BY PHAGOCYTIC CELLS AND SPECIALIZED PROTEINS. IN THE ADAPTIVE IMMUNE RESPONSE, LYMPHOCYTES TARGET SPECIFIC PATHOGENS.

COMPLEMENT SYSTEM

An important part of the innate immune system is the complement system, in which specialized proteins circulating freely in blood plasma target microorganisms. These proteins are ordinarily present as separate molecules, but once activated they act together as a cascade, initiating a complementary chain reaction that attacks and destroys pathogens. Complement proteins are activated by bacterial surface features and respond to infections throughout the body.

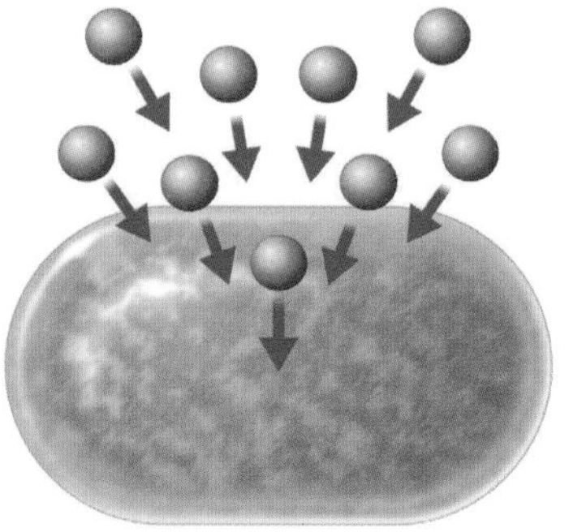

1 APPROACH Bacterial surface proteins activate the complement system, causing proteins in the blood plasma to assemble at the surface of the bacterium.

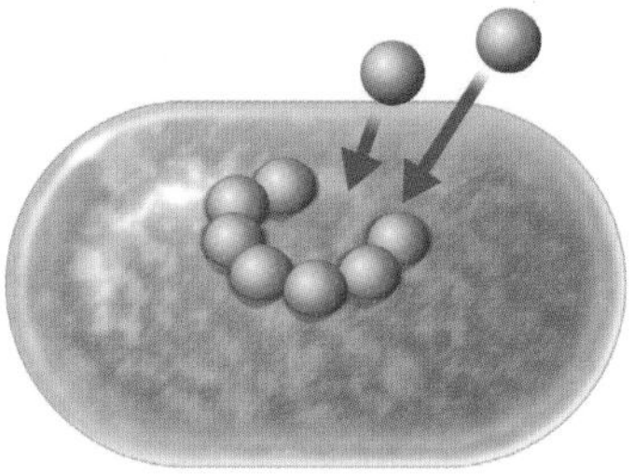

2 MEMBRANE ATTACK The proteins combine to form the "membrane attack complex"—a structure that punches a hole in the surface of the bacterium.

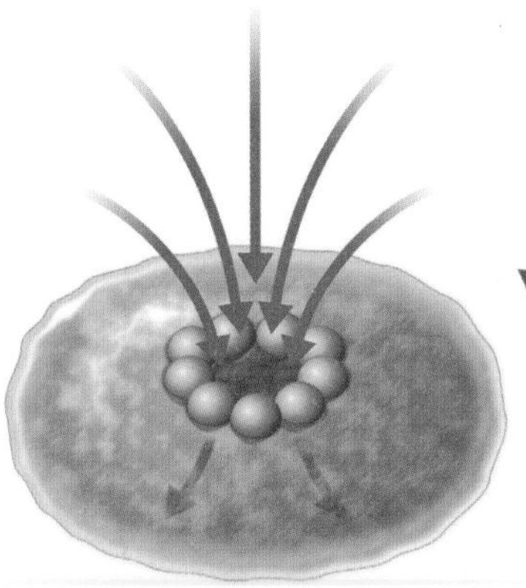

3 PERFORATION The resultant hole allows extracellular fluid to enter the bacterium. This process occurs repeatedly over the surface of the bacterium.

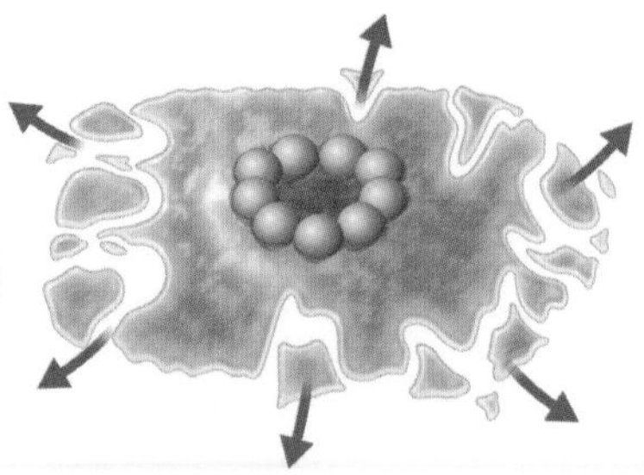

4 RUPTURE The influx of extracellular fluid causes the bacterium to expand. Eventually, so much fluid enters the bacterium that it ruptures.

PHAGOCYTES

Fundamental to the innate immune reponse are white blood cells known as phagocytes (macrophages and neutrophils). Phagocytes, often the first cells to reach an infection site, recognize pathogens via surface contact. Once a microorganism has been identified as an invader, the phagocyte destroys it via phagocytosis, a process of engulfing, absorbing, and digesting the pathogen. Some phagocytes also interact with lymphocytes in the adaptive immune response.

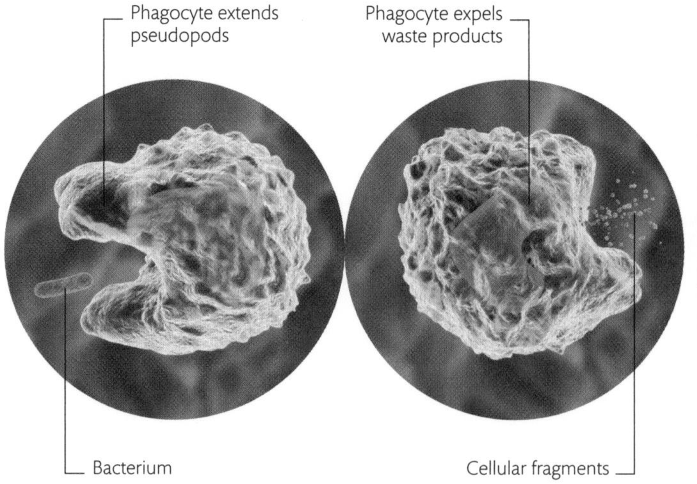

RECOGNITION
The phagocyte recognizes a target bacterium on contact of the two cells' surfaces. Extending projections (pseudopods), the phagocyte engulfs and absorbs the bacterium.

EXPULSION
Within the phagocyte, chemical reactions break down the bacterium. Waste fragments of the bacterial cell that cannot be broken down further are expelled by the phagocyte.

LYMPHOCYTES

The key cells of the adaptive immune response are T and B lymphocytes. The surfaces of these cells have receptors that are programed to recognize specific antigens (foreign proteins). T and B lymphocytes are capable of remembering a specific pathogen and acting quickly if it re-infects. Two types of T lymphocyte—killer, or cytotoxic (attack), cells and helper (coordinating) cells—target infected body cells. B lymphocytes respond to fluid infections.

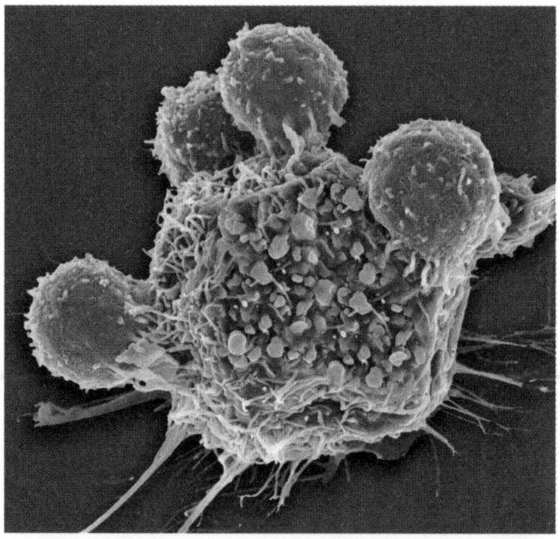

T CELL ATTACK
T lymphocytes are able to target body cells that have become malignant. In this micrograph, four T cells (red) can be seen attacking a cancer cell (gray).

ADAPTIVE RESPONSES

There are two main types of adaptive immune response—cell-mediated and antibody-mediated—both of which depend on the actions of T and B lymphocytes. In cell-mediated immunity, T lymphocytes multiply rapidly on recognition of an antigen, to launch a direct attack on the pathogen or infected cell. In antibody-mediated immunity, B lymphocytes multiply on recognition of an antigen and make gammoglobulins (protein antibodies), which react against antigens. Integral to adaptive immunity is the activity of antigen-presenting cells, including some macrophages, which present antigens to lymphocytes to promote their response.

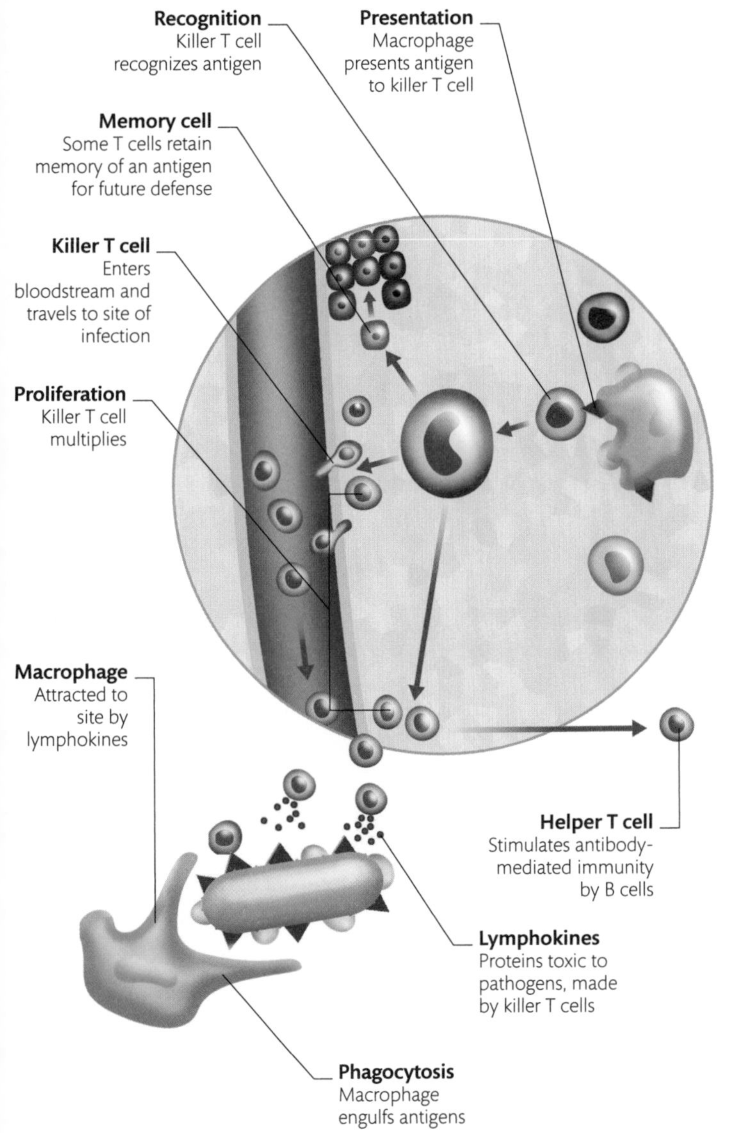
Recognition Killer T cell recognizes antigen
Presentation Macrophage presents antigen to killer T cell
Memory cell Some T cells retain memory of an antigen for future defense
Killer T cell Enters bloodstream and travels to site of infection
Proliferation Killer T cell multiplies
Macrophage Attracted to site by lymphokines
Helper T cell Stimulates antibody-mediated immunity by B cells
Lymphokines Proteins toxic to pathogens, made by killer T cells
Phagocytosis Macrophage engulfs antigens

CELL-MEDIATED IMMUNITY
Presented with an antigen, a killer T cell responds by multiplying. Killer T cells attack pathogens and infected body cells with powerful proteins (lymphokines). Helper T cells activate B cells to assist antibody-mediated immunity, and macrophages to engulf pathogens and debris.

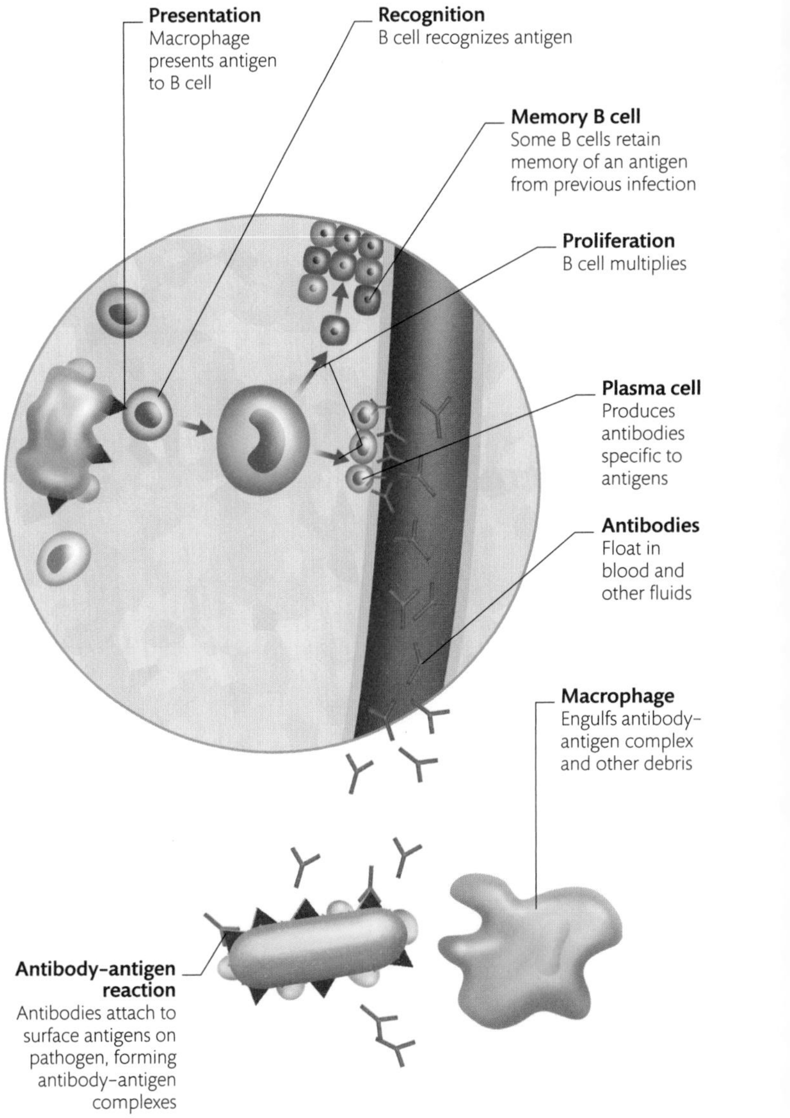
Presentation Macrophage presents antigen to B cell
Recognition B cell recognizes antigen
Memory B cell Some B cells retain memory of an antigen from previous infection
Proliferation B cell multiplies
Plasma cell Produces antibodies specific to antigens
Antibodies Float in blood and other fluids
Macrophage Engulfs antibody-antigen complex and other debris
Antibody-antigen reaction Antibodies attach to surface antigens on pathogen, forming antibody-antigen complexes

ANTIBODY-MEDIATED IMMUNITY
Once activated by the presence of antigens, B cells multiply into plasma and other cell types. The plasma cells produce antibodies, which attach to the surface antigens of specific pathogens. The resultant antibody-antigen complex enables macrophages to destroy the pathogen.

See also pp. 133, 156

HEAD AND NECK

THE HEAD AND NECK HAVE AN EXTENSIVE NETWORK OF LYMPH NODES AND LYMPHATIC VESSELS, AS WELL AS LYMPHOID TISSUE IN THE FORM OF TONSILS.

LYMPH NODES

There is a ring of lymph nodes lying close to the skin where the head meets the neck, from the occipital nodes (against the back of the skull) to the submandibular and submental nodes (which are tucked under the jaw). Superficial nodes lie along the sides and front of the neck, and deep nodes are clustered around the internal jugular vein, under the sternocleidomastoid muscle. Lymph from all other nodes passes into these deep ones, then into the jugular lymphatic trunks, before draining back into the veins in the base of the neck.

LYMPH NODES OF HEAD AND NECK

Color and/or label the structures indicated on the diagram using the key below.

1. Parotid nodes
2. Buccal node
3. Submandibular nodes
4. Mandibular node
5. Submental nodes
6. Prelaryngeal nodes
7. Infrahyoid nodes
8. Pretracheal nodes
9. Paratracheal nodes
10. Jugulo-omohyoid node
11. Superficial cervical nodes
12. Jugulodigastric node
13. Occipital nodes
14. Mastoid nodes

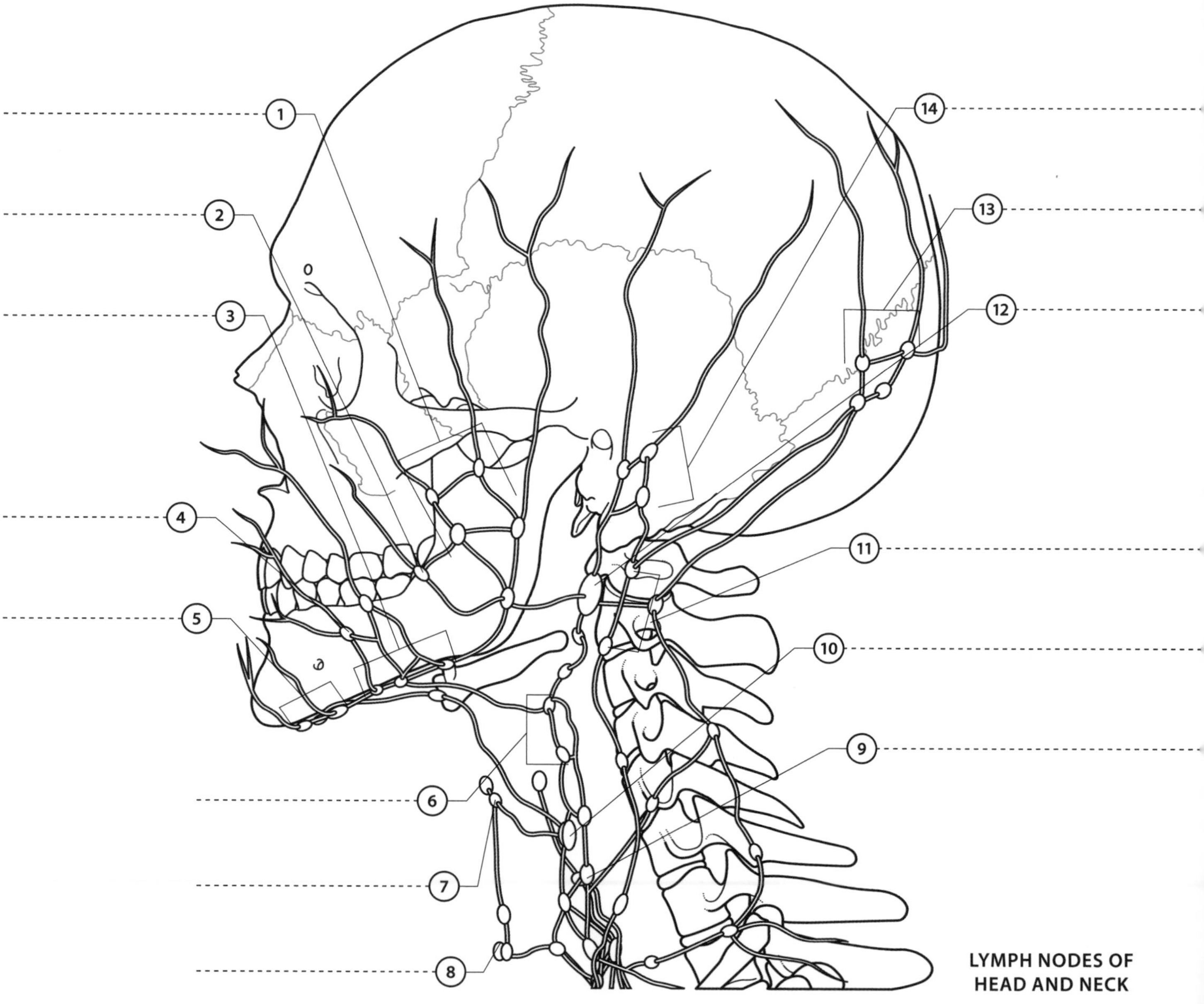

LYMPH NODES OF HEAD AND NECK

TONSILS

The palatine, pharyngeal, and lingual tonsils form a protective ring around the upper parts of the respiratory and digestive tracts. The palatine tonsils lie under the mucosa of the oropharynx; the two are often just called the tonsils. The pharyngeal tonsil—commonly referred to as the adenoid—is in the nasopharynx, near the openings of the Eustachian tubes. The lingual tonsil lies under the mucosa at the back of the tongue.

LOCATION OF TONSILLAR TISSUE

Color and/or label the structures indicated on the diagram using the key below.

1. Nasal cavity
2. Palatine tonsil
3. Tongue
4. Lingual tonsil
5. Larynx
6. Epiglottis
7. Soft palate
8. Opening of pharyngotympanic (Eustachian) tube
9. Pharyngeal tonsil (adenoid)

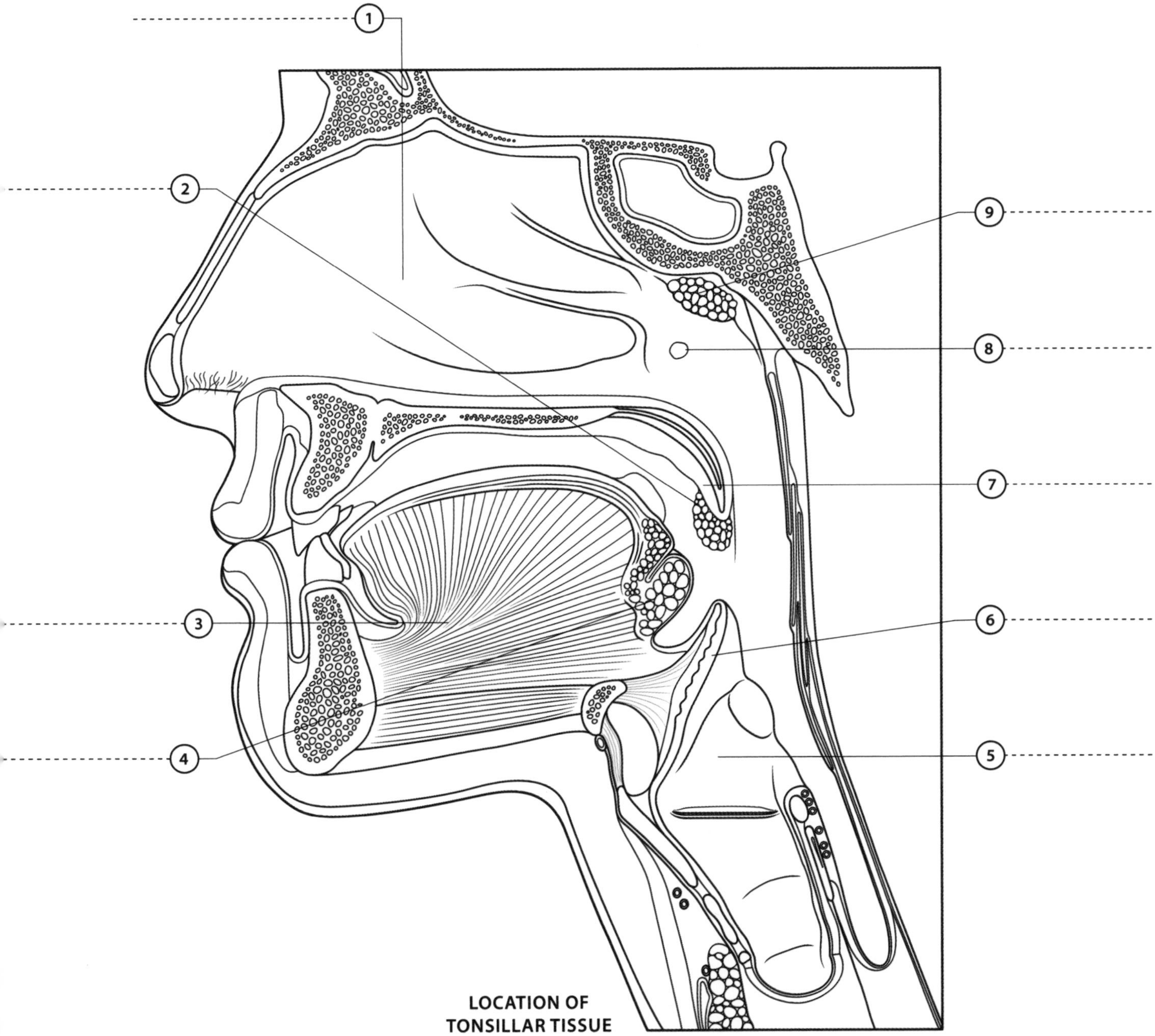

LOCATION OF TONSILLAR TISSUE

See also pp. 118, 134, 136, 166, 174

THORAX, ABDOMEN, AND PELVIS

IN THE THORAX, ABDOMEN, AND PELVIS THE LYMPH NODES GENERALLY OCCUR IN CLUSTERS, MAINLY AROUND THE ARMPITS AND GROIN. THE THYMUS IS SITUATED IN THE THORAX, WHEREAS THE SPLEEN IS IN THE ABDOMEN.

THORAX

Most of the fluid from superficial tissues of the chest drains into axillary nodes, high in the armpits. The complex drainage of the female breast passes to these and to parasternal, supraclavicular, and abdominal nodes. Lymph from deeper tissues drains into nodes in the thorax. Lymph from the thorax's left side ultimately drains into the thoracic duct, whereas lymph from the right side drains into the right lymphatic duct. Both ducts empty into veins at the base of the neck. The thymus, which is largest during childhood, lies behind the sternum.

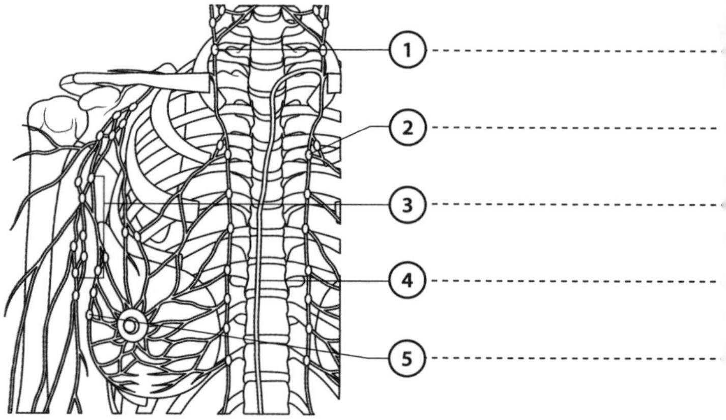

FEMALE THORAX (ANTERIOR)

FEMALE THORAX (ANTERIOR)

Color and/or label the structures indicated on the diagram using the key below.

1. Supraclavicular node
2. Parasternal node
3. Axillary nodes
4. Thoracic duct
5. Paramammary node

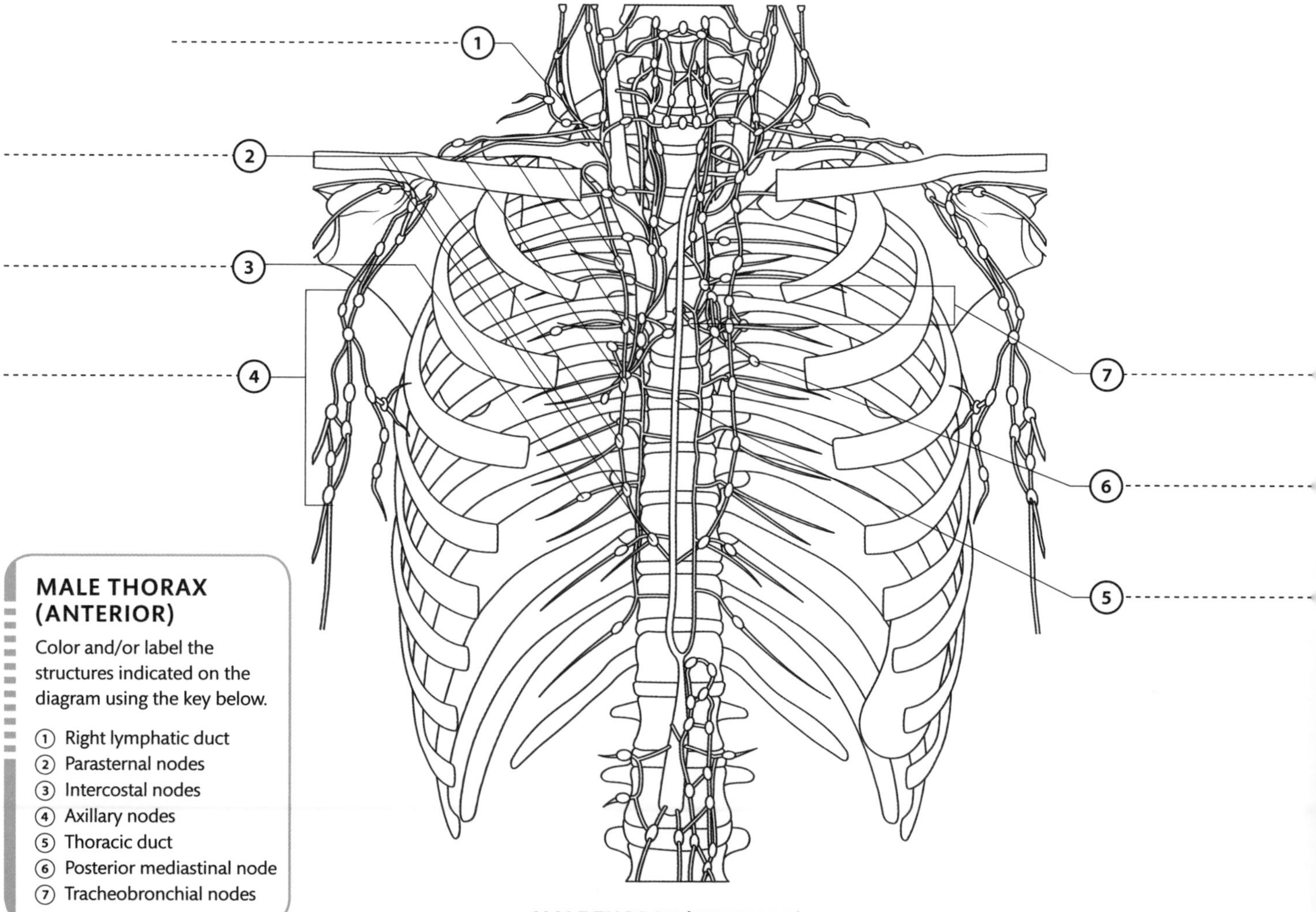

MALE THORAX (ANTERIOR)

Color and/or label the structures indicated on the diagram using the key below.

1. Right lymphatic duct
2. Parasternal nodes
3. Intercostal nodes
4. Axillary nodes
5. Thoracic duct
6. Posterior mediastinal node
7. Tracheobronchial nodes

MALE THORAX (ANTERIOR)

ABDOMEN AND PELVIS

The deep lymph nodes of the abdomen are clustered around arteries. Nodes on each side of the aorta receive lymph from paired structures, such as the muscles of the abdominal wall, the kidneys and suprarenal glands, and the testes or ovaries. Iliac nodes collect lymph from the legs and pelvis. Nodes around the branches on the front of the aorta collect lymph from the gut and abdominal organs. Eventually, all this lymph from the legs, pelvis, and abdomen passes into the cisterna chyli; this narrows to become the thoracic duct, which runs up into the thorax. Most lymph nodes are small, bean-sized structures, but the abdomen also contains an important immune organ: the spleen.

ANTERIOR VIEW

Color and/or label the structures indicated on the diagram using the key below.

1. Thoracic duct
2. Celiac nodes
3. Lateral aortic nodes
4. External iliac nodes
5. Proximal superficial inguinal nodes
6. Distal superficial inguinal nodes
7. Internal iliac nodes
8. Common iliac nodes
9. Mesenteric nodes
10. Cisterna chyli
11. Spleen

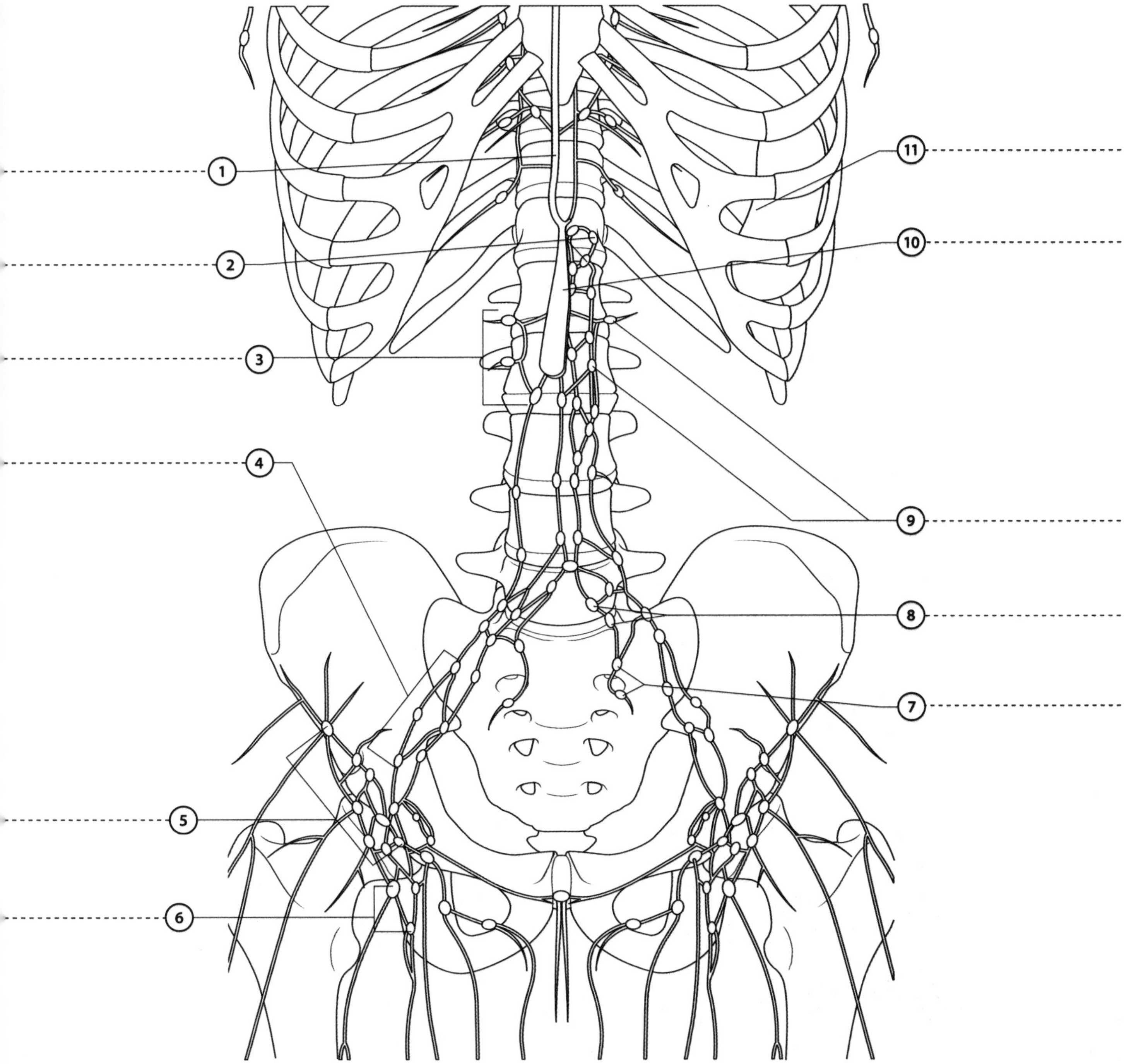

ANTERIOR VIEW

See also pp. 138, 144

SHOULDER, ARM, HIP, AND LEG

THE LIMBS ARE RELATIVELY POORLY SUPPLIED WITH LYMPHATIC VESSELS AND NODES. THE MAIN CLUSTERS OF NODES ARE IN THE ARMPIT AND GROIN, BUT THERE ARE ALSO SMALLER GROUPS NEAR THE ELBOW AND KNEE.

SHOULDER AND ARM

Ultimately, all the lymph from the hand, forearm, and upper arm drains into the axillary nodes in the armpit. However, there are a few nodes, lower in the arm, through which lymph may pass on its way to the axilla. The supratrochlear nodes lie in the subcutaneous fat on the inner arm, above the elbow. They collect lymph from the medial side of the hand and forearm. The infraclavicular nodes, lying along the cephalic vein, below the clavicle, receive lymphatics draining from the thumb and lateral side of the arm and forearm. The axillary nodes drain lymph from the arm and receive it from the chest wall, including the breast.

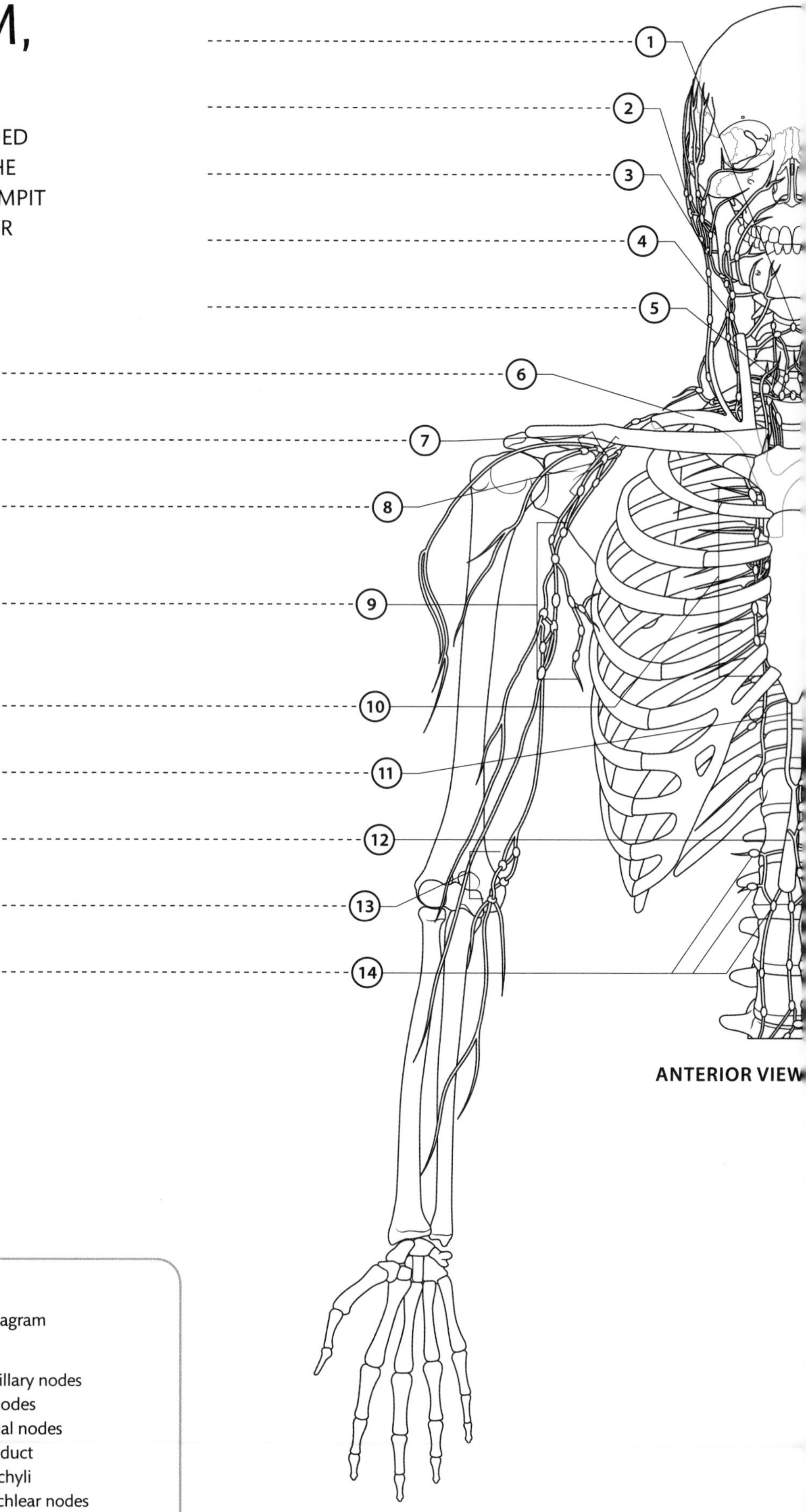

ANTERIOR VIEW

ANTERIOR VIEW

Color and/or label the structures indicated on the diagram using the key below.

1. Prelaryngeal nodes
2. Preauricular nodes
3. Subparotid nodes
4. Deep cervical nodes
5. Pretracheal nodes
6. Right subclavian vein
7. Infraclavicular nodes
8. Apical axillary nodes
9. Axillary nodes
10. Parasternal nodes
11. Thoracic duct
12. Cisterna chyli
13. Supratrochlear nodes
14. Lateral aortic nodes

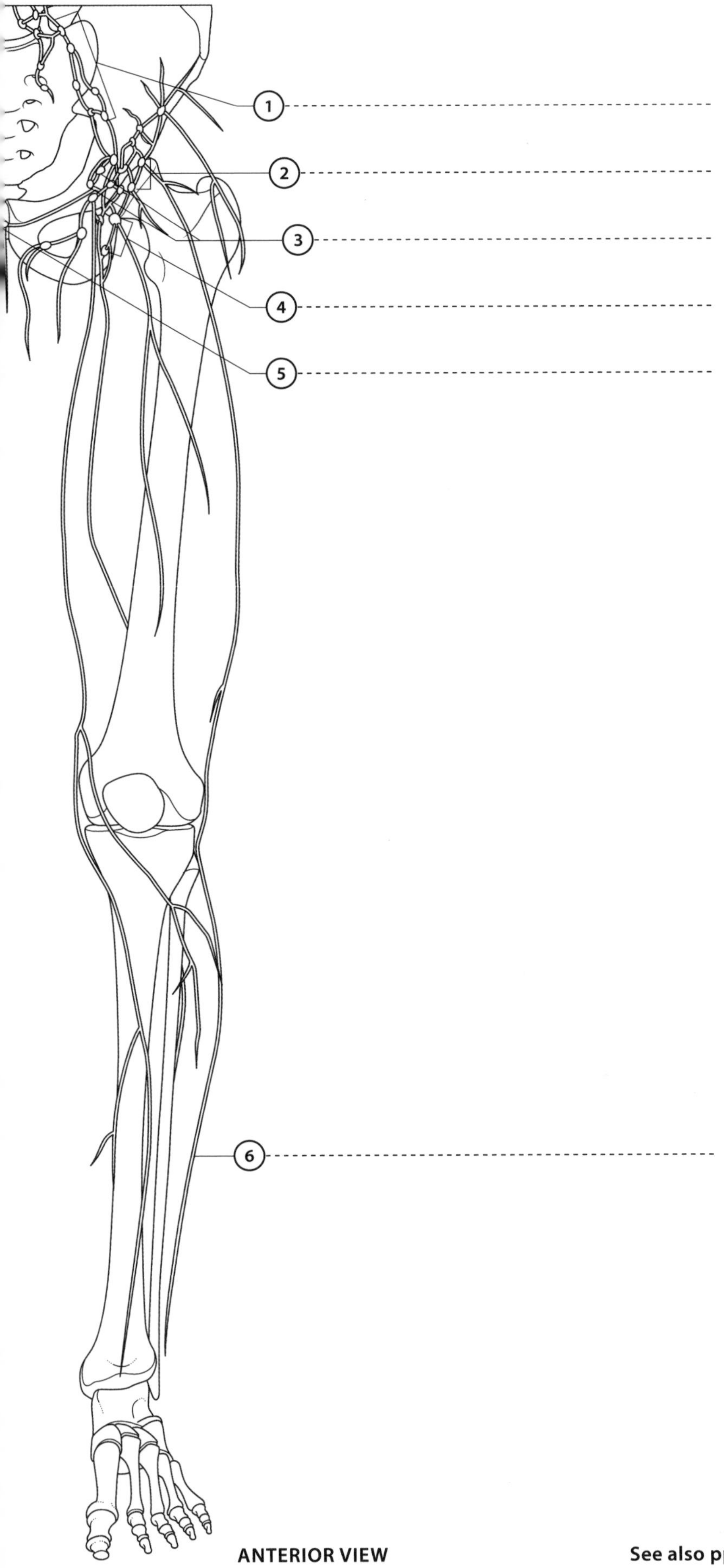

ANTERIOR VIEW

HIP AND LEG

Most lymph from the thigh, leg, and foot passes through the inguinal lymph nodes in the groin. However, lymph from the deep tissues of the buttock passes straight to nodes in the pelvis. Eventually, all the lymph from the leg reaches the lateral aortic nodes on the back wall of the abdomen. As in the arm, there are groups of nodes clustered around points at which superficial veins drain into deep veins. Popliteal nodes are close to the drainage of the small saphenous vein into the popliteal vein, while the superficial inguinal nodes lie close to the great saphenous vein, just before it empties into the femoral vein.

ANTERIOR VIEW

Color and/or label the structures indicated on the diagram using the key below.

1. External iliac nodes
2. Inguinal nodes
3. Deep inguinal nodes
4. Distal superficial inguinal nodes
5. Presymphyseal node
6. Lymphatic of lower leg

See also pp. 40, 46, 70, 72, 74, 76, 102, 104, 146, 148, 150, 152

LYMPHATIC CIRCULATION

THE LYMPHATIC CIRCULATION IS CLOSELY LINKED TO THE BLOOD CIRCULATION AND PLAYS A KEY ROLE IN DRAINING FLUID FROM BODY TISSUES.

MOVEMENT OF LYMPH

Fluid components of blood plasma filter out of the blood through capillary walls and enter the interstitial spaces of body tissues. This interstitial fluid is secreted faster than it can be re-absorbed. Blind-ended channels (initial lymphatics) allow the excess fluid to drain into the lymphatic system, via one-way valves, forming lymph. The initial lymphatics drain into the main lymphatic vessels, which carry the lymph around the body. These vessels have contractile walls that aid the forward movement of lymph, and bicuspid valves that prevent its backflow. Scattered along the lymphatics are lymph nodes, which filter the passing lymph and screen it for signs of infection, using the immune cells in the nodes. Eventually, the lymph drains through the right lymphatic duct and thoracic duct into the internal jugular vein.

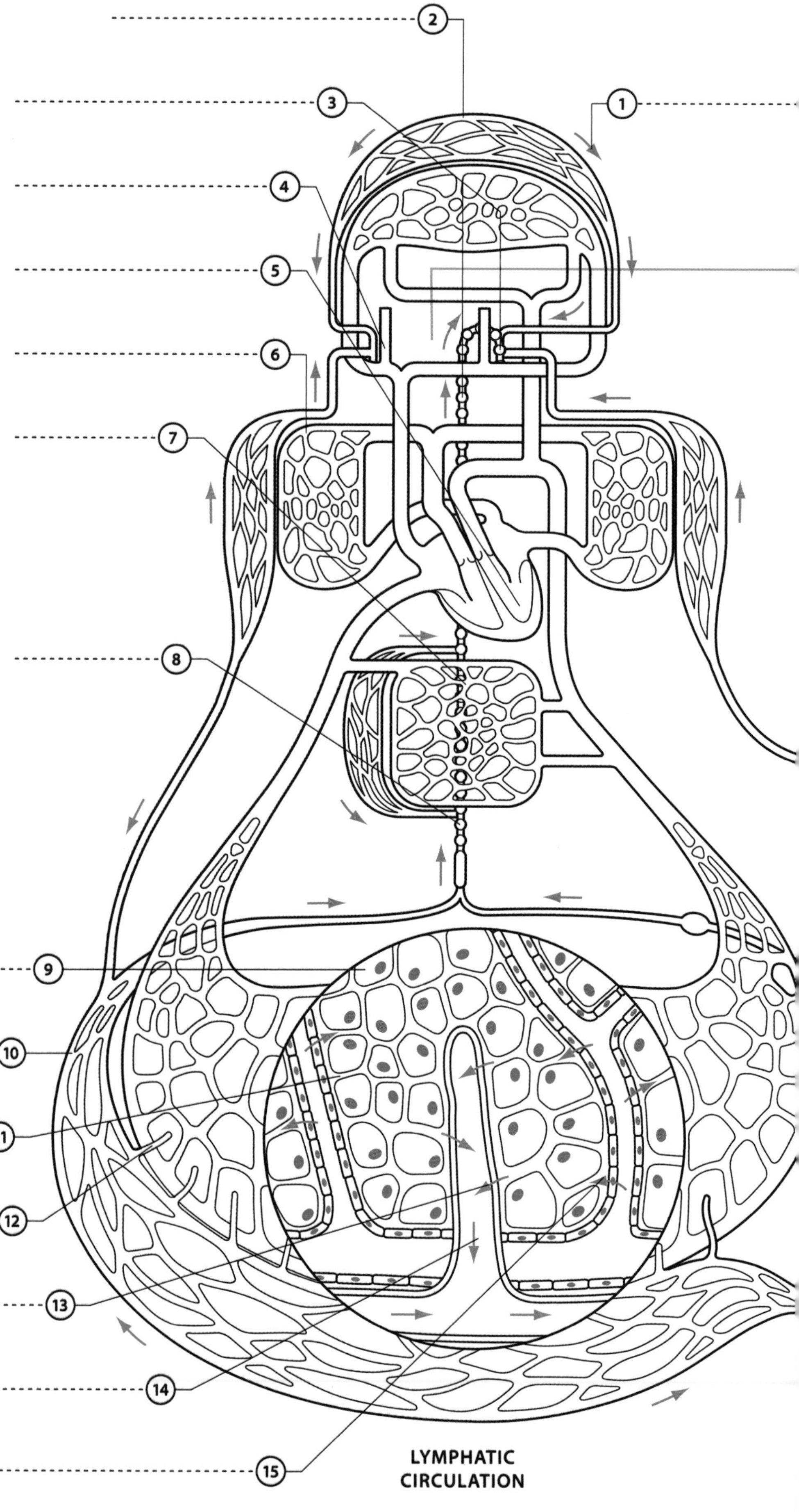

LYMPHATIC CIRCULATION

LYMPHATIC CIRCULATION

Color and/or label the structures indicated on the diagram using the key below. Color lymphatic system structures green and blood vessels red.

1. Flow of lymph
2. Lymphatic vessels of head and upper body
3. Thoracic duct
4. Right lymphatic duct
5. Heart
6. Blood vessels of lung
7. Blood vessels of abdominal cavity
8. Lymphatic vessel in gut
9. Body cell
10. Lymphatic vessels and nodes of lower body
11. Interstitial space
12. Initial lymphatic
13. Interstitial fluid enters initial lymphatic
14. Lymph moves into circulation
15. Plasma filters out of capillary
16. Afferent lymphatic vessel
17. Capsule of lymph node
18. Reticular fibers
19. Blood vessels
20. Efferent lymphatic vessel
21. Inner (paracortical) cortex
22. Outer cortex
23. Valve
24. Right lymphatic duct
25. Thoracic duct
26. Left subclavian vein
27. Flow of lymph
28. Left internal jugular vein
29. Right internal jugular vein
30. Right subclavian vein

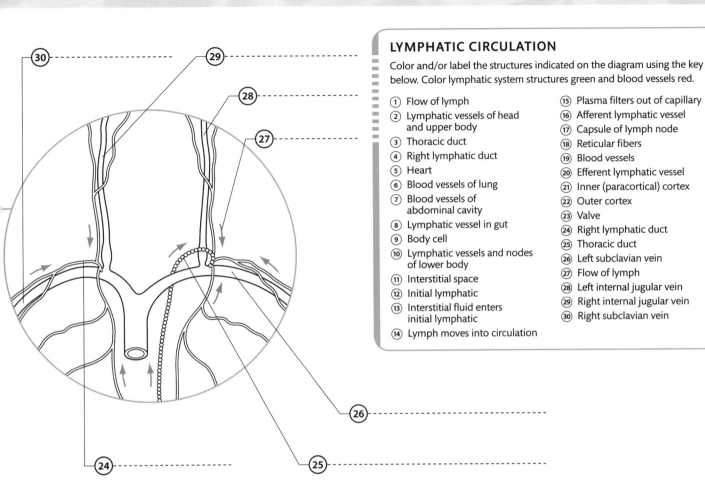

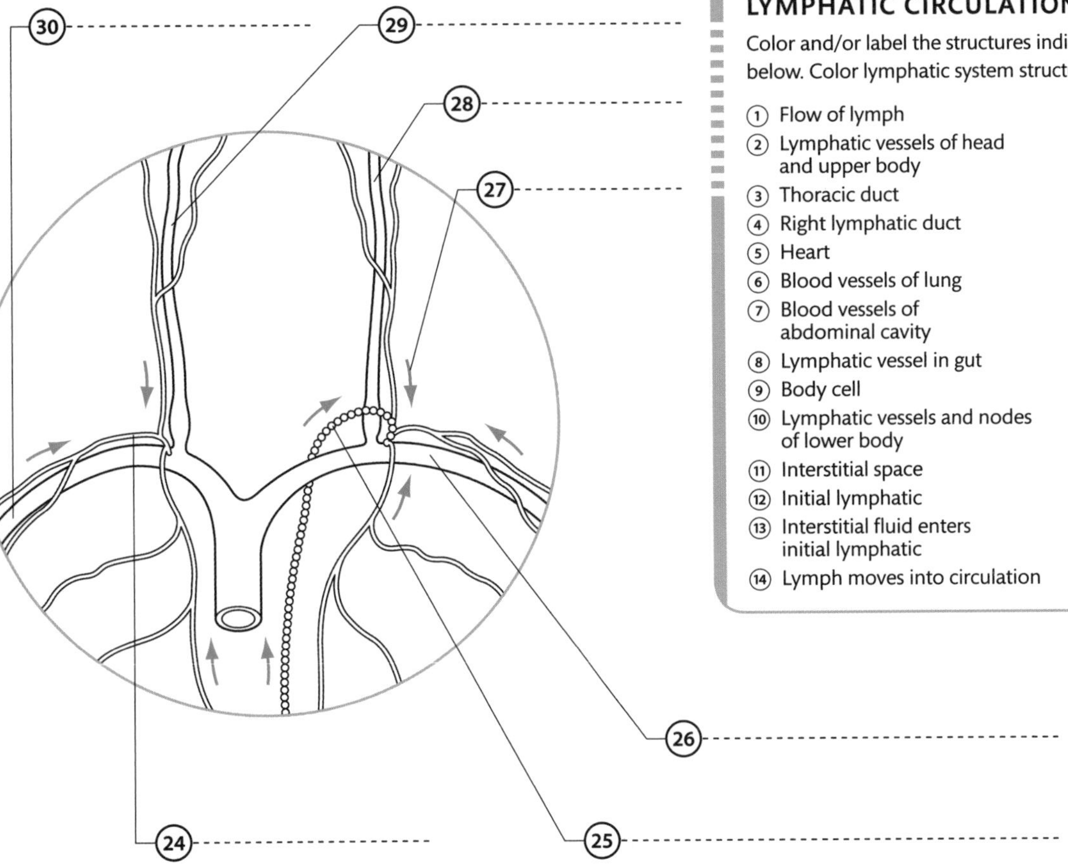

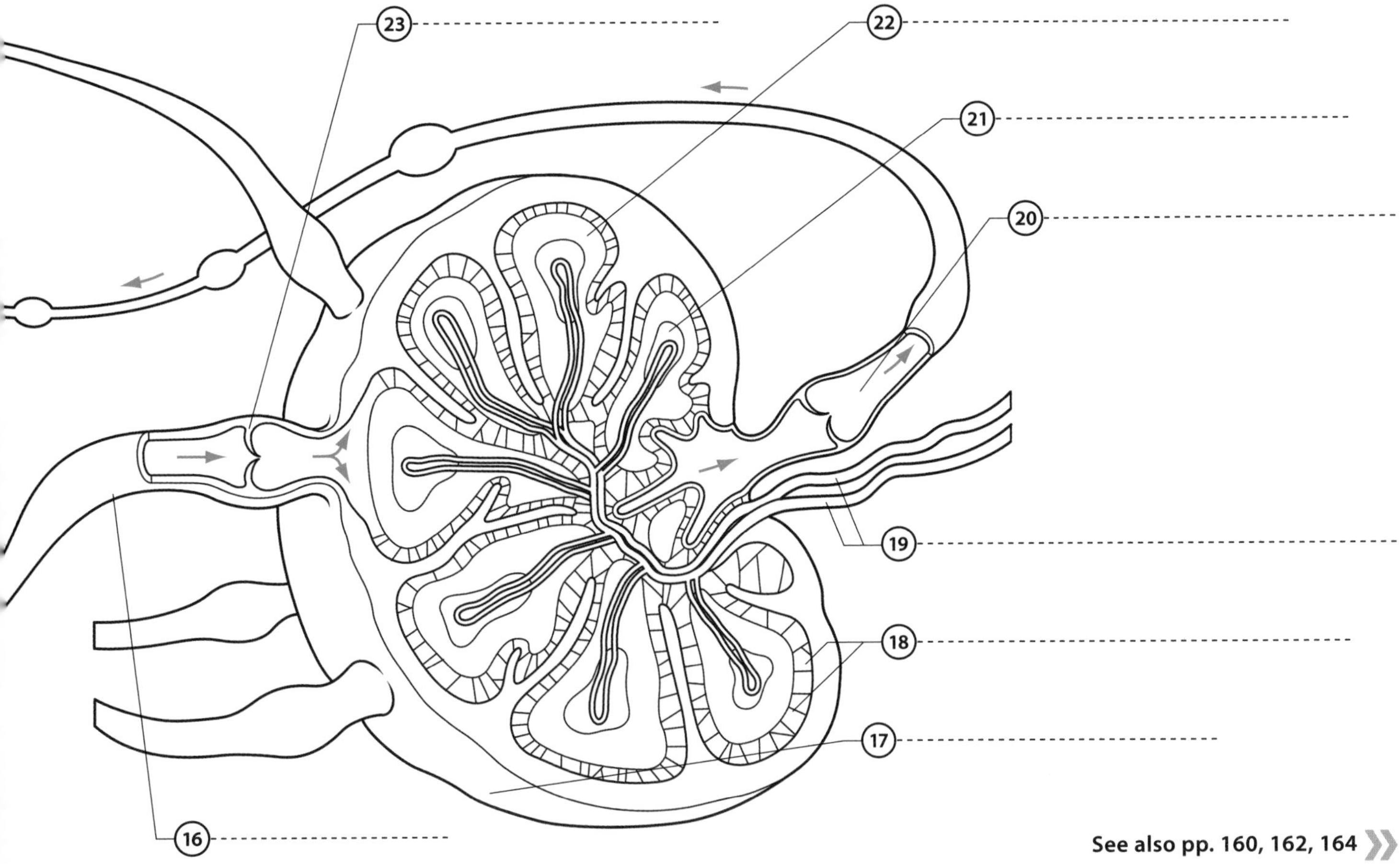

See also pp. 160, 162, 164

DIGESTIVE SYSTEM

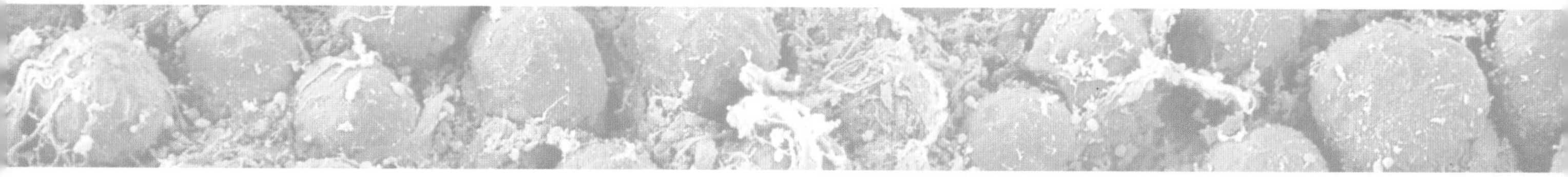

08

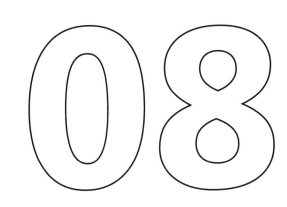

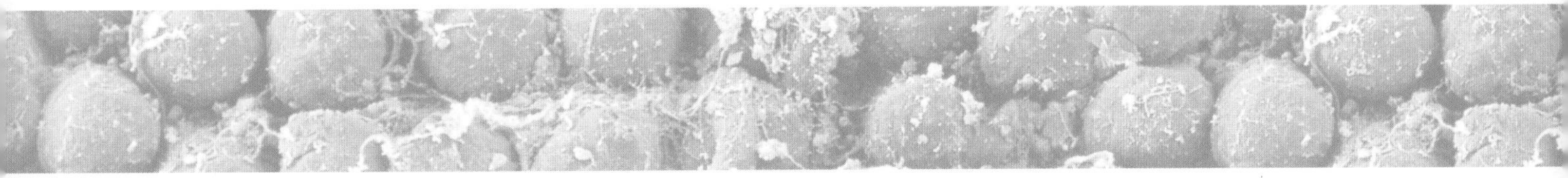

DIGESTIVE SYSTEM 1

THE DIGESTIVE SYSTEM CONSISTS OF THE LONG DIGESTIVE TRACT, OR ALIMENTARY CANAL, AND ASSOCIATED ORGANS, INCLUDING THE LIVER, GALLBLADDER, AND PANCREAS. ALONG THE COURSE OF THE TRACT, FOOD IS BROKEN DOWN AND ITS NUTRIENTS AND WASTE PRODUCTS ARE EXTRACTED.

DIGESTIVE TRACT

The digestive tract starts at the mouth, where the teeth, tongue, and saliva begin the process of breaking down food and forming it into a bolus that can be swallowed. The bolus passes down the pharynx, then travels through the esophagus to the stomach, small intestine, large intestine, and anus. In the small intestine, chemical digestion breaks the food down further into molecules small enough to be absorbed into the bloodstream. What cannot be digested is compacted as feces in the large intestine and eliminated through the anus. The digestive system also includes several glands: the salivary glands; the pancreas, which produces digestive juices; and the body's major nutrient processor, the liver.

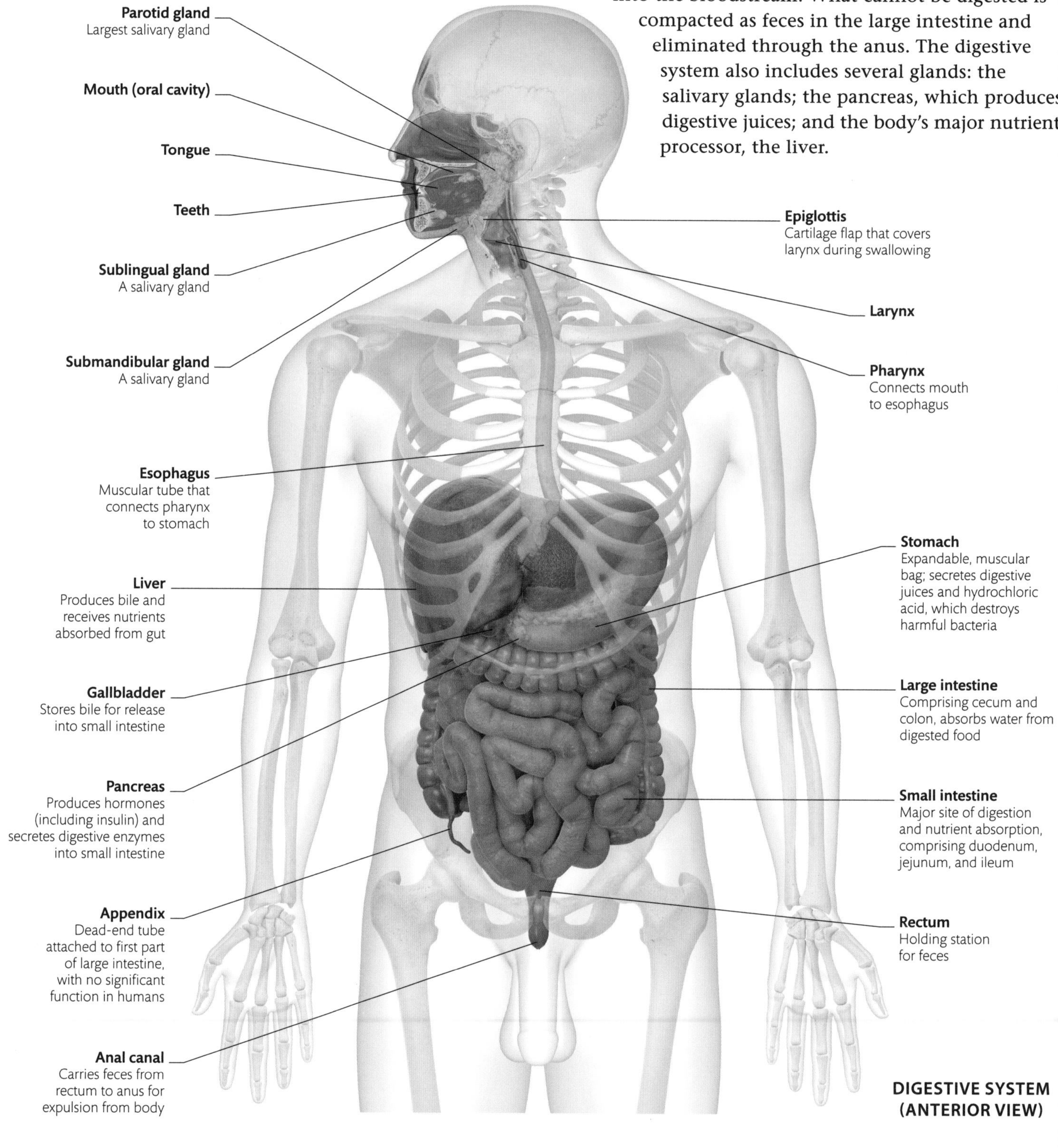

Parotid gland Largest salivary gland
Mouth (oral cavity)
Tongue
Teeth
Sublingual gland A salivary gland
Submandibular gland A salivary gland
Esophagus Muscular tube that connects pharynx to stomach
Liver Produces bile and receives nutrients absorbed from gut
Gallbladder Stores bile for release into small intestine
Pancreas Produces hormones (including insulin) and secretes digestive enzymes into small intestine
Appendix Dead-end tube attached to first part of large intestine, with no significant function in humans
Anal canal Carries feces from rectum to anus for expulsion from body
Epiglottis Cartilage flap that covers larynx during swallowing
Larynx
Pharynx Connects mouth to esophagus
Stomach Expandable, muscular bag; secretes digestive juices and hydrochloric acid, which destroys harmful bacteria
Large intestine Comprising cecum and colon, absorbs water from digested food
Small intestine Major site of digestion and nutrient absorption, comprising duodenum, jejunum, and ileum
Rectum Holding station for feces

DIGESTIVE SYSTEM (ANTERIOR VIEW)

PERITONEUM

The peritoneum is the largest serous membrane in the body. A complex, two-layered membrane, it produces a fluid that reduces friction between organs in the abdominal and pelvic cavities. The parietal peritoneum lines the abdominal wall; the fused slinglike layers of the visceral peritoneum, known as mesenteries, suspend organs within the abdomen, and carry nerves and blood vessels to them. The greater omentum is a specialized double-fold of fatty peritoneum hanging from the stomach.

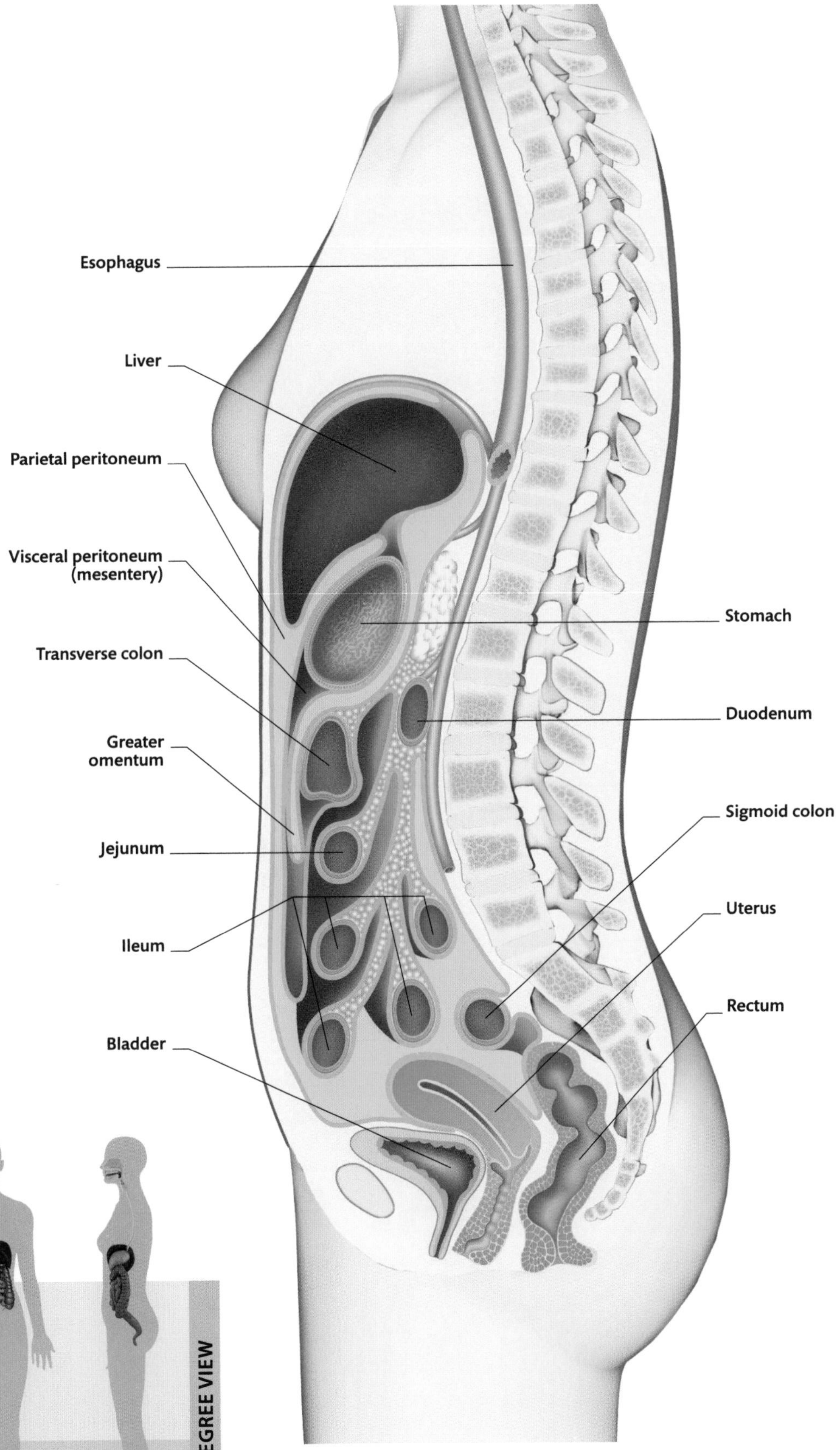

PERITONEUM (SAGITTAL SECTION, FEMALE)

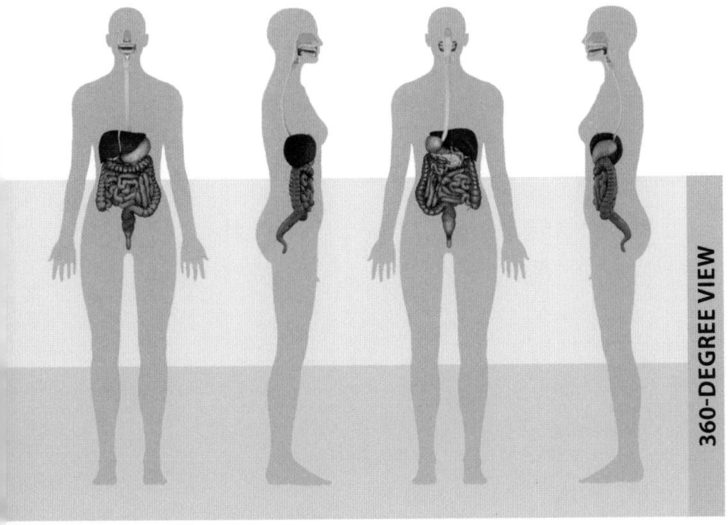

See also p. 172

DIGESTIVE SYSTEM 2

THE DIGESTIVE PROCESS PROVIDES NUTRIENTS AS RAW MATERIALS FOR METABOLISM—THE BODY'S INTERNAL BIOCHEMICAL REACTIONS AND PROCESSES. MOST NUTRIENTS ARE PROCESSED IN THE LIVER BEFORE BEING RELEASED INTO THE CIRCULATION.

BREAKDOWN OF NUTRIENTS

Nutrients encompass all substances that are useful to the body. These include complex chemicals broken down to release energy, chiefly carbohydrates and fats; proteins, which are mainly used for building the structural parts of cells; and vitamins and minerals, which ensure healthy body function. The digestive system absorbs the nutrients into the blood and lymph at different stages along the tract. Blood flow from the major absorption sites of the intestines is along the hepatic portal vein to the liver, the body's chief processor of nutrients. According to the body's needs, the liver breaks down some nutrients into smaller, simpler molecules, stores others, and releases others into the circulation.

FINAL STAGES OF DIGESTION

The colon is the last main site for breakdown and uptake of nutrients, including minerals, salts, and some vitamins. A considerable amount of water, mainly from the digestive juices, is also re-absorbed. Fiber, such as pectin and cellulose, gives bulk to the digestive remnants as they are compressed into feces awaiting expulsion. Fiber also helps to delay absorption of some molecules, including sugars, and so spreads out their uptake over an extended period. In addition, fiber binds with some fatty substances, such as cholesterol, and helps to prevent their over-absorption.

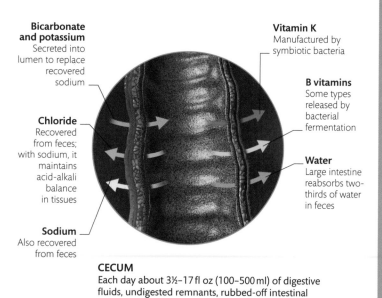

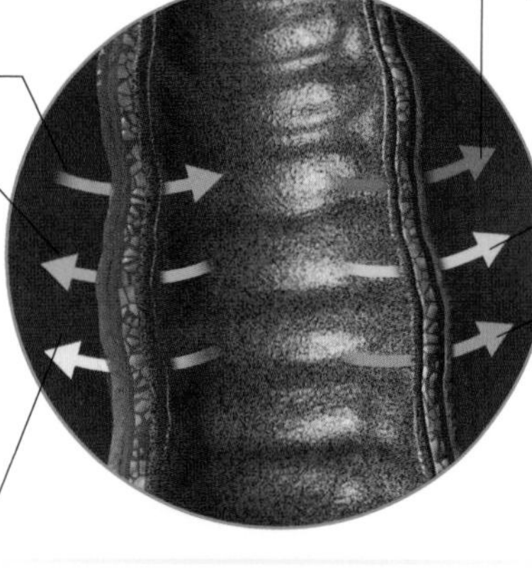
Bicarbonate and potassium Secreted into lumen to replace recovered sodium
Chloride Recovered from feces; with sodium, it maintains acid-alkali balance in tissues
Sodium Also recovered from feces
Vitamin K Manufactured by symbiotic bacteria
B vitamins Some types released by bacterial fermentation
Water Large intestine reabsorbs two-thirds of water in feces

CECUM
Each day about 3½–17 fl oz (100–500 ml) of digestive fluids, undigested remnants, rubbed-off intestinal linings, and other matter enters the first chamber of the large intestine, the cecum.

FATE OF NUTRIENTS

The digestive process takes on average, 12–24 hours. Food remains in the stomach for 2–4 hours, and in the small intestine for 1–5 hours. The final stages of digestion and waste compaction in the large intestine may take 12 hours. Different breakdown products are available for absorption at different times.

	MOUTH	STOMACH	SMALL INTESTINE	LARGE INTESTINE
PROTEINS		Hydrochloric acid and pepsin break protein into peptide chains	Peptidases snip peptides into amino acids for absorption	
CARBOHYDRATES	Salivary amylase begins starch digestion during chewing	Stomach acid inactivates salivary amylase	Enzymes, such as pancreatic amylase, yield simple sugars	
FATS (LIPIDS)		Gastric lipase splits lipids into fatty acids and monoglycerides	Pancreatic lipase products enter lacteals	
FIBER				Soluble fiber broken down, not absorbed
SOLUBLE				
INSOLUBLE				
WATER		Small amounts absorbed by stomach lining	Absorbed by small intestine lining	Most water absorbed by large intestine
FAT-SOLUBLE VITAMINS (A, D, K)			Emulsified by bile salts and absorbed	Further absorption; K manufacture by gut bacteria
WATER-SOLUBLE VITAMINS (B, C)			Dissolve and are absorbed relatively easily	Continued absorption
MINERALS				Most minerals dissolve easily as inorganic salts for uptake in the small intestine and colon
IRON				
SODIUM				
CALCIUM				

HOW FOOD IS USED

The three major food components—carbohydrates, proteins, and fats—yield different breakdown products. Carbohydrates (starches and sugars) can be reduced to the simple sugar glucose; proteins are cut into polypeptide chains, peptides, and finally into single amino acids; fats (lipids) are reduced to fatty acids and glycerol. The major use of glucose is as the body's most adaptable and readiest source of energy. Uses of fatty acids include forming the bi-lipid membranes around and inside cells. Amino acids are reassembled into the body's own proteins, both structural (collagen, keratin, and similar tough substances) and functional (enzymes). However, the body can adapt and divert nutrients to different uses as conditions dictate.

ENERGY PRODUCTION
The simple sugar glucose is the energy source used by all body cells to power their life processes. Fats, or in starvation circumstances proteins, also suffice, either from the liver or mobilized from storage in tissues.

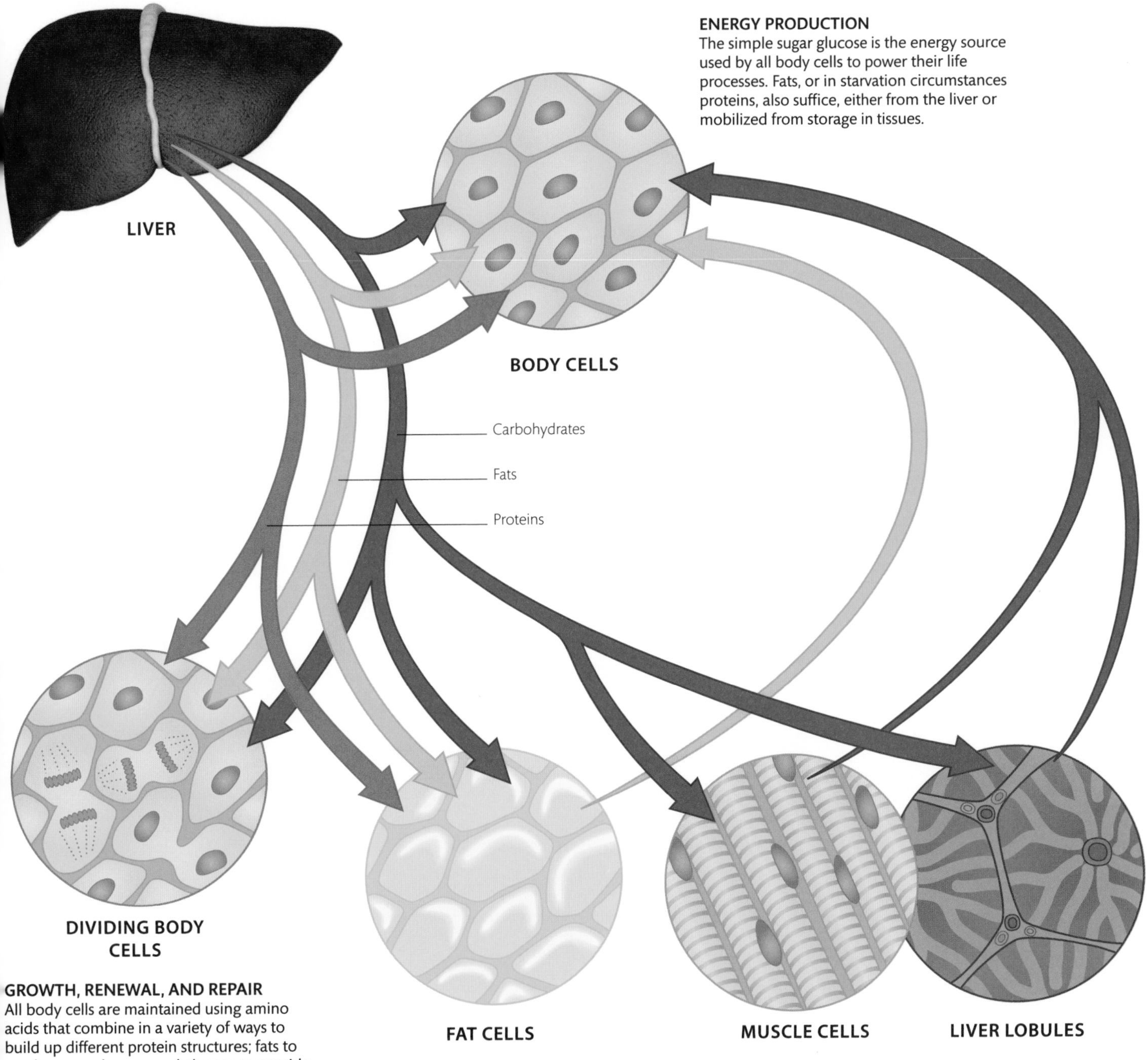

GROWTH, RENEWAL, AND REPAIR
All body cells are maintained using amino acids that combine in a variety of ways to build up different protein structures; fats to produce membranes; and glucose to provide energy. Cells dividing for growth or repair require increased supplies of these nutrients.

ENERGY STORAGE (FAT CELLS, MUSCLE CELLS, LIVER)
Surplus glucose is converted into glycogen, which can be stockpiled in the liver and muscle cells. Fatty acids are a concentrated energy store. They are derived either directly from dietary fats, from conversion of excess amino acids, or from conversion of glucose.

See also p. 170 »

HEAD AND NECK 1

THE HEAD AND NECK CONTAIN THE FIRST PARTS OF THE DIGESTIVE TRACT AND ARE WHERE THE PROCESSES OF MECHANICAL AND CHEMICAL BREAKDOWN OF FOOD BEGIN.

MOUTH AND THROAT

In the mouth, the teeth cut and grind up food, and the salivary glands secrete saliva. The tongue manipulates the food, and also has taste buds that help to quickly distinguish between appetizing food and food that may contain potentially harmful toxins. During swallowing, the tongue pushes up against the hard palate, the airway is sealed off, and the muscular pharynx contracts in a wave to push the bolus of food down into the esophagus.

SAGITTAL VIEW

Color and/or label the structures indicated on the diagram using the key below.

1. Tongue
2. Lower incisor
3. Upper incisor
4. Upper lip
5. Lower lip
6. Submandibular duct
7. Sublingual gland
8. Geniohyoid
9. Mylohyoid
10. Submandibular gland
11. Hyoid bone
12. Pre-epiglottic fat body
13. Epiglottis
14. Vocal fold
15. Esophagus
16. Laryngopharynx
17. Oropharynx
18. Nasopharynx
19. Parotid gland
20. Hard palate

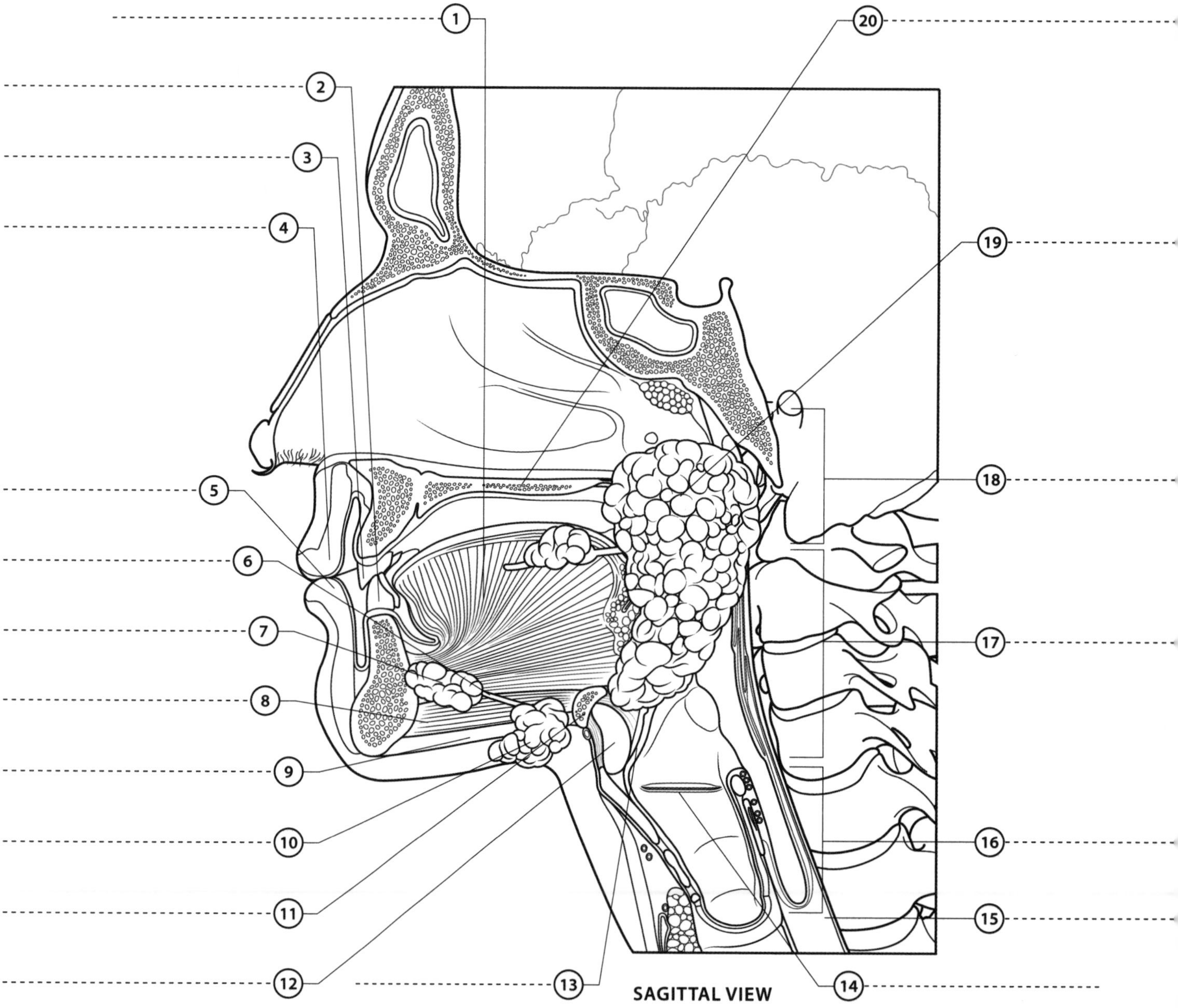

SAGITTAL VIEW

SALIVARY GLANDS

There are three pairs of salivary glands: the parotids, in front of and just below each ear; the submandibulars, on the inner ides of the mandible; and the sublinguals in the floor of the mouth, below the tongue. In addition, numerous small accessory glands are found in the mucous membranes lining he mouth and tongue. Saliva contains digestive enzymes hat begin the chemical breakdown of food, and lubricates food to make chewing and swallowing easier.

SALIVARY GLANDS

Color and/or label the structures indicated on the diagram using the key below.

1. Parotid gland
2. Accessory parotid gland
3. Nasal cavity
4. Parotid (Stensen's) duct
5. Uvula
6. Tongue
7. Tooth (central incisor)
8. Sublingual duct
9. Sublingual gland
10. Submandibular duct
11. Submandibular gland

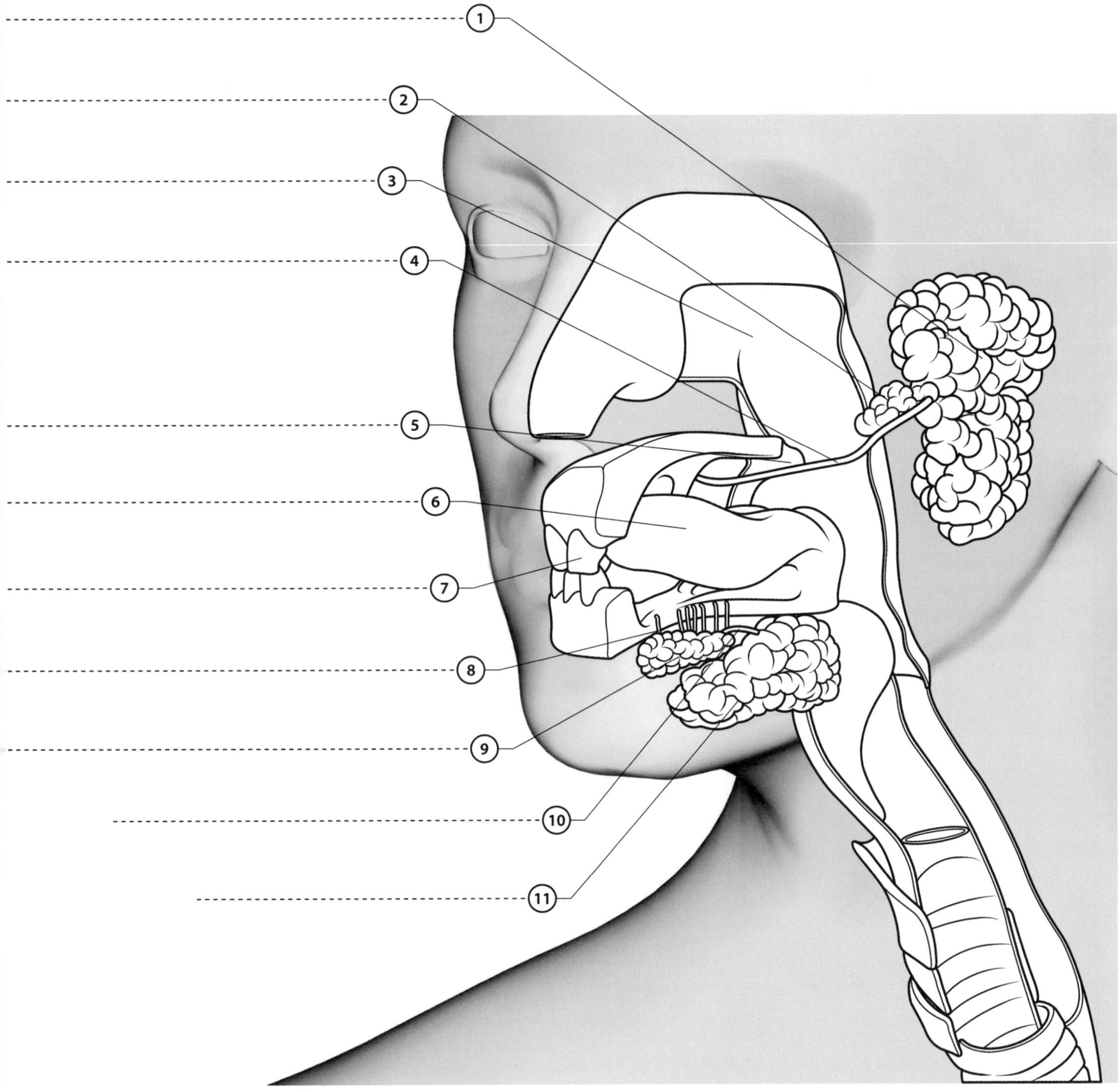

See also pp. 28, 58, 60, 118, 134, 160

HEAD AND NECK 2

THE MAIN DIGESTIVE SYSTEM STRUCTURES IN THE MOUTH ARE THE TEETH AND TONGUE, USED FOR THE INITIAL PROCESSING AND SWALLOWING OF FOOD, AND THE SALIVARY GLANDS.

TEETH

The complete set of adult teeth comprises a total of 32 teeth. Each jaw has 16 teeth: four incisors, two canines, four premolars, and six molars.

ADULT TEETH

Color and/or label the structures indicated on the diagram using the key below.

1. Incisors
2. Canine
3. Premolars
4. First molar
5. Second molar
6. Third molar (wisdom tooth)
7. Second molar
8. First molar
9. Premolars
10. Canine
11. Incisors

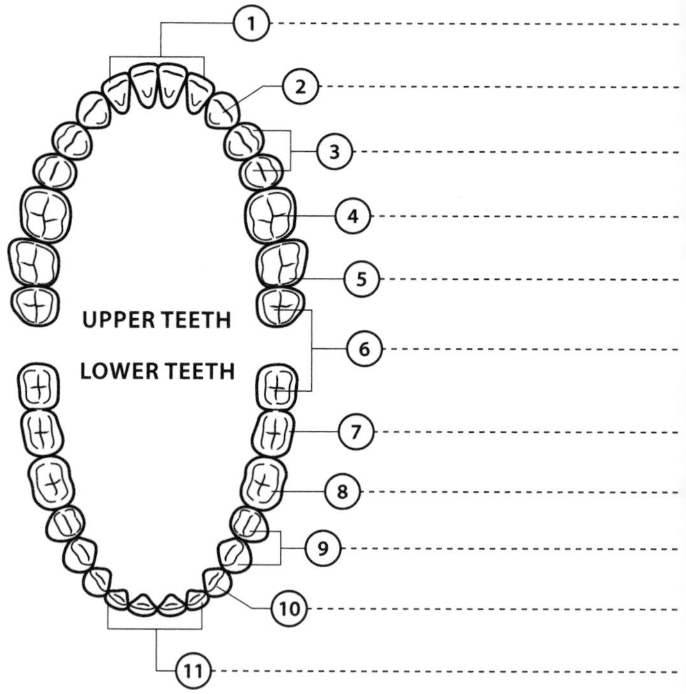

ADULT TEETH

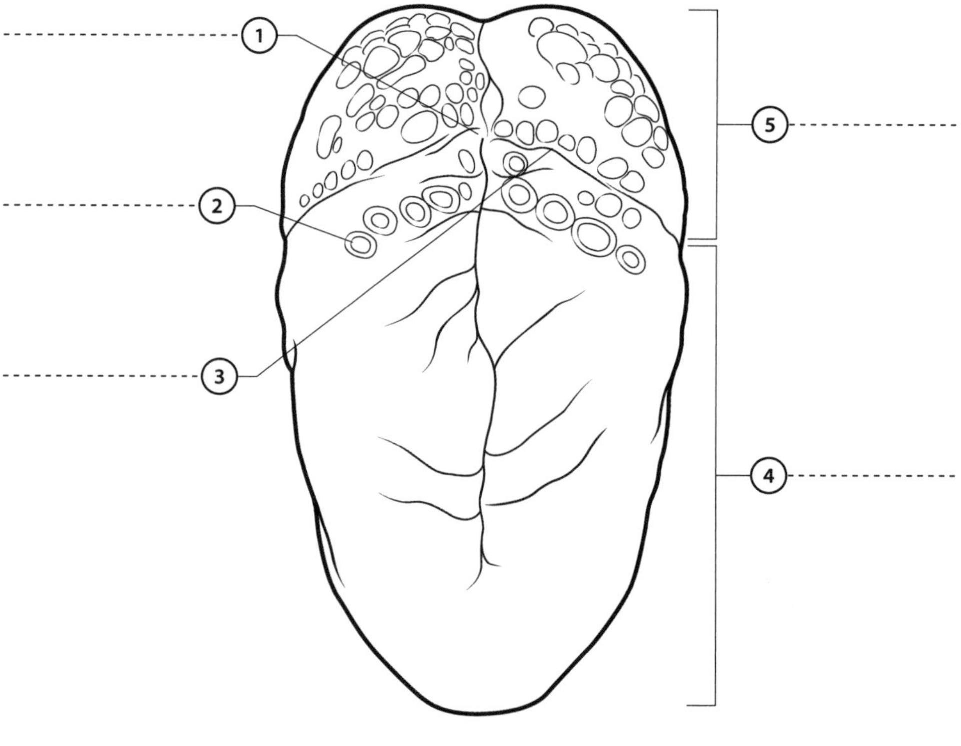

TONGUE

TONGUE

The tongue consists largely of muscle and is covered with mucous membrane. The upper surface of the tongue is covered with small protruberances called papillae. Filiform papillae are tiny and hair-shaped, giving the tongue a velvety texture. Foliate papillae are leaf-shaped and form a series of ridges on each side of the back of the tongue. Fungiform papillae are mushroom-shaped and bear taste buds. Vallate papillae occur at the back of the tongue; each is surrounded by a furrow that contains taste buds.

TONGUE

Color and/or label the structures indicated on the diagram using the key below.

1. Foramen cecum
2. Vallate papilla
3. Sulcus terminalis
4. Oral part of the tongue
5. Pharyngeal part of tongue

TOOTH STRUCTURE

Although teeth vary in shape and size, they have the same structure. Each consists of a hard outer shell surrounding a cavity of soft tissue, known as pulp, which contains nerves and blood vessels. The crown—the exposed part of the shell—is coated in a tough layer of enamel, beneath which is a layer of yellowish substance, similar to ivory, called dentine. The dentine and pulp form long roots that extend into the jawbone and are covered by the firm, fleshy gums.

SECTION THROUGH TOOTH

Color and/or label the structures indicated on the diagram using the key below.

1. Crown
2. Neck
3. Root
4. Blood vessel
5. Nerve
6. Jawbone (maxilla or mandible)
7. Cementum
8. Periodontal ligament
9. Pulp
10. Gum (gingiva)
11. Dentine
12. Enamel

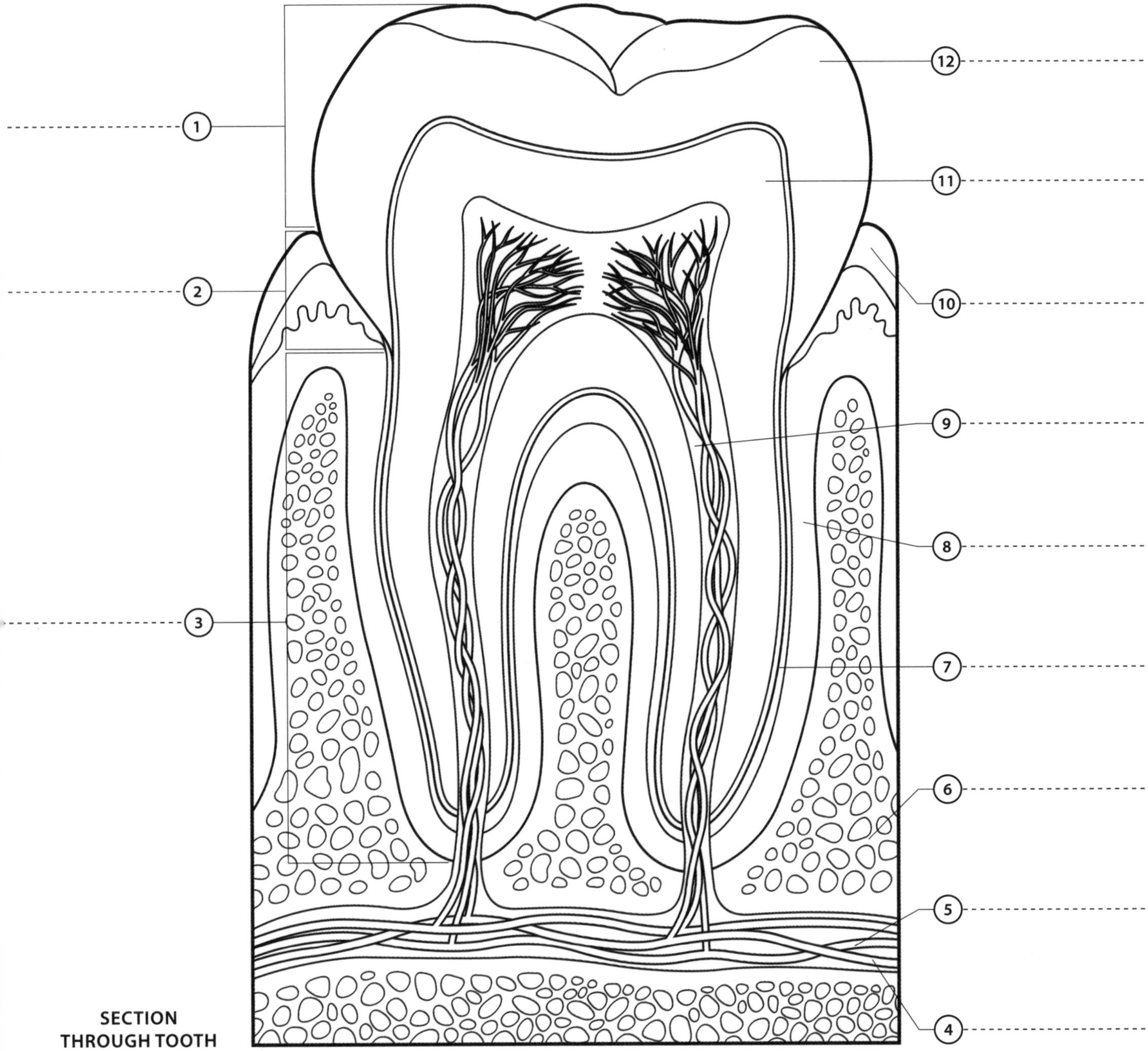

SECTION THROUGH TOOTH

See also p. 174

THORAX, ABDOMEN, AND PELVIS 1

MOST OF THE SPACE IN THE ABDOMEN AND PELVIS IS OCCUPIED BY THE DIGESTIVE TRACT AND ITS ASSOCIATED ORGANS.

STOMACH

The stomach is a J-shaped muscular bag linking the esophagus to the duodenum, the first part of the small intestine. In the stomach, gastric glands secrete hydrochloric acid, which destroys potentially harmful microorganisms, and the enzyme pepsin, which initiates the breakdown of protein in food. Peristalsis, generated by contractions of muscle in the stomach wall, churns the food into a liquid (called chyme) and pushes it to the pyloric sphincter at the stomach's exit.

STOMACH

Color and/or label the structures indicated on the diagram using the key below.

1. Esophagus
2. Longitudinal muscle layer
3. Pyloric sphincter
4. Duodenum
5. Secretory cell of gastric gland
6. Muscularis
7. Submucosa
8. Mucosa
9. Gastric gland
10. Gastric pit
11. Gastric mucosa (stomach lining)
12. Secretion
13. Rugae
14. Oblique muscle layer
15. Circular muscle layer

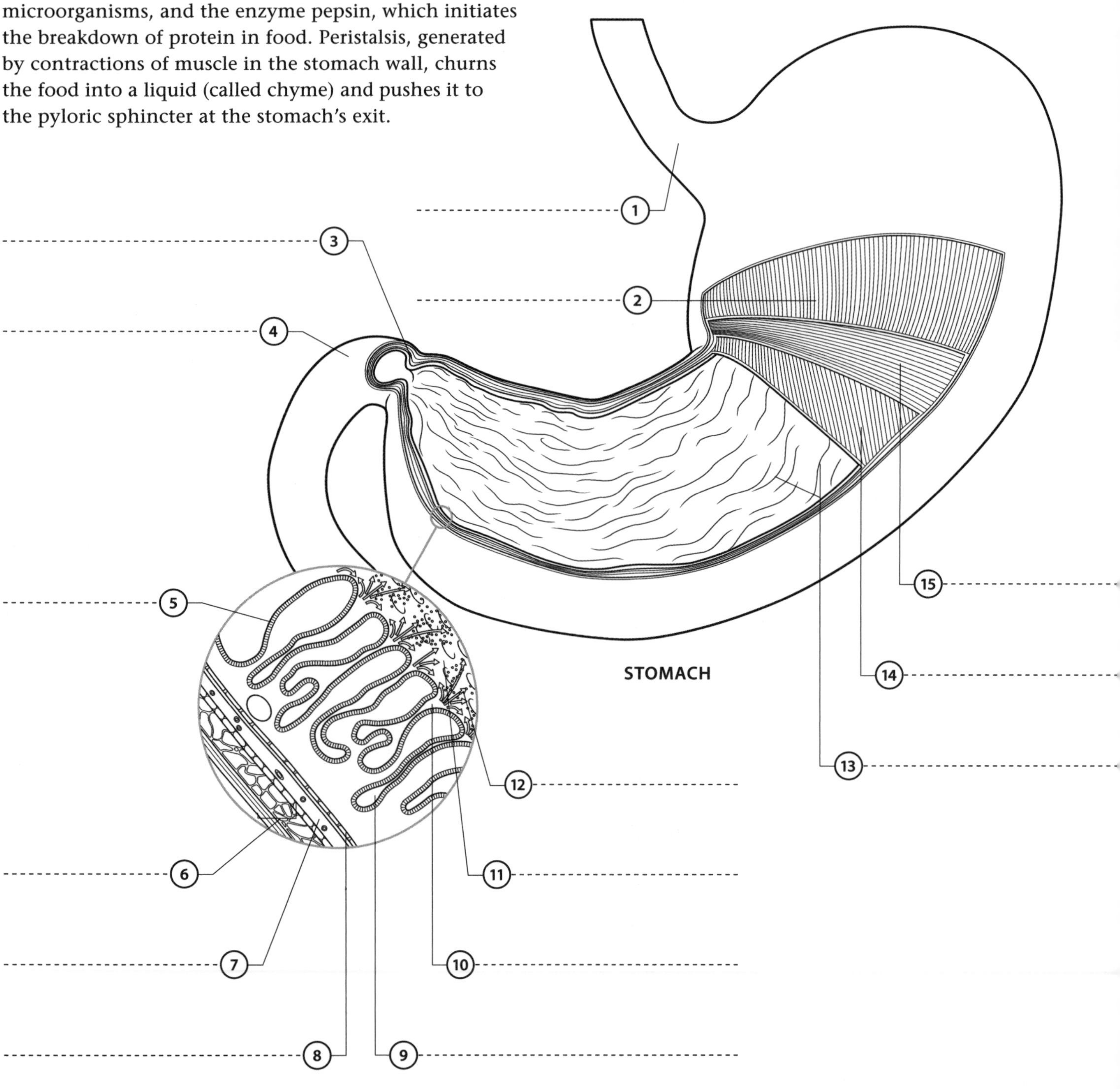

STOMACH

ANATOMY

The stomach and liver occupy the upper abdomen, and the coils of the small intestine take up much of the middle of the abdominal cavity. The large intestine forms an M shape, starting at the cecum low down on the right. The ascending colon runs up the right flank and tucks under the liver. The transverse colon hangs below the liver and stomach, and the descending colon runs down the left side, becoming the sigmoid colon, which runs down into the pelvis to become the rectum.

ANTERIOR VIEW

Color and/or label the structures indicated on the diagram using the key below.

1. Esophagus
2. Liver
3. Gallbladder
4. Transverse colon
5. Ascending colon
6. Ileum
7. Cecum
8. Appendix
9. Rectum
10. Anal canal
11. Sigmoid colon
12. Descending colon
13. Jejunum
14. Stomach
15. Pancreas

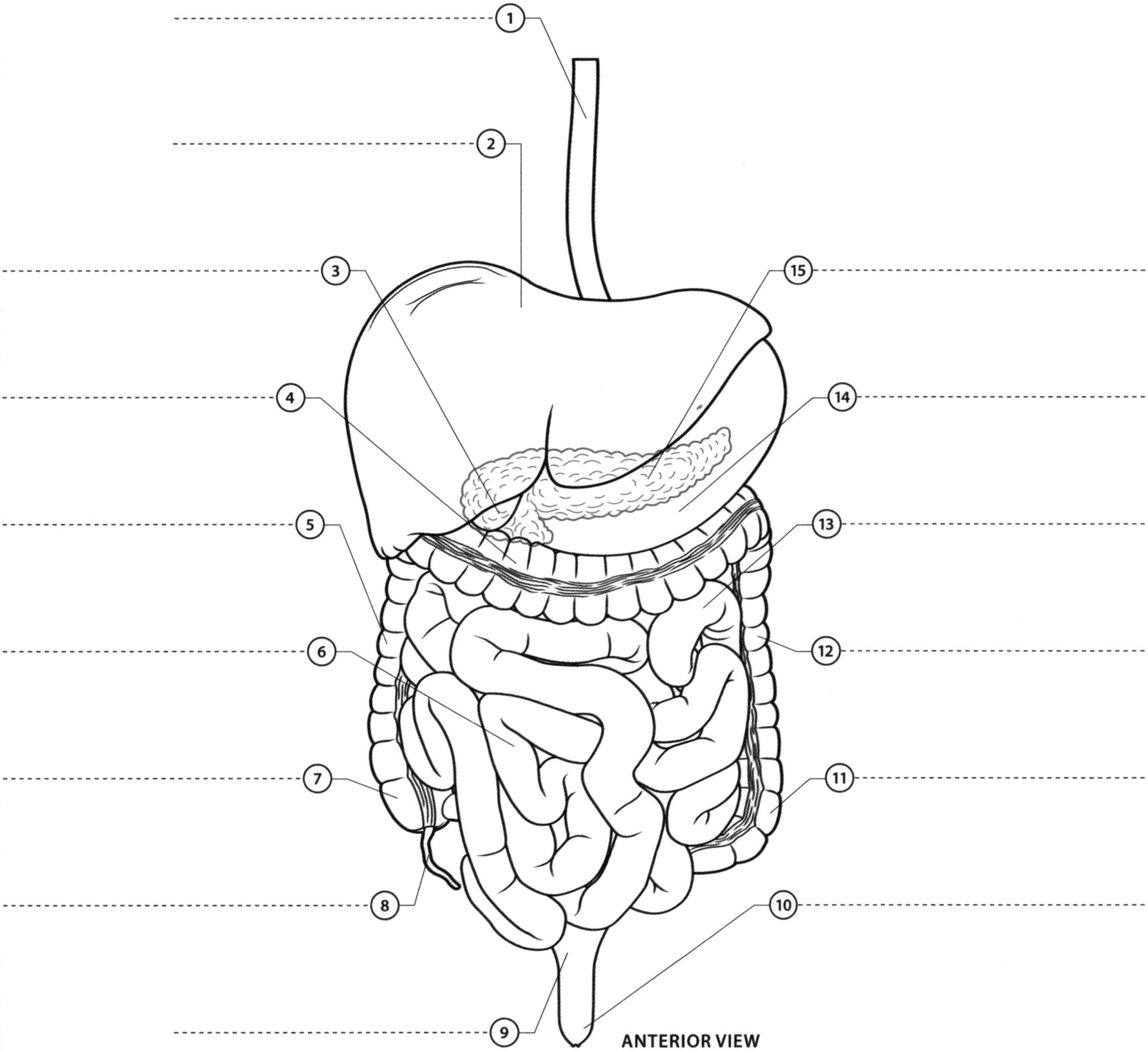

ANTERIOR VIEW

See also pp. 36, 38, 62, 66, 100, 120, 138, 144, 162, 204, 218, 224

THORAX, ABDOMEN, AND PELVIS 2

THE SMALL AND LARGE INTESTINES ARE THE LONGEST SECTIONS OF THE DIGESTIVE TRACT. THE SMALL INTESTINE IS ABOUT 17 FT (5 M) LONG, AND THE LARGE INTESTINE ABOUT 5 FT (1.5 M).

SECTION OF SMALL INTESTINE

Color and/or label the structures indicated on the diagram using the key below.

① Serosa
② Muscularis
③ Submucosa
④ Mucosa
⑤ Villus

SMALL INTESTINE

The small intestine consists of the duodenum, jejunum, and ileum. In the small intestine, digestion continues, nutrients are absorbed by intestinal villi, and peristalsis moves the liquid food on to the large intestine. The wall of the small intestine has four layers. From outside in, these are the serosa; the muscularis, which has longitudinal and circular smooth muscle fibers; the submucosa, containing blood and lymphatic vessels; and the mucosa, which is covered by tiny, fingerlike villi.

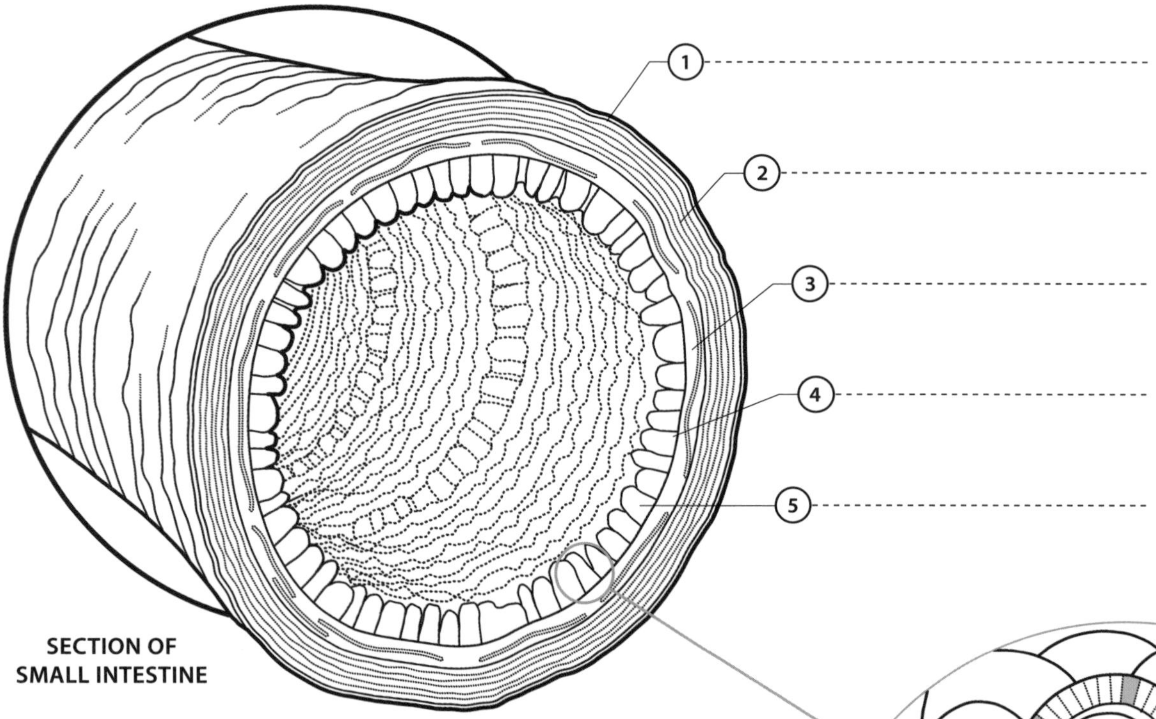

SECTION OF SMALL INTESTINE

INTESTINAL VILLI

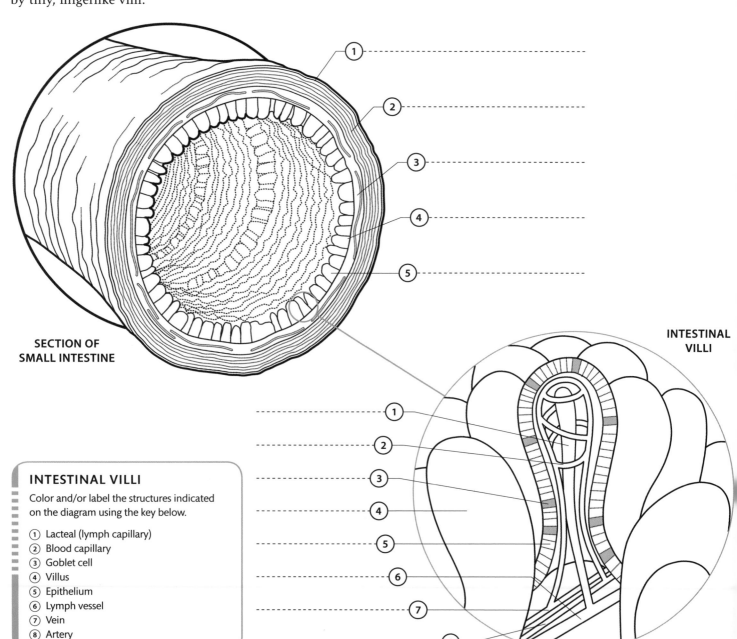

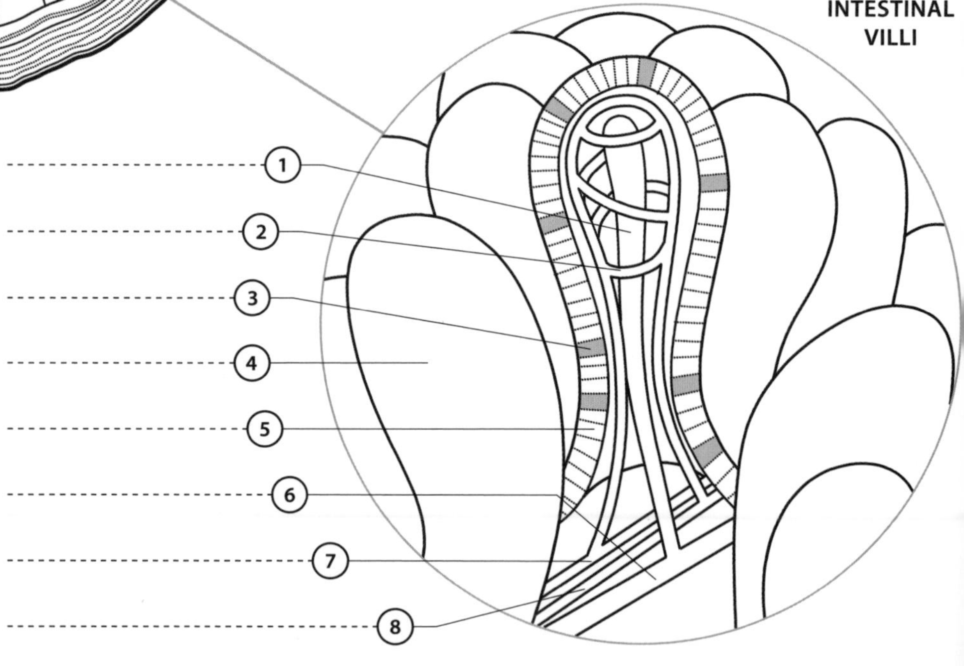

INTESTINAL VILLI

Color and/or label the structures indicated on the diagram using the key below.

① Lacteal (lymph capillary)
② Blood capillary
③ Goblet cell
④ Villus
⑤ Epithelium
⑥ Lymph vessel
⑦ Vein
⑧ Artery

LARGE INTESTINE

The large intestine comprises three main regions: the cecum, colon, and rectum. The cecum is a short pouch that links the small intestine to the colon. The colon absorbs water, sodium, and chloride from the liquid chyme, turning it into semisolid feces. The rectum, which is normally empty (except just before and during defecation) ends in the anal canal, the opening of which is surrounded by two anal sphincters. Like the small intestine, the colon wall has four layers: serosa, muscularis, submucosa, and mucosa. The mucosa secretes mucus, which lubricates the passage of feces.

SECTION OF LARGE INTESTINE

Color and/or label the structures indicated on the diagram using the key below.

1. Longitudinal muscle (taenia coli)
2. Circular muscle
3. Submucosa
4. Serosa
5. Mucosa

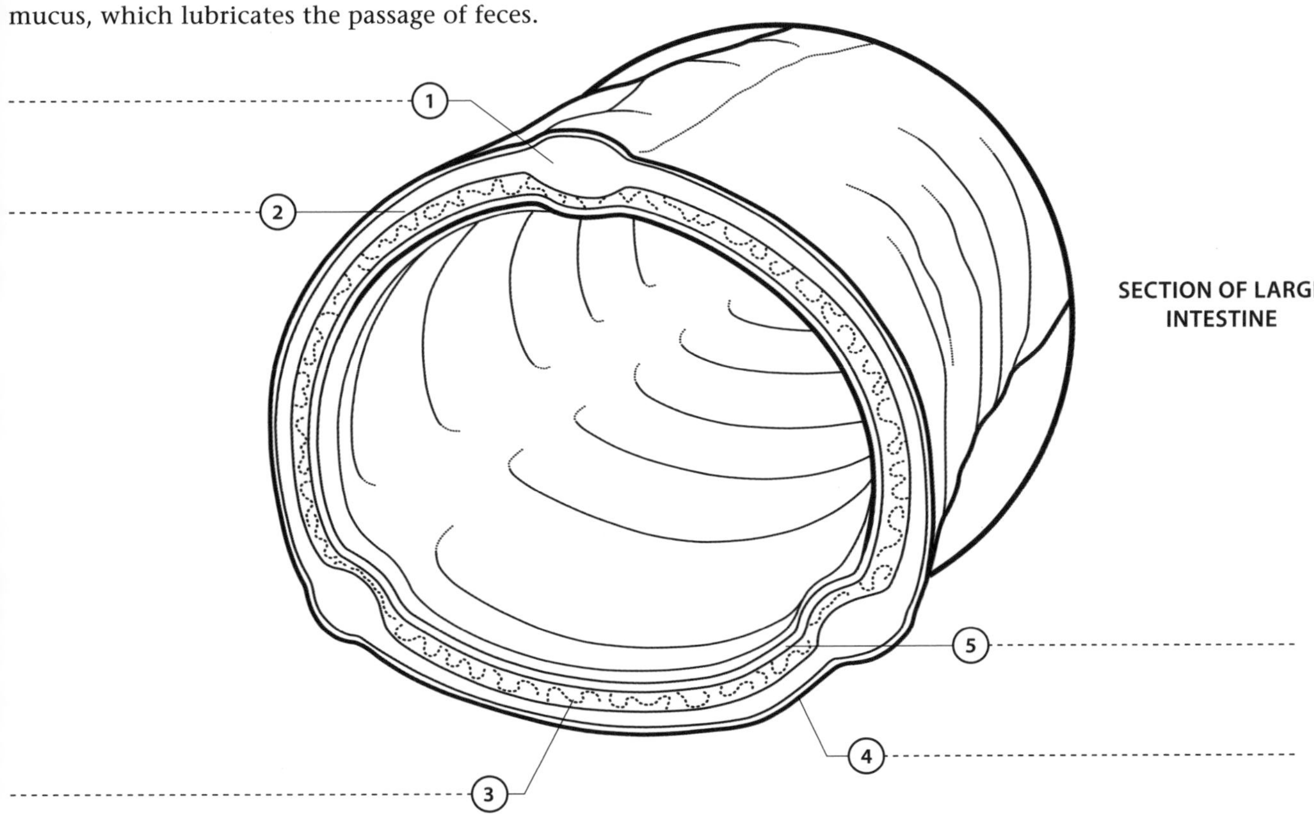

SECTION OF LARGE INTESTINE

SECTION THROUGH RECTUM

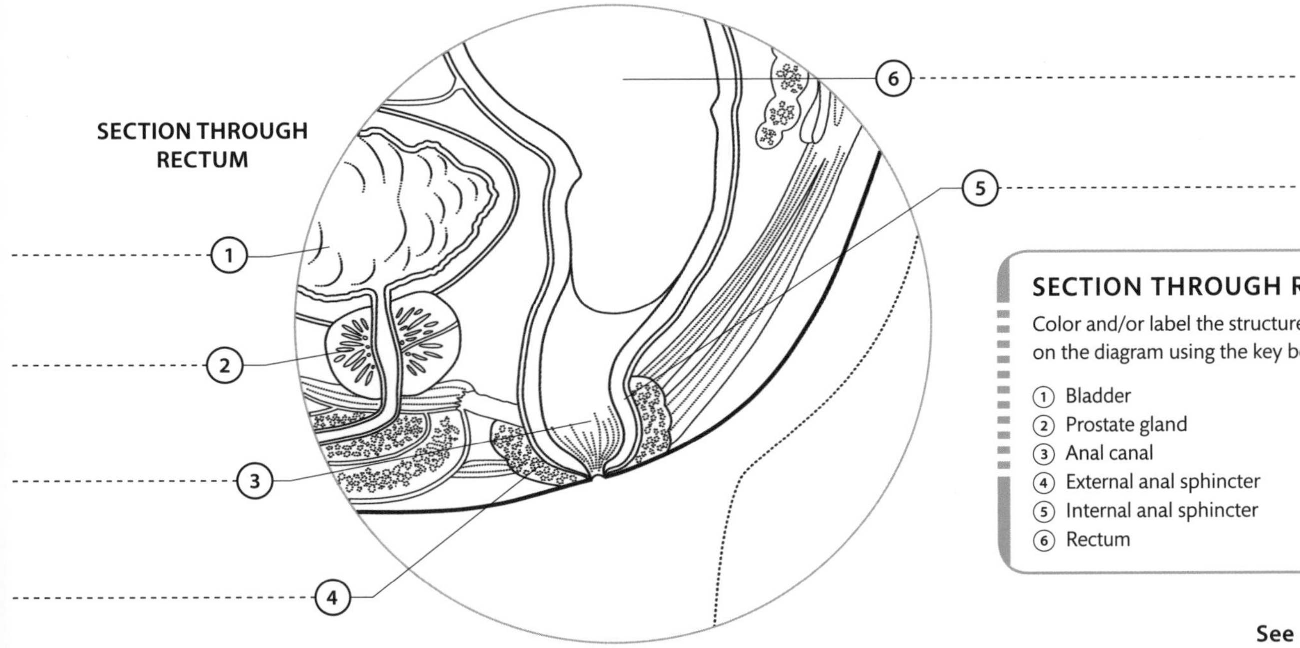

SECTION THROUGH RECTUM

Color and/or label the structures indicated on the diagram using the key below.

1. Bladder
2. Prostate gland
3. Anal canal
4. External anal sphincter
5. Internal anal sphincter
6. Rectum

See also p. 178 »

THORAX, ABDOMEN, AND PELVIS 3

THE LIVER IS THE LARGEST INTERNAL ORGAN AND PLAYS A VITAL ROLE IN NUMEROUS METABOLIC PROCESSES, MANY OF THEM RELATED TO DIGESTION. THE PANCREAS PRODUCES DIGESTIVE ENZYMES AND ALSO HORMONES.

LIVER

Weighing up to 6 lb (3 kg), the liver fills the upper right abdomen below the diaphragm. It produces bile, which is stored in the gallbladder, and receives nutrients from the gut via the hepatic portal vein. (The liver has two blood supplies: the hepatic artery, delivering oxygenated blood, and the hepatic portal vein.) As well as processing the nutrients, it detoxifies substances such as alcohol and drugs, helps to regulate glucose levels, stores various vitamins and minerals, plays a role in the immune system, and recycles red blood cells. At a microscopic level, the liver's structural units (lobules) are made up of hepatocytes, tiny branches of the hepatic artery and vein, and bile ducts. The lobules filter incoming blood into constituents destined for bile ducts, storage, or waste disposal.

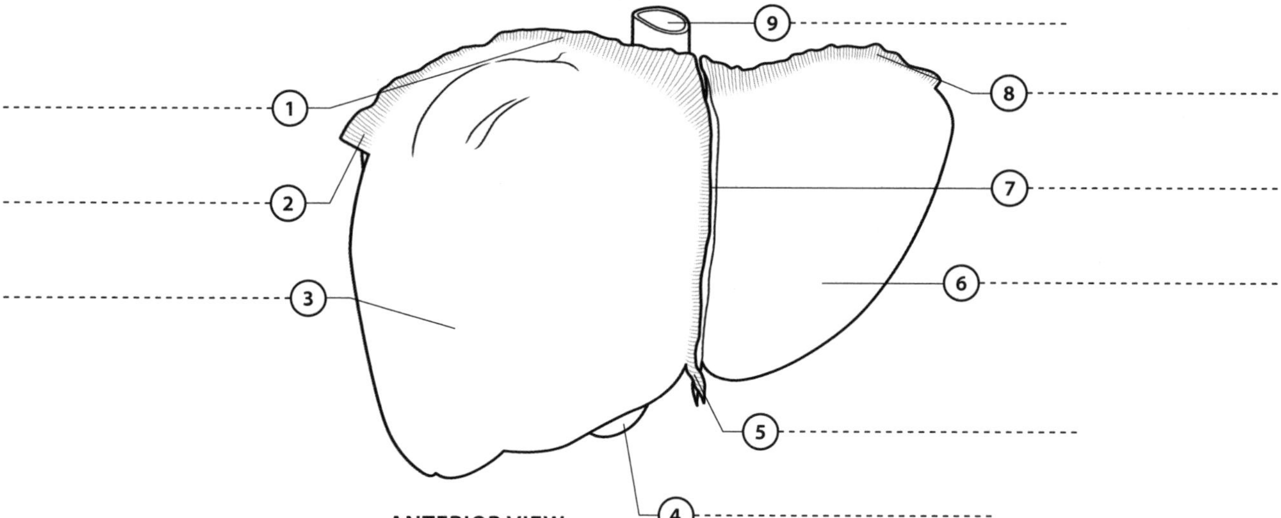

ANTERIOR VIEW

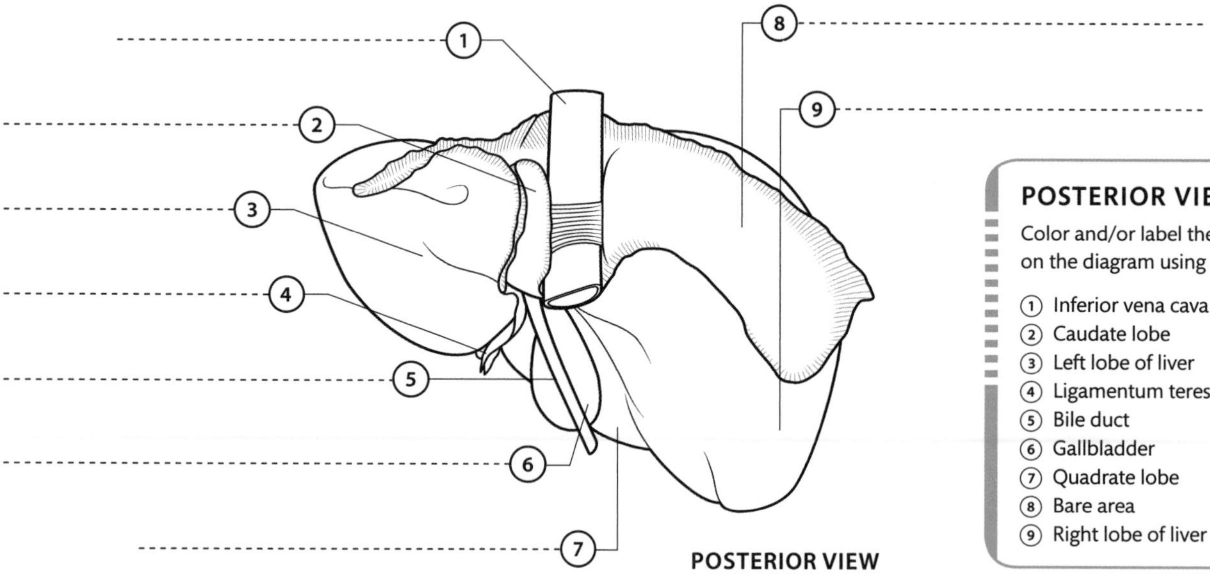

POSTERIOR VIEW

ANTERIOR VIEW

Color and/or label the structures indicated on the diagram using the key below.

1. Coronary ligament (cut)
2. Right triangular ligament (cut)
3. Right lobe of liver
4. Gallbladder
5. Ligamentum teres (cut)
6. Left lobe of liver
7. Falciform ligament (cut)
8. Left triangular ligament (cut)
9. Inferior vena cava

POSTERIOR VIEW

Color and/or label the structures indicated on the diagram using the key below.

1. Inferior vena cava
2. Caudate lobe
3. Left lobe of liver
4. Ligamentum teres
5. Bile duct
6. Gallbladder
7. Quadrate lobe
8. Bare area
9. Right lobe of liver

LIVER LOBULES

Color and/or label the structures indicated on the diagram using the key below.

1. Artery
2. Central vein
3. Bile duct
4. Vein
5. Exterior of lobule
6. Central vein

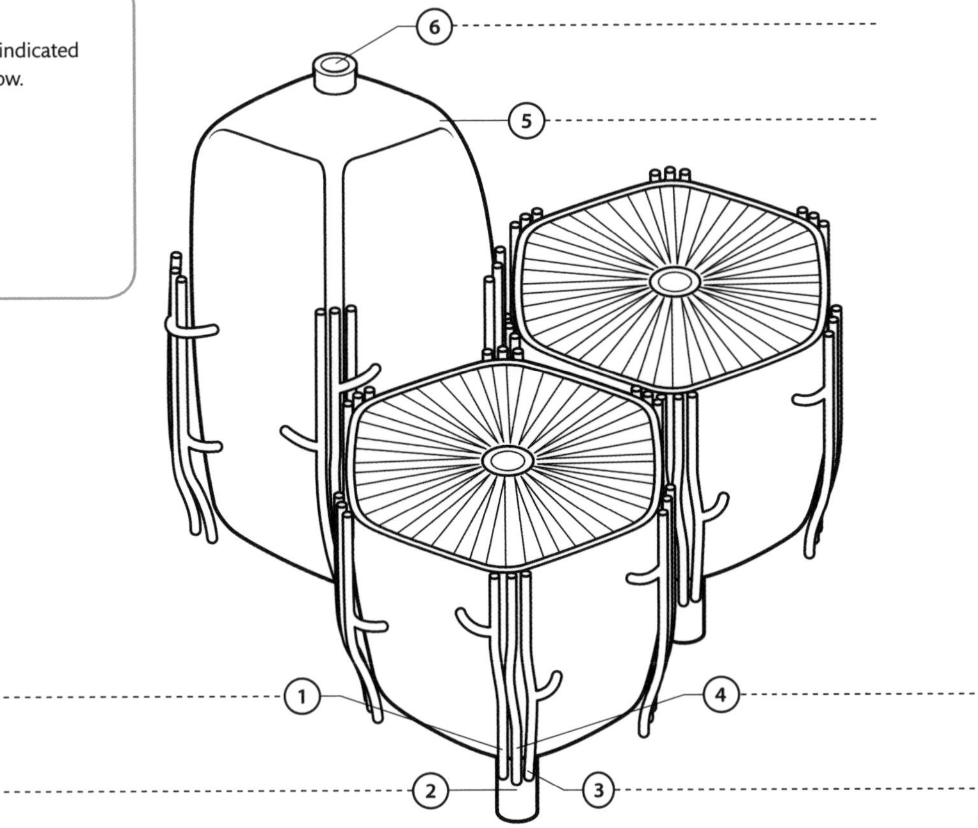

LIVER LOBULES

INSIDE A LOBULE

Color and/or label the structures indicated on the diagram using the key below.

1. Sinusoid receives blood from hepatic portal vein and hepatic artery
2. Branch of hepatic artery
3. Flow of oxygen-rich blood
4. Flow of nutrient-rich blood
5. Flow of bile
6. Hepatocytes process blood and make bile
7. Branch of bile duct conveys bile from liver
8. Branch of hepatic artery
9. Branch of portal vein
10. Central vein conveys processed blood away from liver

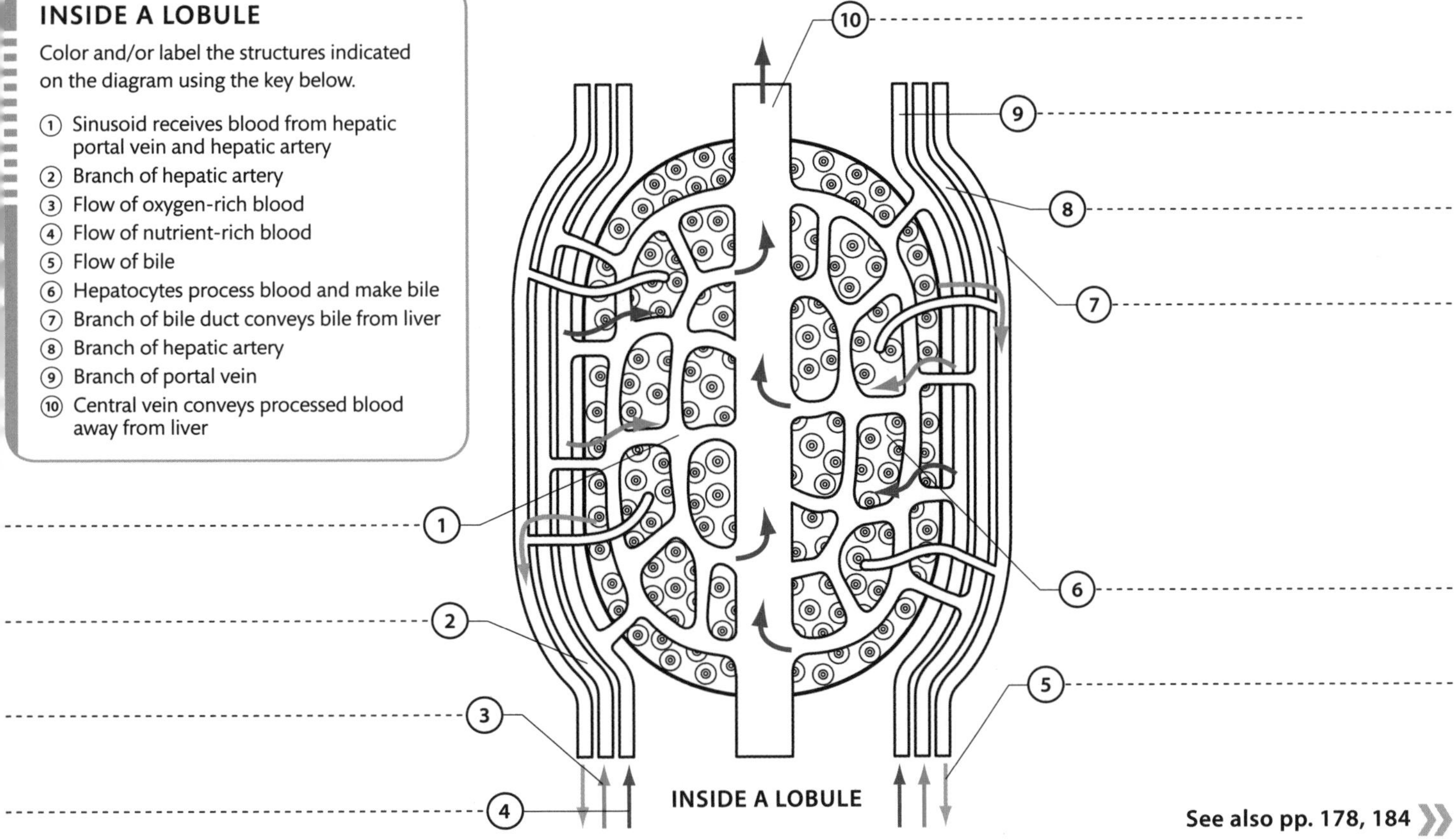

INSIDE A LOBULE

See also pp. 178, 184

THORAX, ABDOMEN, AND PELVIS 4

THE GALLBLADDER IS SITUATED UNDER THE LIVER'S RIGHT LOBE. THE PANCREAS LIES UNDER THE LIVER AND BELOW THE STOMACH.

HEPATIC PORTAL CIRCULATION

The liver receives two blood supplies. The hepatic artery delivers oxygenated blood. The hepatic portal vein supplies nutrient-rich (but oxygen-poor) blood from the digestive tract, before this blood returns to the heart. Veins from several organs, including the intestines, pancreas, stomach, and spleen, drain into the hepatic portal vein. After processing by the liver, the blood passes into the hepatic veins, which drain into the inferior vena cava.

PORTAL VEINS AND TRIBUTARIES

Color and/or label the structures indicated on the diagram using the key below.

1. Inferior vena cava
2. Hepatic vein
3. Liver
4. Hepatic portal vein
5. Large intestine (colon)
6. Superior mesenteric vein
7. Jejunal and ileal veins
8. Superior rectal vein
9. Middle rectal vein
10. Inferior rectal vein
11. Inferior mesenteric vein
12. Splenic vein
13. Right gastric vein
14. Left gastric vein
15. Esophageal vein

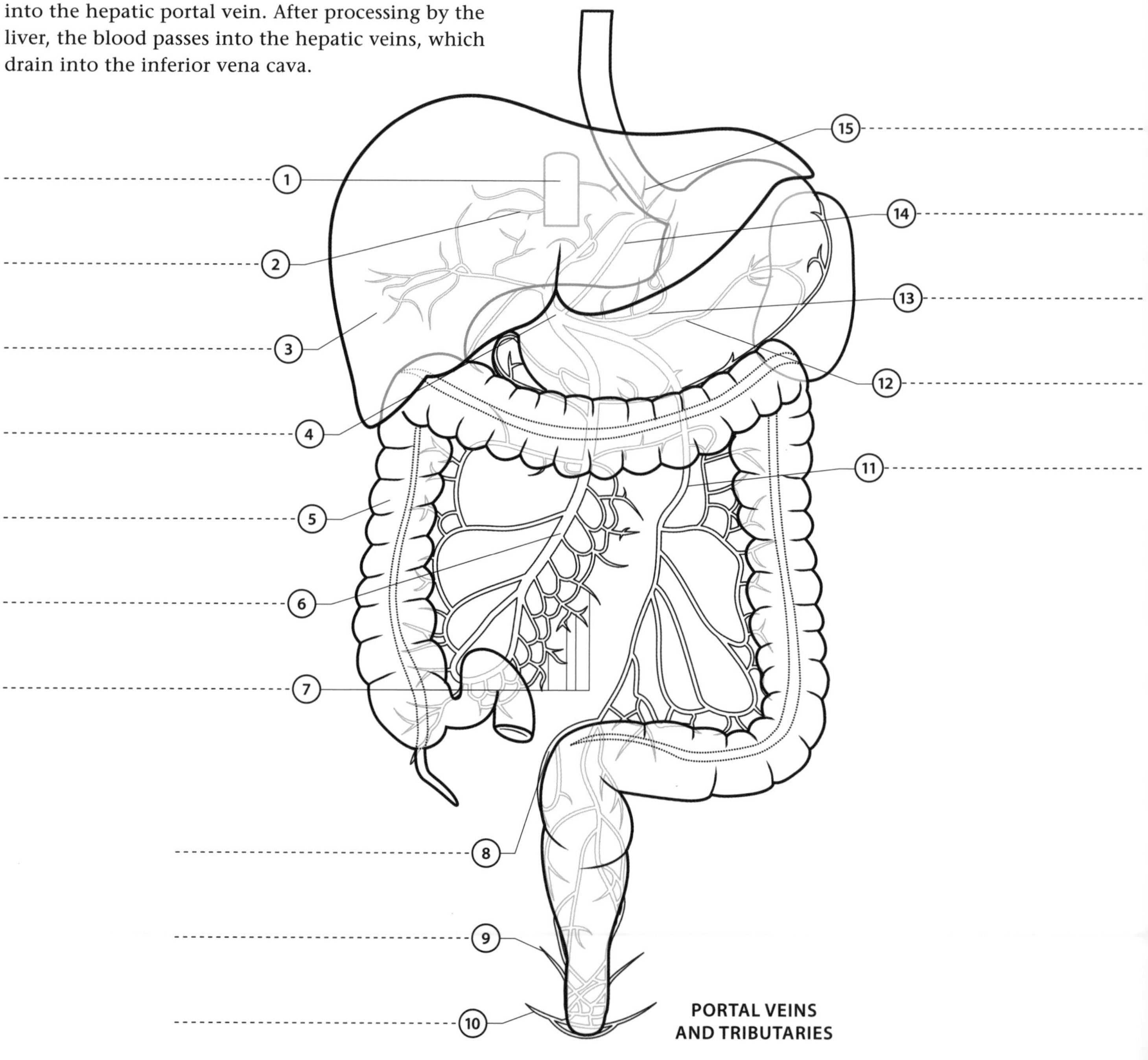

PORTAL VEINS AND TRIBUTARIES

GALLBLADDER AND PANCREAS

The gallbladder stores and concentrates bile produced by the liver. Bile from the liver's two lobes passes along the left and right hepatic ducts, then along the common hepatic and cystic ducts to the gallbladder. Bile is released after a meal, when it passes along the cystic duct and bile duct to enter the duodenum. The pancreas produces hormones (such as insulin, which regulates blood glucose levels), which are secreted into the blood, and pancreatic juice, which contains digestive enzymes and enters the duodenum via the main and accessory pancreatic ducts.

GALLBLADDER AND PANCREAS

Color and/or label the structures indicated on the diagram using the key below.

1. Cystic duct
2. Neck of gallbladder
3. Body of gallbladder
4. Fundus of gallbladder
5. Neck of pancreas
6. Duodenum
7. Accessory pancreatic duct
8. Main pancreatic duct
9. Head of pancreas
10. Uncinate process of pancreas
11. Tail of pancreas
12. Body of pancreas
13. Bile duct
14. Common hepatic duct
15. Left hepatic duct
16. Right hepatic duct

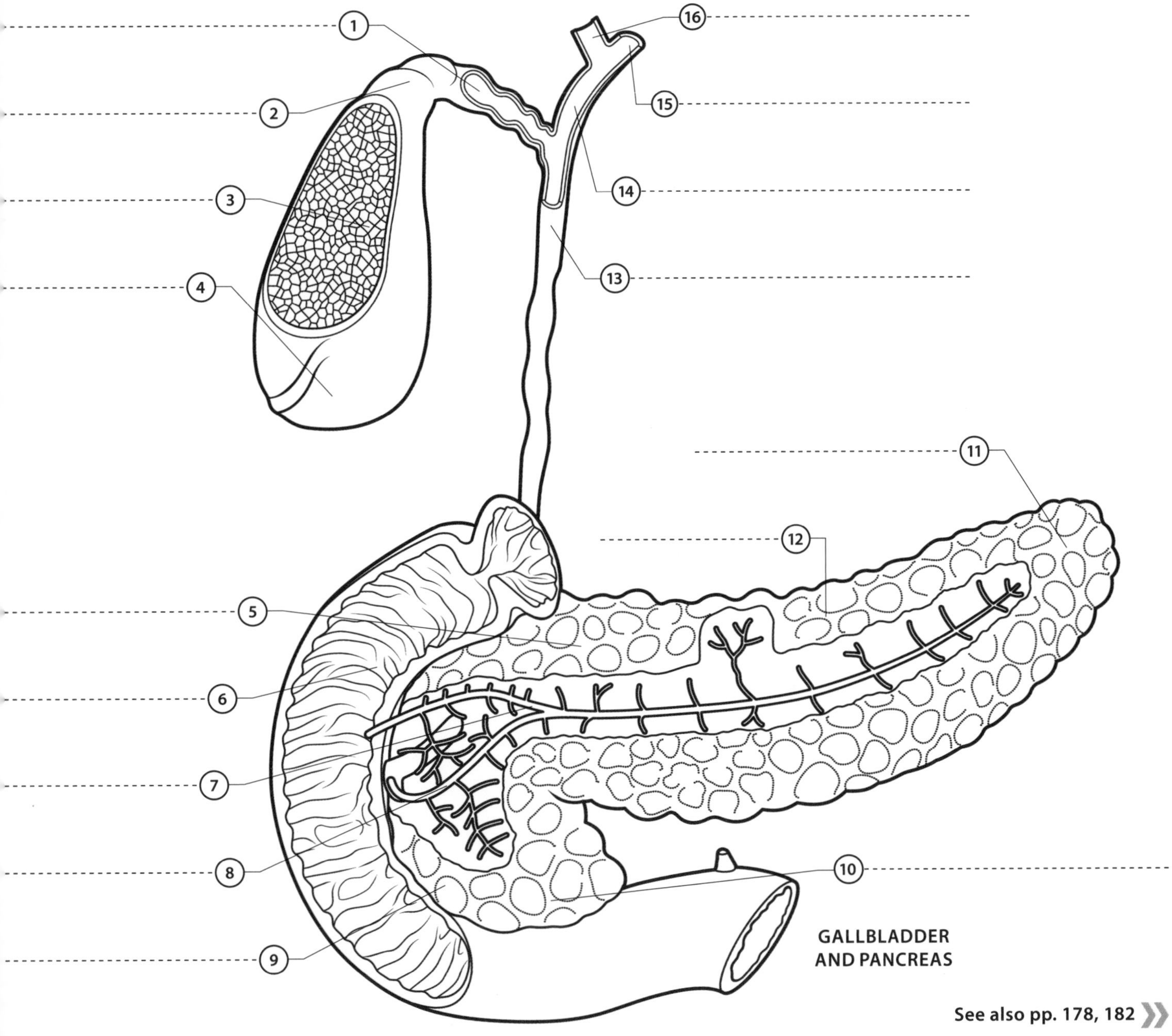

GALLBLADDER AND PANCREAS

See also pp. 178, 182

ENDOCRINE SYSTEM

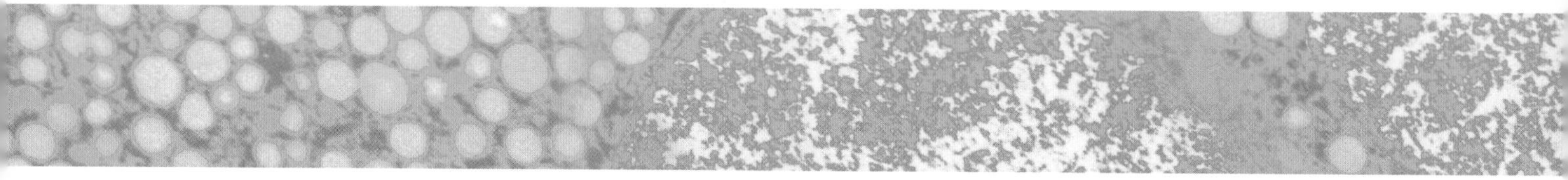

09

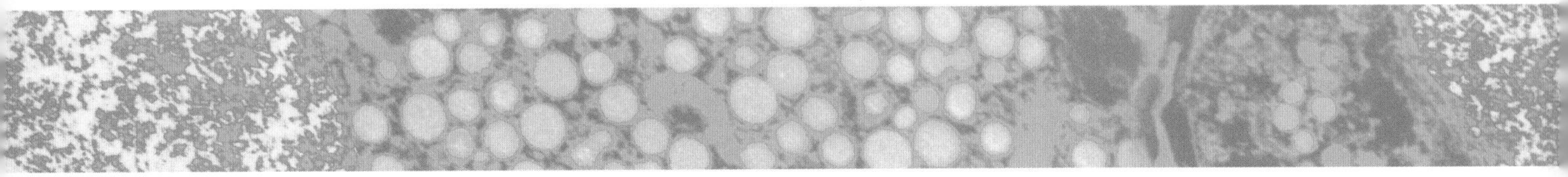

ENDOCRINE SYSTEM 1

THE ENDOCRINE SYSTEM USES CHEMICAL SIGNALS, IN THE FORM OF HORMONES, TO CONTROL AND COORDINATE BODY FUNCTIONS IN MUCH THE SAME WAY AS THE NERVOUS SYSTEM USES ELECTRICAL SIGNALS. BOTH SYSTEMS INTEGRATE IN THE BRAIN AND COMPLEMENT ONE ANOTHER.

FUNCTION OF ENDOCRINE SYSTEM

Hormones are made by the endocrine system, mainly in glands, such as the thyroid, but also in tissues within certain organs, such as the testes. Endocrine glands secrete their hormones directly into the blood, by which means they reach every target cell in the body. Each hormone has a specific molecular shape that slots only into receptors on its target tissues or organs. Hormones regulate processes such as the breakdown of chemical substances in metabolism, fluid balance and urine production, the body's growth and development, and sexual reproduction. Hormone output from a gland can be influenced by several factors, including levels of chemicals in the blood and input from the nervous system.

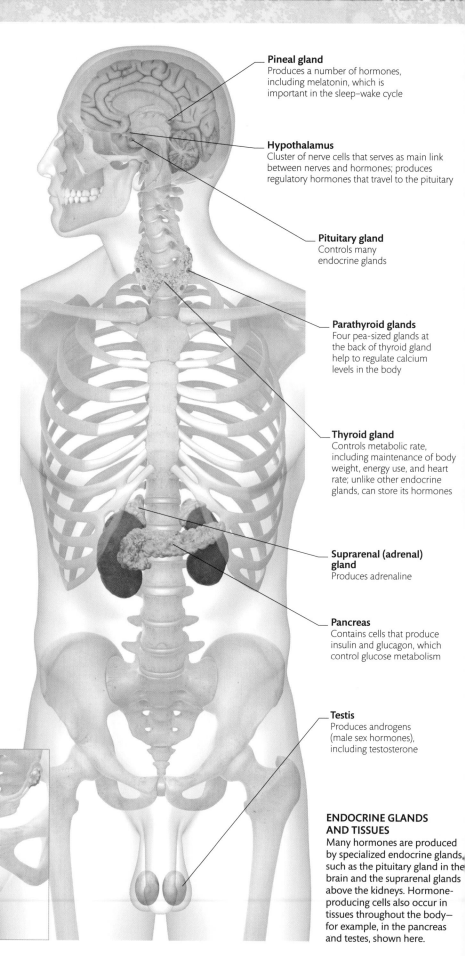

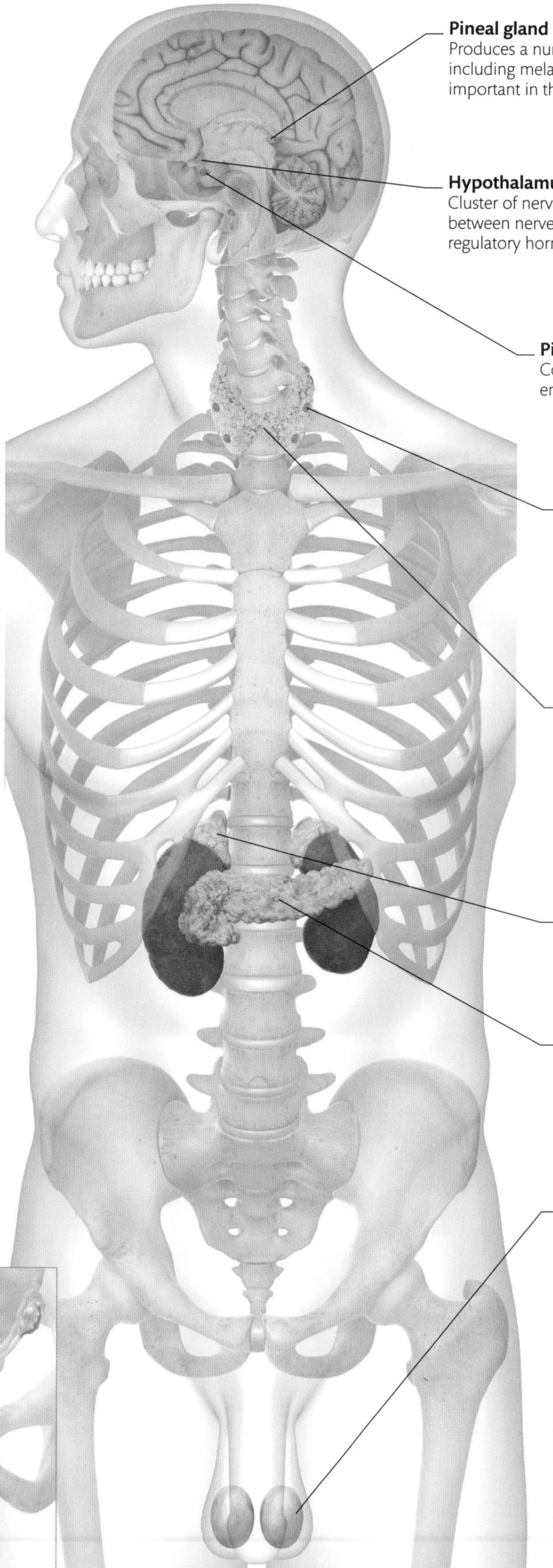
Pineal gland
Produces a number of hormones, including melatonin, which is important in the sleep-wake cycle
Hypothalamus
Cluster of nerve cells that serves as main link between nerves and hormones; produces regulatory hormones that travel to the pituitary
Pituitary gland
Controls many endocrine glands
Parathyroid glands
Four pea-sized glands at the back of thyroid gland help to regulate calcium levels in the body
Thyroid gland
Controls metabolic rate, including maintenance of body weight, energy use, and heart rate; unlike other endocrine glands, can store its hormones
Suprarenal (adrenal) gland
Produces adrenaline
Pancreas
Contains cells that produce insulin and glucagon, which control glucose metabolism
Testis
Produces androgens (male sex hormones), including testosterone

ANTERIOR

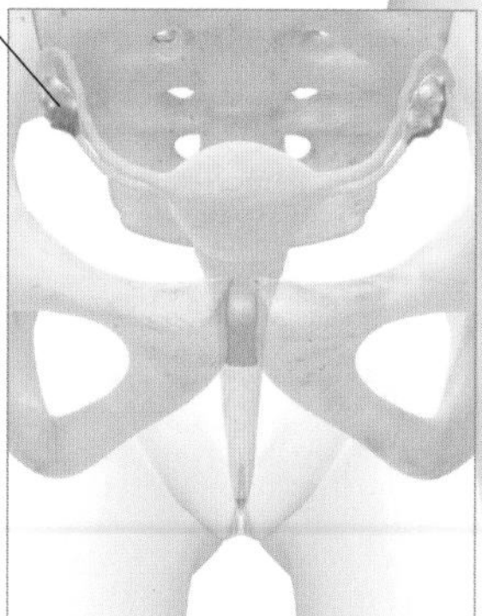
Ovary

FEMALE

FEMALE SEX HORMONES
The two ovaries manufacture the female sex hormones: estrogen, which stimulates ripening of the ova; and progesterone, which stimulates thickening of the uterine wall. Levels of these hormones fluctuate during the course of the menstrual cycle.

ENDOCRINE GLANDS AND TISSUES
Many hormones are produced by specialized endocrine glands, such as the pituitary gland in the brain and the suprarenal glands above the kidneys. Hormone-producing cells also occur in tissues throughout the body—for example, in the pancreas and testes, shown here.

DUAL-PURPOSE HORMONE PRODUCERS

Many organs and tissues in the body that primarily have another function also produce hormones. They include the kidneys, heart, skin, adipose tissue, and gastrointestinal tract, and they are as important as the purely endocrine glands in controlling vital functions. Hormones from the kidneys and heart help control blood pressure and stimulate production of red blood cells. Skin produces the hormone cholecalciferol, a precursor form of vitamin D. Endocrine cells lining the gastrointestinal tract secrete a number of different hormones, most of which play a role in the digestive process. Some of these gastrointestinal hormones, the incretins, affect many other body tissues in addition to those in the gut. Incretins stimulate insulin production in the pancreas, enhance bone formation, help to promote energy storage, and, by targeting the brain, suppress appetite. Another hormone that also affects appetite is leptin, produced by adipose tissue.

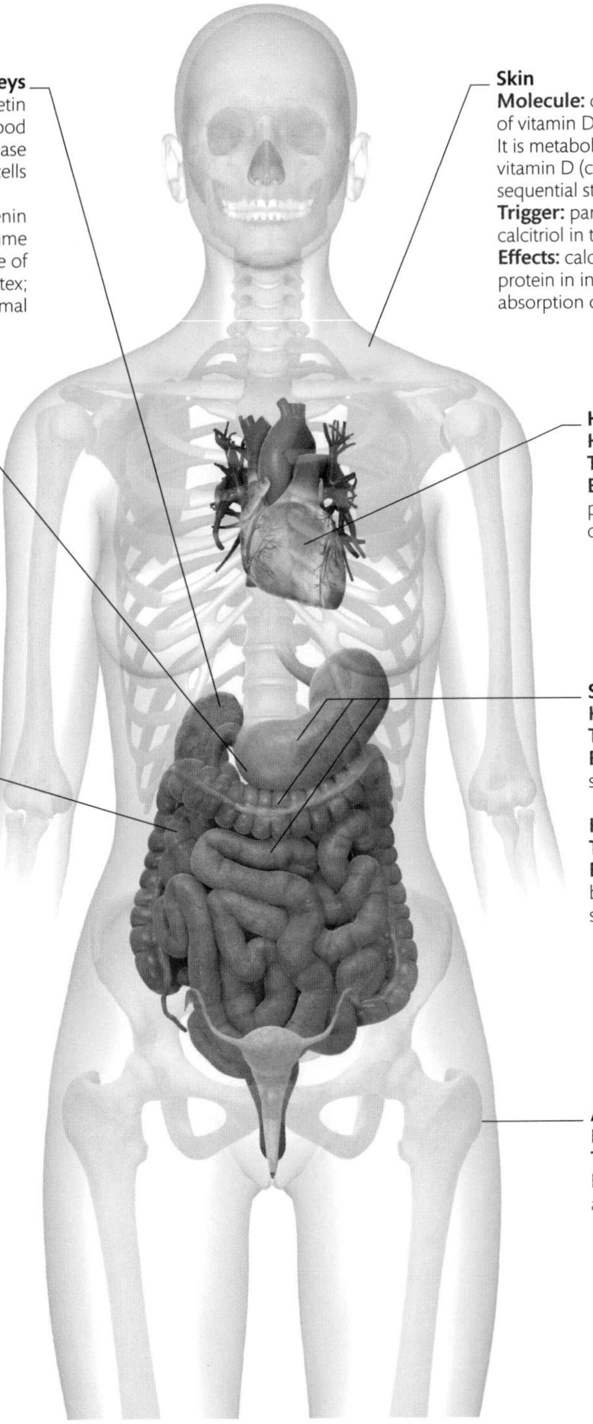
Kidneys
Hormone: erythropoietin
Trigger: low level of oxygen in blood
Effects: stimulates bone marrow to increase production of red blood cells
Hormone: renin
Trigger: low blood pressure or blood volume
Effects: initiates mechanism for release of aldosterone from suprarenal cortex; returns blood pressure to normal
Skin
Molecule: cholecalciferol (a biologically inactive form of vitamin D) is generated on exposure to UV radiation. It is metabolized to the biologically active form of vitamin D (calcitriol), which is a steroid hormone, in two sequential steps, first in the liver and then in the kidneys
Trigger: parathyroid hormone stimulates production of calcitriol in the kidneys
Effects: calcitriol induces synthesis of a calcium-binding protein in intestinal epithelial cells and so helps absorption of calcium from food in to the blood
Stomach
Hormone: gastrin
Trigger: response to food
Effects: stimulates gastric acid secretion
Hormone: ghrelin
Trigger: long period without food
Effects: appears to stimulate appetite and eating; stimulates growth hormone secretion
Heart
Hormone: atrial natriuretic hormone
Trigger: high pressure in heart
Effects: signals kidneys to increase production of urine and inhibits release of aldosterone, reducing blood pressure
Stomach, duodenum, and colon
Hormone: motilin
Trigger: associated with fasting
Effects: stimulates movements of stomach and small intestine
Hormone: incretins
Trigger: fat and glucose in small intestine
Effects: stimulate insulin secretion, enhance bone formation, promote energy storage, suppress appetite
Duodenum
Hormone: intestinal gastrin
Trigger: response to food
Effects: stimulates gastric acid secretion and movements of gastrointestinal tract
Hormone: secretin
Trigger: acid environment
Effects: stimulates release of bicarbonate-rich juice from pancreas and bile ducts; inhibits production of gastric acid in stomach
Hormone: cholecystokinin
Trigger: response to fats in food
Effects: stimulates secretion of enzymes in pancreas, and contraction and emptying of gallbladder to allow bile and pancreatic enzymes to enter duodenum
Adipose tissue
Hormone: leptin
Trigger: released following uptake of nutrients
Effects: regulates appetite, energy expenditure, and food intake

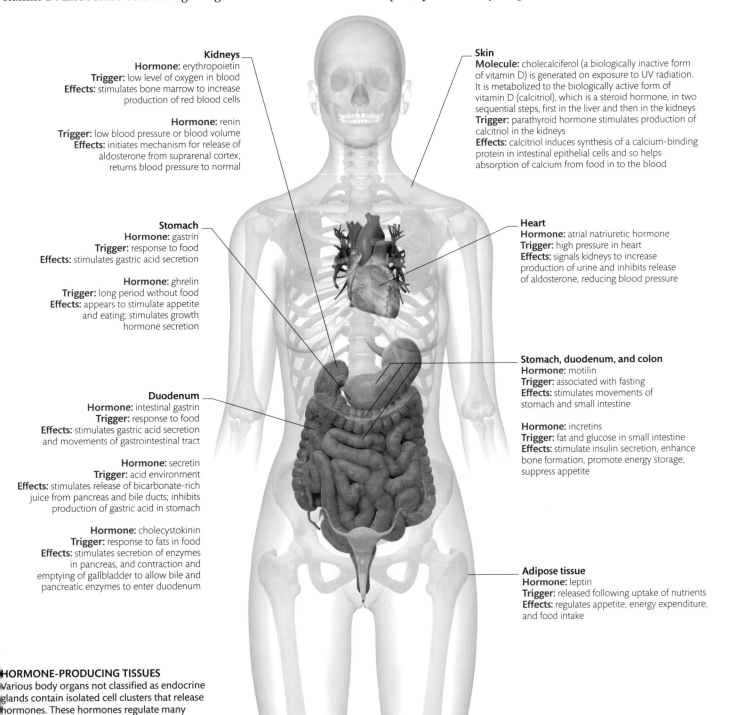

HORMONE-PRODUCING TISSUES
Various body organs not classified as endocrine glands contain isolated cell clusters that release hormones. These hormones regulate many important processes in the body.

See also p. 190 »

ENDOCRINE SYSTEM 2

THE RELEASE OF HORMONES FROM ENDOCRINE CELLS IS STIMULATED BY VARIOUS MEANS, AND BLOOD LEVELS OF MOST HORMONES ARE REGULATED BY FEEDBACK LOOPS. THERE ARE TWO MAIN TYPES OF HORMONES—WATER-SOLUBLE AND FAT-SOLUBLE—WHICH USE DIFFERENT MECHANISMS TO PRODUCE A BIOCHEMICAL REACTION IN THEIR TARGET CELLS.

TRIGGERS FOR HORMONE RELEASE

Factors stimulating the production and release of hormone vary. Some endocrine glands are stimulated by the presence of certain minerals or nutrients in the blood. For example, low blood levels of calcium stimulate the parathyroid gland to release parathyroid hormone, while insulin, made in the pancreas, is released in response to rising glucose levels. Many endocrine glands respond to hormones produced by other endocrine glands. These include hormones produced by the hypothalamus, which stimulate the anterior pituitary gland to produce its hormones. In turn, these pituitary hormones stimulate other target glands; for example, adrenocorticotropic hormone stimulates the cortex of the suprarenal gland to produce corticosteroid hormones.

Hormonal stimulation leads to the rhythmic release of hormones, so that the levels rise and fall in a particular pattern. In a few cases, release of hormones is triggered by signals from the nervous system. An example is the medulla of the suprarenal gland, which releases adrenaline when stimulated by nerve fibers from the sympathetic nervous system. With this type of stimulation, hormone release occurs in bursts rather than rhythmically.

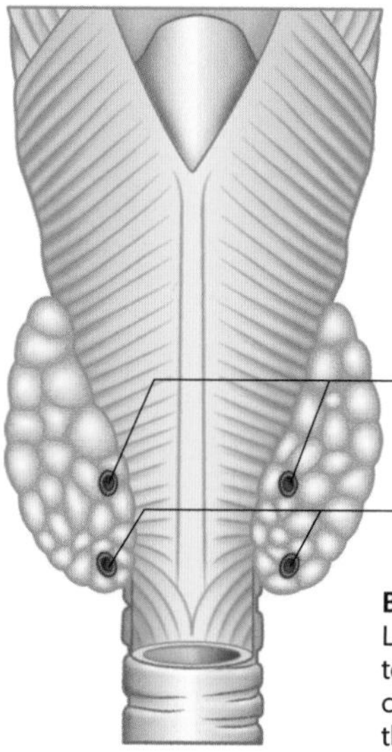

BLOOD LEVEL RESPONSE
Low blood calcium prompts the parathyroid to release parathyroid hormone, which raises calcium levels. The release of calcitonin from the thyroid is inhibited.

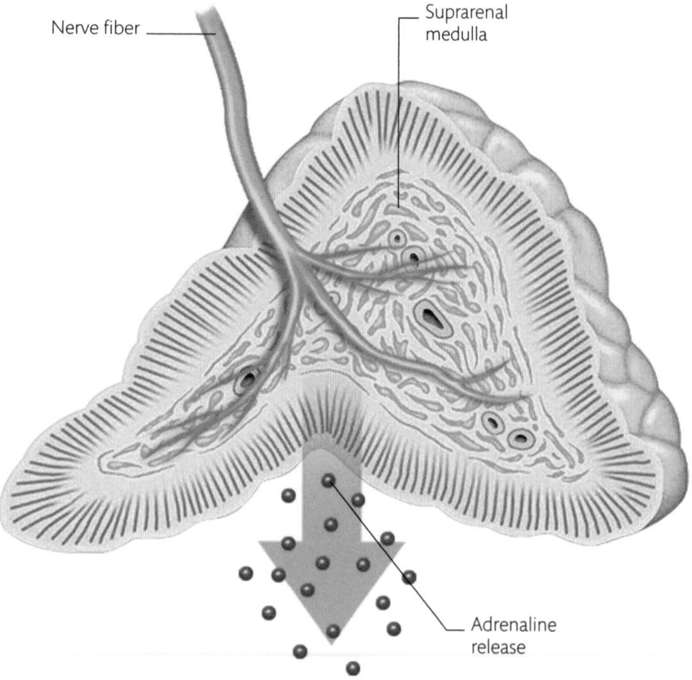

NERVOUS STIMULATION
Nerve fibers of the sympathetic nervous system, signaled by the hypothalamus, stimulate the suprarenal medulla to release adrenaline in times of stress.

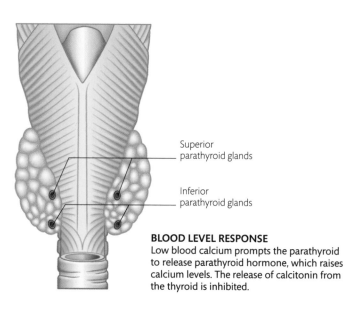

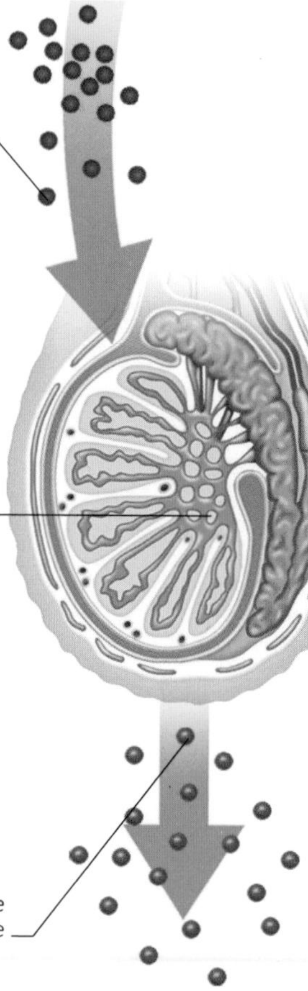

RESPONSE TO HORMONES
Gonadotropin hormones from the pituitary gland stimulate the ovaries and testes to secrete more sex hormones. In the testis, shown here, this is testosterone.

HORMONE REGULATION

Hormones are powerful and affect target organs at low concentrations. The duration of their action is limited—from seconds to several hours—so blood levels need to be kept within an optimal range. This stabilization of hormone levels in blood (homeostasis) is regulated by feedback mechanisms. The amount of a particular hormone in the bloodstream is detected and the information is passed on to a control unit—often the hypothalamus—pituitary complex in the brain. If the level of a hormone increases beyond normal, the control unit responds by reducing hormone production. If the hormone level decreases below normal, the control unit stimulates production.

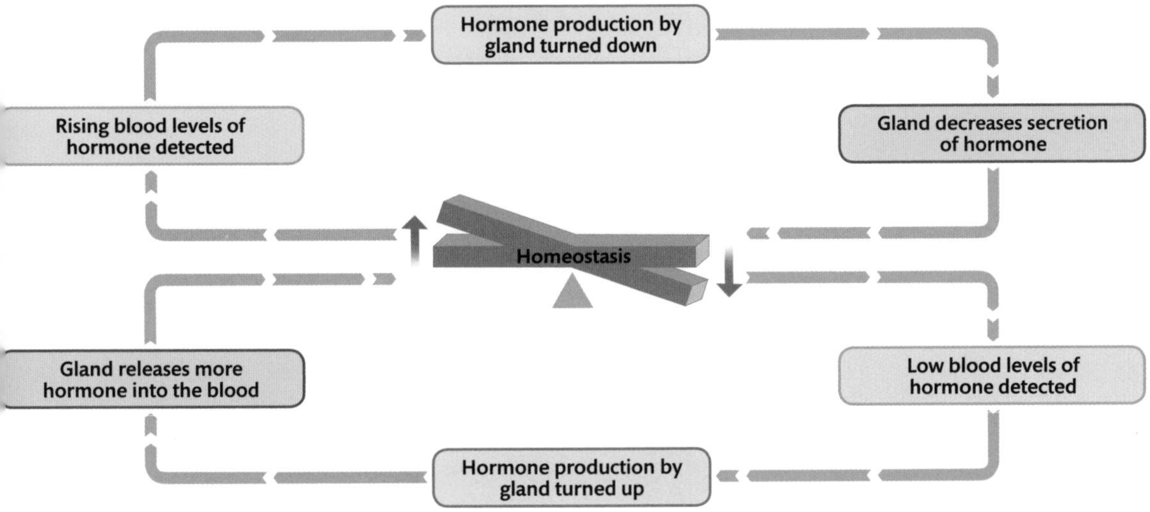

REGULATION FLOWCHART
The maintenance of stable blood hormone levels, or homeostasis, is controlled by feedback mechanisms. Hormone levels are monitored and if they become too low or too high, production is increased or decreased, respectively.

WATER-SOLUBLE HORMONES

Most hormones are built from amino acids, the basic structural units of proteins, and are water-soluble. These hormones are unable to pass through target cell membranes, which have fatty layers. To have an effect on target cells, water-soluble hormones bind to receptors on the cell surface.

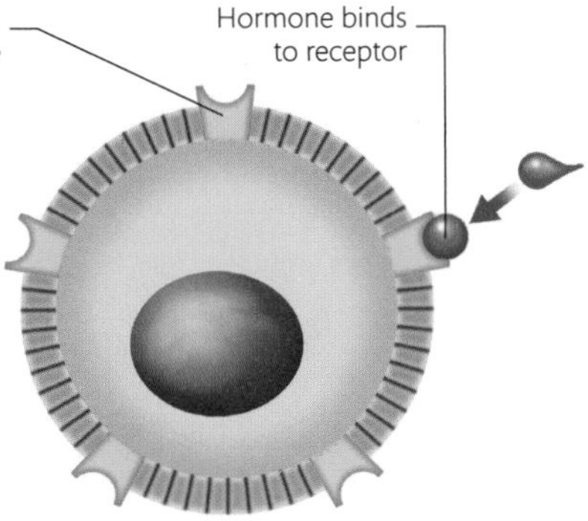

1 RECEPTOR BINDING
The hormone recognizes a receptor protruding from the surface of the target cell and binds to it. This triggers a chemical response inside the cell.

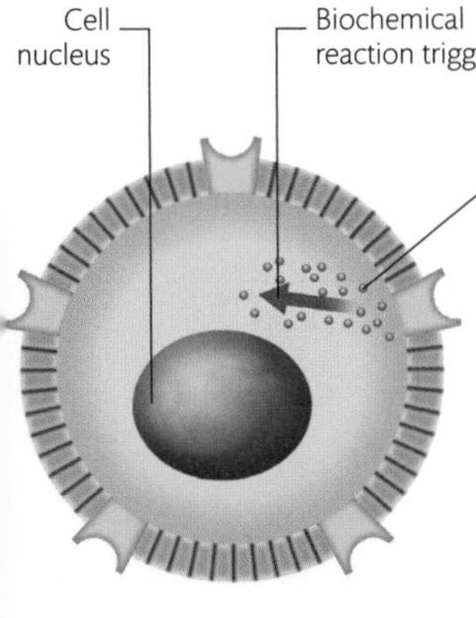

2 ACTIVATION
Enzymes inside the cell are activated, altering the biochemical activity of the cell. The rates of normal cell processes are either increased or decreased.

FAT-SOLUBLE HORMONES

Hormones that are fat-soluble are usually made from cholesterol. These hormones are able to pass through a target cell membrane and produce their effect by binding with receptors inside the cell. Fat-soluble hormones include the sex hormones and thyroid hormone.

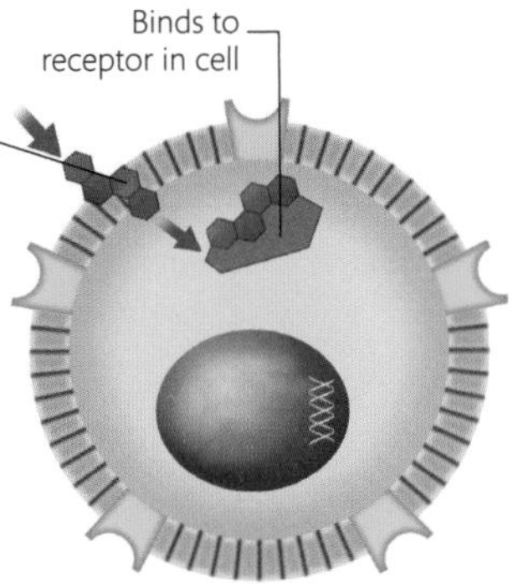

1 BINDING IN CELL
The hormone diffuses through the cell membrane and binds to a mobile receptor within the cell, which is activated by the process.

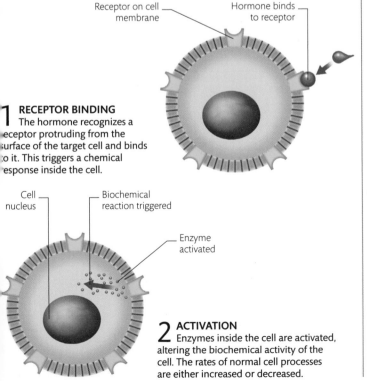

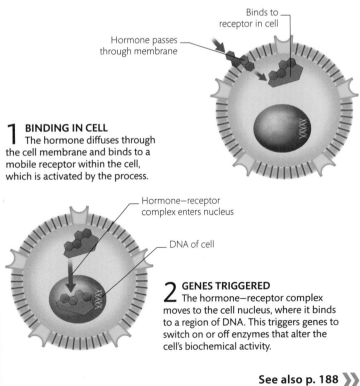

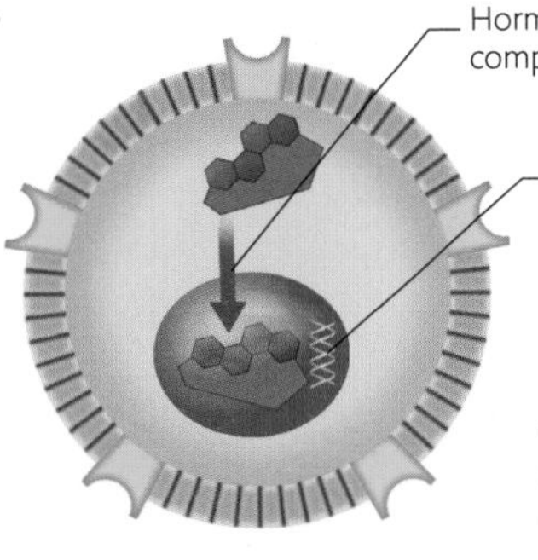

2 GENES TRIGGERED
The hormone–receptor complex moves to the cell nucleus, where it binds to a region of DNA. This triggers genes to switch on or off enzymes that alter the cell's biochemical activity.

See also p. 188 »

HEAD AND NECK 1

THE HEAD AND NECK CONTAIN SEVERAL KEY ENDOCRINE GLANDS, INCLUDING THE PITUITARY, THE PINEAL, THE THYROID AND PARATHYROID GLANDS, AND THE HYPOTHALAMUS.

ANATOMY

The pineal is a tiny gland about 5⁄16in (8 mm) long, in the middle of the brain. It has links to the visual pathway and secretes the hormone melatonin, which is important in regulating circadian rhythms. The thyroid is butterfly-shaped and is situated in the front of the neck, with the four small parathyroid glands embedded toward the rear of its "wings." The hypothalamus is situated at the base of the brain; the pituitary is connected to it by a short stalk.

ANTEROLATERAL VIEW

Color and/or label the structures indicated on the diagram using the key below.

① Pineal gland
② Hypothalamus
③ Pituitary gland
④ Thyroid gland

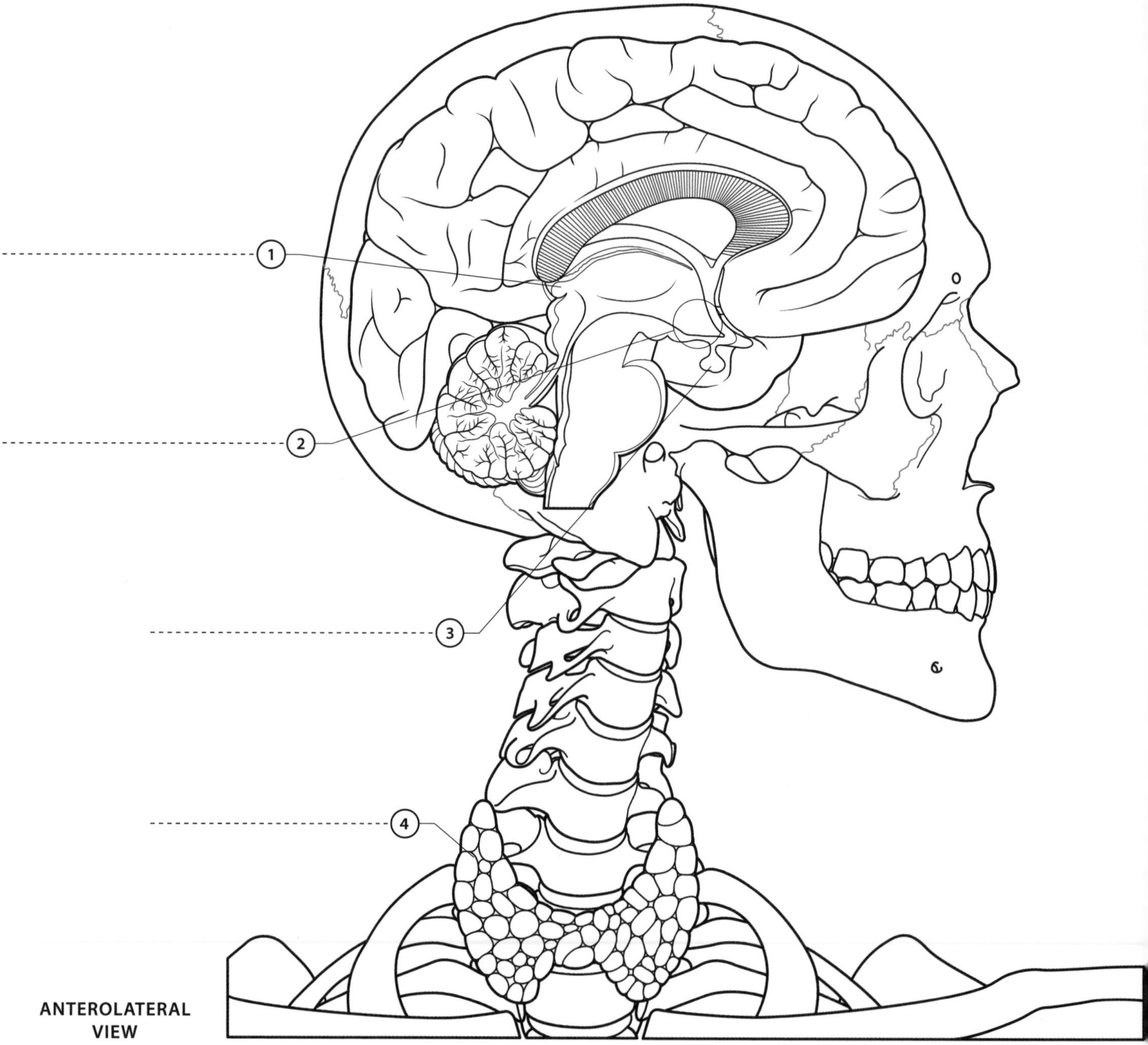

ANTEROLATERAL VIEW

HYROID AND PARATHYROID GLANDS

he thyroid gland produces two important hormones, T3 riiodothyronine) and T4 (thyroxine), which are collectively nown as thyroid hormone (TH). Almost every cell in the ody has receptors for TH. The hormone has widespread fects: it regulates the basal metabolic rate and affects mperature regulation, carbohydrate and fat metabolism, owth and development, reproduction, and heart function. he thyroid also produces calcitonin, which inhibits lcium loss from bones. The parathyroid glands produce arathyroid hormone (PTH), which is the major regulator f calcium levels in the blood.

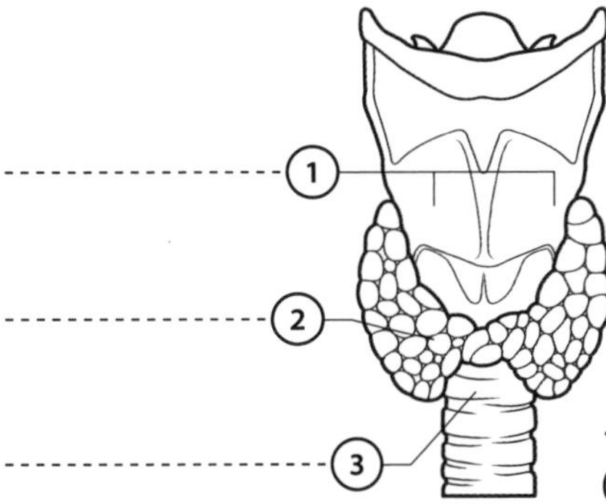

THYROID GLAND (ANTERIOR VIEW)

THYROID GLAND (ANTERIOR VIEW)

Color and/or label the structures indicated on the diagram using the key below.

1. Thyroid cartilage
2. Thyroid gland
3. Tracheal cartilage

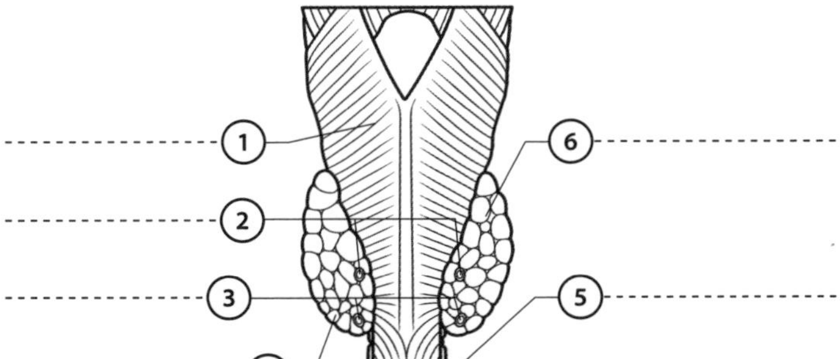

PARATHYROID GLANDS

PARATHYROID GLANDS

Color and/or label the structures indicated on the diagram using the key below.

1. Pharyngeal constrictor muscle
2. Superior parathyroid glands
3. Inferior parathyroid glands
4. Inferior left lobe of thyroid gland
5. Tracheal cartilage
6. Superior right lobe of thyroid gland

THYROID HORMONE REGULATION

The production and release of thyroid hormones (TH) is stimulated by thyrotropin-releasing hormone (TRH) from the hypothalamus and thyroid-stimulating hormone (TSH) from the anterior lobe of the pituitary. Blood levels of TH feed back to the pituitary and hypothalamus to stimulate or inhibit their release of TSH and TRH.

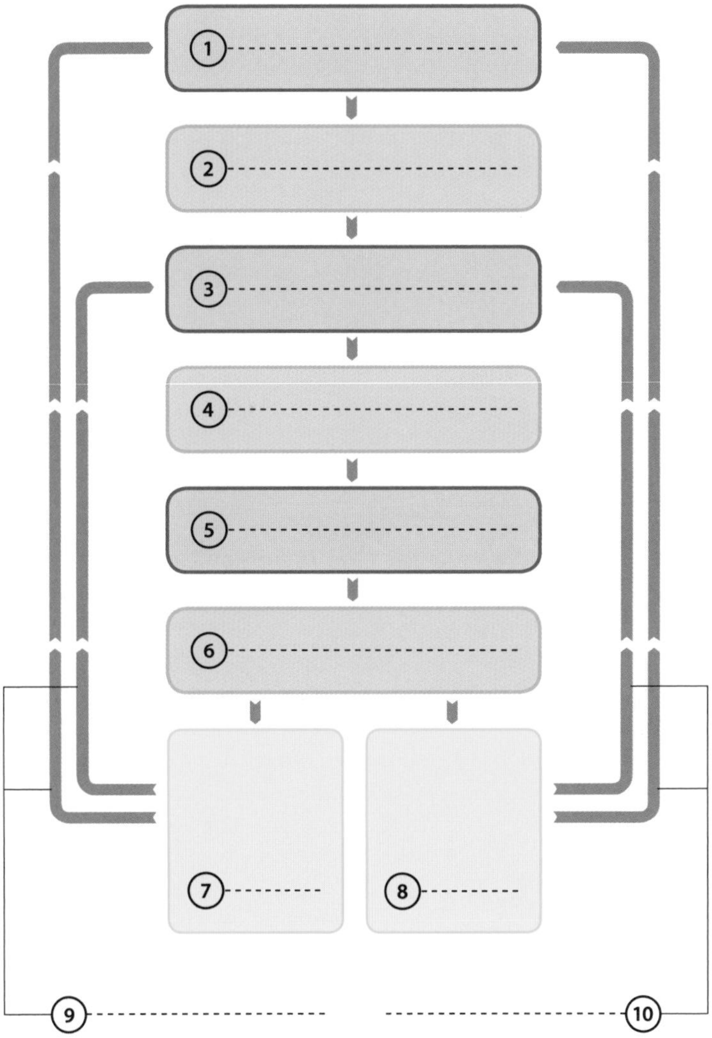

THYROID HORMONE REGULATION

THYROID HORMONE REGULATION

Insert the structures/events onto the flowchart using the key below.

1. Hypothalamus
2. TRH secreted by hypothalamus
3. Pituitary gland
4. TSH secreted by pituitary gland
5. Thyroid gland
6. Thyroid hormone (TH)
7. Decreased levels of hormones in bloodstream
8. Increased levels of hormones in bloodstream
9. Stimulation
10. Inhibition

See also pp. 188, 190, 194 »

HEAD AND NECK 2

THE PITUITARY GLAND IS THE MOST INFLUENTIAL GLAND IN THE ENDOCRINE SYSTEM. IT IS LINKED TO THE HYPOTHALAMUS, WHICH ACTS AS AN INTERMEDIARY BETWEEN THE ENDOCRINE SYSTEM AND THE AUTONOMIC NERVOUS SYSTEM.

PITUITARY GLAND

The anterior lobe forms the bulk of the pituitary and consists of glandular tissue that manufactures seven hormones: thyroid-stimulating hormone (TSH), adrenocorticotropic hormone (ACTH), follicle-stimulating hormone (FSH), luteinizing hormone (LH), growth hormone (GH), melanocyte-stimulating hormone (MSH), and prolactin. The release of these hormones is regulated by the hypothalamus, which secretes releasing or inhibiting hormones. The posterior lobe of the pituitary stores two hormones—oxytocin and antidiuretic hormone (ADH)—which are made in the hypothalamus and pass down axons to the posterior lobe of the pituitary. These hormones are released in response to nerve impulses from the hypothalamus.

PITUITARY AND HYPOTHALAMUS

Color and/or label the diagram using the key below.

1. MSH
2. Skin
3. Anterior lobe of pituitary
4. ACTH
5. Suprarenal gland
6. Thyroid
7. TSH
8. Bone and general growth
9. GH
10. Ovary or testis
11. LH and FSH
12. Vein
13. Prolactin
14. Mammary glands
15. Uterine muscles
16. Oxytocin
17. Kidney tubules
18. ADH
19. Posterior lobe of pituitary
20. Artery
21. Pituitary stalk
22. Axon
23. Neurosecretory cell in hypothalamus
24. Hypothalamus

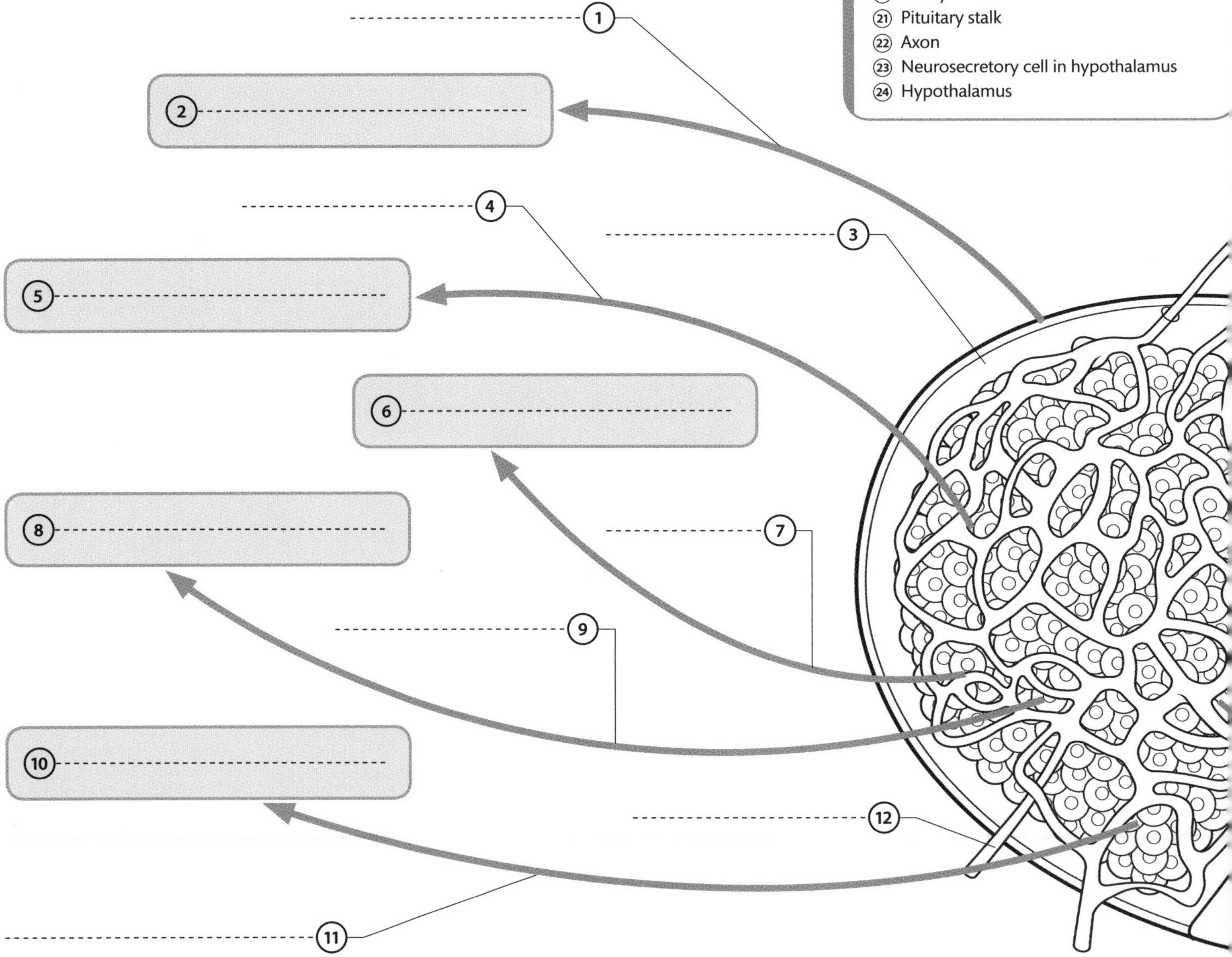

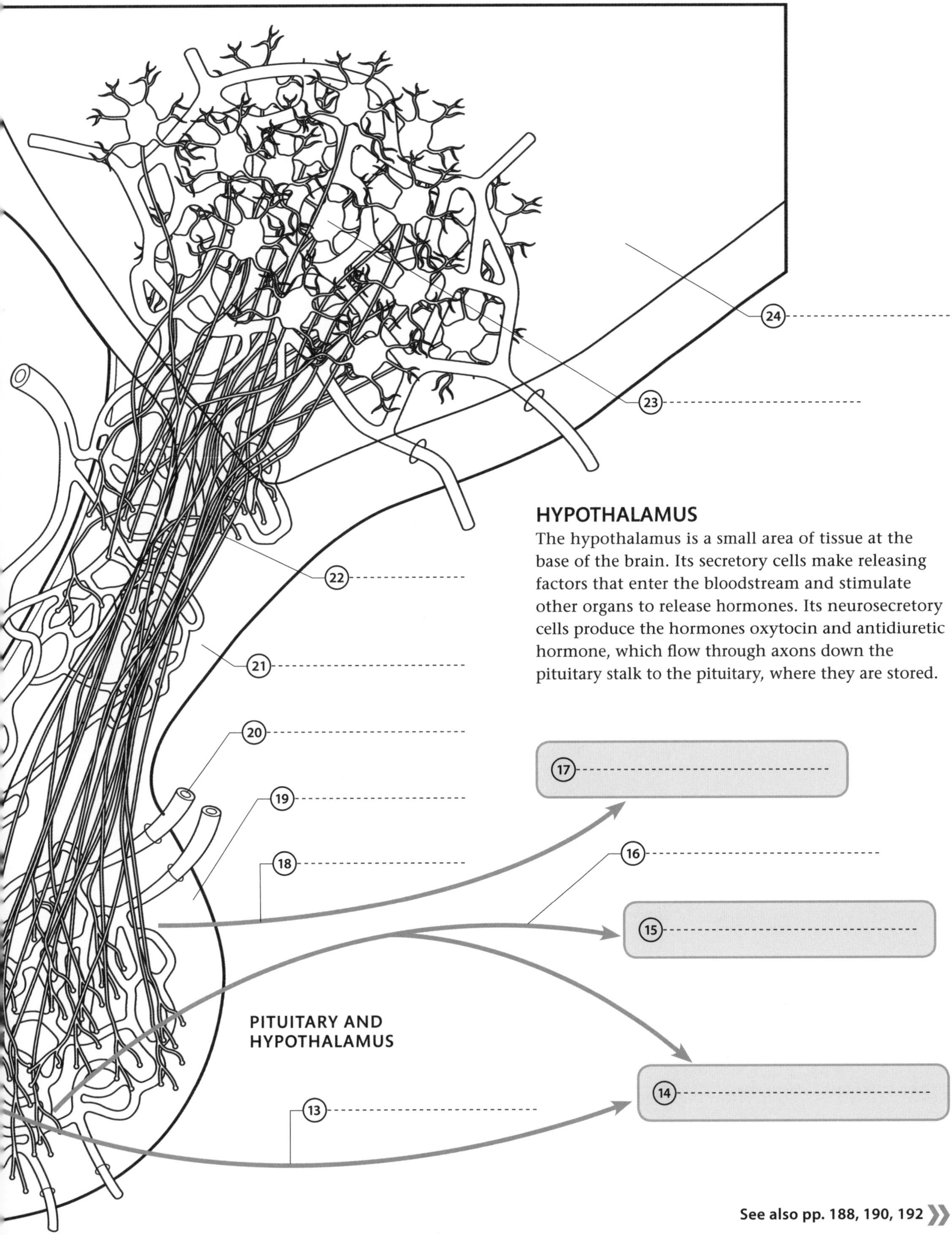

HYPOTHALAMUS

The hypothalamus is a small area of tissue at the base of the brain. Its secretory cells make releasing factors that enter the bloodstream and stimulate other organs to release hormones. Its neurosecretory cells produce the hormones oxytocin and antidiuretic hormone, which flow through axons down the pituitary stalk to the pituitary, where they are stored.

See also pp. 188, 190, 192

ABDOMEN AND PELVIS 1

THE PANCREAS IS BOTH A DIGESTIVE ORGAN, PRODUCING ENZYME-CONTAINING PANCREATIC JUICE, AND AN ENDOCRINE ORGAN, PRODUCING HORMONES THAT REGULATE BLOOD GLUCOSE LEVELS.

PANCREAS

The pancreas lies under the liver and behind the stomach. Most of its cells are acinar cells, which produce digestive enzymes. Scattered among these cells are about a million pancreatic islets (also called islets of Langerhans), clusters of cells that produce hormones. There are four types of hormone-producing cell. Beta cells make insulin, which enhances transport of glucose into cells, thereby lowering blood glucose levels. Alpha cells secrete glucagon, which stimulates the liver to release glucose, thereby raising blood glucose levels. Delta cells secrete somatostatin, which regulates alpha and beta cells. F cells secrete pancreatic peptide, which inhibits secretion of bile and pancreatic digestive enzymes.

PANCREAS

Color and/or label the structures indicated on the diagram using the key below.

1. Neck of pancreas
2. Duodenum
3. Accessory duct
4. Bile duct
5. Main pancreatic duct
6. Head of pancreas
7. Uncinate process of pancreas
8. Body of pancreas
9. Tail of pancreas

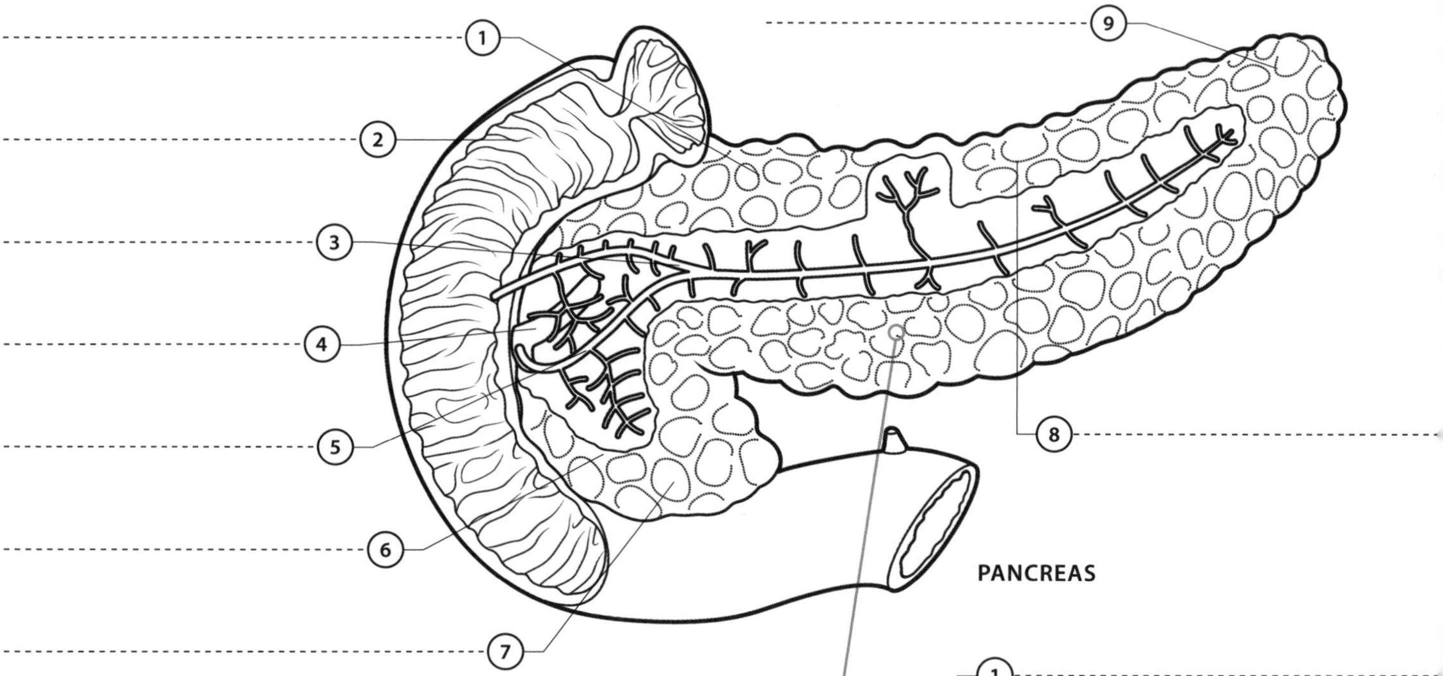

PANCREAS

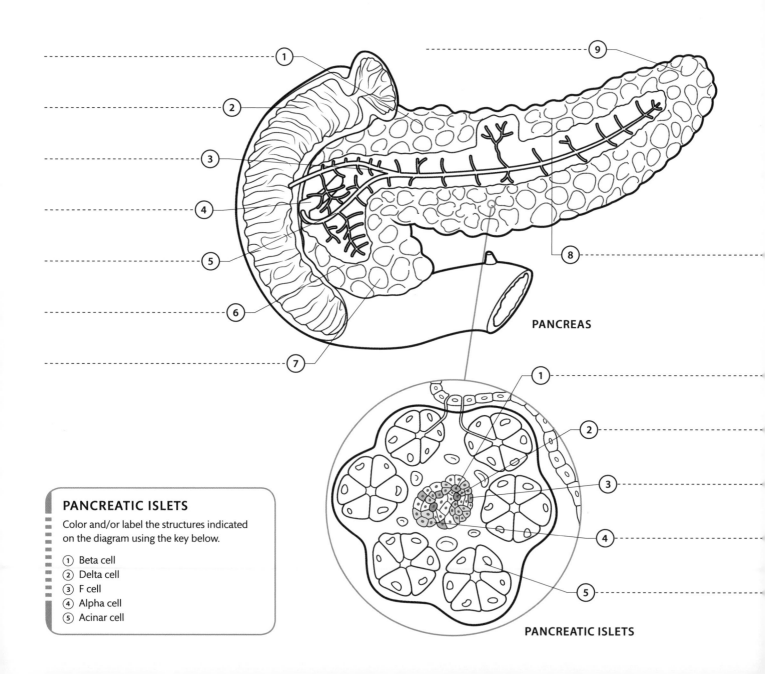

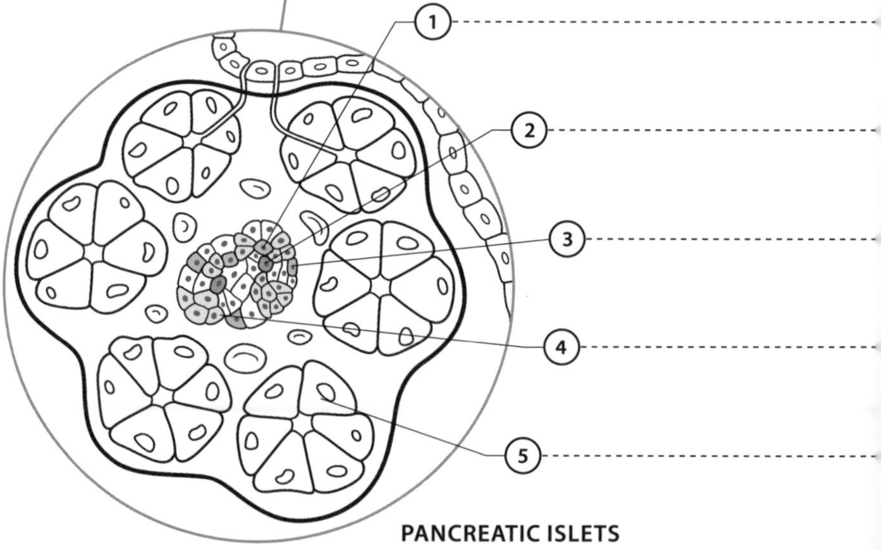

PANCREATIC ISLETS

PANCREATIC ISLETS

Color and/or label the structures indicated on the diagram using the key below.

1. Beta cell
2. Delta cell
3. F cell
4. Alpha cell
5. Acinar cell

LOOD GLUCOSE REGULATION

he body needs to regulate blood glucose levels to ensure that cells receive nough glucose (their main source of energy). When blood glucose levels fall elow normal, alpha cells in the pancreas release glucagon, which acts on the ver and muscles, causing them to break down stored glycogen to form glucose, hich is then released into the bloodstream. When blood glucose levels are o high, beta cells in the pancreas release insulin, stimulating the liver and uscles to convert glucose into glycogen for storage. Some excess glucose is ombined with fatty acids and stored as fat.

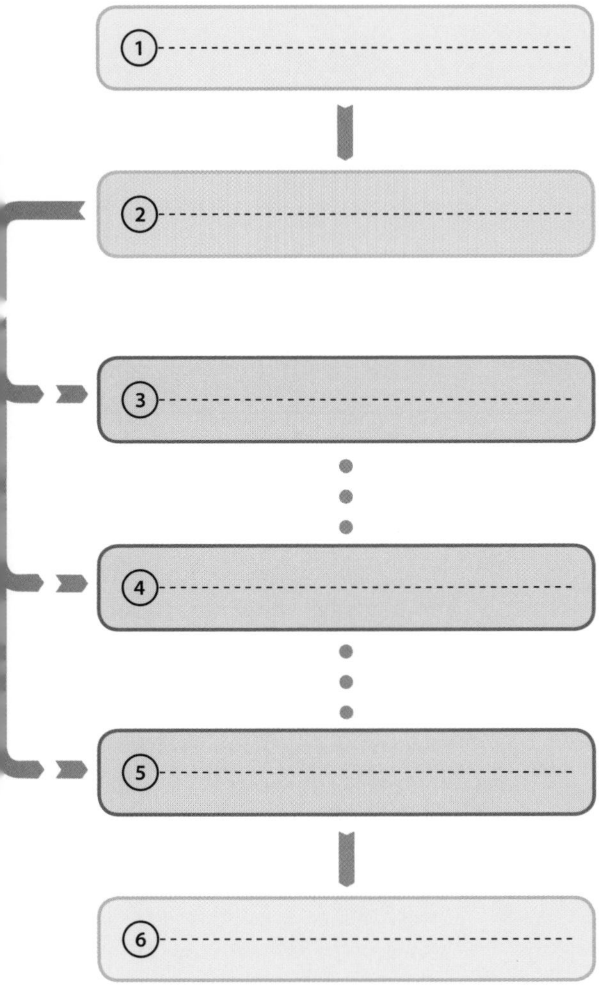

REGULATING HIGH BLOOD GLUCOSE

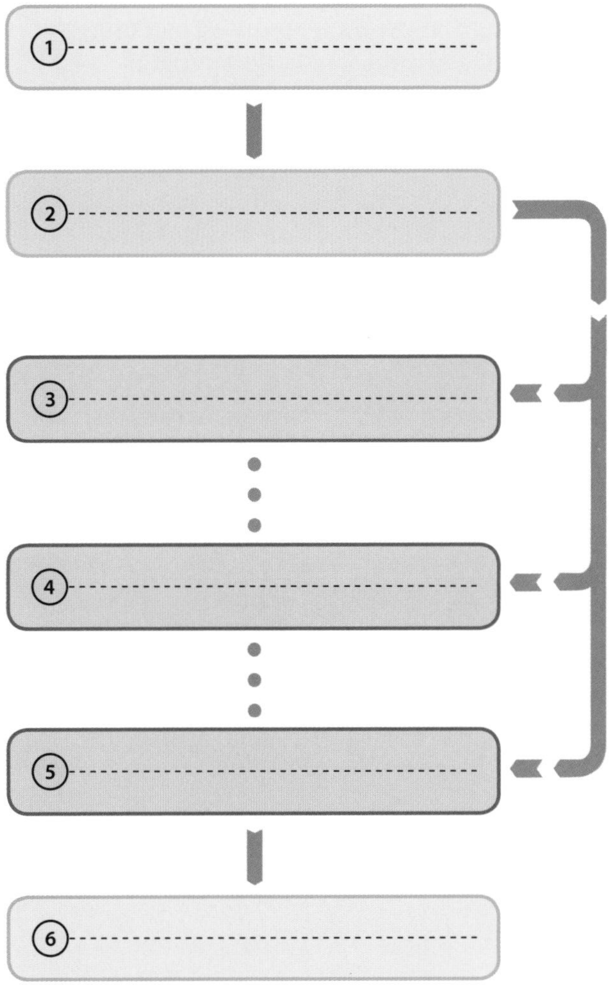

REGULATING LOW BLOOD GLUCOSE

REGULATING HIGH BLOOD GLUCOSE

Insert the events indicated on the flowchart using the key below.

1. Blood glucose level rises (e.g. after eating)
2. Beta cells in pancreas release insulin
3. Glucose stored in liver
4. Glucose stored in muscles
5. Glucose stored as fat
6. Blood glucose level stabilized

REGULATING LOW BLOOD GLUCOSE

Insert the events indicated on the flowchart using the key below.

1. Blood glucose level falls (e.g. between meals)
2. Alpha cells in pancreas release glucagon
3. Liver releases glucose
4. Muscles release glucose
5. If necessary, fats broken down for energy
6. Blood glucose level stabilized

See also pp. 184, 188 »

ABDOMEN AND PELVIS 2

THE SUPRARENAL GLANDS (ALSO KNOWN AS THE ADRENAL GLANDS) SIT ON TOP OF THE KIDNEYS AND PRODUCE THREE DIFFERENT GROUPS OF HORMONES. THE GONADS—OVARIES AND TESTES—PRODUCE SEX HORMONES.

SUPRARENAL GLANDS

The outer suprarenal cortex produces mineral corticosteroids, glucocorticosteroids, and androgens. The main mineral corticosteroid is aldosterone, which regulates the sodium–potassium balance and helps to control blood pressure. The main glucocorticosteroid is cortisol, which affects metabolism and helps the body to resist stress. Androgens produced by the suprarenals are relatively weak in their effects, compared with those produced by the ovaries and testes from late puberty. The inner suprarenal medulla produces epinephrine and norepinephrine, which are involved in the stress response.

SUPRARENAL GLAND

Color and/or label the structures indicated on the diagram using the key below.

1. Blood vessel
2. Cortex
3. Medulla
4. Fat pad
5. Kidney

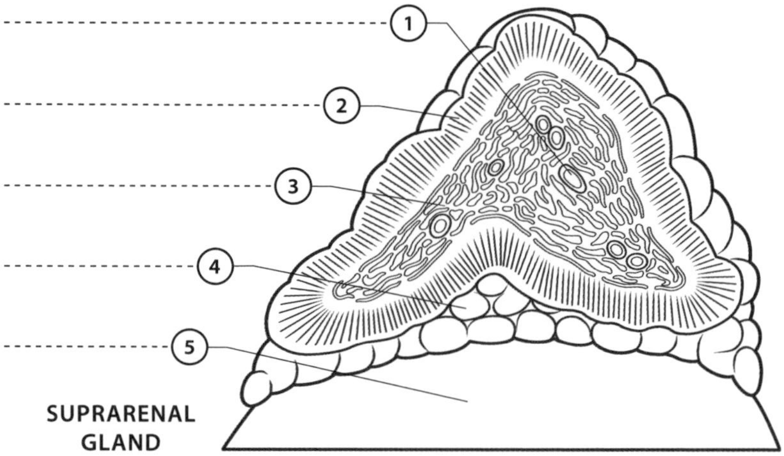

SUPRARENAL GLAND

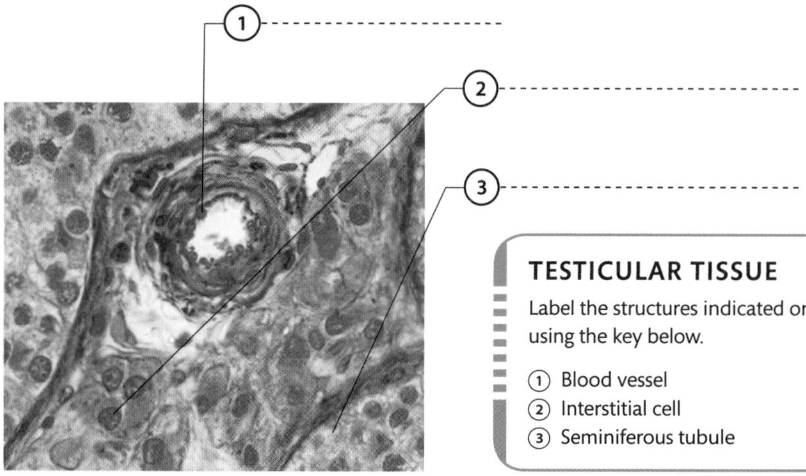

TESTICULAR TISSUE

TESTICULAR TISSUE

Label the structures indicated on the diagram using the key below.

1. Blood vessel
2. Interstitial cell
3. Seminiferous tubule

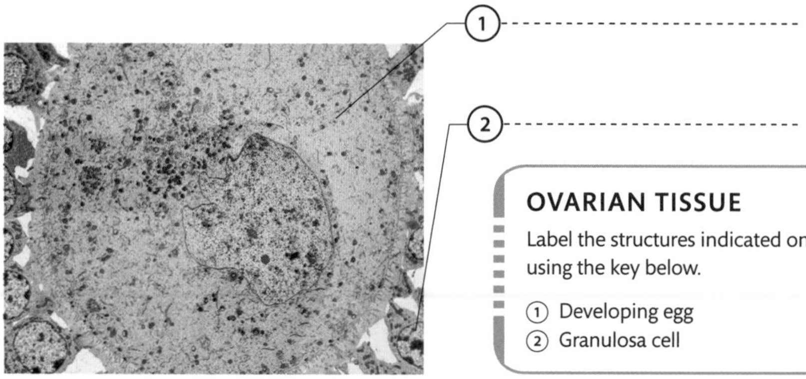

OVARIAN TISSUE

OVARIAN TISSUE

Label the structures indicated on the diagram using the key below.

1. Developing egg
2. Granulosa cell

OVARIES AND TESTES

The female ovaries and male testes produce eggs and sperm, respectively. They also produce sex hormones, the most important of which are estrogens and progesterone in females, and testosterone in males. In the ovaries, granulosa cells secrete estrogen, and progesterone is produced mainly by the corpus luteum after ovulation. In the testes, testosterone is produced by interstitial cells in the connective tissue between the seminiferous tubules. The ovaries and testes also secrete inhibin, which inhibits secretion of FSH by the anterior pituitary. Additionally, the ovaries produce relaxin, which prepares the body for childbirth.

STRESS RESPONSE

When stress is detected, nerve impulses from the hypothalamus activate the sympathetic nervous system, which initiates a fight-or-flight response, preparing the body for action. Hormones from the suprarenal medulla prolong the response. Next the body tries to respond to the stressful situation. This reaction is initiated mainly by hypothalamic-releasing hormones, which trigger the anterior pituitary to release growth hormone and other hormones that stimulate the thyroid and suprarenal cortex to secret their hormones. These mobilize the synthesis of proteins for energy and repair.

STRESS RESPONSE

Color and/or label the diagram using the key below.

1. Blood vessels in brain dilate
2. Pupil dilates
3. Thyroid gland releases T3 and T4 to increase glucose utilization
4. Airways and blood vessels in lungs dilate
5. Suprarenal glands release cortisol, epinephrine, and norepinephrine
6. Kidneys reduce urine output
7. Blood vessels in skin constrict, hair stands on end, and sweat pores open
8. Blood vessels in skeletal muscles dilate
9. Bladder sphincter constricts
10. Intestinal movements diminish
11. Digestive activity in stomach decreases
12. Spleen contracts
13. Liver releases glucose
14. Heart rate and force of heart beat increase
15. Anterior pituitary gland releases growth hormone, and hormones that stimulate the thyroid and suprarenal cortex
16. Hypothalamus triggers fight-or-flight response and stimulates suprarenal medulla; releases hormones that stimulate the anterior pituitary

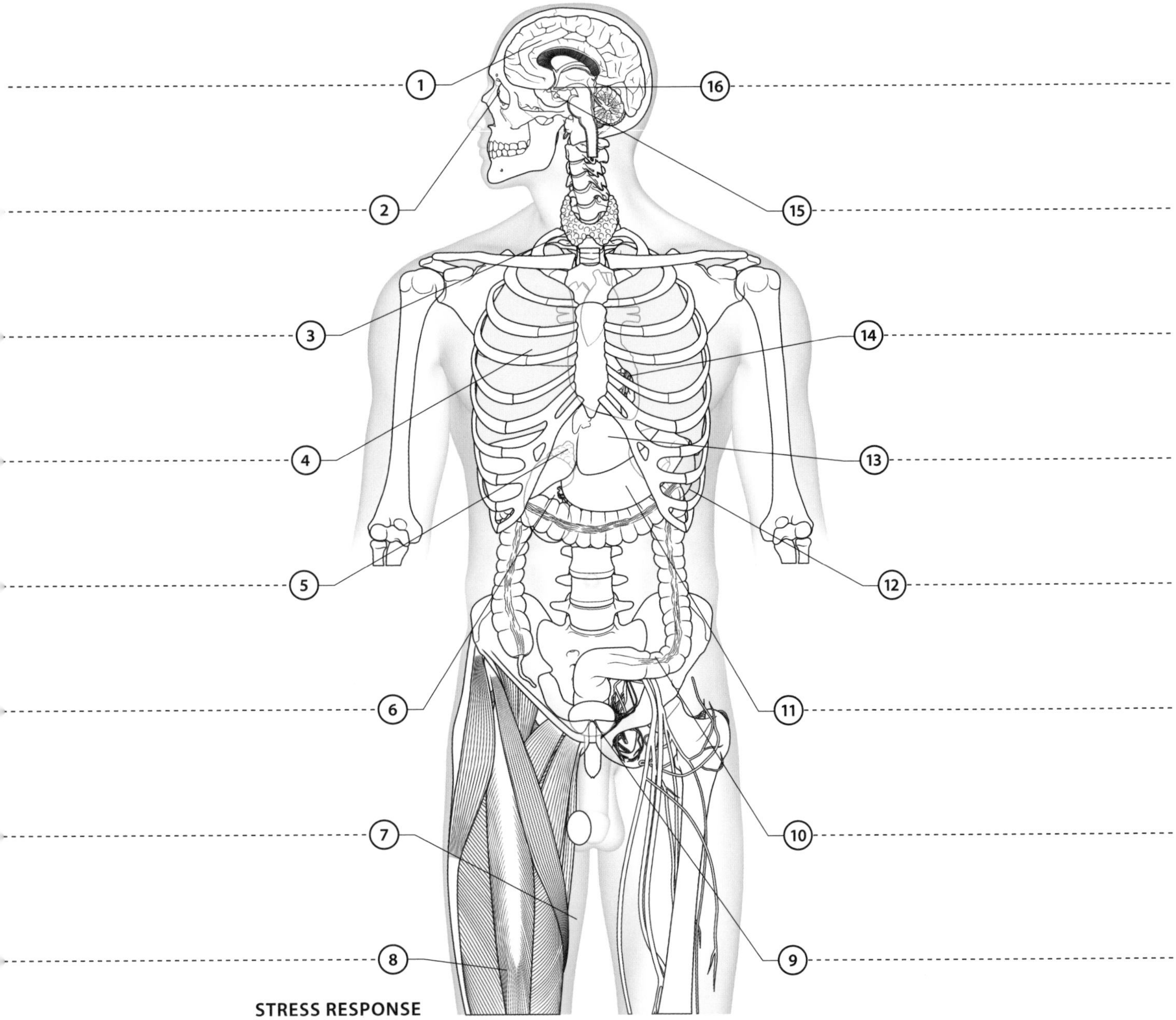

STRESS RESPONSE

See also pp. 188, 212, 214, 222, 226, 228

URINARY SYSTEM

10

URINARY SYSTEM

THE FUNCTIONS OF THE URINARY SYSTEM ARE TO MAINTAIN THE BALANCE OF BODY FLUIDS, REMOVE WASTE PRODUCTS FROM THE BLOOD, AND EXPEL EXCESS WATER FROM THE BODY. THE DIFFERENCES BETWEEN MALE AND FEMALE SYSTEMS ARE ILLUSTRATED HERE.

URINARY SYSTEM

The urinary system comprises the kidneys, ureters, bladder, and urethra. The kidneys lie high up in the abdomen, on its back wall, with the upper part of each tucked under the twelfth rib. These organs contain microscopic units that filter the blood, removing wastes produced as a result of chemical reactions in body cells. The filtering process regulates the level of water in the body and also maintains the correct balance of fluids and substances such as salts within the body. A ureter connected to each kidney transports wastes and excess water in the form of urine to the bladder. The bladder expands as it fills, until nerve impulses trigger the impulse to urinate. Urine is expelled from the bladder via the urethra.

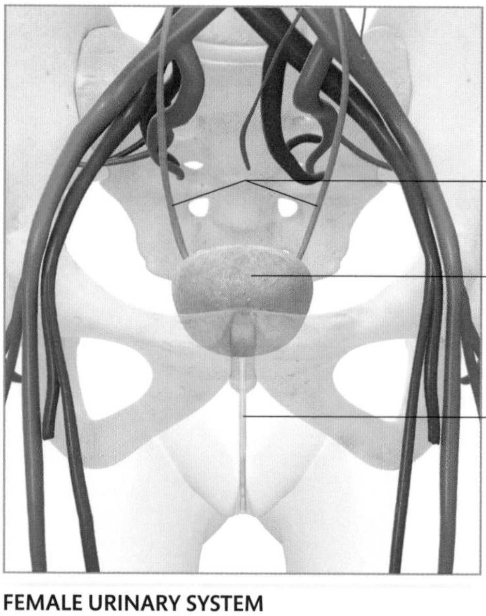

FEMALE URINARY SYSTEM
In females, the urethra is short, around $1\frac{1}{2}$ in (4 cm) long. It passes through the muscles of the pelvic floor and a muscular sphincter before opening between the clitoris and the vagina.

MALE URINARY SYSTEM
The male urethra runs from the bladder, through the prostate gland, to the tip of the penis. It also functions as a pathway for sperm from the testes during ejaculation.

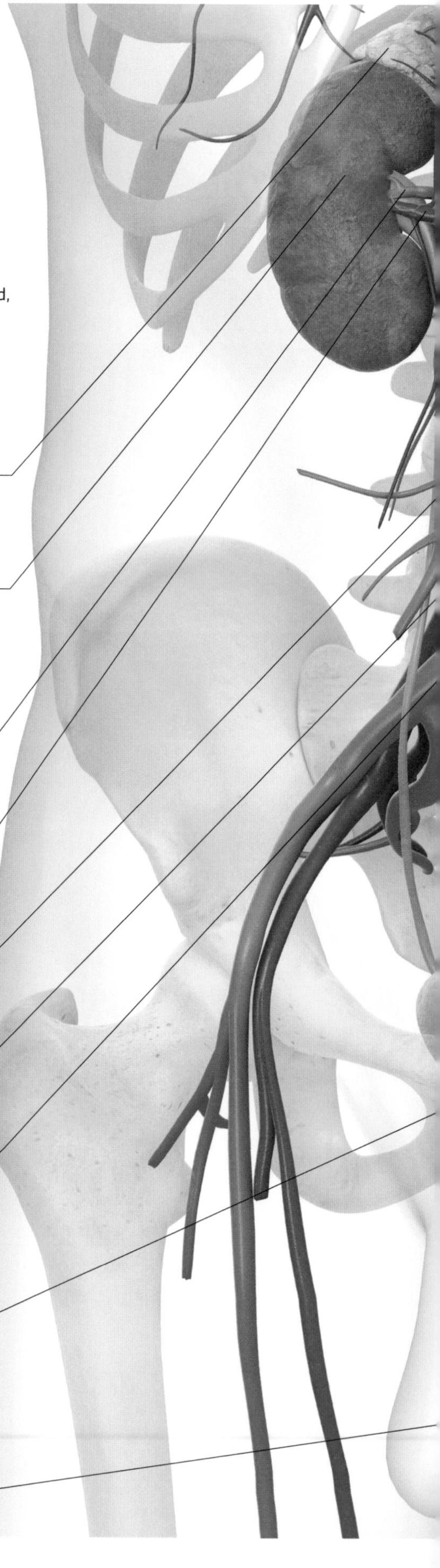

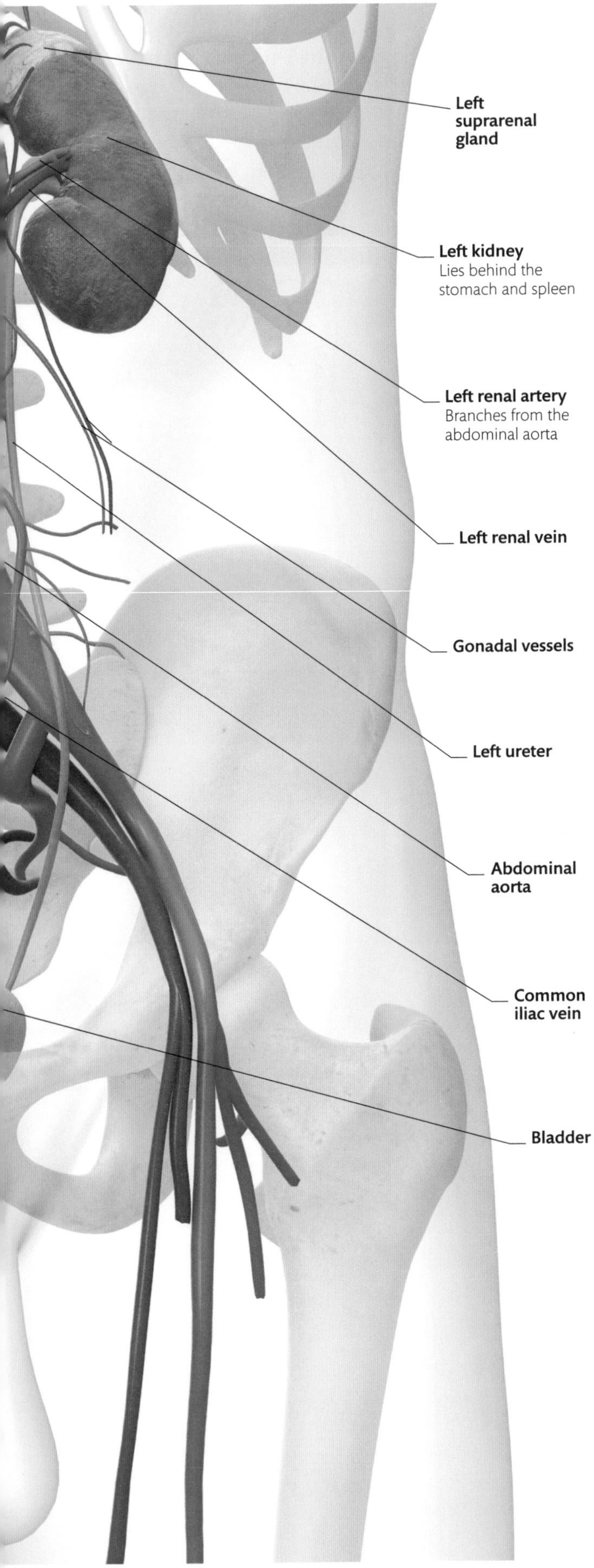
Left suprarenal gland
Left kidney
Lies behind the stomach and spleen
Left renal artery
Branches from the abdominal aorta
Left renal vein
Gonadal vessels
Left ureter
Abdominal aorta
Common iliac vein
Bladder

FLUID BALANCE

The body's fluid content is maintained by balancing intake with excretion. The osmolarity (concentration) of body fluids is detected in the brain by nerve cells called osmoreceptors. If osmolarity rises, signaling dehydration, antidiuretic hormone (ADH) is secreted from the pituitary gland and acts on the kidney to increase re-absorption of water and decrease urine output. If water intake is increased, osmolarity falls and ADH output is reduced, leading to decreased fluid re-absorption in the kidney and increased urine volume. When the body is sufficiently hydrated, urine is a pale straw color. Darker urine usually signals a need for increased water intake.

THE PROCESS OF THIRST
Although the kidney can conserve body water, it cannot replace it. Thirst, prompted by increased osmolarity, reduced body fluid volume, and symptoms such as a dry mouth, signals the need to increase fluid intake.

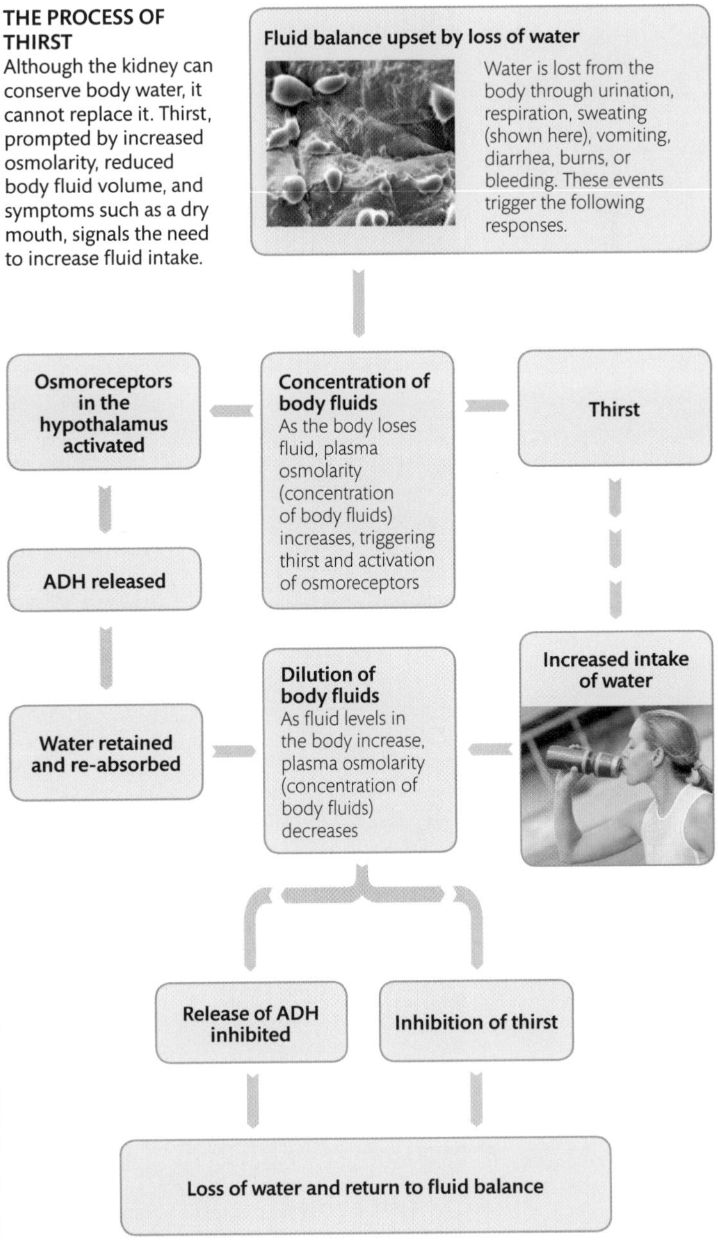
Fluid balance upset by loss of water
Water is lost from the body through urination, respiration, sweating (shown here), vomiting, diarrhea, burns, or bleeding. These events trigger the following responses.
Osmoreceptors in the hypothalamus activated
Concentration of body fluids
As the body loses fluid, plasma osmolarity (concentration of body fluids) increases, triggering thirst and activation of osmoreceptors
Thirst
ADH released
Increased intake of water
Water retained and re-absorbed
Dilution of body fluids
As fluid levels in the body increase, plasma osmolarity (concentration of body fluids) decreases
Release of ADH inhibited
Inhibition of thirst
Loss of water and return to fluid balance

See also pp. 204, 206, 208 »

ABDOMEN AND PELVIS 1

THE ORGANS OF THE URINARY SYSTEM—THE KIDNEYS, URETERS, BLADDER, AND URETHRA—LIE WITHIN THE ABDOMEN AND PELVIS, ALTHOUGH THE MALE URETHRA EXTENDS OUTSIDE THE PELVIC CAVITY INTO THE PENIS.

ANATOMY

The kidneys lie high up on the back wall of the abdomen, tucked under the twelfth ribs. They filter the blood carried to them via the renal arteries, and help to maintain the body's fluid balance. The urine they produce flows into the narrow, muscular ureters and then to the bladder in the pelvis. The bladder is a muscular bag that can expand to hold up to about 1 pint (0.5 litres) of urine, which is passed out of the body through the urethra.

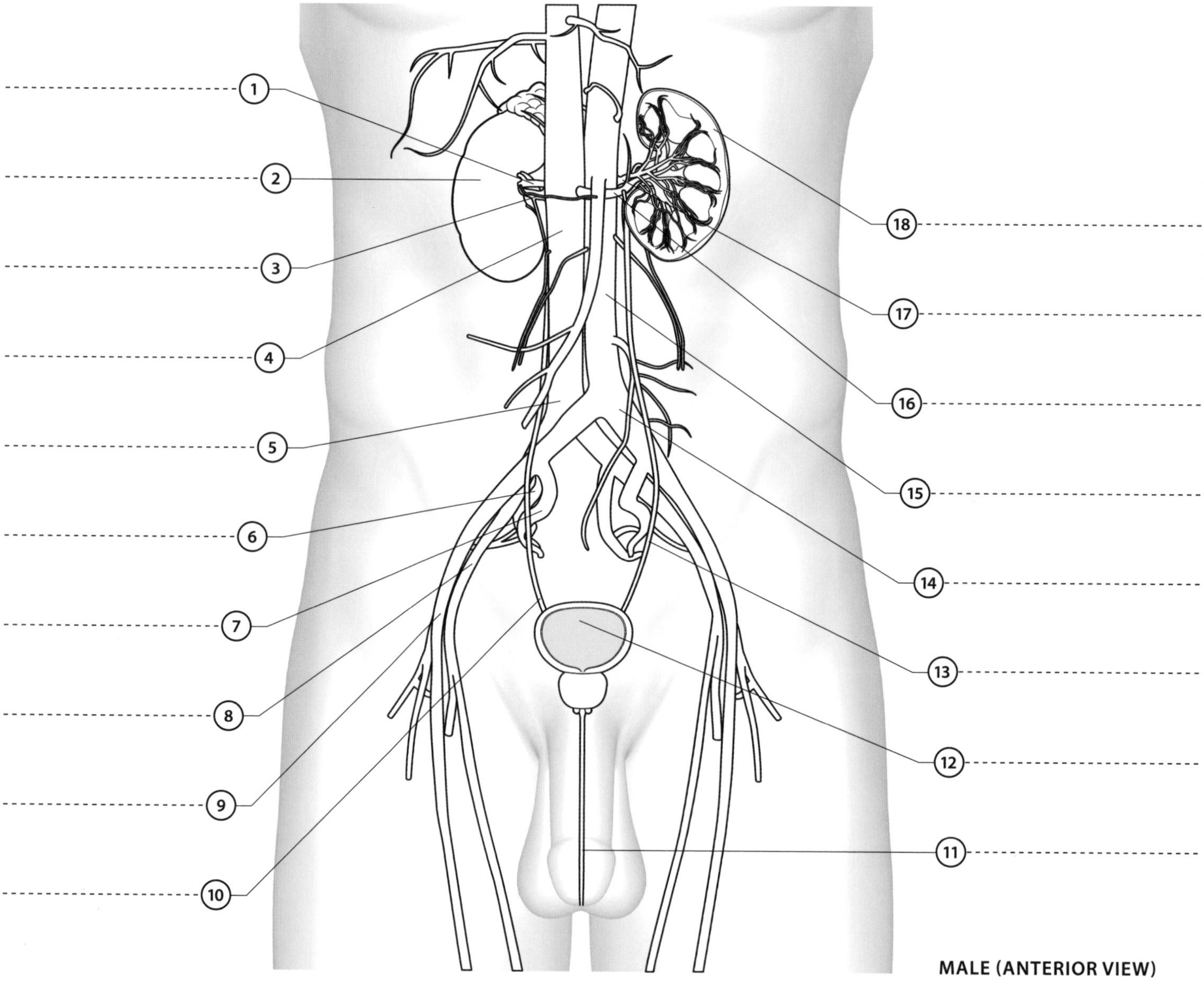

MALE (ANTERIOR VIEW)

MALE (ANTERIOR VIEW)

Color and/or label the structures indicated on the diagram using the key below.

1. Right renal artery
2. Right kidney
3. Right renal vein
4. Inferior vena cava
5. Right common iliac vein
6. Right internal iliac vein
7. Right internal iliac artery
8. Right external iliac vein
9. Right external iliac artery
10. Right ureter
11. Urethra
12. Bladder
13. Left ureter
14. Left common iliac artery
15. Abdominal aorta
16. Left renal vein
17. Left renal artery
18. Left kidney (sectioned)

SEXUAL DIFFERENCES

The lower urinary tract is different in males and females. In males, the urethra is longer—about 8 in (20 cm) in length—and passes through the prostate gland, carrying either urine or semen to the opening at the tip of the penis. The female bladder sits under the uterus, and the urethra—which is about 1.5 in (4 cm) long—carries urine to the opening in front of the vagina.

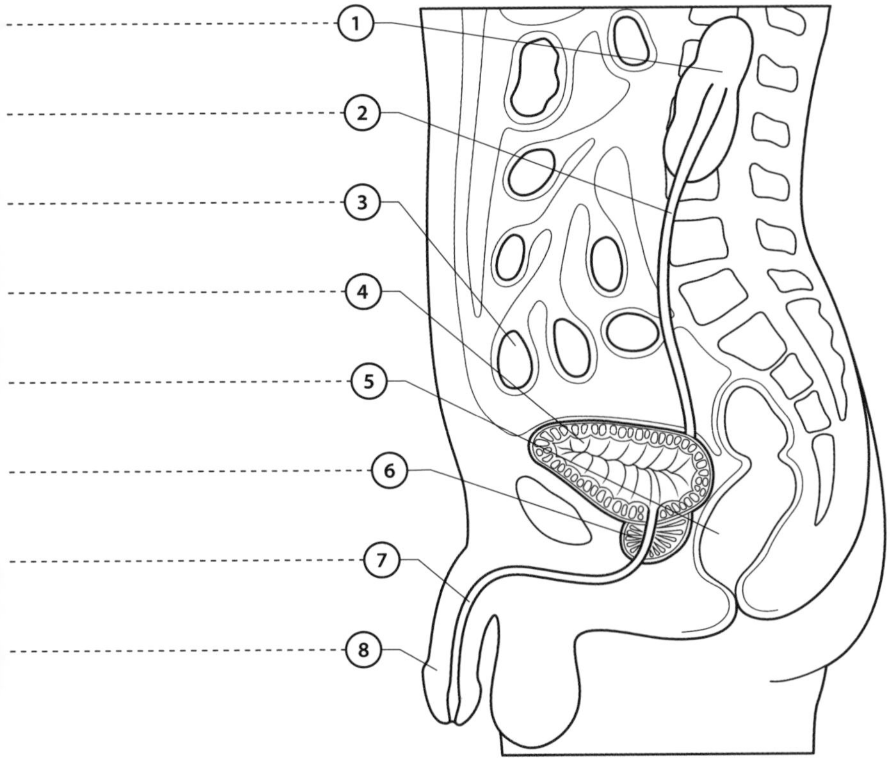

MALE (LATERAL VIEW)

MALE (LATERAL VIEW)

Color and/or label the structures indicated on the diagram using the key below.

1. Kidney
2. Ureter
3. Loop of intestine
4. Bladder
5. Rectum
6. Prostate gland
7. Urethra
8. Penis

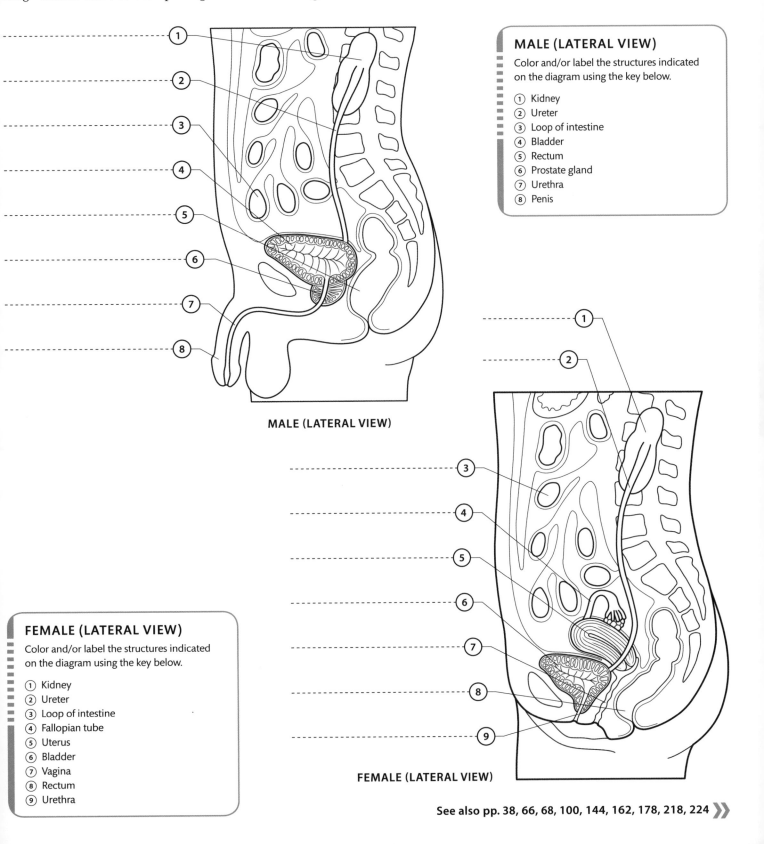

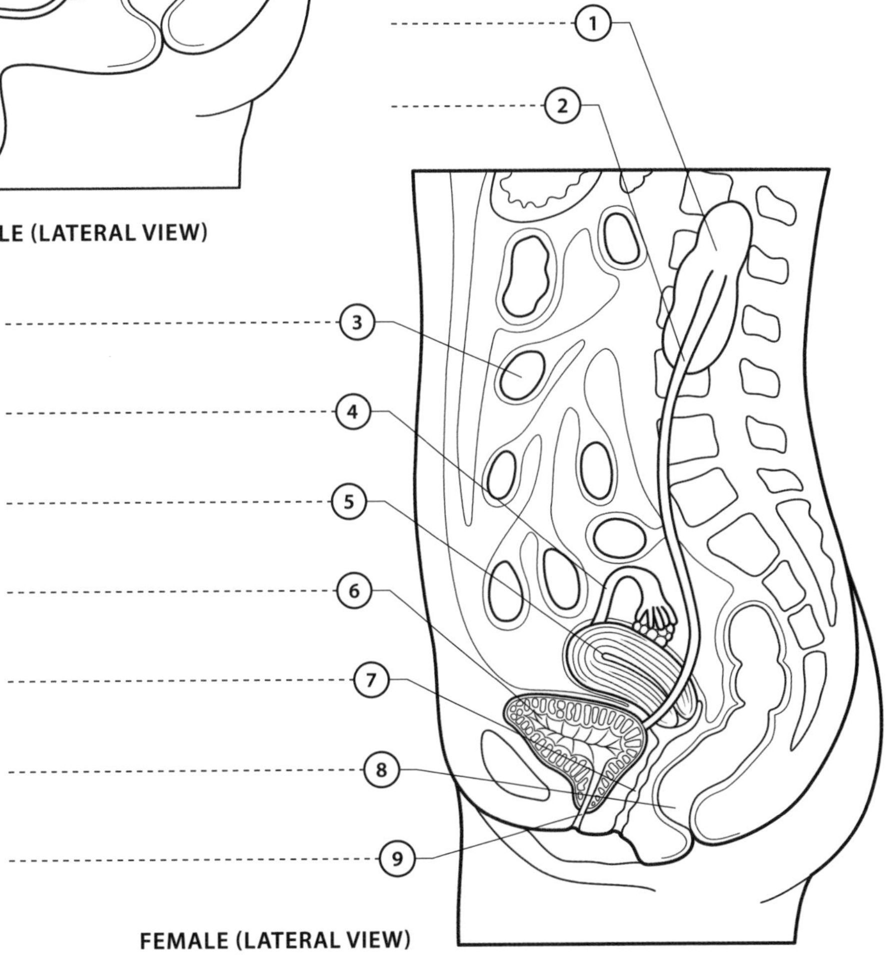

FEMALE (LATERAL VIEW)

FEMALE (LATERAL VIEW)

Color and/or label the structures indicated on the diagram using the key below.

1. Kidney
2. Ureter
3. Loop of intestine
4. Fallopian tube
5. Uterus
6. Bladder
7. Vagina
8. Rectum
9. Urethra

See also pp. 38, 66, 68, 100, 144, 162, 178, 218, 224 »

ABDOMEN AND PELVIS 2

THE KIDNEYS REGULATE THE BODY'S FLUID BALANCE, FILTER WASTE PRODUCTS AND TOXINS FROM THE BLOOD, AND REGULATE BLOOD ACIDITY.

KIDNEY STRUCTURE AND FUNCTION

The cortex of each kidney contains about a million nephrons. These are filtration units, made up of a glomerulus and a tubule. The glomerulus consists of a capillary network surrounded by a glomerular (Bowman's) capsule. The tubule is a looped tube connected to the glomerulus. Together, they filter the blood plasma, re-absorbing water and useful ions from the filtrate and producing urine as an excretory product. Loops from the nephrons dip down into the medulla of the kidney, where the amount of salt and water in the urine is controlled. About 85 percent of nephrons are cortical (short-looped); the rest are juxtamedullary (long-looped). Collecting ducts carry the outflow of the nephrons to the renal pelvis, from where urine flows into the ureter and bladder for excretion. As well as producing urine, the kidneys also secrete hormones: erythropoietin, which stimulates red blood cell production, and renin, which helps to regulate blood presssure.

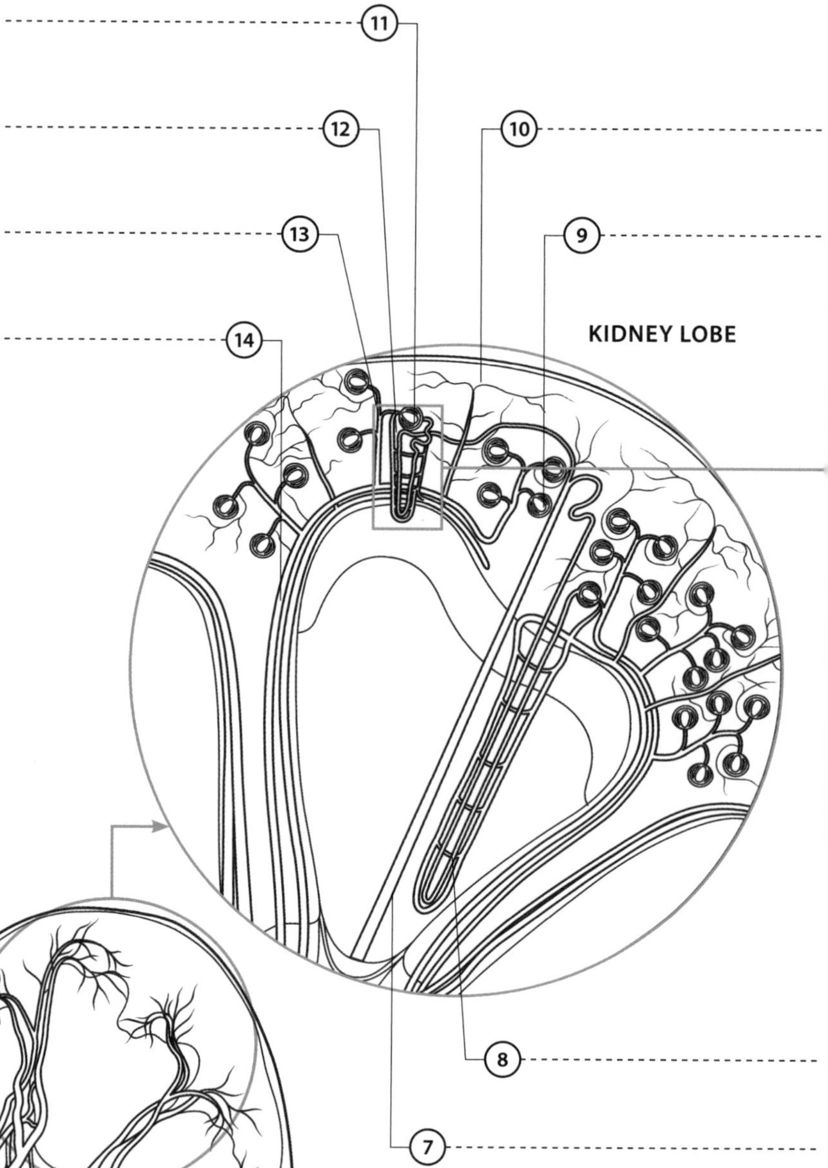

KIDNEY LOBE

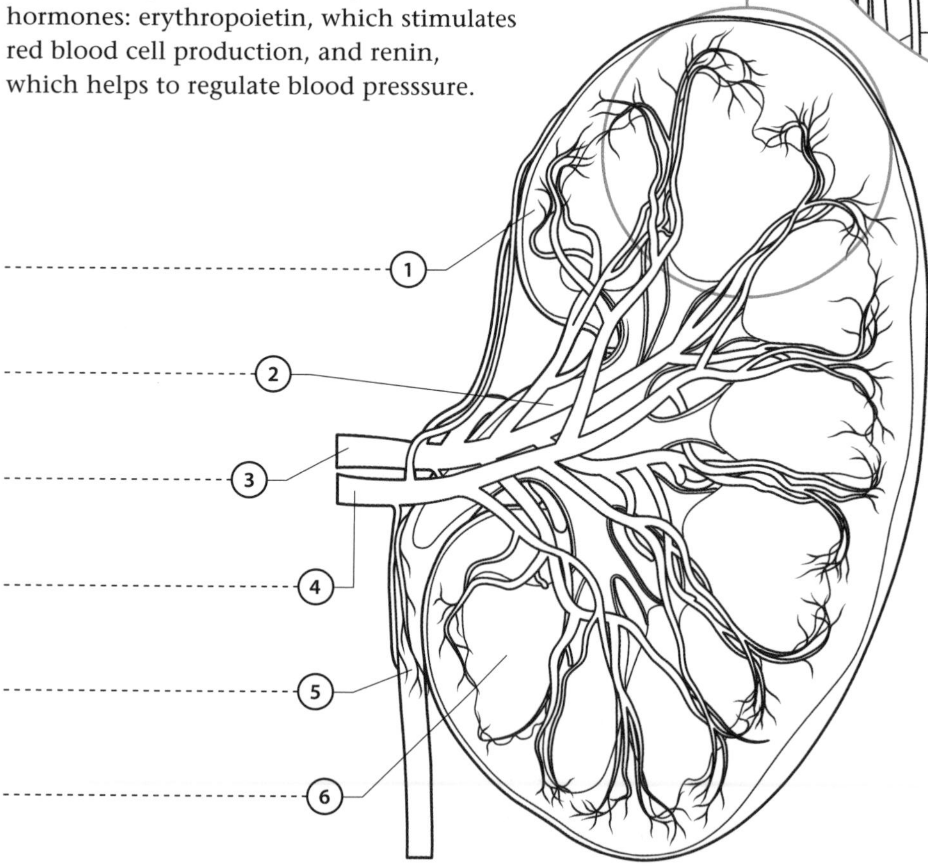

KIDNEY HEMISECTION

KIDNEY HEMISECTION AND KIDNEY LOBE

Color and/or label the structures indicated on the diagram using the key below.

1. Renal cortex
2. Renal pelvis
3. Renal artery
4. Renal vein
5. Ureter
6. Renal medulla
7. Urine-collecting duct
8. Capillary
9. Juxtamedullary nephron
10. Renal cortex
11. Glomerulus
12. Tubule
13. Cortical nephron
14. Blood vessel

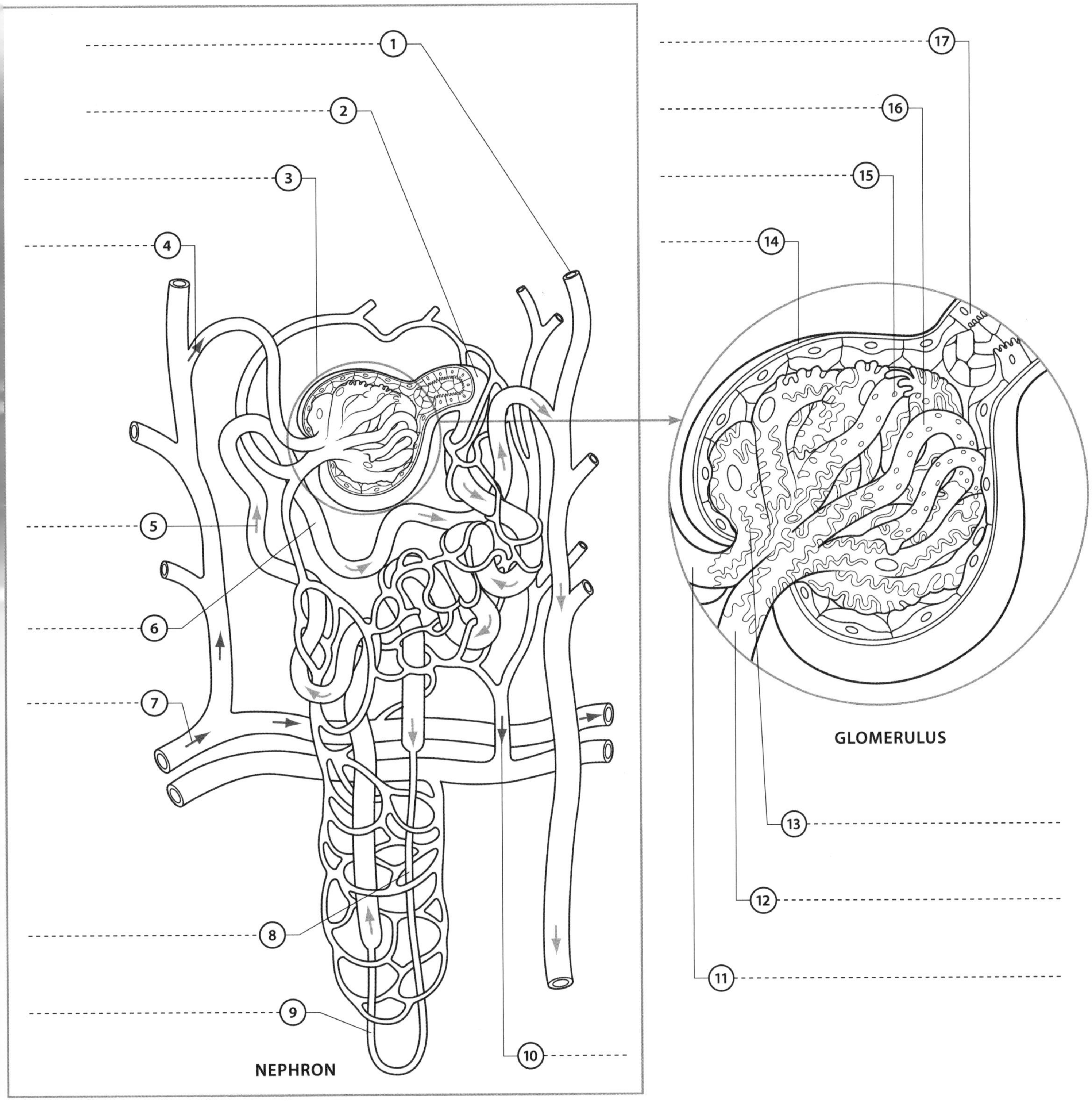

NEPHRON AND GLOMERULUS

Color and/or label the structures indicated on the diagram using the key below.

1. Urine-collecting duct
2. Proximal convoluted tubule
3. Glomerulus
4. Blood enters nephron
5. Flow of filtrate
6. Distal convoluted tubule
7. Blood flow
8. Descending limb of loop of Henle
9. Ascending limb of loop of Henle
10. Filtered blood leaves the nephron
11. Afferent arteriole to glomerulus
12. Efferent arteriole from glomerulus
13. Filtration slit between podocytes
14. Bowman's capsule
15. Fenestration (pore)
16. Glomerular capillary
17. Proximal convoluted tubule

See also pp. 202, 204, 208

URINE PRODUCTION

TOGETHER, THE KIDNEYS FILTER ABOUT 316 PINTS (180 LITRES) OF BLOOD PLASMA EACH DAY, PRODUCING 1.75–3.5 PINTS (1–2 LITRES) OF URINE AS AN EXCRETORY PRODUCT.

MAKING URINE

The glomerulus of each nephron receives blood under high pressure from the renal artery. The pressure squeezes blood through its membranes so that water and small molecules pass through but larger cells and molecules (such as proteins) are retained in the blood. Each glomerulus is inside a Bowman's capsule, which conveys the filtrate to a proximal convoluted tubule. This tubule is the first part of a tube that then runs down into the medulla in a loop—the loop of Henle—and back up the distal convoluted tubule to join tubules from other nephrons passing into the collecting ducts. In the proximal tubule, glucose is re-absorbed into the blood. In the loop of Henle, most of the water is re-absorbed back into the capillaries that surround it. In the distal tubule, most of the salts are re-absorbed. What remains is urine, containing urea and other waste products, which flows into the urine-collecting duct.

MAKING URINE

Color and/or label the diagram using the key below.

1. Blood filtered by glomerulus
2. Bowman's capsule
3. Blood enters nephron
4. Filtrate
5. Proximal convoluted tubule
6. Re-absorption of glucose, water, and some salts
7. Re-absorption of water into blood
8. Loop of Henle
9. Re-absorption of salts into blood
10. Filtered blood leaves nephron
11. Urine
12. Urine-collecting duct
13. Distal convoluted tubule

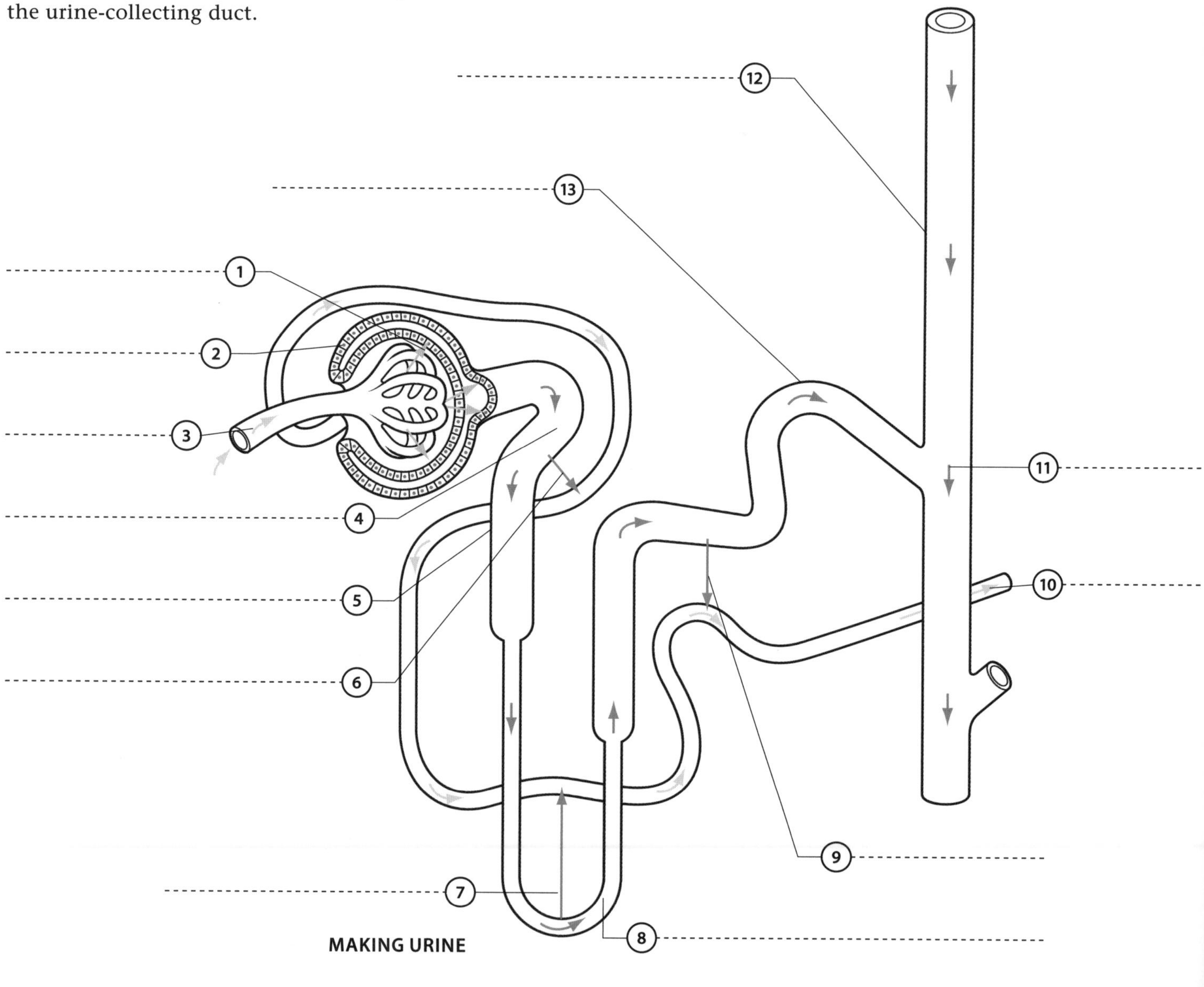

MAKING URINE

JRINATION

Jrine enters the bladder via the two ureters. Valves at he entry points into the bladder prevent reflux back up the ıreters. At the exit from the bladder there are two sphincters hat prevent urine from draining continuously into the ırethra. The internal sphincter opens and closes automatically ɔut the external sphincter is under voluntary control. When he bladder is empty, the detrusor muscle in its walls is relaxed and both sphincters are closed. As the bladder fills, the walls stretch, prompting a small reflex contraction in the detrusor muscle and triggering the urge to urinate. This urge can be resisted voluntarily by keeping the external sphincter closed. When it is convenient to urinate, the external sphincter and pelvic floor muscles are consciously relaxed, and the detrusor muscle contracts, expelling urine out of the bladder.

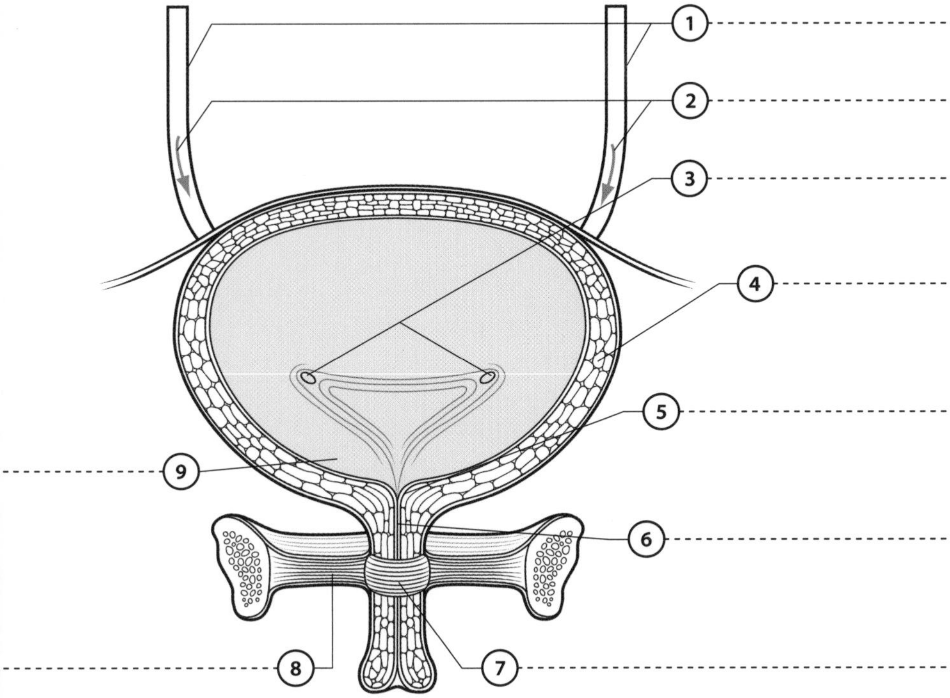

BLADDER FILLS

BLADDER FILLS

Color and/or label the diagram using the key below.

1. Ureters
2. Urine flows into bladder
3. Ureteric orifices with valves
4. Detrusor muscle stretched
5. Internal sphincter closed
6. Urethra
7. External sphincter closed
8. Pelvic floor muscle
9. Urine in bladder

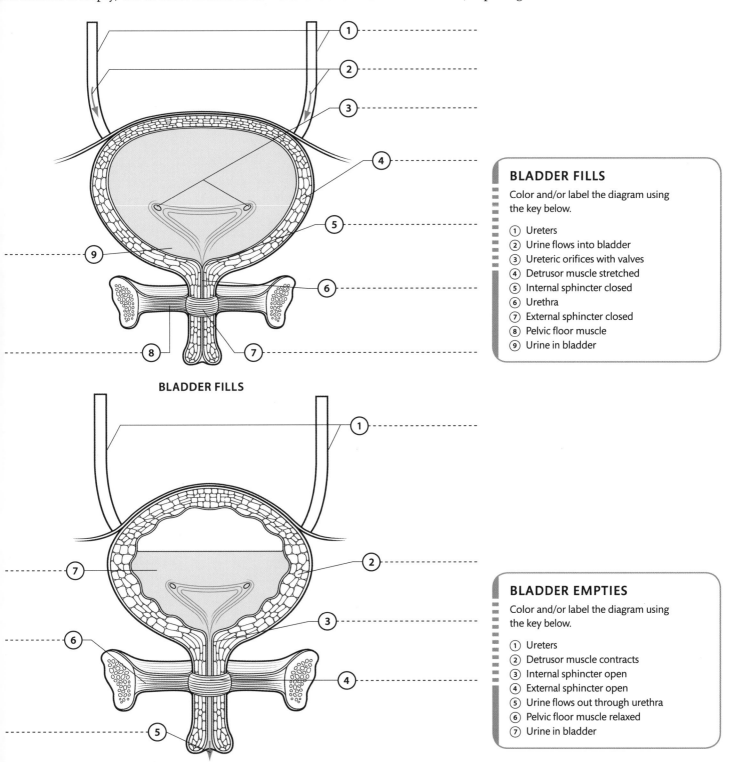

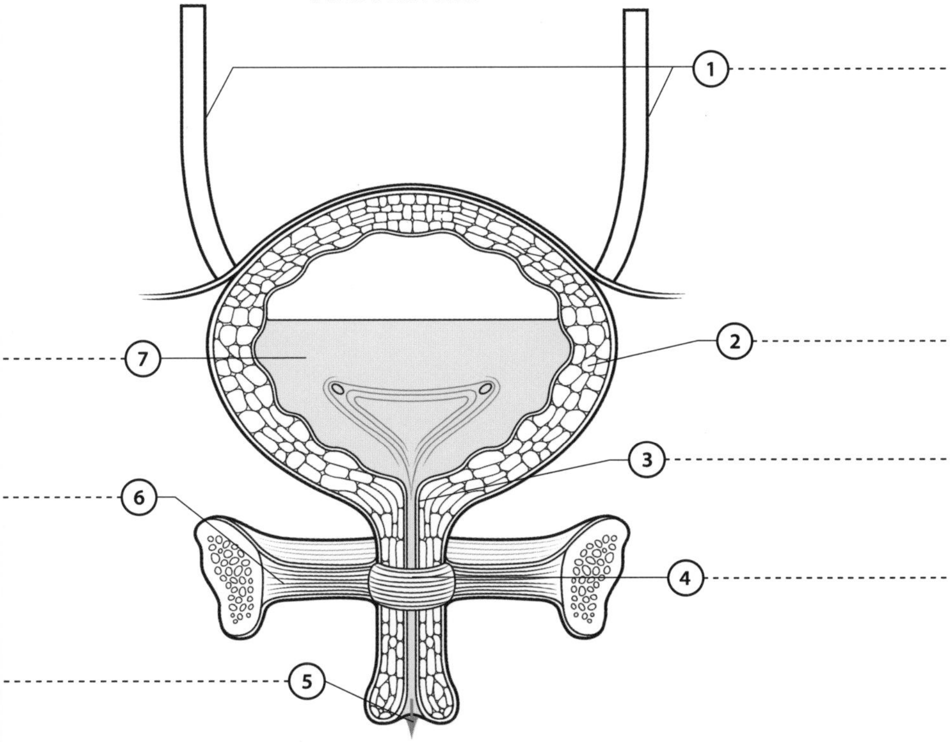

BLADDER EMPTIES

BLADDER EMPTIES

Color and/or label the diagram using the key below.

1. Ureters
2. Detrusor muscle contracts
3. Internal sphincter open
4. External sphincter open
5. Urine flows out through urethra
6. Pelvic floor muscle relaxed
7. Urine in bladder

See also pp. 204, 206 »

REPRODUCTIVE SYSTEM

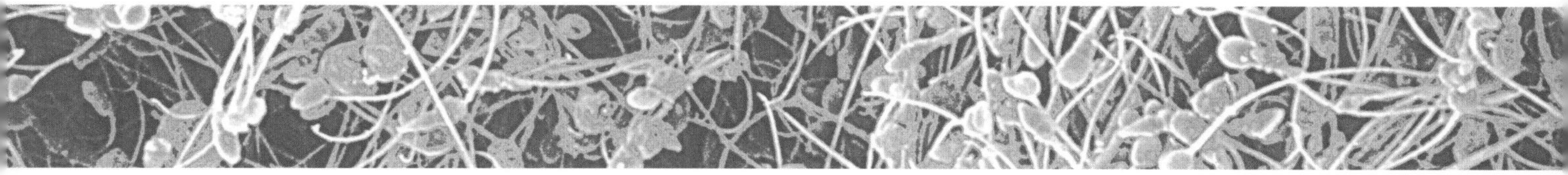

11

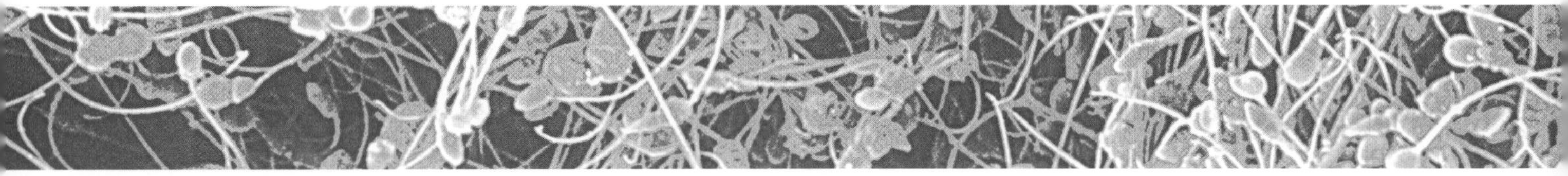

REPRODUCTIVE SYSTEM 1

MOST ORGANS IN THE BODY ARE SIMILAR IN MEN AND IN WOMEN, BUT THOSE OF THE REPRODUCTIVE SYSTEM DIFFER GREATLY BETWEEN THE SEXES. THIS IS THE ONLY SYSTEM THAT DOES NOT FUNCTION UNTIL PUBERTY.

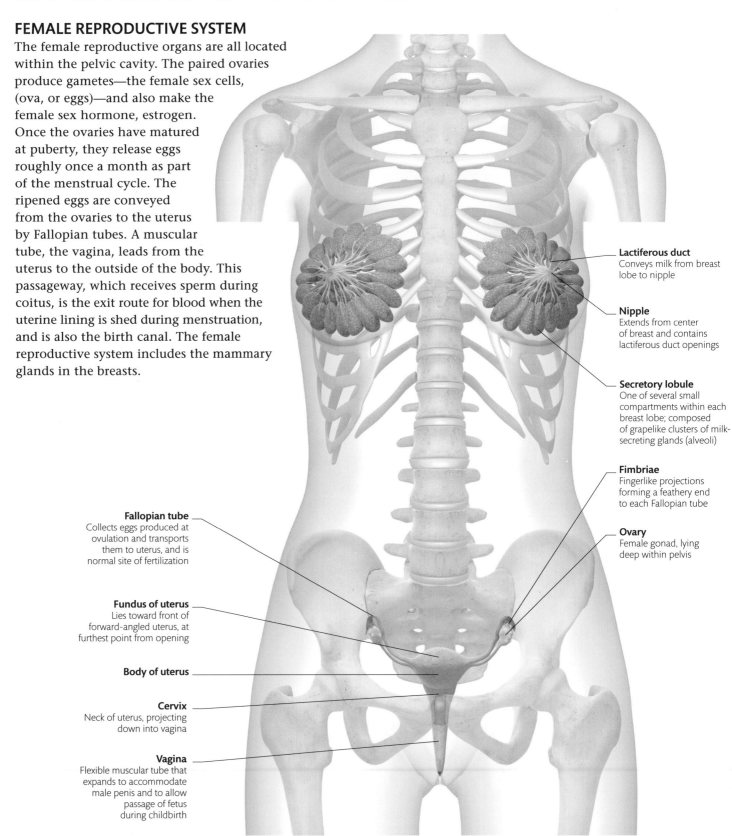

FEMALE REPRODUCTIVE SYSTEM

The female reproductive organs are all located within the pelvic cavity. The paired ovaries produce gametes—the female sex cells, (ova, or eggs)—and also make the female sex hormone, estrogen. Once the ovaries have matured at puberty, they release eggs roughly once a month as part of the menstrual cycle. The ripened eggs are conveyed from the ovaries to the uterus by Fallopian tubes. A muscular tube, the vagina, leads from the uterus to the outside of the body. This passageway, which receives sperm during coitus, is the exit route for blood when the uterine lining is shed during menstruation, and is also the birth canal. The female reproductive system includes the mammary glands in the breasts.

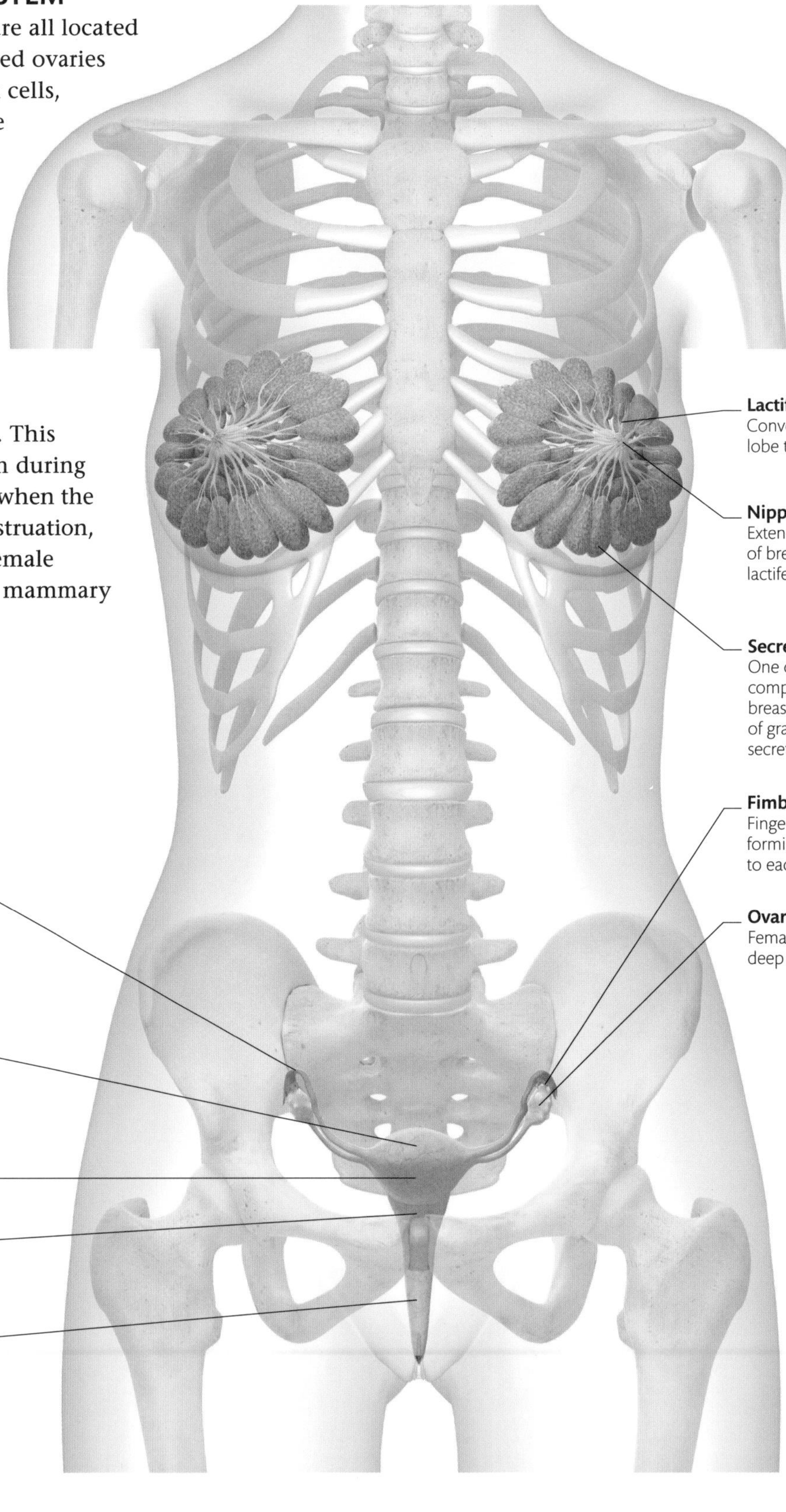

Lactiferous duct
Conveys milk from breast lobe to nipple
Nipple
Extends from center of breast and contains lactiferous duct openings
Secretory lobule
One of several small compartments within each breast lobe; composed of grapelike clusters of milk-secreting glands (alveoli)
Fimbriae
Fingerlike projections forming a feathery end to each Fallopian tube
Ovary
Female gonad, lying deep within pelvis
Fallopian tube
Collects eggs produced at ovulation and transports them to uterus, and is normal site of fertilization
Fundus of uterus
Lies toward front of forward-angled uterus, at furthest point from opening
Body of uterus
Cervix
Neck of uterus, projecting down into vagina
Vagina
Flexible muscular tube that expands to accommodate male penis and to allow passage of fetus during childbirth

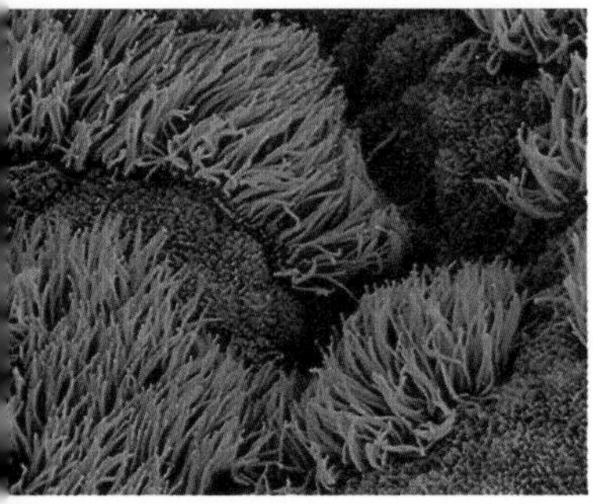

INSIDE A FALLOPIAN TUBE
The cells lining the tube can be seen in this (false-color) electron micrograph. Secretory cells (purple) lubricate the surface and cilia (pink) waft a current of fluid down the tube, carrying the ovum to the uterus.

ENDOMETRIUM
This electron micrograph shows the endometrium, the lining of the uterus. Thickened and folded, the vascular glandular tissue is ready to receive a fertilized ovum.

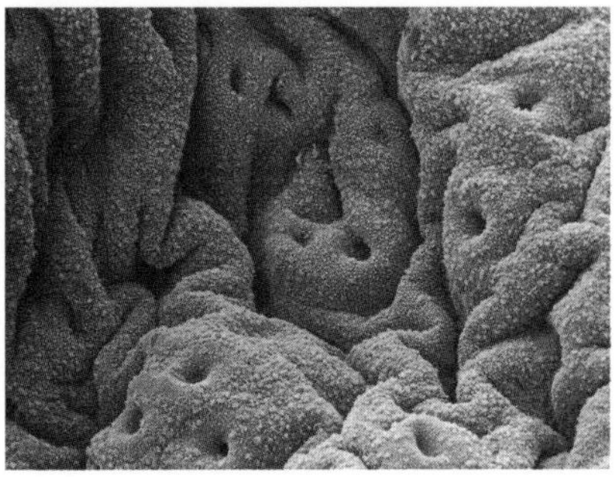

'ALLOPIAN TUBES

xtending from the uterus to each ovary, the Fallopian ubes transport ripened eggs released from the ovaries. 'heir muscular walls are lined with both secretory cells nd with cells bearing cilia. Once an egg has been released rom an ovary and enters a Fallopian tube, a combination f muscular contractions and beating cilia propel it down he tube toward the uterus. The tubes also transport perm traveling upward from the uterus and are the normal ite for fertilization.

UTERUS

The uterus is located behind the bladder and in front of the intestines. This hollow, thick-walled organ consists mainly of muscle that expands to accommodate the developing fetus during pregnancy. It is lined with glandular tissue, the endometrium. In response to hormones released during the menstrual cycle, the endometrium retains fluid and increases in thickness in preparation for potential pregnancy. If a fertilized ovum does not implant in the uterus, the endometrium is shed as menstrual blood.

ΛALE REPRODUCTIVE SYSTEM

n males, the reproductive organs comprise the penis, the estes, a pair of sperm-carrying tubes (the vasa deferentia), ccessory sex glands (the seminal vesicles and the prostate), nd the urethra. The testes, which produce sperm and the nale sex hormone, testosterone, hang outside the pelvic avity in the scrotum, where they can maintain the correct emperature for sperm production.

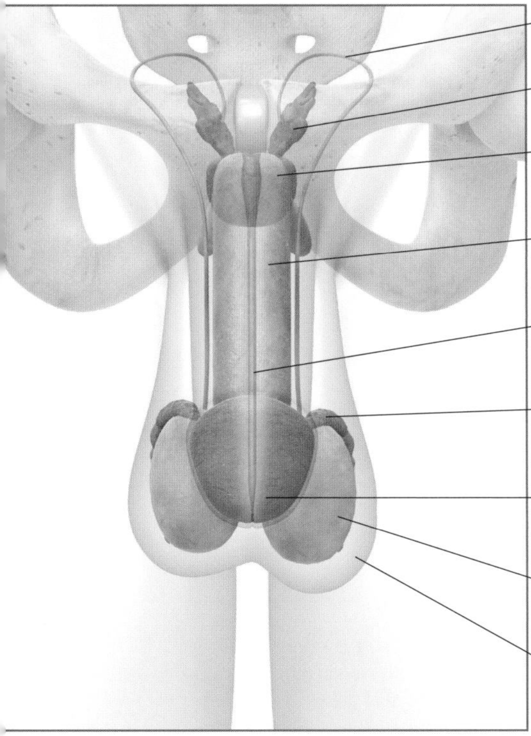

Vas deferens
Carries sperm from testis to urethra
Seminal vesicle
Contributes fluid to semen
Prostate gland
Accessory gland at base of bladder; contributes some fluid to semen
Shaft of penis
Consists of erectile tissue that engorges with blood during erection
Urethra
Conveys sperm and urine through penis
Epididymis
Coiled tube on back of testis where sperm are stored and mature
Glans penis
Sensitive enlarged end of penis
Testis
Male gonad; hangs outside body in scrotum
Scrotum
Pouch of skin and muscle encasing testes

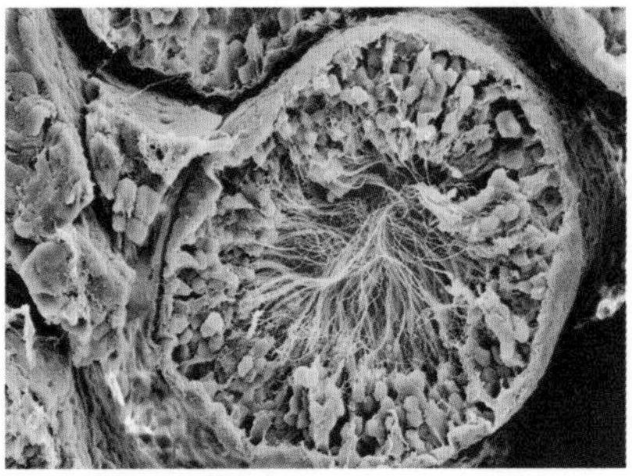

SEMINIFEROUS TUBULE
This cross section of a seminiferous tubule shows sperm taking shape as they move toward the center of the tubule. Approaching the final stages of development, they are developing long tails, clearly seen here.

TESTES

By birth, or within the next few months, the testes have usually descended to the surface of the body, where they hang in the scrotum. To regulate temperature, the scrotum elevates the testes to preserve warmth and lowers them away from the body to cool them. From puberty onward, sperm production—spermatogenesis—takes place in the seminiferous tubules, the delicate coiled tubes within each testis.

See also pp. 214, 216, 218, 220, 222, 224, 226, 228 »

REPRODUCTIVE SYSTEM 2

FROM PUBERTY ONWARDS, THE FUNCTION OF THE REPRODUCTIVE ORGANS IN BOTH SEXES IS REGULATED BY HORMONES. IN FEMALES, DECLINING LEVELS OF CERTAIN HORMONES IN LATER LIFE RESULT IN MENOPAUSE.

PUBERTY

At puberty, hormonal changes stimulate development of the sex organs and enable reproduction, as well as triggering increase in height and other physical changes. The age of onset of puberty is highly variable. In males, puberty begins around age 12 or 13, with signs of development becoming apparent by age 14. The testicles and penis enlarge, and sperm production begins; other changes include hair growth in the pubic area and armpits, increase of muscle bulk, deepening voice, and growth of facial hair. Puberty in males is usually completed by age 17 or 18. In females, puberty starts earlier, at around 10 or 11 years, with physical changes often appearing around age 13. Usually the first signs are breast development and growth of body hair, followed by the onset of menstruation. Female puberty is normally complete by about age 16.

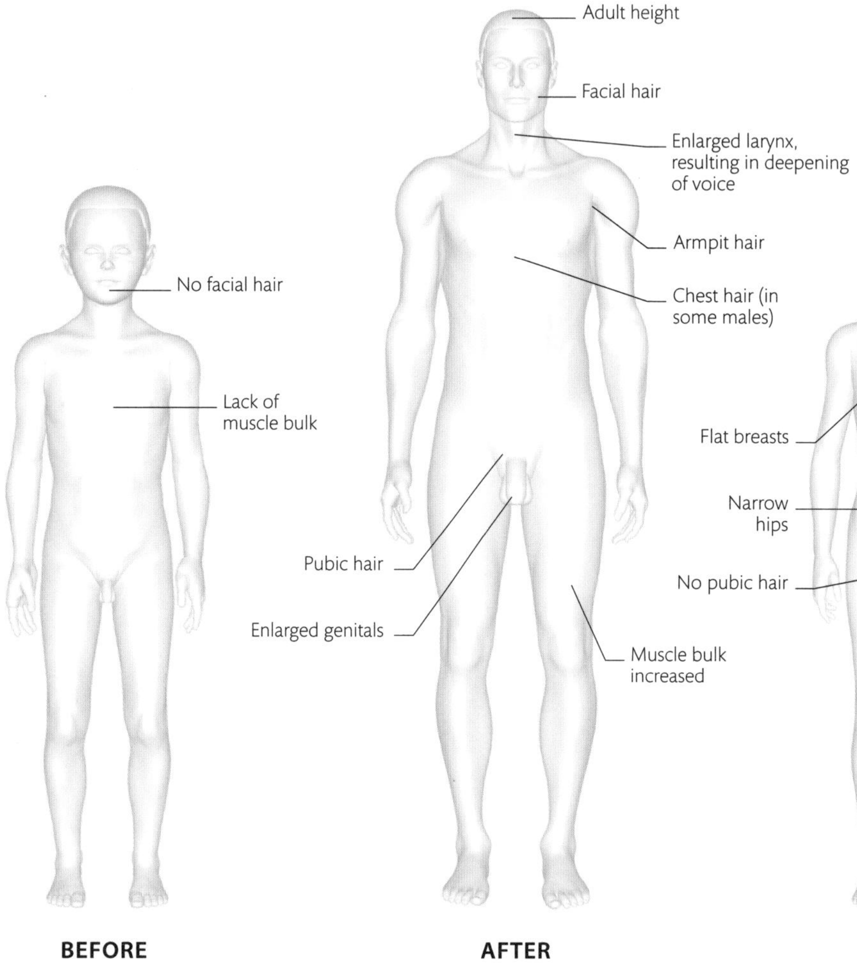

PHYSICAL CHANGES IN MALES
In developing males, the genitals enlarge, the scrotum darkens, and hair grows in the armpits, the pubic region, and sometimes the trunk. Appearance of facial hair is one of the last changes to occur during puberty.

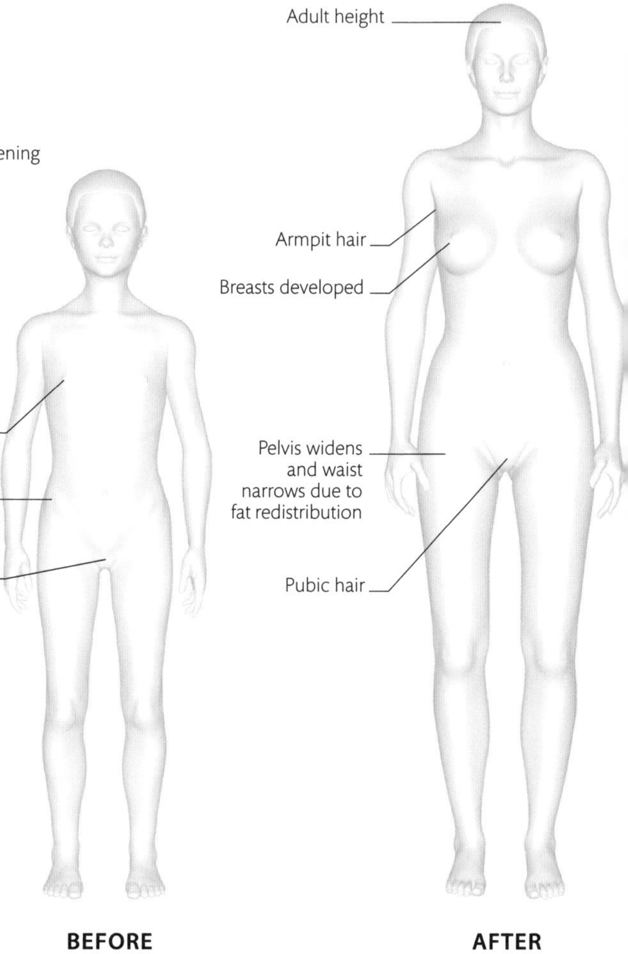

PHYSICAL CHANGES IN FEMALES
Changes associated with female puberty include development of the breasts, widening of the pelvis, and growth of pubic and underarm hair. An extra layer of fat develops under the skin.

MALE HORMONE CONTROL

Hormone production is often regulated by feedback, when the amount of a substance in a system controls how much of it is made. Sperm production and male hormones are controlled by feedback loops involving the testes, hypothalamus, and pituitary gland. The pituitary controls testis function by producing follicle-stimulating hormone (FSH) and luteinizing hormone (LH). Pituitary production depends on gonadotropin-releasing hormone (GnRH) from the hypothalamus. The process is governed by negative feedback—high levels of testosterone act on the pituitary to slow the release of LH and FSH.

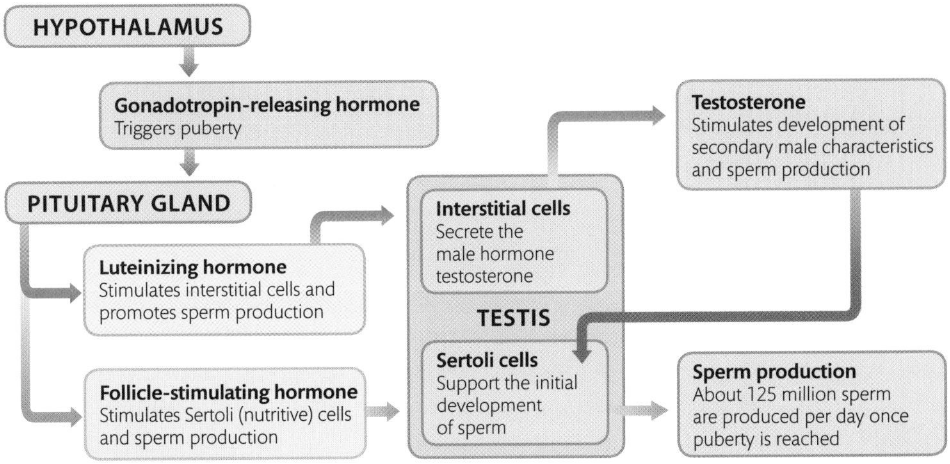

MALE HORMONE CHART
Production of sex hormones in males is "switched on" at puberty. The secretion of GnRH in the hypothalamus initiates a series of hormonal messages that prompt physical development and, later, production of sperm.

FEMALE HORMONE CONTROL

Female hormones, like those of males, are tightly regulated by the hypothalamus and pituitary gland. The biological clock that regulates the female menstrual cycle consists of the rhythmic release of gonadotropin-releasing hormone (GnRH) from the hypothalamus. GnRH regulates release of luteinizing hormone (LH) and follicle-stimulating hormone (FSH) in the anterior pituitary gland. These hormones also send feedback to the hypothalamus and the pituitary. Disturbances in the release of GnRH or in the pituitary gland can result in inadequate production of FSH and LH, and abnormal function of the ovaries.

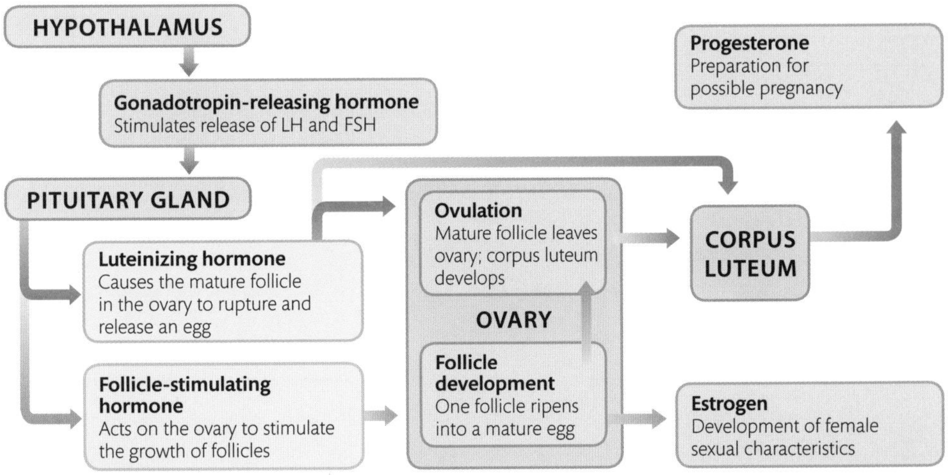

FEMALE HORMONE CHART
Activity of the ovaries and production of female sex hormones is controlled by the secretion of hormones in the hypothalamus and pituitary. Complex interaction between hormones regulates the menstrual cycle.

HORMONE CHANGES DURING MENOPAUSE

Menopause occurs between the ages of 45 and 55, when a woman's ovaries stop responding to follicle-stimulating hormone (FSH) and produce less of the female sex hormones estrogen and progesterone. This drop in hormone levels brings an end to ovulation and menstruation.

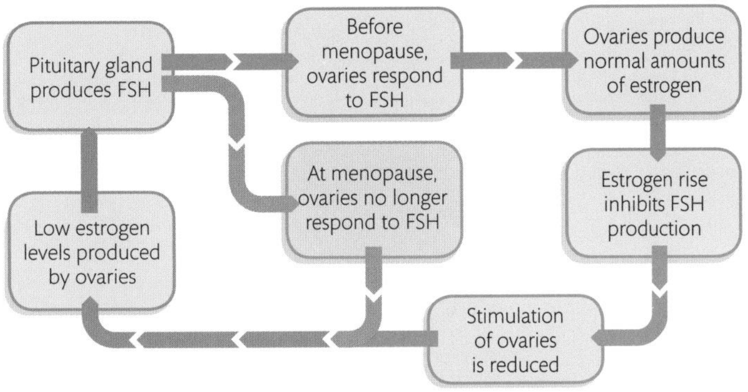

MENOPAUSAL CHANGES
At menopause, the ovaries no longer respond to FSH and produce little estrogen. The pituitary gland reacts by increasing its production of FSH (which may give rise to menopausal symptoms).

See also pp. 188, 194, 212, 228 »

FEMALE THORAX

THE BREASTS, OR MAMMARY GLANDS, ARE AN IMPORTANT PART OF THE REPRODUCTIVE SYSTEM IN FEMALES. MALES ALSO HAVE BREASTS, BUT THEY ARE MUCH SMALLER AND SERVE NO REPRODUCTIVE FUNCTION.

BREASTS

The breasts develop at puberty, when they grow due to the increased production of glandular tissue and fat. They lie on the pectoralis major muscle on each side, and each extends upward and outward on the chest wall to form an axillary tail in the armpit. Each breast contains 15–20 lobes, and each lobe, in turn, contains several lobules, where milk is produced. The lobes are connected to the nipple by lactiferous ducts, with each duct draining one lobe. Just before they enter the nipple, the lactiferous ducts expand slightly in a lactating breast to form a lactiferous sinus. Surrounding each nipple is the areola, which becomes darker during pregnancy.

ANTERIOR VIEW

Color and/or label the structures indicated on the diagram using the key below.

1. Nipple
2. Areola
3. Axillary tail
4. Secretory lobule
5. Lactiferous duct

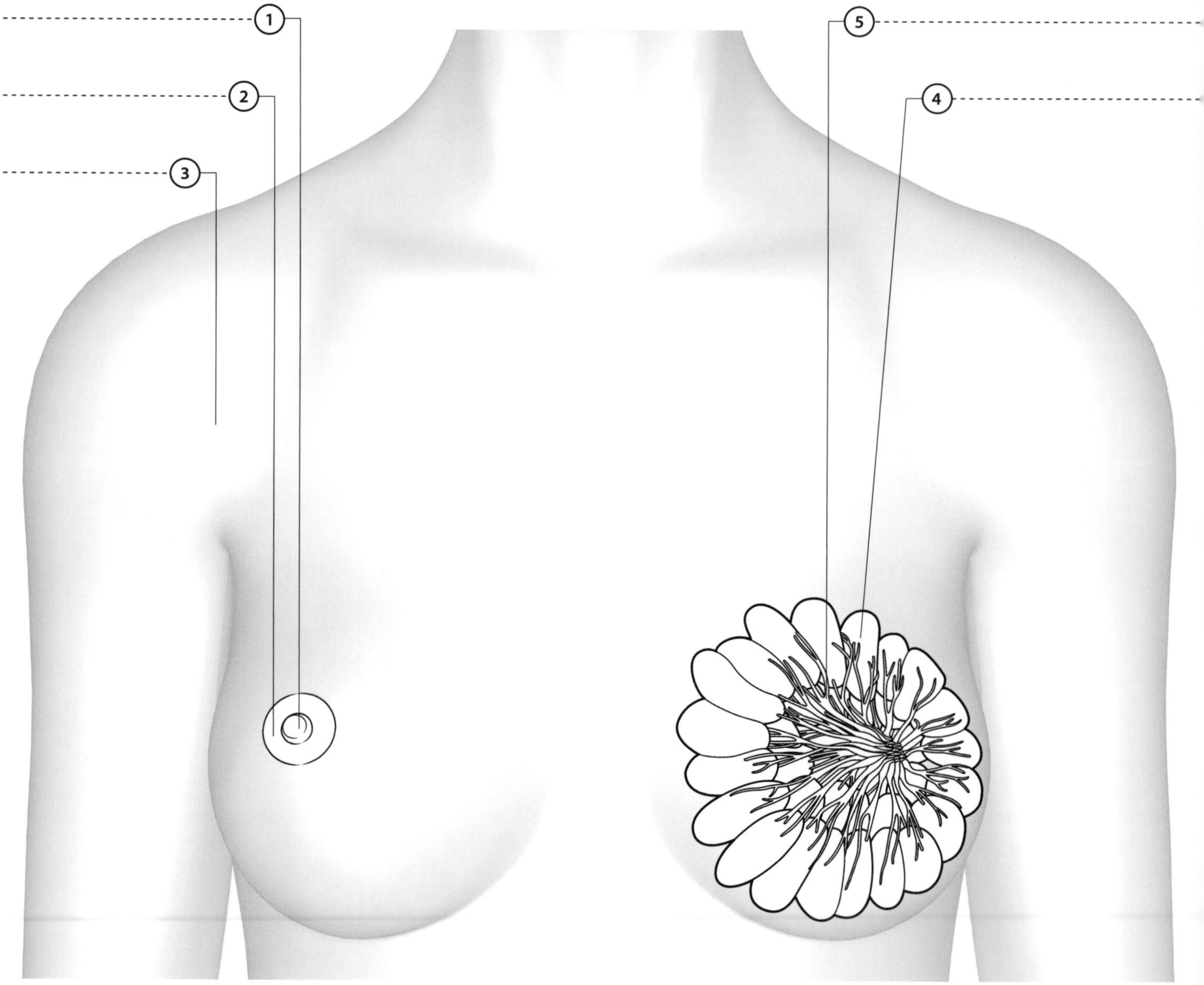

ANTERIOR VIEW

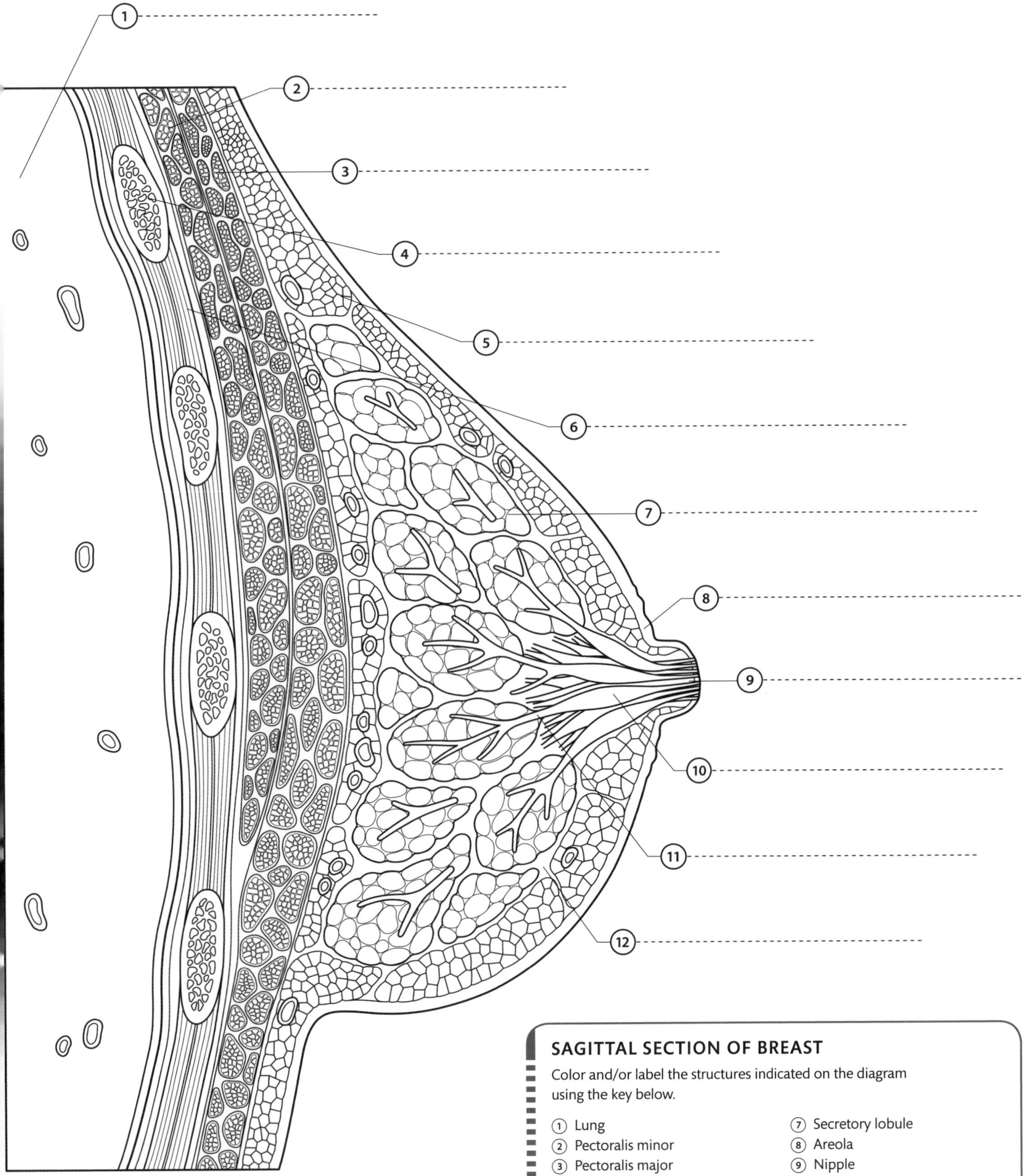

SAGITTAL SECTION OF BREAST

SAGITTAL SECTION OF BREAST

Color and/or label the structures indicated on the diagram using the key below.

1. Lung
2. Pectoralis minor
3. Pectoralis major
4. Rib
5. Superficial fascia
6. Intercostal muscle
7. Secretory lobule
8. Areola
9. Nipple
10. Lactiferous sinus
11. Lactiferous duct
12. Stroma

See also pp. 36, 62, 138, 162, 178 »

MALE ABDOMEN AND PELVIS 1

IN MALES, THE ABDOMEN AND PELVIS CONTAIN ONLY PART OF THE REPRODUCTIVE TRACT; THE PENIS AND SCROTUM LIE OUTSIDE THE BODY.

ACCESSORY GLANDS

The seminal vesicles, prostate, and bulbourethral glands (also called Cowper's glands) are collectively known as the accessory glands. They add their secretions to sperm by the contraction of their muscular sheaths during ejaculation. Fluid from the seminal vesicles makes up about 60 percent of semen by volume, and contains sugar, vitamin C, and prostaglandins. Prostate secretions account for about 30 percent of semen and include enzymes, fatty acids, cholesterol, and salts to adjust the semen's acid–alkali balance. The bulbourethral glands make up five percent of semen, and their secretions also neutralize the acid–alkali balance of traces of urine in the urethra.

POSTERIOR VIEW OF GLANDS

Color and/or label the structures indicated on the diagram using the key below.

1. Ureter
2. Seminal vesicle
3. Urethra
4. Bulbourethral gland
5. Prostate gland
6. Ejaculatory duct
7. Vas deferens
8. Bladder

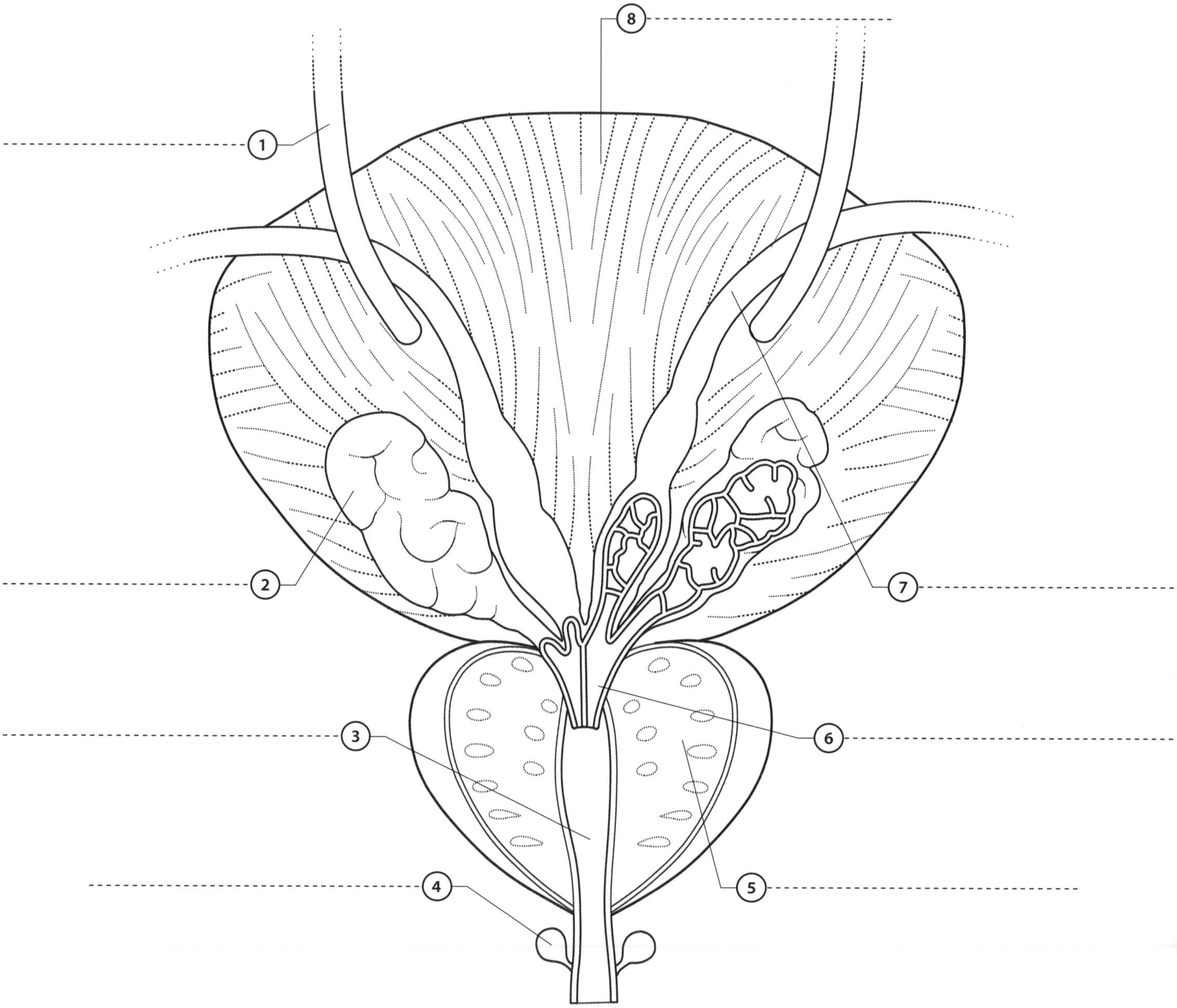

POSTERIOR VIEW OF GLANDS

MALE REPRODUCTIVE ORGANS

The male reproductive organs include the penis, testes, and various structures concerned with storage and transportation of sperm. The two oval-shaped testes are situated outside the body in the scrotum, where they can be kept at the optimum temperature for sperm production—about 5°F (3°C) lower than body temperature. The testes not only produce sperm, they also manufacture testosterone. From each testis, sperm pass into the epididymis for maturation. They are stored in the epididymides until they are either broken down and re-absorbed or ejaculated—forced by movement of fluid from the accessory glands down the vas deferens.

SAGITTAL SECTION

Color and/or label the structures indicated on the diagram using the key below.

1. Bladder
2. Vas deferens
3. Pubic symphysis
4. Penis
5. Corpus cavernosum
6. Corpus spongiosum
7. Glans penis
8. External urethral orifice
9. Scrotum
10. Testis
11. Epididymis
12. Urethra
13. External urethral sphincter
14. Rectum
15. Seminal vesicle
16. Prostate gland

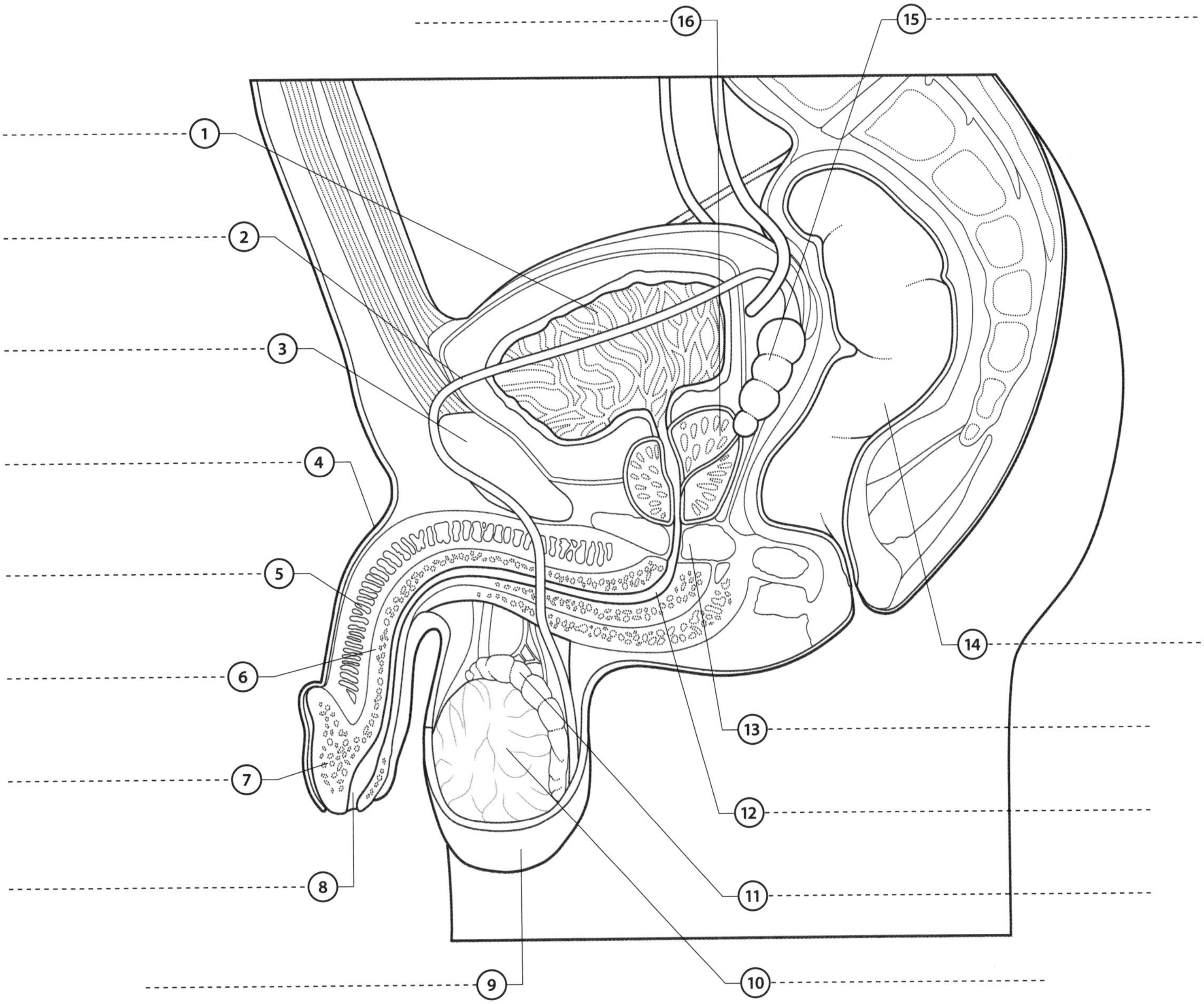

SAGITTAL SECTION

See also pp. 220, 222, 230

MALE ABDOMEN AND PELVIS 2

THE SOLE FUNCTION OF THE TESTES IS SPERM PRODUCTION, WHEREAS THE PENIS HAS DUAL URINARY AND REPRODUCTIVE FUNCTIONS AND THEREFORE CONTAINS THE URETHRA AS WELL AS ERECTILE TISSUE.

TESTES AND PENIS

Each testis contains about 200–300 lobules, and each lobule contains 1–3 tightly packed seminiferous tubules, in which sperm are generated. A series of interconnecting tubes—the rete testis—links the seminiferous tubules with the efferent ductules, which convey seminal fluid from the testis to the epididymis. From the epididymis, the seminal fluid passes into the vas deferens, up to the seminal vesicle, and then through the prostate to the urethra. In the penis, the urethra is contained within the corpus spongiosum, on either side of which are the corpora cavernosa.

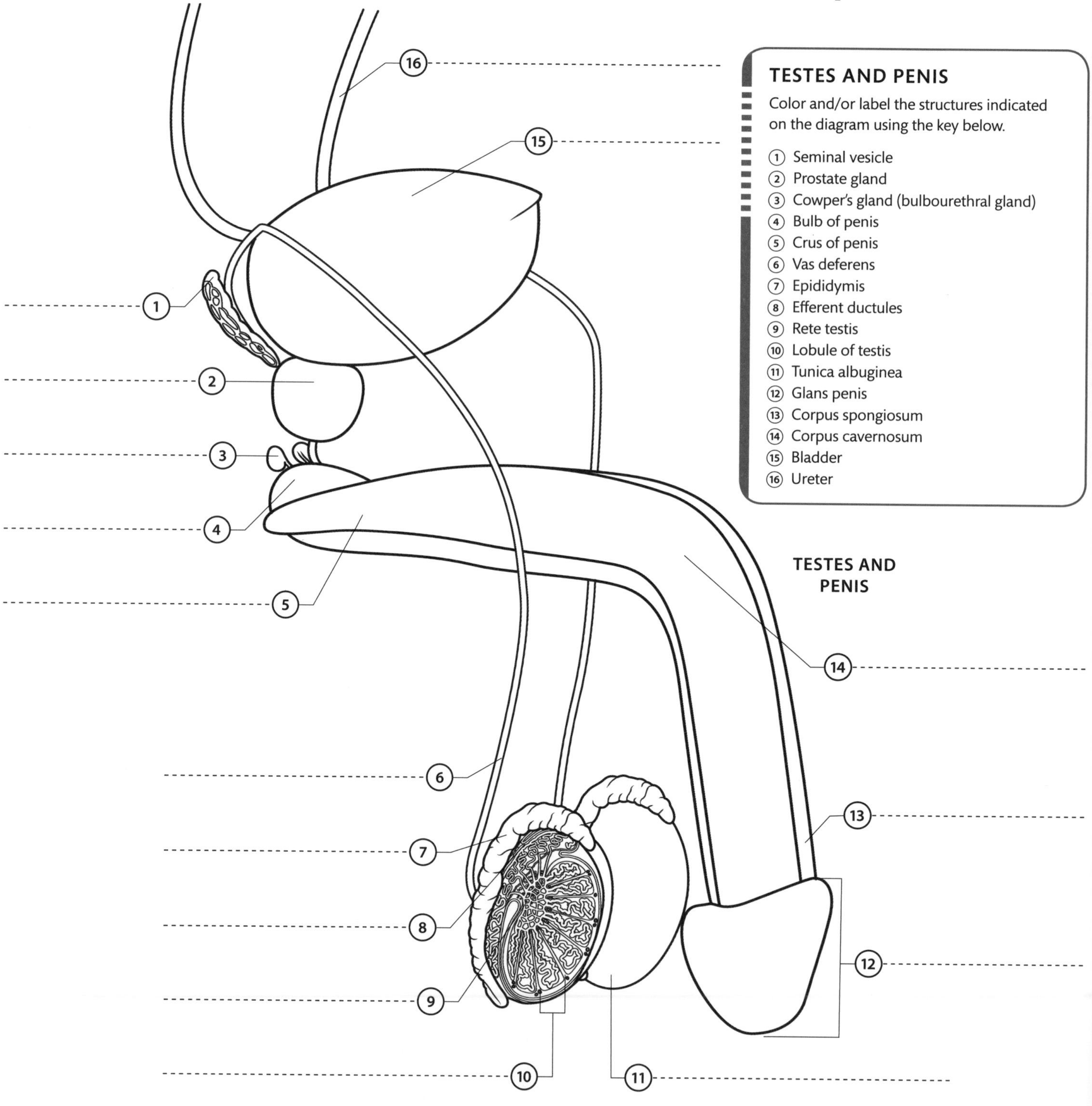

TESTES AND PENIS

TESTES AND PENIS

Color and/or label the structures indicated on the diagram using the key below.

1. Seminal vesicle
2. Prostate gland
3. Cowper's gland (bulbourethral gland)
4. Bulb of penis
5. Crus of penis
6. Vas deferens
7. Epididymis
8. Efferent ductules
9. Rete testis
10. Lobule of testis
11. Tunica albuginea
12. Glans penis
13. Corpus spongiosum
14. Corpus cavernosum
15. Bladder
16. Ureter

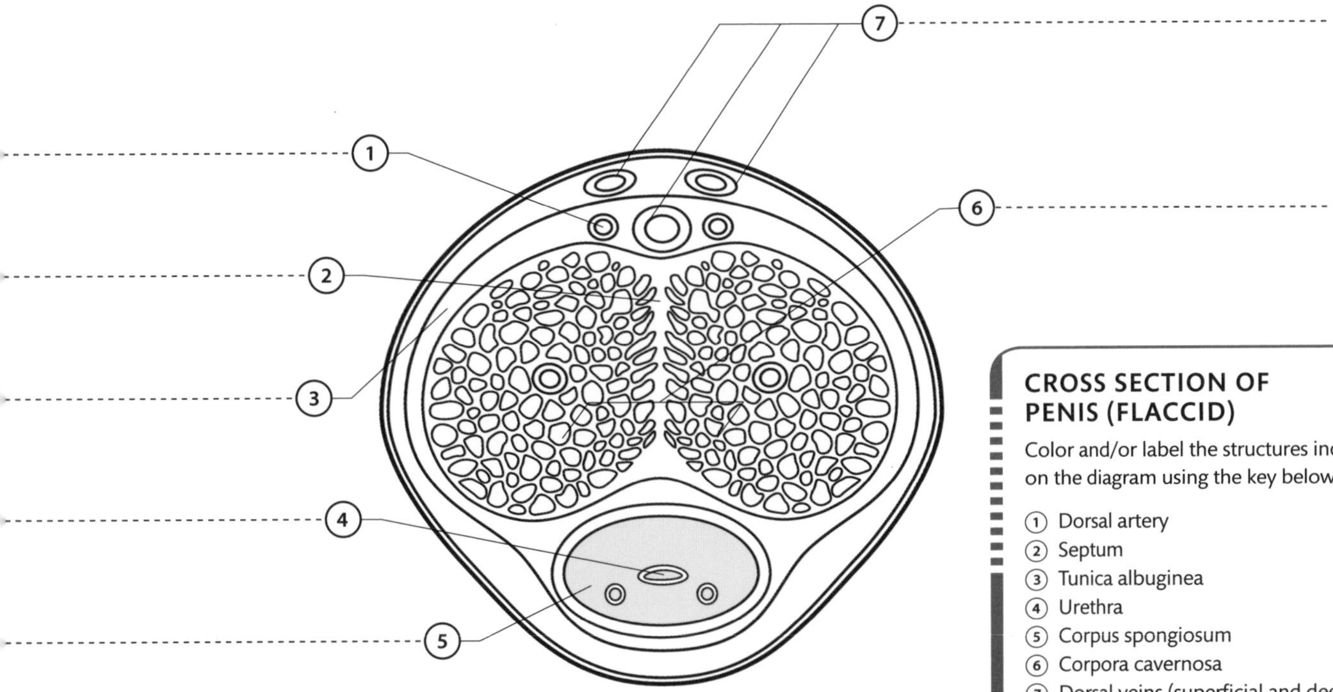

CROSS SECTION OF PENIS (FLACCID)

CROSS SECTION OF PENIS (FLACCID)

Color and/or label the structures indicated on the diagram using the key below.

1. Dorsal artery
2. Septum
3. Tunica albuginea
4. Urethra
5. Corpus spongiosum
6. Corpora cavernosa
7. Dorsal veins (superficial and deep)

PENILE ERECTION

During an erection, the smooth muscle in the corpora cavernosa relaxes, allowing rapid inflow of blood into the trabecular arteries, filling the cavernous spaces. The resulting distension obstructs the veins draining the erectile tissue and produces tumescence.

CROSS SECTION OF PENIS (ERECT)

Color and/or label the structures indicated on the diagram using the key below.

1. Dorsal veins compressed
2. Corpora cavernosa fill with blood
3. Corpus spongiosum
4. Deep arteries dilated

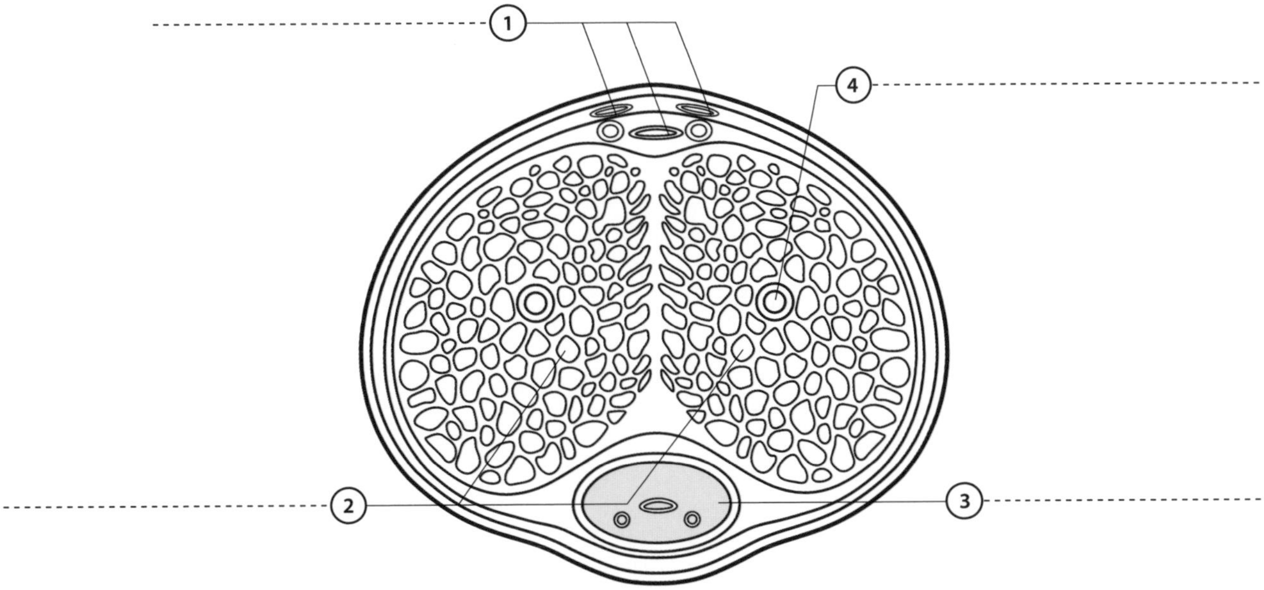

CROSS SECTION OF PENIS (ERECT)

See also pp. 218, 222 »

MALE ABDOMEN AND PELVIS 3

THE SCROTUM CONTAINS THE TESTES, WHERE SPERMATOGENESIS OCCURS, GENERATING SEVERAL HUNDRED MILLION SPERM EVERY DAY FROM PUBERTY TO OLD AGE—AS MANY AS ABOUT 12 TRILLION SPERM IN A LIFETIME.

SCROTUM

In the scrotum, each testis is covered by a thin tissue layer, the tunica vaginalis, around which is a layer of connective tissue (fascia). The dartos muscle relaxes in hot weather to drop the testes away from the body to help keep them cool. In cold weather, the dartos muscle contracts to draw the testes up nearer the body. The spermatic cord, which suspends each testis within the scrotum, contains the testicular artery and vein, lymph vessels, nerves, and the vas deferens.

SCROTAL SECTION

Color and/or label the structures indicated on the diagram using the key below.

1. Vas deferens
2. Epididymis
3. Efferent ductule
4. Head of epididymis
5. Rete testis
6. Septum of testis
7. Seminiferous tubule

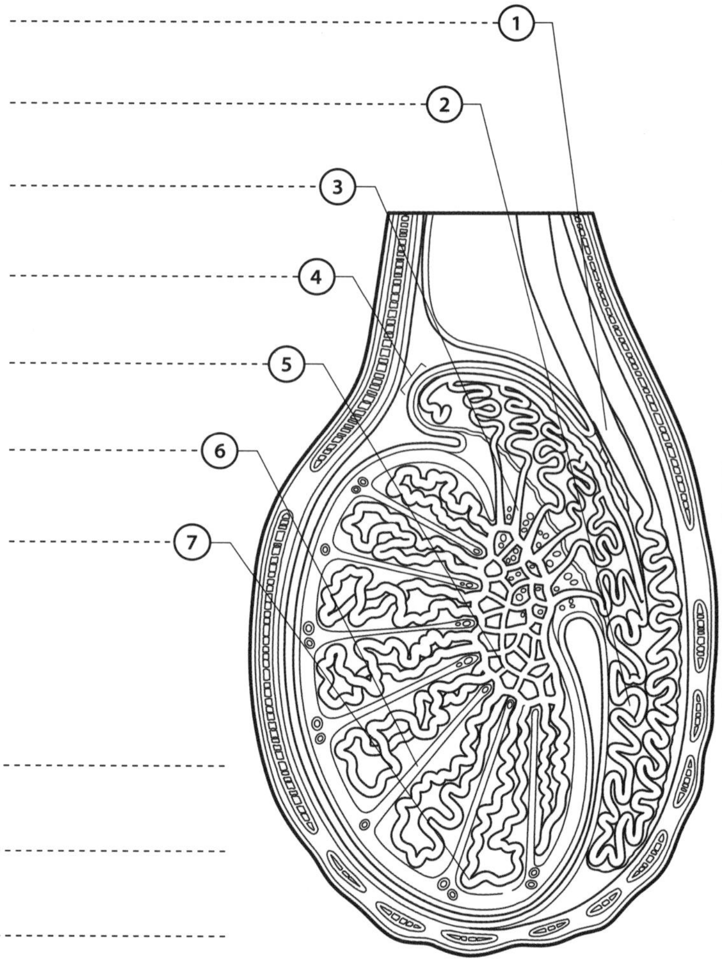

SCROTAL SECTION

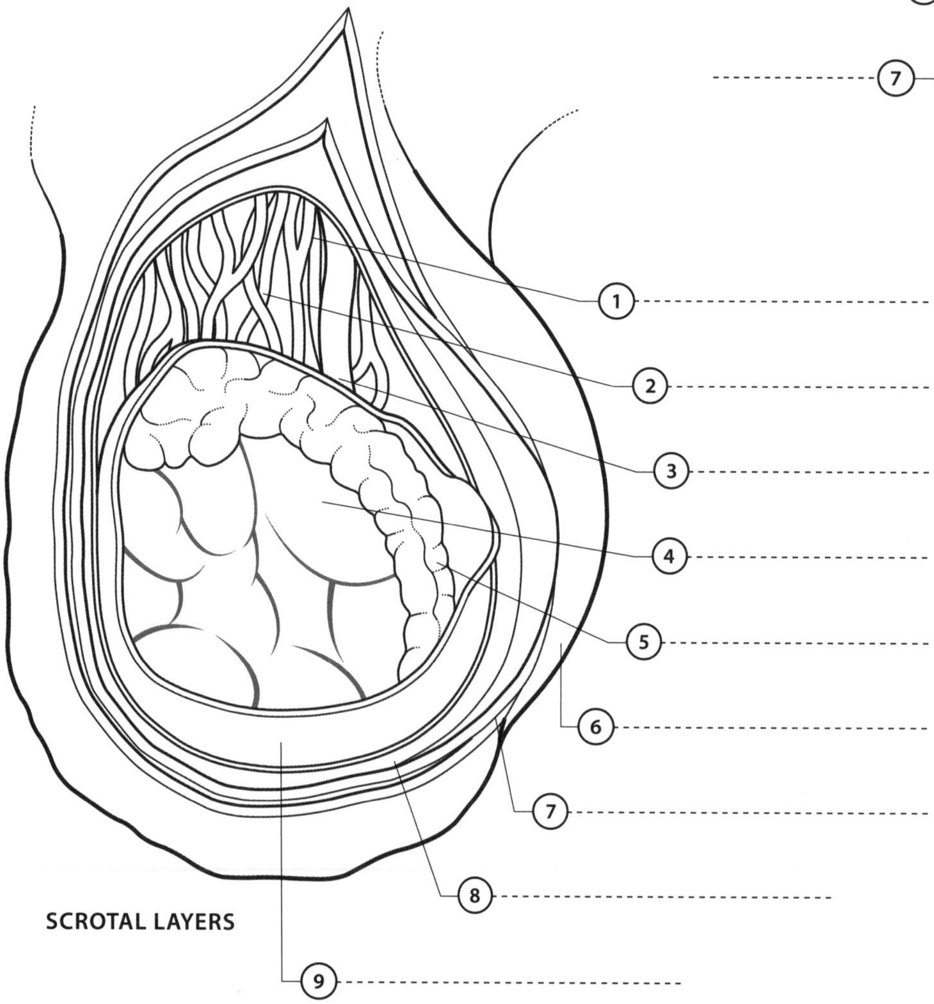

SCROTAL LAYERS

SCROTAL LAYERS

Color and/or label the structures indicated on the diagram using the key below.

1. Vein
2. Artery
3. Vas deferens
4. Testis (lobules)
5. Epididymis
6. Scrotal skin
7. Dartos muscle
8. Fascia
9. Tunica vaginalis

SPERM PRODUCTION

The seminiferous tubules in the testis contain immature male germ cells (spermatogonia). These initially multiply by mitosis to produce primary spermatocytes, each with the diploid number of chromosomes (46). Each primary spermatocyte undergoes a first meiotic division, producing two secondary spermatocytes, each with the haploid number of chromosomes (23). Both cells then undergo a second meiotic division (but remain haploid), producing two spermatids each. Finally, the spermatids undergo spermiogenesis, a process in which the spermatids develop tails and form mature sperm cells. While in the seminiferous tubules, the developing sperm are nourished and protected by Sertoli cells.

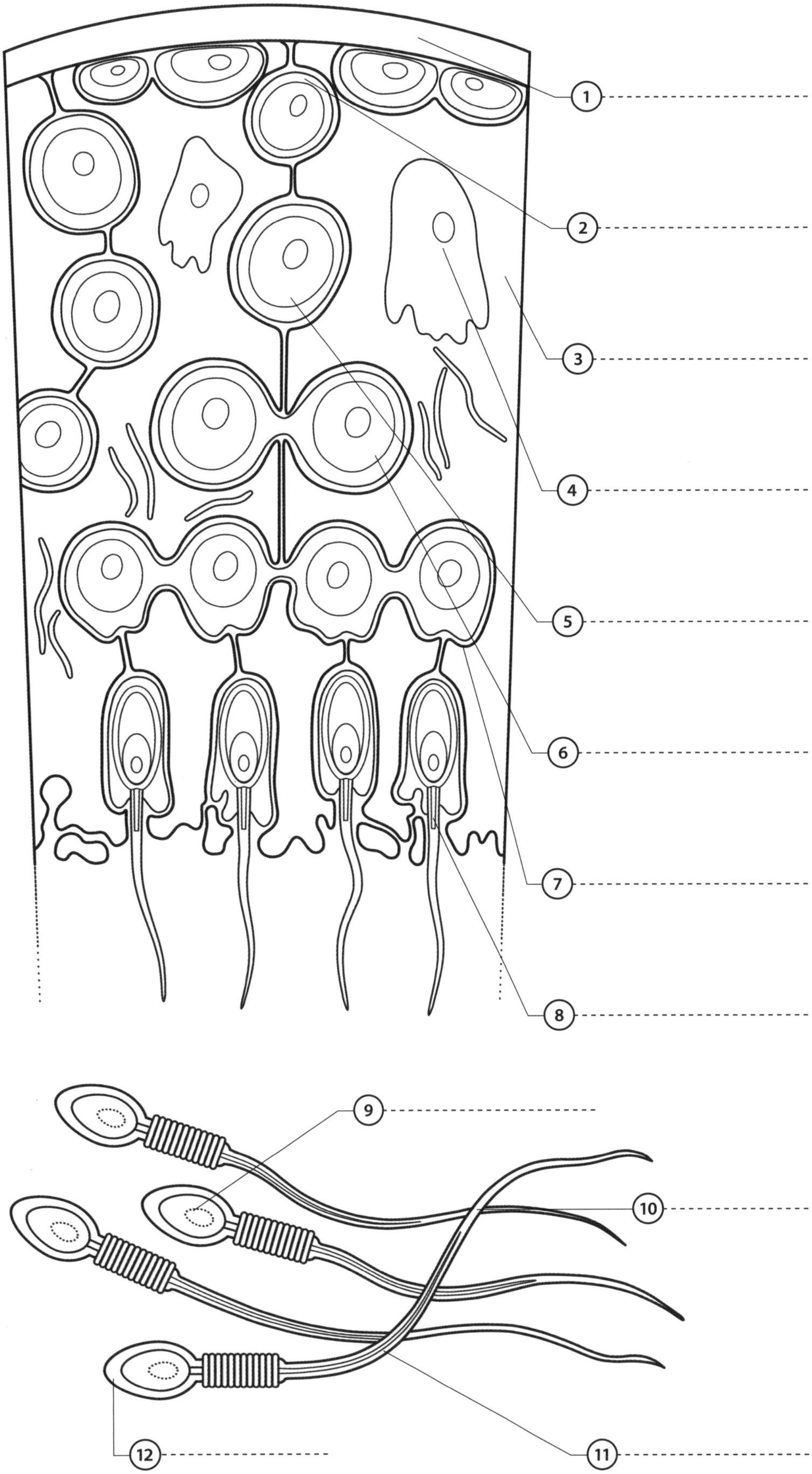

SPERM PRODUCTION

SPERM PRODUCTION

Color and/or label the structures indicated on the diagram using the key below.

1. Membrane of seminiferous tubule
2. Spermatogonium
3. Sertoli cell
4. Nucleus of Sertoli cell
5. Primary spermatocyte
6. Secondary spermatocyte
7. Early spermatid
8. Late spermatid
9. Head
10. Tail
11. Mature sperm
12. Acrosome

See also pp. 218, 220, 230 »

FEMALE ABDOMEN AND PELVIS 1

UNLIKE IN MALES, THE REPRODUCTIVE ORGANS IN FEMALES ARE SITUATED ENTIRELY INSIDE THE BODY, EXCEPT FOR THE EXTERNAL GENITALIA.

FEMALE REPRODUCTIVE ORGANS

The female reproductive organs include the ovaries, Fallopian tubes, uterus, and vagina. The ovaries are deep within the pelvis and contain follicles in various stages of the ovarian cycle. Mature eggs are collected by a pair of Fallopian tubes, which connect to the muscular, thick-walled uterus. The neck of the uterus (cervix) leads to the vagina, an expandable receptacle for sperm, exit for menstrual blood, and birth passage.

SAGITTAL SECTION

Color and/or label the structures indicated on the diagram using the key below.

1. Endometrium
2. Myometrium
3. Uterine cavity
4. Round ligament
5. Vesicouterine pouch
6. Bladder
7. Pubic symphysis
8. Clitoris
9. Urethra
10. Anterior fornix of vagina
11. Vagina
12. Cervix of uterus
13. Posterior fornix of vagina
14. Anus
15. Rectum
16. Uterus
17. Rectouterine pouch
18. Fimbria of Fallopian tube
19. Ovary
20. Fallopian tube

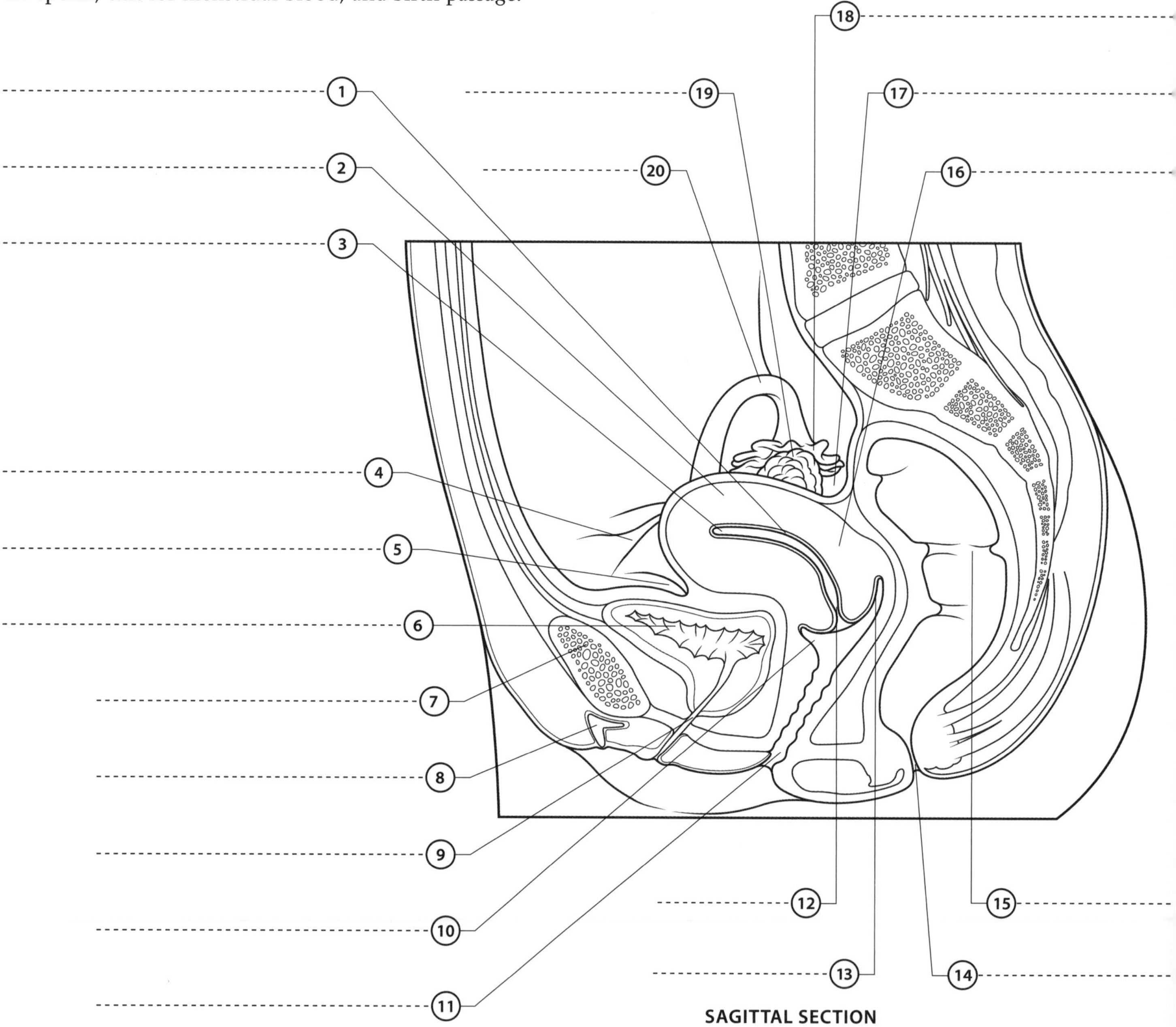

SAGITTAL SECTION

FEMALE PERINEUM

The female external genitalia are collectively known as the vulva. They are sited under the mons pubis, a mound of fatty tissue that covers the pubic symphysis. Outermost in the vulva are the flaplike labia majora, with the smaller, foldlike labia minora within them. The labia majora contain fatty and connective tissue, sebaceous glands, smooth muscles, and sensory nerve endings. At puberty, their exposed surfaces grow hairs. Within the vulva are the openings of the vagina and urethra. At the front end of the labia minora is the clitoris, a sensitive, erectile organ that contains two corpora cavernosa, which become engorged with blood during sexual arousal.

FEMALE PERINEUM

Color and/or label the structures indicated on the diagram using the key below.

1. Vaginal orifice
2. Glans clitoris
3. External urethral orifice
4. Ischiocavernosus
5. Bulb of vestibule
6. Labium minus
7. Labium majus
8. Bulbospongiosus
9. Anus

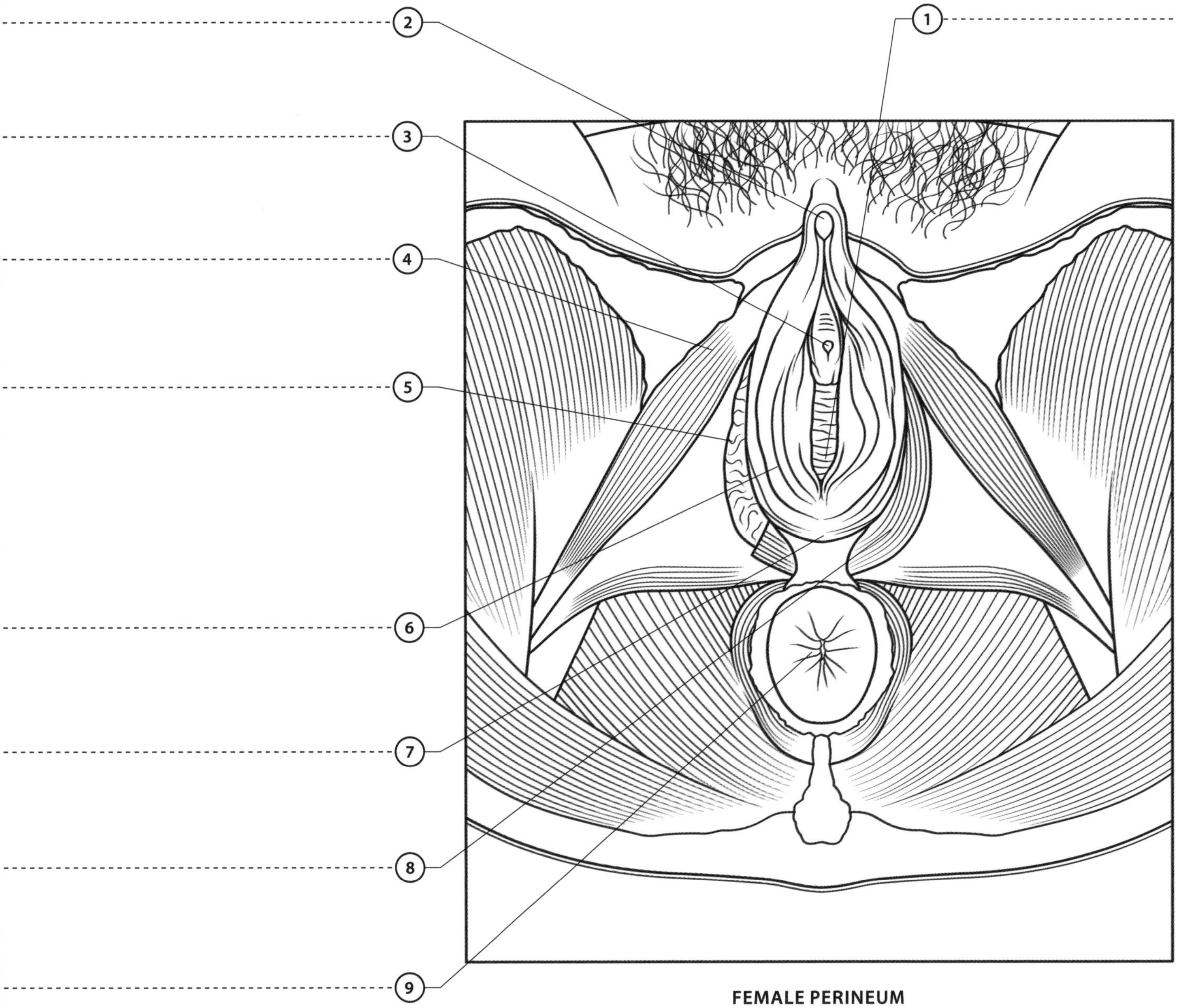

FEMALE PERINEUM

See also pp. 38, 66, 68, 100, 144, 162, 178, 198, 204, 226, 228

FEMALE ABDOMEN AND PELVIS 2

THE FEMALE REPRODUCTIVE ORGANS CONSIST OF THE UTERUS, OVARIES, AND FALLOPIAN TUBES, AND THE VAGINA.

UTERUS, CERVIX, AND VAGINA

The uterus is a hollow, thick-walled, muscular organ that lies between the bladder and rectum. It consists of two main parts: the body and the cervix. In a non-pregnant woman, the body is about 2 in (5 cm) long. The Fallopian tubes (oviducts) open into the sides of the uterine body. Running from each ovary to each side of the uterine body are the ovarian ligaments. The cervix is about 1 in (2.5 cm) long (in a non-pregnant woman); and its lower end protrudes into the vagina.

UTERUS, CERVIX, OVARIES, AND VAGINA

Color and/or label the structures indicated on the diagram using the key below.

1. Body of uterus
2. Ampulla of Fallopian tube
3. Infundibulum of Fallopian tube
4. Ovary
5. Isthmus of Fallopian tube
6. Cavity of uterus
7. Lateral fornix of vagina
8. Cervical canal
9. Vagina
10. Cervix
11. Ovarian ligament
12. Corpus albicans
13. Corpus luteum
14. Fimbriae of Fallopian tube
15. Mature follicle
16. Secondary follicle
17. Primary follicle

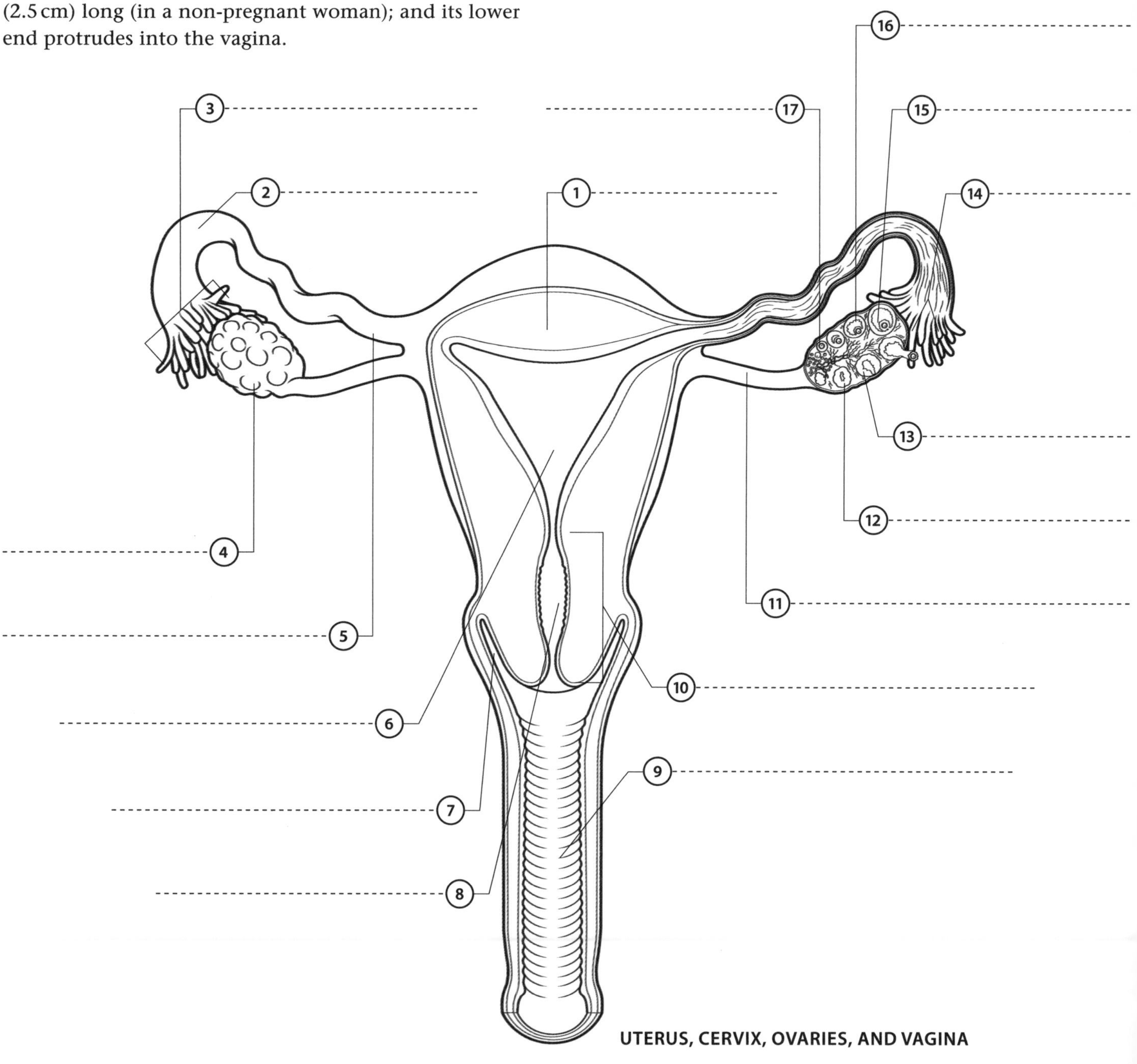

UTERUS, CERVIX, OVARIES, AND VAGINA

OLLICULAR DEVELOPMENT

mmature ova (eggs) are protected within layers of cells alled ovarian follicles. The smallest, primordial follicles ave a single layer of cells. Each month, some of these levelop to become mature (Graafian) follicles. Just before vulation, one mature follicle moves toward the surface of he ovary and bursts through to release its egg. Its remnant orms the corpus luteum, which, if the egg is not fertilized, hrinks to a small white body called the corpus albicans. f the egg is fertilized, the corpus luteum persists and roduces progesterone, which prompts thickening of the ndometrium in preparation for receiving the fertilized gg. At birth, girls have about a million follicles in each vary. These degenerate to about 350,000 by puberty, nd 1,500 by menopause.

FOLLICULAR DEVELOPMENT

Color and/or label the structures indicated on the diagram using the key below.

1. Primordial follicles
2. Enlarging primordial follicle
3. Primary follicle
4. Secondary (developing) follicles
5. Mature (Graafian) follicle
6. Released egg
7. Rupturing follicle
8. Corpus luteum
9. Corpus luteum degenerates if egg is unfertilized
10. Corpus albicans
11. Ovarian blood vessels

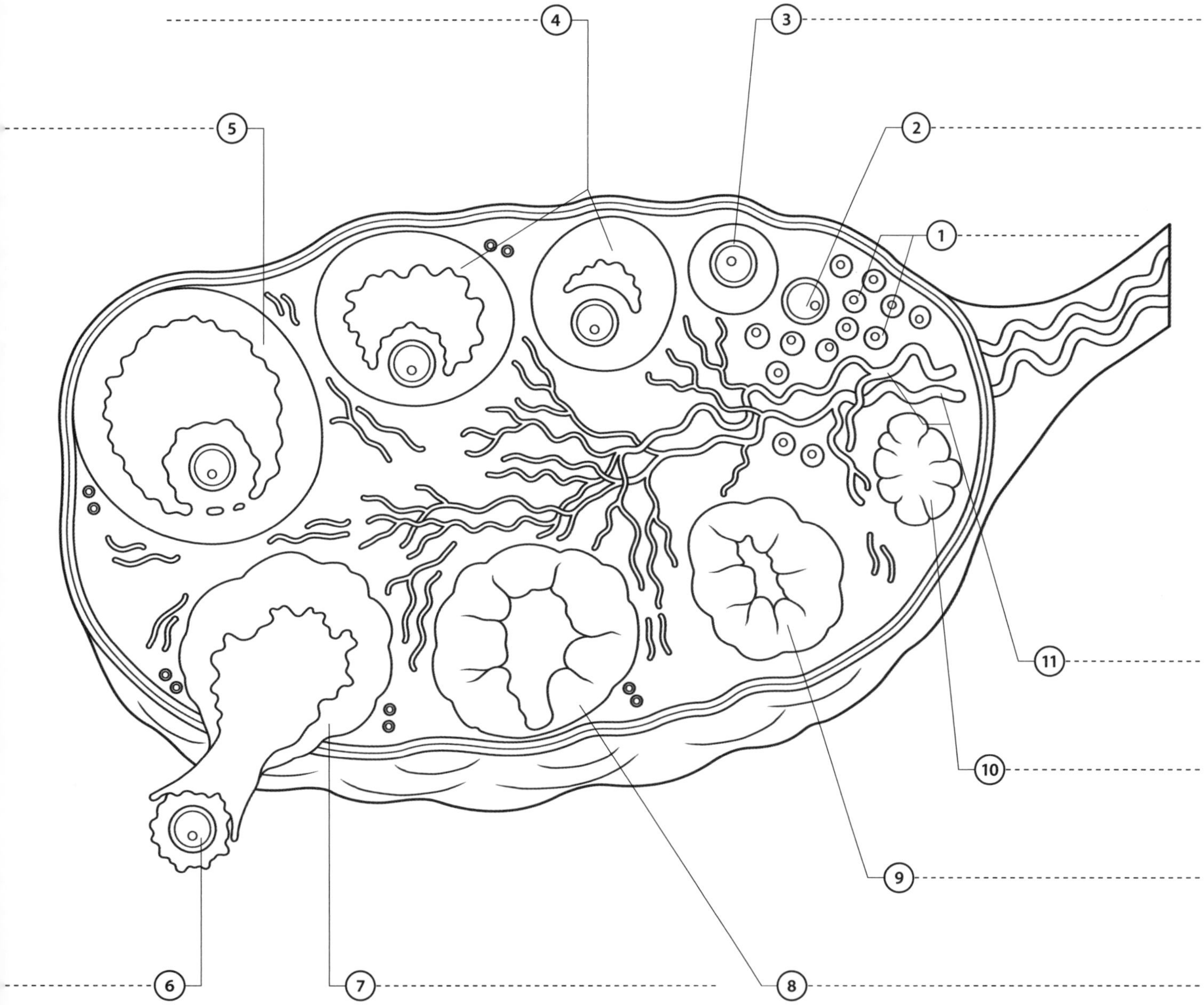

FOLLICULAR DEVELOPMENT

See also pp. 224, 228, 230, 232

FEMALE ABDOMEN AND PELVIS 3

THE OVARIES RELEASE AN EGG AT MONTHLY INTERVALS. IF THE EGG IS NOT FERTILIZED, IT IS SHED WITH THE ENDOMETRIUM AT MENSTRUATION. IF THE EGG IS FERTILIZED, IT IMPLANTS IN THE UTERUS AND DEVELOPS INTO AN EMBRYO (SEE PP. 232–233).

OVULATION

Each month 10 or more follicles in each ovary start to ripen, but usually only one egg is released from either the right or left ovary—the right is favored about 60 percent of the time. The egg passes into the Fallopian tube and, if it is not fertilized, continues along the tube to the uterus and is shed from the body along with the endometrium during the next menstrual period. If the egg is fertilized in the Fallopian tube, the resulting cell mass may implant in the wall of the uterus and develop into an embryo.

OVULATION

Color and/or label the diagram using the key below.

1. Path of egg
2. Ovarian ligament
3. Fimbriae of Fallopian tube
4. Released egg
5. Egg travels down Fallopian tube
6. Fallopian tube
7. Egg reaches uterus
8. Endometrium
9. Myometrium

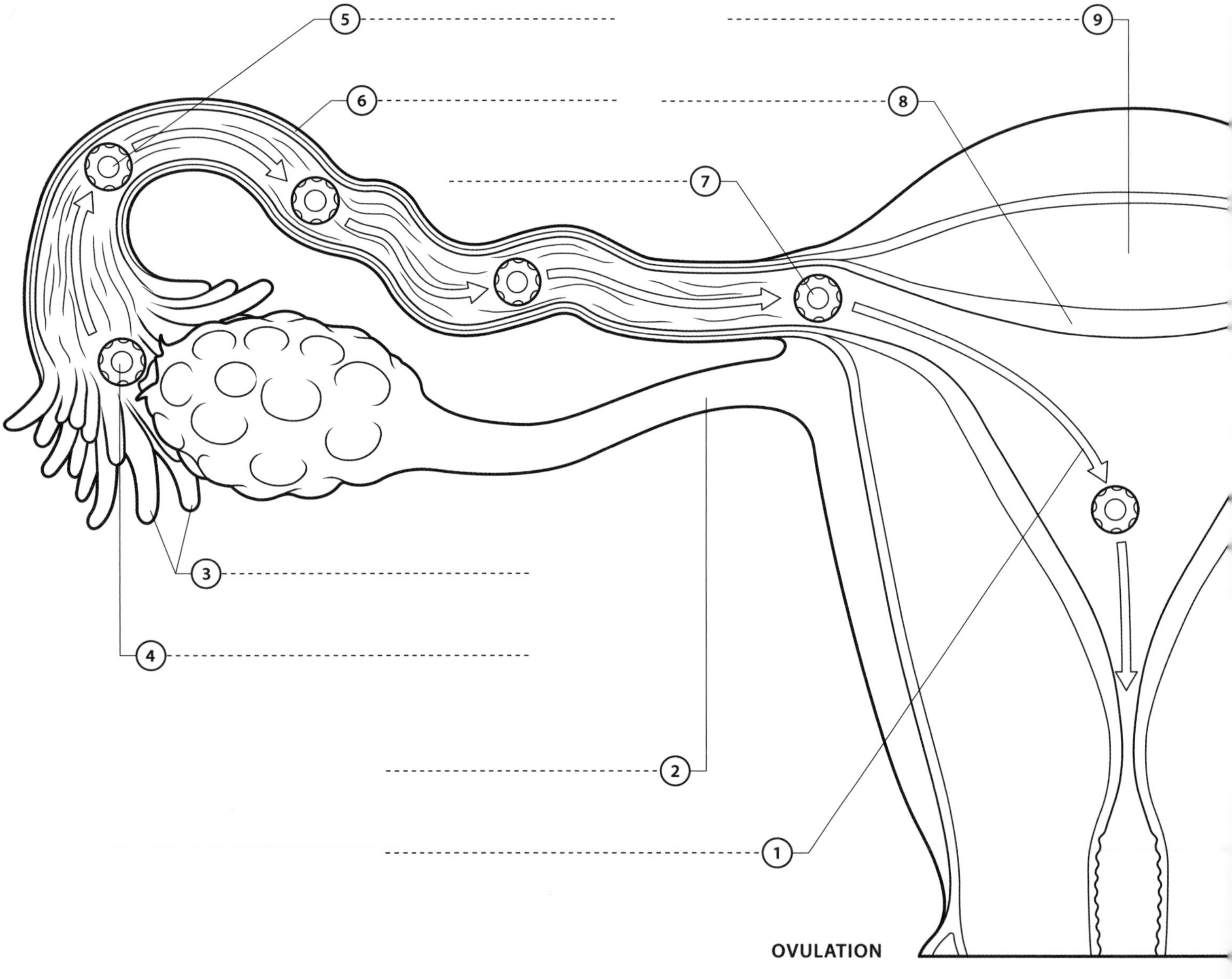

OVULATION

MENSTRUAL CYCLE

The menstrual cycle is controlled by two hormones from the pituitary gland: follicle-stimulating hormone (FSH) and luteinizing hormone (LH). FSH causes the ovarian follicles to ripen and produce estrogen. When estrogen levels are high enough, a surge of LH from the pituitary prompts final maturation of the egg and its release from the ovary. The estrogen also stimulates endometrial thickening, which is temporarily maintained by progesterone from the corpus luteum. However, the endometrium is shed as hormone levels fall. After ovulation, as estrogen levels fall, FSH production increases to repeat the cycle.

MENSTRUAL CYCLE

Color and/or label the diagram using the key below.

1. Menstruation
2. Preovulation
3. Ovulation
4. Postovulation
5. Follicle-stimulating hormone (FSH)
6. Estrogen
7. Luteinizing hormone (LH)
8. Progesterone
9. Menstruation
10. Endometrial thickening
11. Uterine blood vessels
12. Menstruation

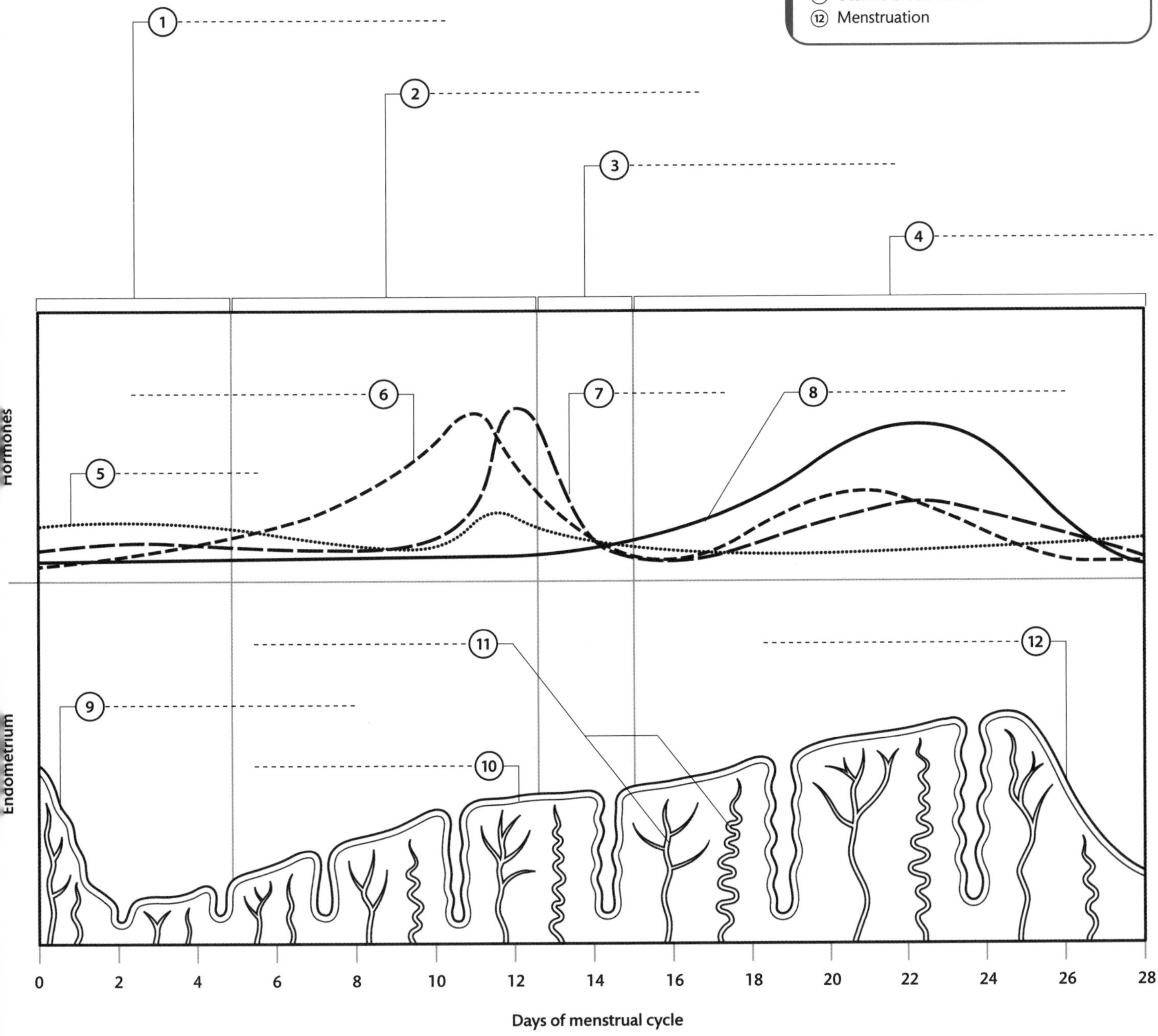

MENSTRUAL CYCLE

See also pp. 224, 226 »

SEXUAL INTERCOURSE

HUMAN REPRODUCTION INVOLVES THE FUSION OF FEMALE AND MALE GERM CELLS (SPERM AND EGGS), EACH CONTAINING HALF OF THE GENETIC INFORMATION REQUIRED TO CREATE A FETUS THAT CAN DEVELOP INTO A NEW HUMAN BEING.

SEX

Sexual arousal in both sexes leads to progressive engorgement of the genital organs as blood flow increases, along with muscle tension, heart rate, and blood pressure. The penis becomes erect, and the woman's clitoris and labia increase in size. The vagina lengthens and its walls secrete lubricating fluid to enable the penis to enter and ejaculate semen high up in the vagina, near the opening of the cervix

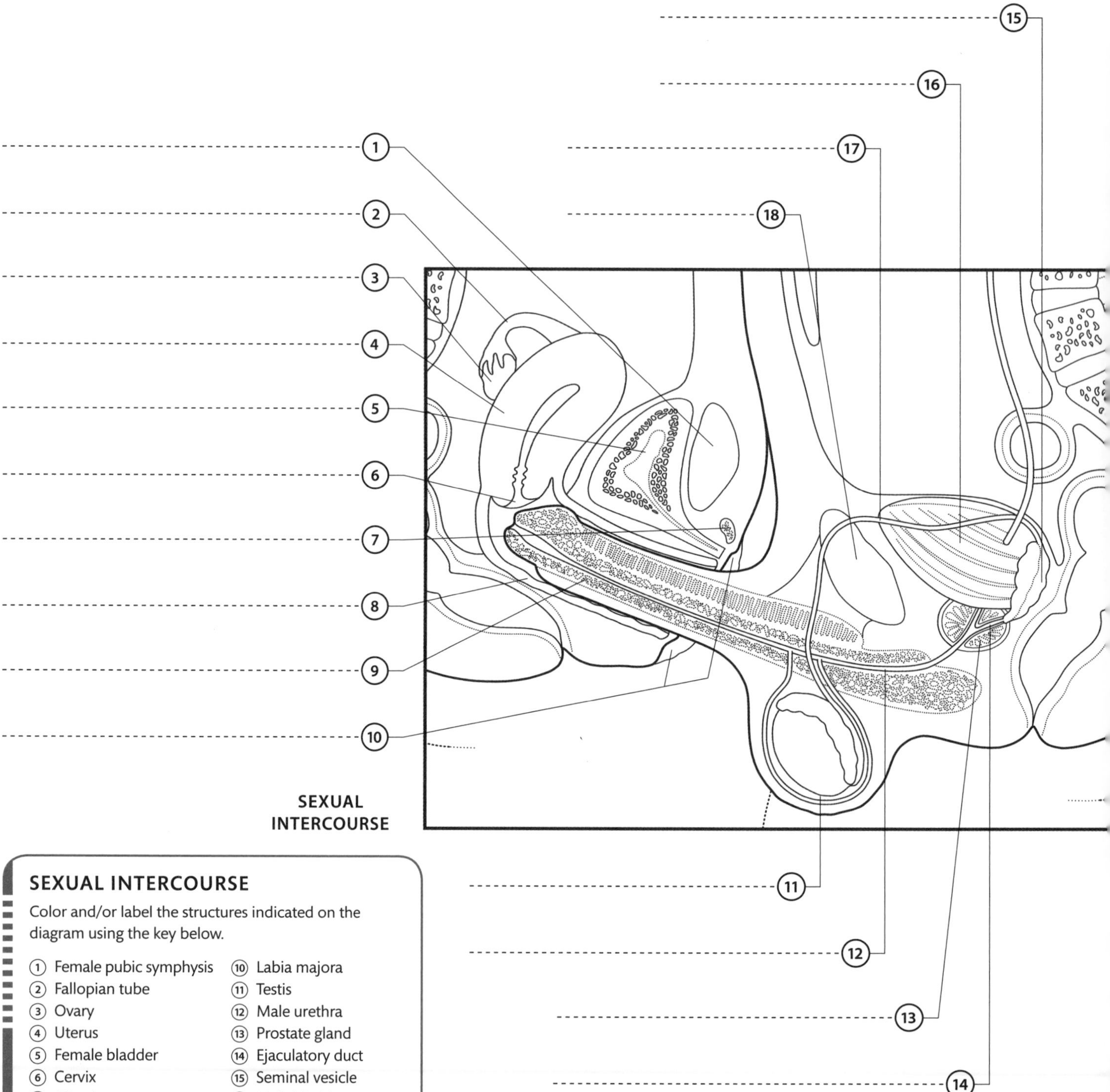

SEXUAL INTERCOURSE

SEXUAL INTERCOURSE

Color and/or label the structures indicated on the diagram using the key below.

1. Female pubic symphysis
2. Fallopian tube
3. Ovary
4. Uterus
5. Female bladder
6. Cervix
7. Clitoris
8. Vagina
9. Penis
10. Labia majora
11. Testis
12. Male urethra
13. Prostate gland
14. Ejaculatory duct
15. Seminal vesicle
16. Male bladder
17. Vas deferens
18. Male pubic symphysis

PATH OF SPERM

Sperm are produced in the seminiferous tubules and pass into the epididymis, where they mature to become motile and fertile. From there, they pass into the vas deferens and then the ejaculatory duct, receiving nutrients and fluid from the seminal vesicle en route. As the semen enters the urethra, the prostate gland contributes fluid. The Cowper's gland (bulbourethral gland) also secretes fluid to lubricate the urethra and flush out any urine before ejaculation. At ejaculation, about 100–300 million sperm enter the vagina, but only about 100,000 reach the uterus; fewer still—about 200—reach the Fallopian tubes, and only one sperm can fertilize an egg.

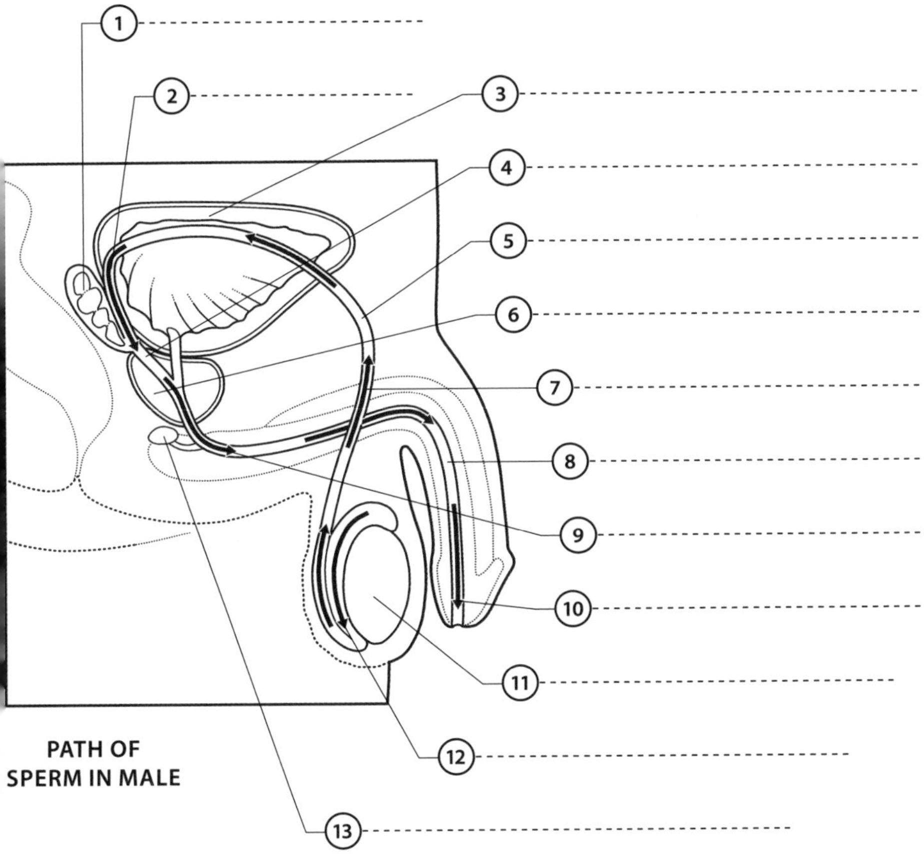

PATH OF SPERM IN MALE

PATH OF SPERM IN MALE

Color and/or label the diagram using the key below.

1. Seminal vesicle
2. Sperm enters ejaculatory duct
3. Bladder
4. Ejaculatory duct
5. Vas deferens
6. Prostate gland
7. Sperm travel along vas deferens
8. Urethra
9. Sperm enters urethra
10. Sperm ejaculated
11. Testis
12. Sperm leave epididymis
13. Cowper's gland

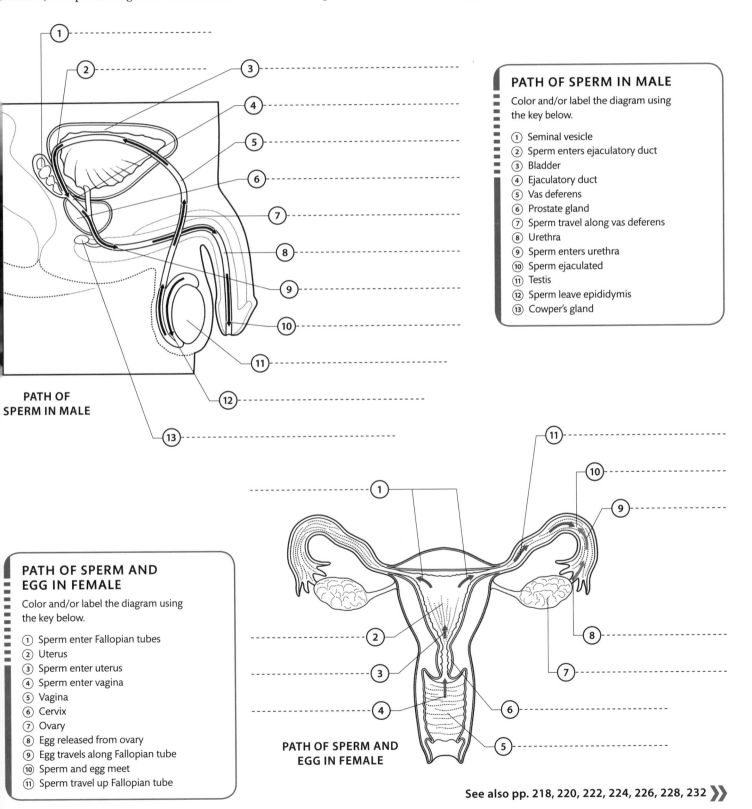

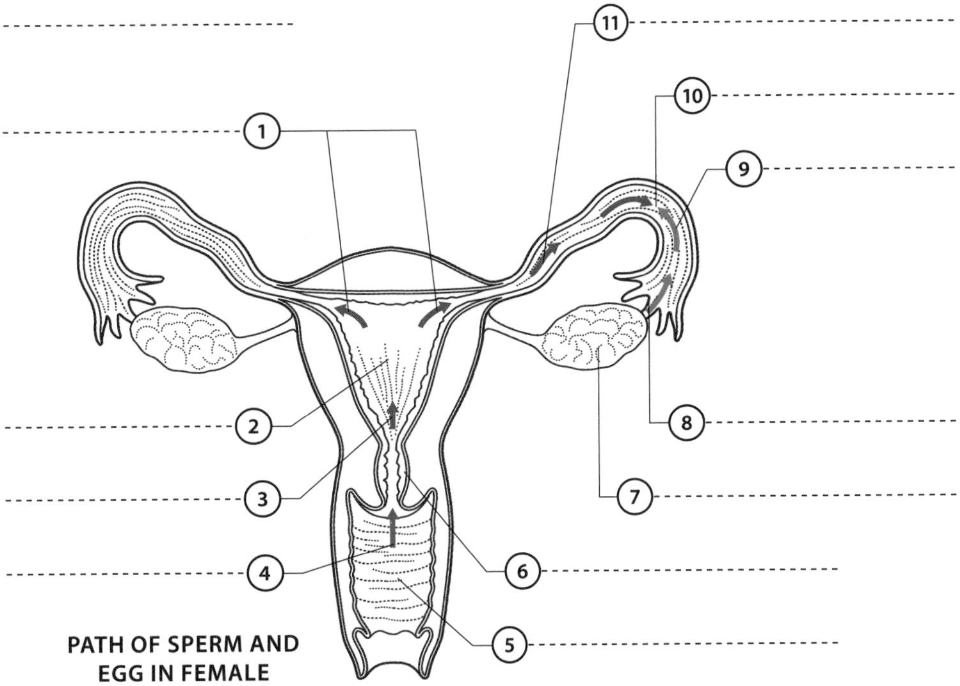

PATH OF SPERM AND EGG IN FEMALE

PATH OF SPERM AND EGG IN FEMALE

Color and/or label the diagram using the key below.

1. Sperm enter Fallopian tubes
2. Uterus
3. Sperm enter uterus
4. Sperm enter vagina
5. Vagina
6. Cervix
7. Ovary
8. Egg released from ovary
9. Egg travels along Fallopian tube
10. Sperm and egg meet
11. Sperm travel up Fallopian tube

See also pp. 218, 220, 222, 224, 226, 228, 232 »

CONCEPTION TO EMBRYO

AFTER AN EGG AND SPERM FUSE AT FERTILIZATION, THE EMBRYONIC CELLS DIVIDE REPEATEDLY TO FORM A BLASTOCYST, WHICH IMPLANTS IN THE UTERINE LINING.

FERTILIZATION AND IMPLANTATION

The first sperm to reach the egg in the Fallopian tube binds to its surface, releasing enzymes that help it to break through the egg's protective coating. The egg releases its own enzymes to block any other sperm from entering. The successful sperm is then absorbed into the egg and loses its tail. The nuclei of the egg and sperm fuse, enabling their genetic information to join together. The fertilized egg (now called a zygote) then continues to travel down the Fallopian tube, undergoing repeated cell divisions to become first a morula and then, about six days after fertilization, a fluid-filled cell cluster called a blastocyst. By this time, the blastocyst has reached the uterus, in which it floats for about two days before settling on the endometrium. The outer cells (trophoblast) of the blastocyst invade the endometrium and develop into the placenta; the inner cell mass develops into the embryo.

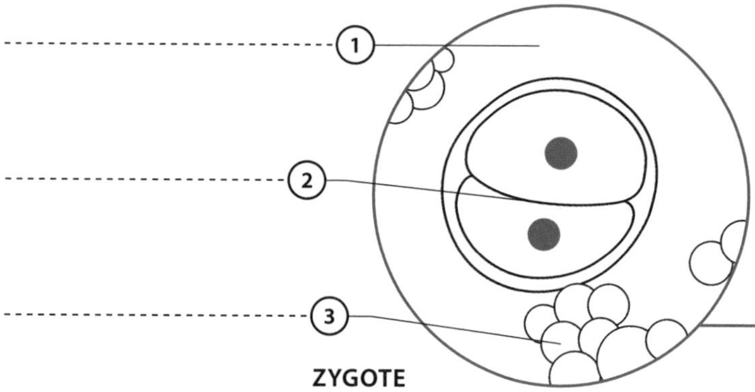

ZYGOTE

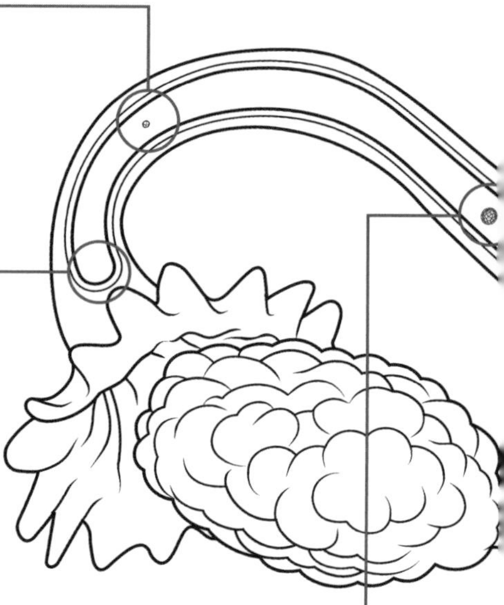

ZYGOTE

Color and/or label the structures indicated on the diagram using the key below.

1. Fallopian tube
2. First cleavage
3. Goblet cells

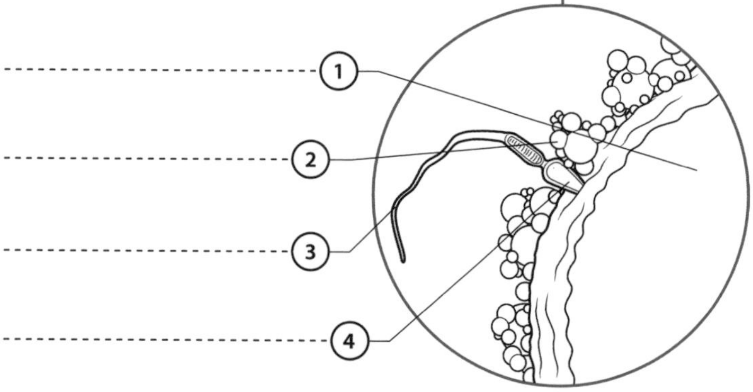

FERTILIZATION

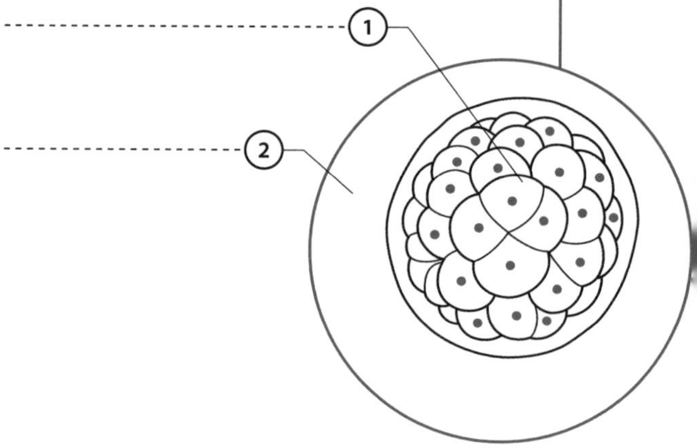

MORULA

FERTILIZATION

Color and/or label the structures indicated on the diagram using the key below.

1. Ovum (egg cell)
2. Corona cell
3. Tail of sperm
4. Sperm head

MORULA

Color and/or label the structures indicated on the diagram using the key below.

1. Morula
2. Fallopian tube lining

BLASTOCYST

Color and/or label the structures indicated on the diagram using the key below.

1. Blastocoele
2. Inner cell mass
3. Trophoblast
4. Endometrium
5. Maternal blood vessels
6. Uterine gland

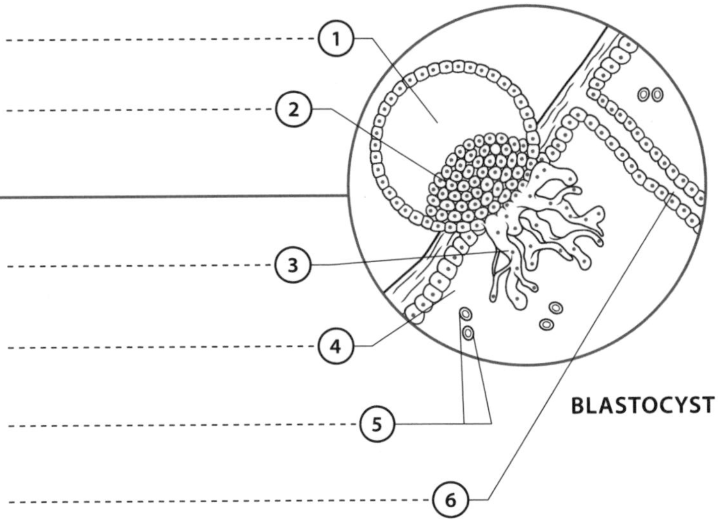

BLASTOCYST

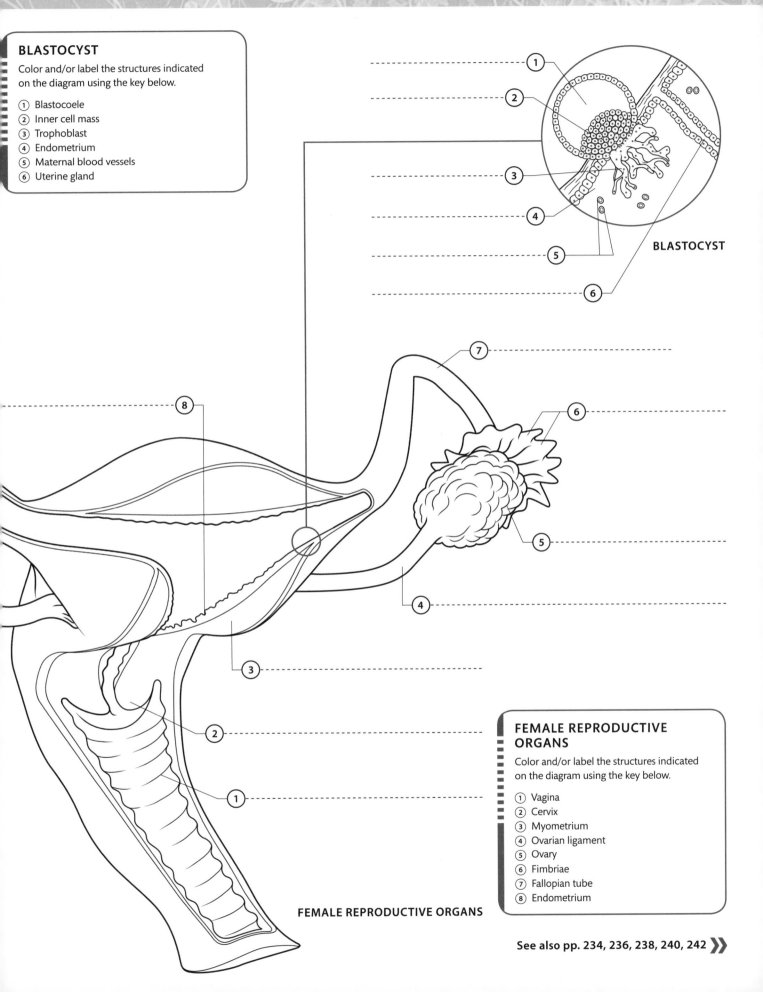

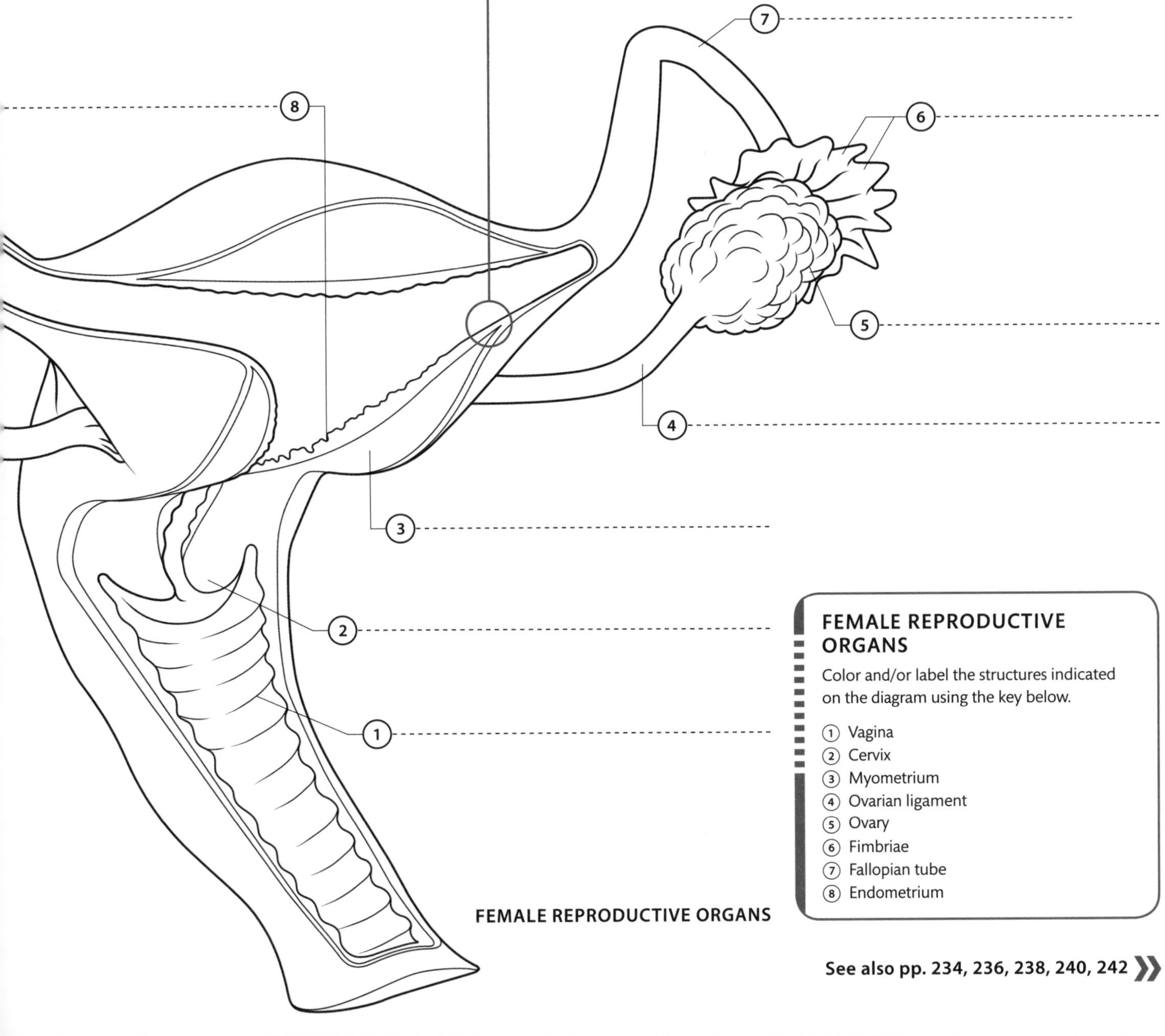

FEMALE REPRODUCTIVE ORGANS

FEMALE REPRODUCTIVE ORGANS

Color and/or label the structures indicated on the diagram using the key below.

1. Vagina
2. Cervix
3. Myometrium
4. Ovarian ligament
5. Ovary
6. Fimbriae
7. Fallopian tube
8. Endometrium

See also pp. 234, 236, 238, 240, 242 »

EMBRYONIC DEVELOPMENT

FROM FERTILIZATION UNTIL THE EIGHTH WEEK OF PREGNANCY THE EMBRYO GROWS RAPIDLY FROM A BALL OF CELLS INTO A MASS OF DISTINCT TISSUES, WHICH DEVELOP INTO ORGANS WITHIN A RECOGNIZABLY HUMAN FORM.

EMBRYO

After implantation in the endometrium, the embryonic cells start to differentiate into specific cell types. Within the inner cell mass of the blastocyst, an embryonic disk forms, consisting of three primary germ layers (ectoderm, endoderm, and mesoderm); these will go on to develop into all the structures in the body. By the time the embryo is two weeks old, cell differentiation has started. By around three weeks, the neural tube has formed, which will become the spinal cord and brain, and heart muscle fibers begin to develop in a simple tubal structure that pulsates. By five weeks, other major organs and the limbs have started to develop, and neural tissue is forming specialist sensory areas, such as the eye. By eight weeks, the embryo has an obvious human shape, all the basic internal organs have formed, skeletal cartilage is starting to turn into bone, and spontaneous movements are occurring. After the end of the eighth week the embryo is referred to as a fetus.

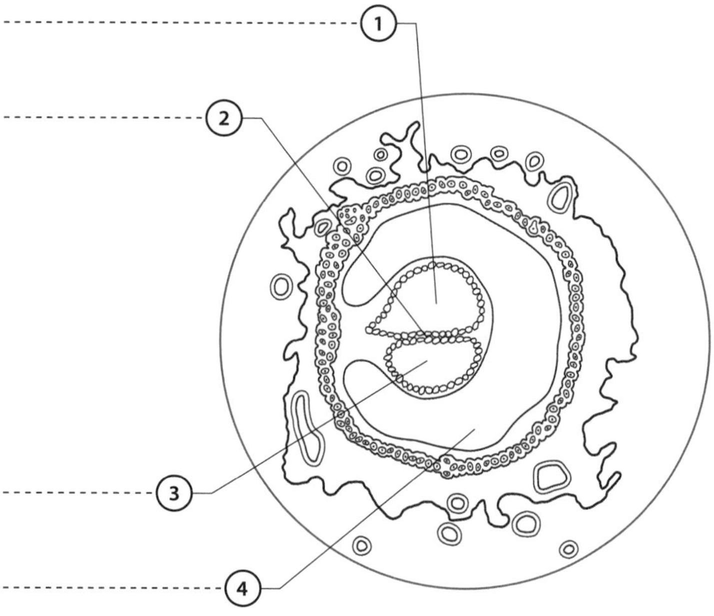

2 WEEKS

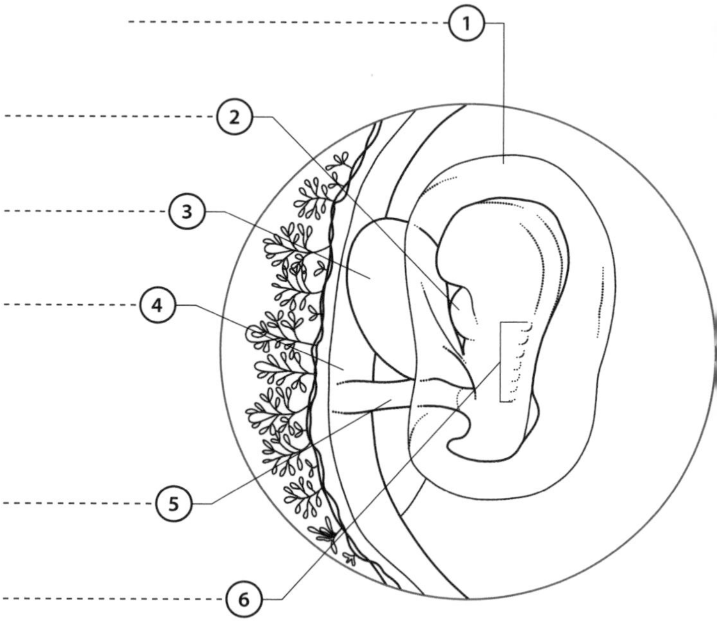

3 WEEKS

2 WEEKS

Label and/or color the diagram using the key below.

1. Yolk sac (nourishes embryo until placenta starts to function)
2. Embryonic disk
3. Amniotic cavity (will become amniotic sac)
4. Chorionic cavity

3 WEEKS

Label and/or color the diagram using the key below.

1. Fluid-filled amniotic sac (cocoons the growing embryo)
2. Muscle fibers have formed a structure that will become the heart
3. Yolk sac
4. Developing placenta
5. Umbilical cord
6. Segmenting tissue down the embryo's back will become the spine

DEVELOPMENT OF THE PLACENTA

The placenta develops in several stages. First, the outer layer of blastocyst cells becomes the trophoblast, which taps into the blood vessels of the maternal endometrium. This forms the placental bed across which nutrients, oxygen, and wastes pass. The trophoblast layer develops fingerlike projections (chorionic villi) that grow out into the maternal blood sinuses. Fetal blood vessels then invade the chorionic villi. By the fifth month, the placenta has become established and a network of villi protrudes into maternal blood-filled chambers (lacunae).

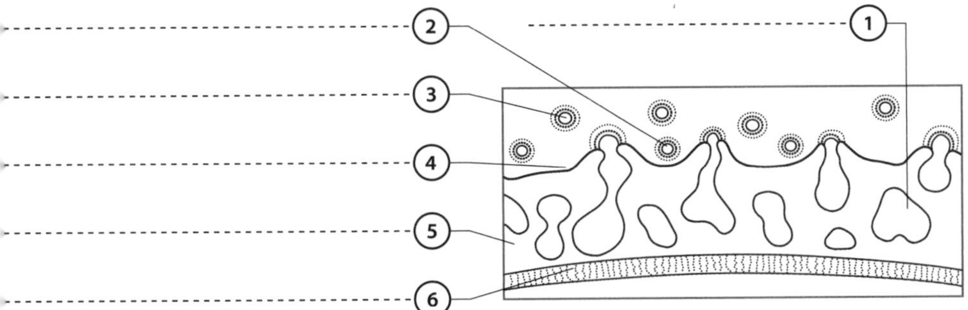

TROPHOBLAST PROLIFERATES

TROPHOBLAST PROLIFERATES

Color and/or label the structures indicated on the diagram using the key below.

1. Maternal blood sinus
2. Maternal artery
3. Maternal vein
4. Endometrium
5. Trophoblast
6. Embryonic cells

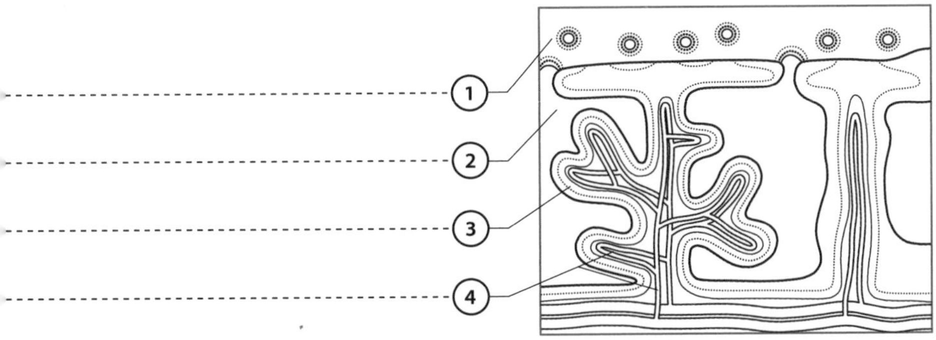

PLACENTA ESTABLISHED

PLACENTA ESTABLISHED

Color and/or label the structures indicated on the diagram using the key below.

1. Endometrium
2. Maternal blood chamber (lacuna)
3. Chorionic villus
4. Fetal blood vessels

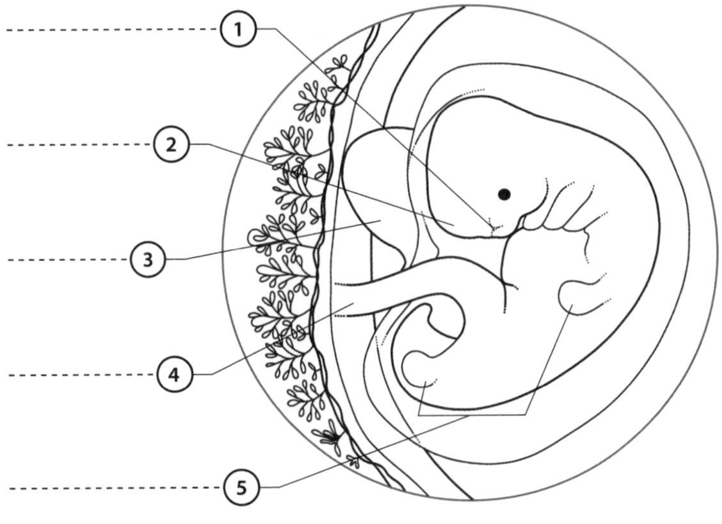

5 WEEKS

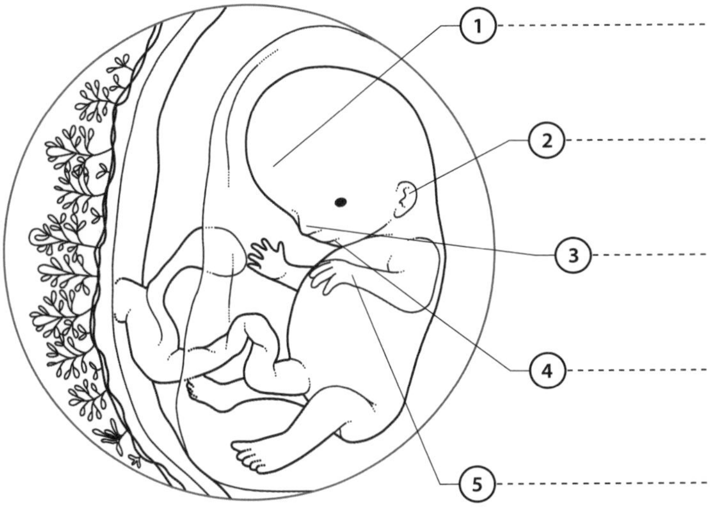

8 WEEKS

5 WEEKS

Label and/or color the diagram using the key below.

1. Nostrils appear as shallow pits
2. Prominent forehead bulges as brain develops
3. Yolk sac shrinking
4. Placenta established
5. Limb buds lengthening

8 WEEKS

Label and/or color the diagram using the key below.

1. Head has lifted off chest
2. Outer ear fully formed
3. Nose protrudes from face
4. Mouth and lips nearly fully developed
5. Wrist has formed

See also pp. 232, 236, 238, 240, 242 »

FETAL DEVELOPMENT 1

FROM EIGHT WEEKS UNTIL DELIVERY, THE FETUS GROWS RAPIDLY IN SIZE AND WEIGHT, AND ITS BODY SYSTEMS DEVELOP UNTIL THEY ARE SUFFICIENTLY MATURE TO SUSTAIN THE BABY ONCE IT HAS BEEN BORN.

EARLY FETAL DEVELOPMENT

By the time the embryo has become a fetus, it has developed a clearly human form. At 11 weeks, the fetus weighs about 1½oz (45g) and is 3½in (9cm) long. It is active and the brain and nervous system are sufficiently developed to sense pressure on the hands and feet, but its eyes remain closed. By 14 weeks, the fetus can swallow; the kidneys are functioning and breathing movements occur. By 19 weeks, the fetus is highly active. Its heart and blood vessel systems are fully developed and, in girls, the ovaries have descended from the abdomen to the pelvis. The skin is covered with fine hair (lanugo) and greasy vernix. At 22 weeks, most body systems are sufficiently developed to cope with independence from the mother, although the lungs are not yet fully mature.

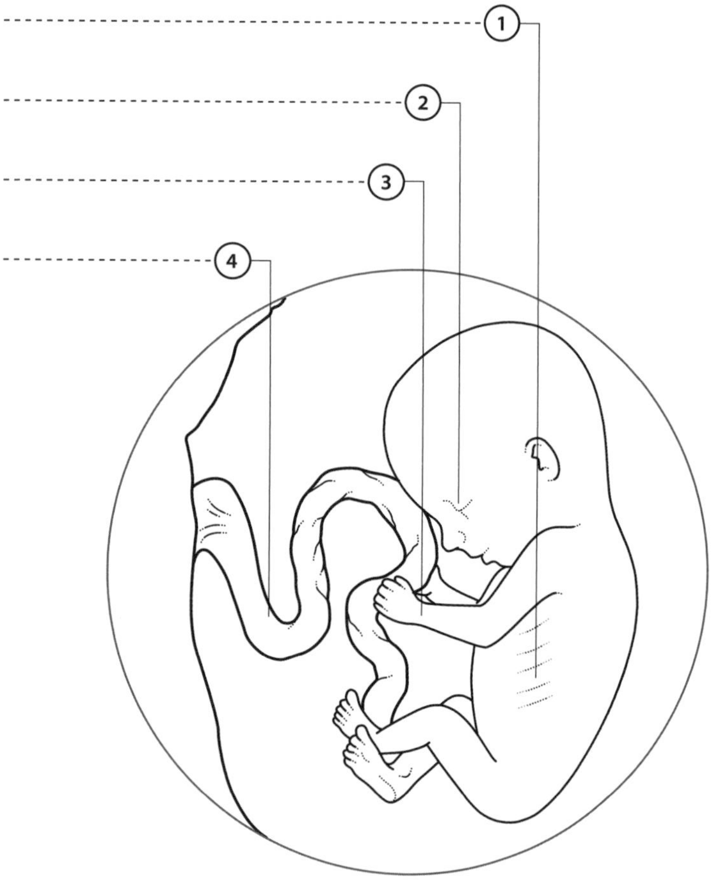

11 WEEKS

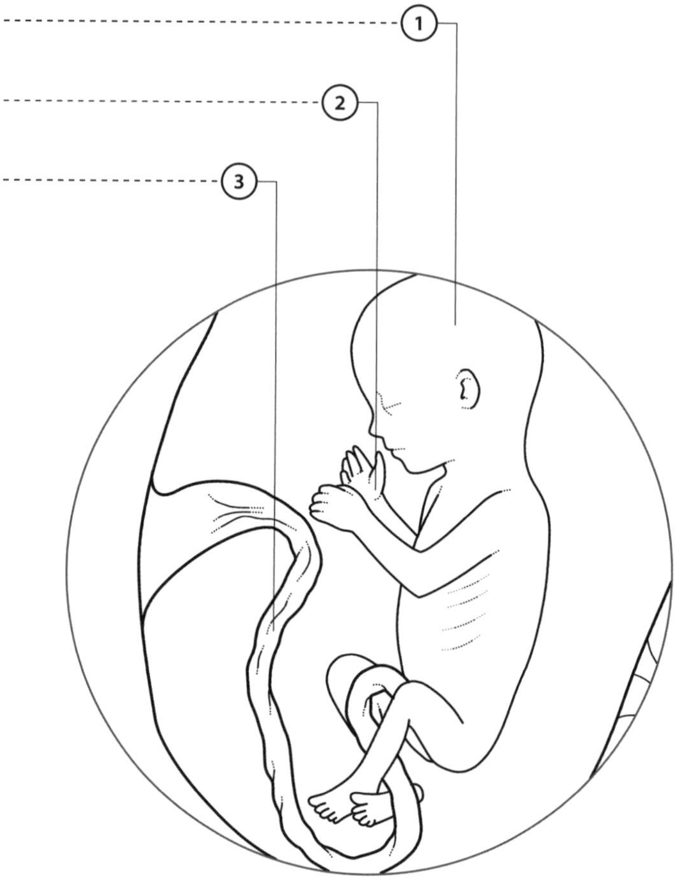

14 WEEKS

11 WEEKS

Label and/or color the diagram using the key below.

1. Body has no underlying fat; bones appear prominent
2. Eyes have moved to the front of the face but remain closed
3. Limbs are lengthening rapidly
4. Umbilical cord

14 WEEKS

Label and/or color the diagram using the key below.

1. In the brain, nerve cells are growing from central to outer areas
2. Greater hand mobility means that the baby is able to suck its thumb
3. Umbilical cord

HOW THE PLACENTA WORKS

The placenta supplies the fetus with oxygen and nutrients (such as glucose, amino acids, and minerals) and removes wastes such as carbon dioxide. The placenta also secretes hormones, including estrogen, progesterone, and human chorionic gonadotropin (hCG). Maternal antibodies can cross the placenta in late pregnancy, giving the fetus passive immunity to infections, and the placenta also prevents the mother's immune system from identifying the fetus as foreign and attacking it.

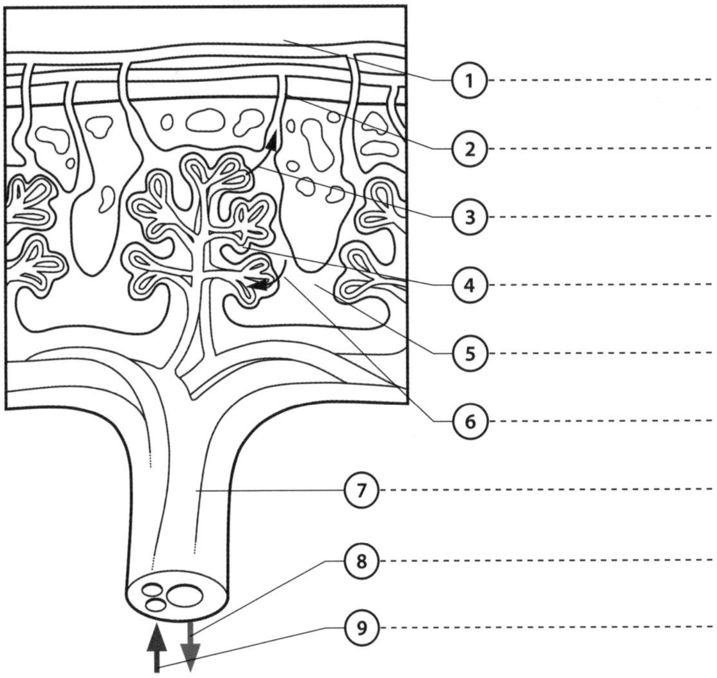

HOW THE PLACENTA WORKS

HOW THE PLACENTA WORKS

Color and/or label the diagram using the key below.

1. Uterine muscle
2. Maternal blood vessel
3. Flow of wastes
4. Fetal blood vessel
5. Maternal blood in intervillous space
6. Flow of nutrients
7. Umbilical cord
8. Blood flow to the fetus
9. Blood flow from the fetus

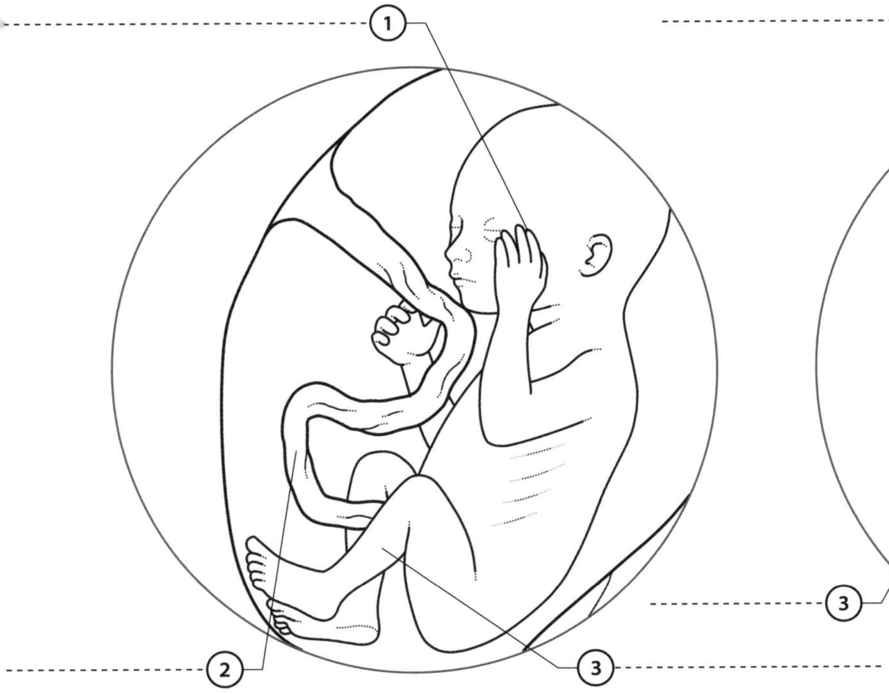

19 WEEKS

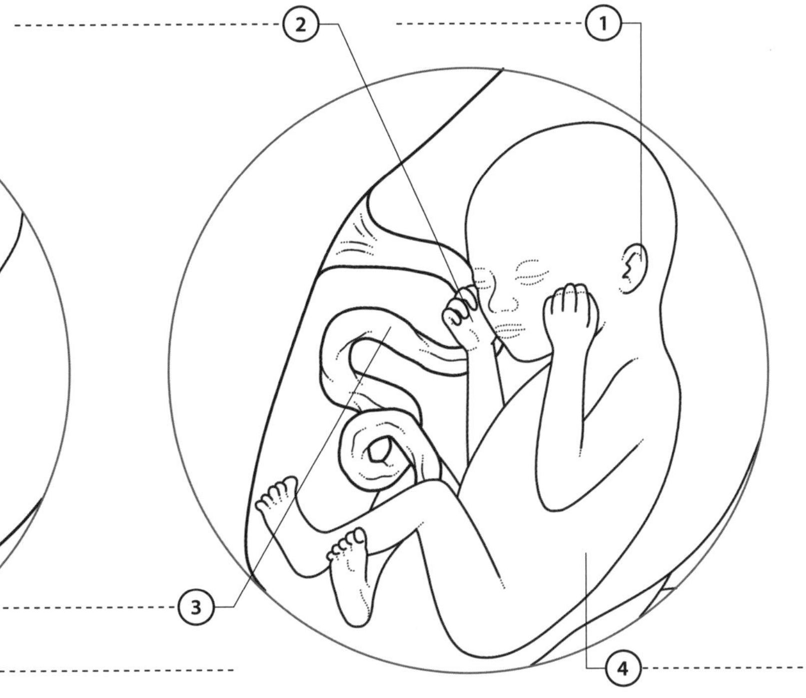

22 WEEKS

19 WEEKS

Label and/or color the diagram using the key below.

1. Fingernails have started to grow
2. Umbilical cord
3. Skin covered with fine hair, called lanugo, and greasy vernix

22 WEEKS

Label and/or color the diagram using the key below.

1. Inner ear organs have matured enough to send nerve signals to the brain
2. Hands are very active
3. Umbilical cord
4. Layers of body fat are being stored beneath the skin

See also pp. 232, 234, 238, 240, 242 »

FETAL DEVELOPMENT 2

DEVELOPMENT DURING THE LATER STAGES IS MOSTLY A PROCESS OF CONSOLIDATION, AS THE FETUS' ORGANS HAVE FORMED BUT NEED TO MATURE.

LATER FETAL DEVELOPMENT

During the final three months the fetus continues to refine its activities and functions, including movement, breathing, swallowing, and urination. The bowels show rhythmic activity but contain a plug of sterile material called meconium (comprising amniotic fluid, skin cells, lanugo, and vernix) that is not usually passed until delivery. At about 26 weeks, the fetus is about 13 in (33 cm) long and weighs about 2 lb (850 g). Its eyes are still closed and will not open for another week or two, when the eyelids have separated. By 30 weeks, the cells that line the lungs start to secrete surfactant, which will help them to inflate when the baby takes its first breath. In boys, the testes will have moved down from the abdomen and will descend into the scrotum. At about 36 weeks, the fetus weighs about 4 lb (1.9 kg) and has accumulated fat deposits. The vernix and lanugo begin to disappear. At full term (40 weeks), the fetus fills the entire uterine space and its organs are fully mature.

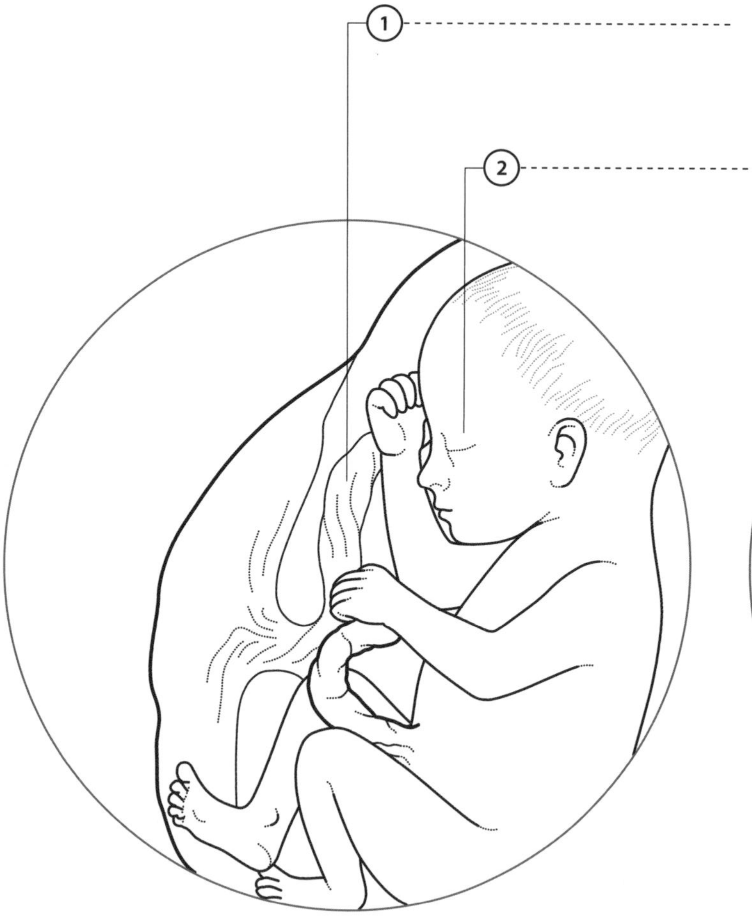

26 WEEKS

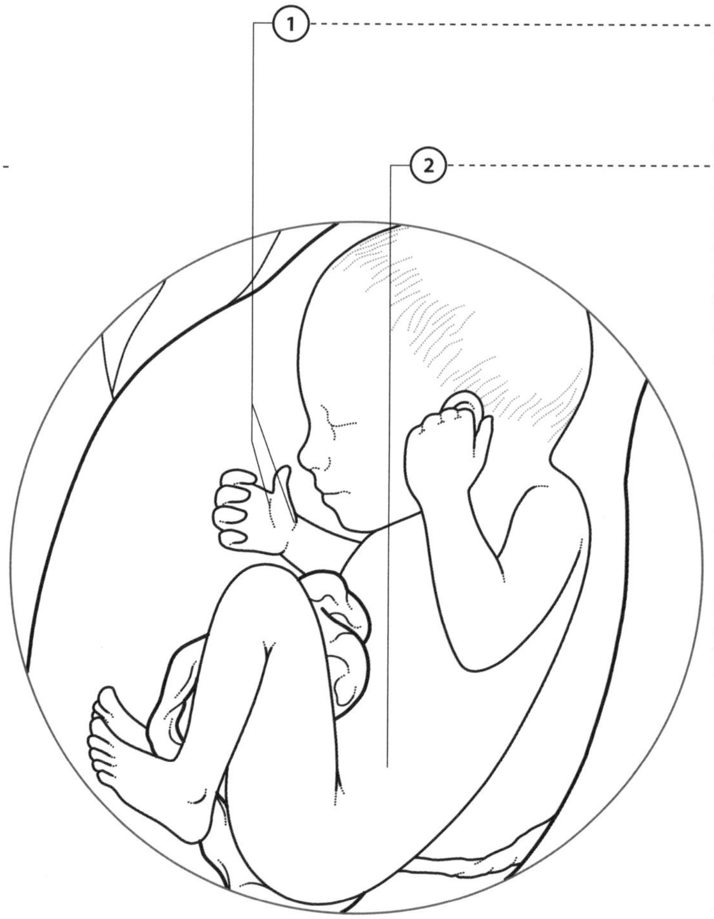

30 WEEKS

26 WEEKS

Label and/or color the diagram using the key below.

1. Blood flow through the umbilical cord regulates the baby's temperature
2. Eyelashes and eyebrows are growing thicker and longer

30 WEEKS

Label and/or color the diagram using the key below.

1. Creases can be seen in the skin of the wrists and the palms of the hands
2. Increased fat layers have rounded out the baby's body

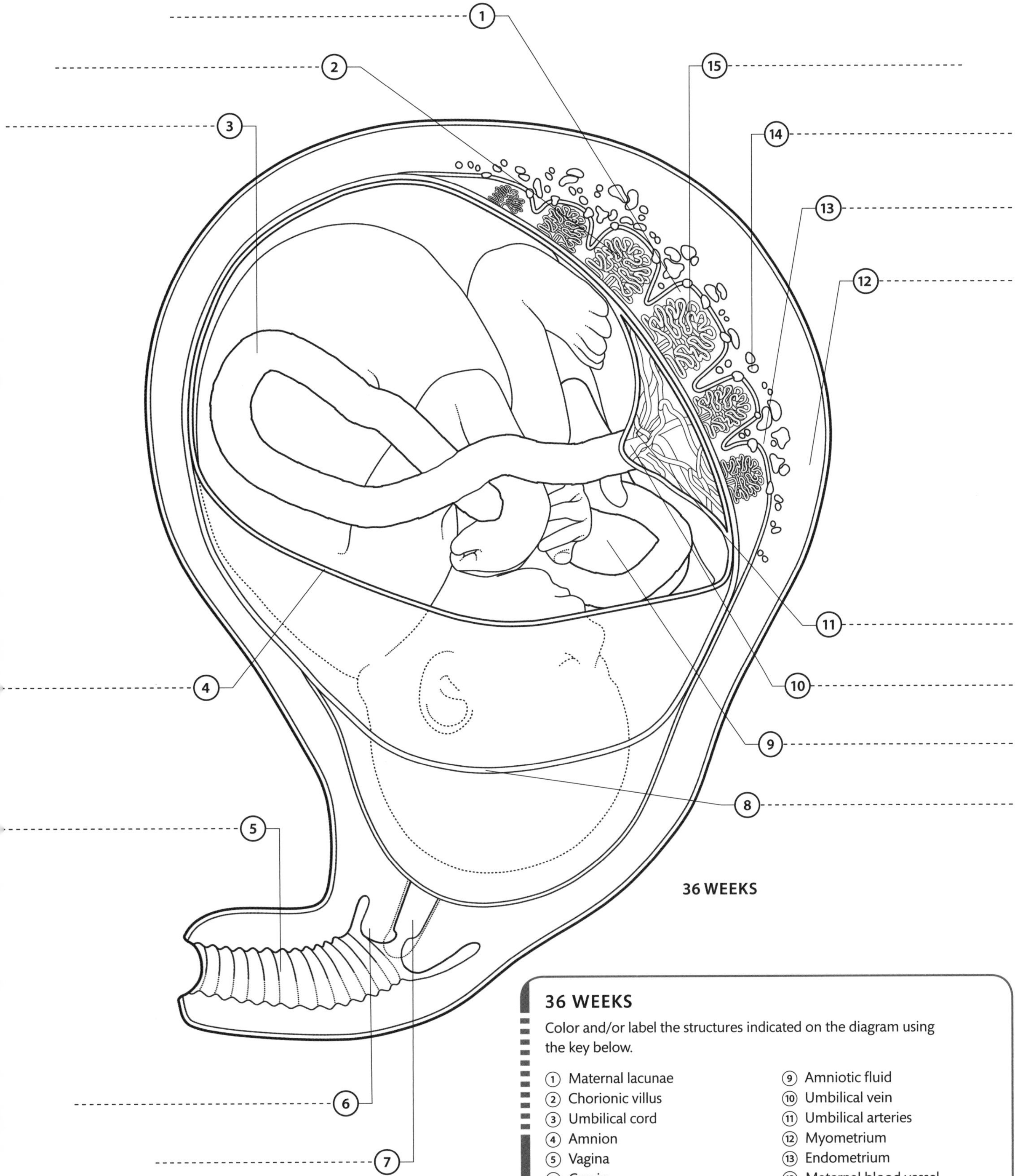

36 WEEKS

Color and/or label the structures indicated on the diagram using the key below.

1. Maternal lacunae
2. Chorionic villus
3. Umbilical cord
4. Amnion
5. Vagina
6. Cervix
7. Cervical plug
8. Chorion
9. Amniotic fluid
10. Umbilical vein
11. Umbilical arteries
12. Myometrium
13. Endometrium
13. Maternal blood vessel
15. Fetal blood vessels

See also pp. 232, 234, 236, 240, 242 »

PREGNANCY

WEEKS OF PREGNANCY ARE DATED FROM THE LAST DAY OF THE WOMAN'S LAST MENSTRUAL PERIOD. PREGNANCY LASTS FOR 40 WEEKS AND IS DIVIDED INTO THIRDS (TRIMESTERS), EACH LASTING ABOUT THREE MONTHS.

TRIMESTERS OF PREGNANCY

During the first trimester (0–12 weeks of pregnancy), nausea is common, the breasts may enlarge, and there is an increased need to urinate. The heart rate rises and food transit through the gut slows, which may result in heartburn or constipation. During the second trimester (13–24 weeks), any nausea usually subsides and food cravings may be experienced. Weight gain is rapid. Back pain and stretch marks on the abdomen are common, and the increased circulation may cause nosebleeds and bleeding gums. In the third trimester (25–40 weeks), the abdomen reaches its maximum protrusion, and the navel may bulge out. Leg cramps and swelling of the hands and feet may occur. Braxton-Hicks contractions ("false labor") often begin in the weeks leading up to labor. The illustrations show some of the common effects of pregnancy in the mother during each trimester.

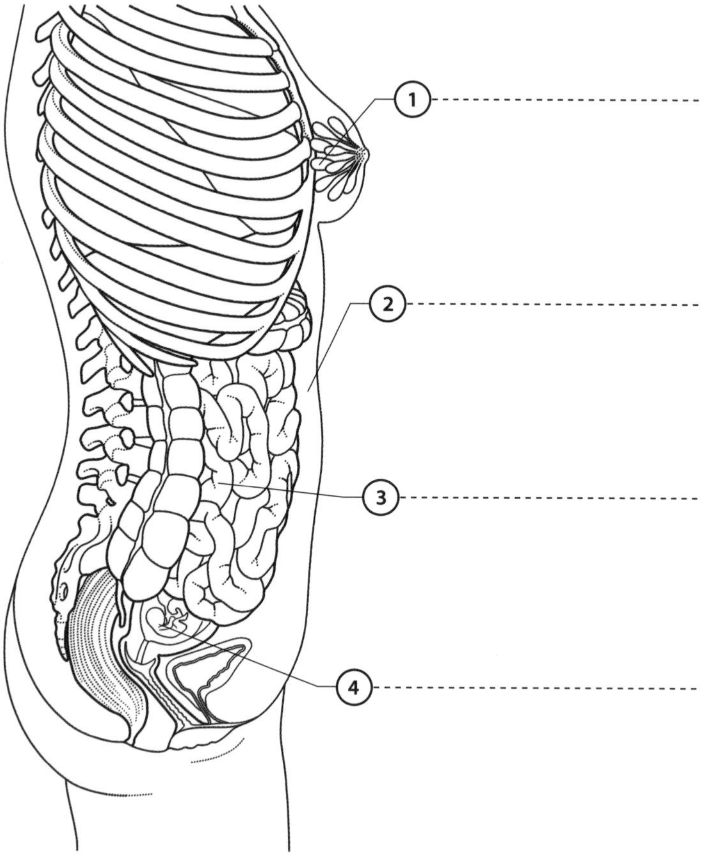

0–12 WEEKS

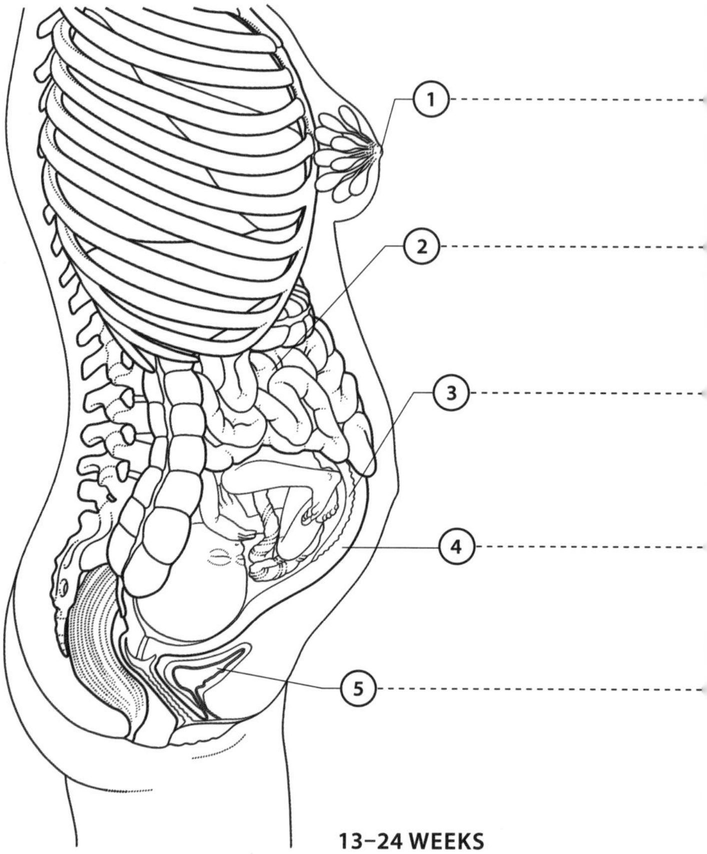

13–24 WEEKS

0–12 WEEKS

Label and/or color the diagram using the key below.

1. Mammary lobules enlarge
2. Waistline may start to thicken
3. Intestines
4. Growing fetus encased in amniotic fluid

13–24 WEEKS

Label and/or color the diagram using the key below.

1. Nipples may darken in response to pregnancy hormones
2. Intestines are compressed by enlarging uterus
3. Placenta is fully formed by 20 weeks
4. Enlarging uterus
5. Bladder becomes slightly compressed

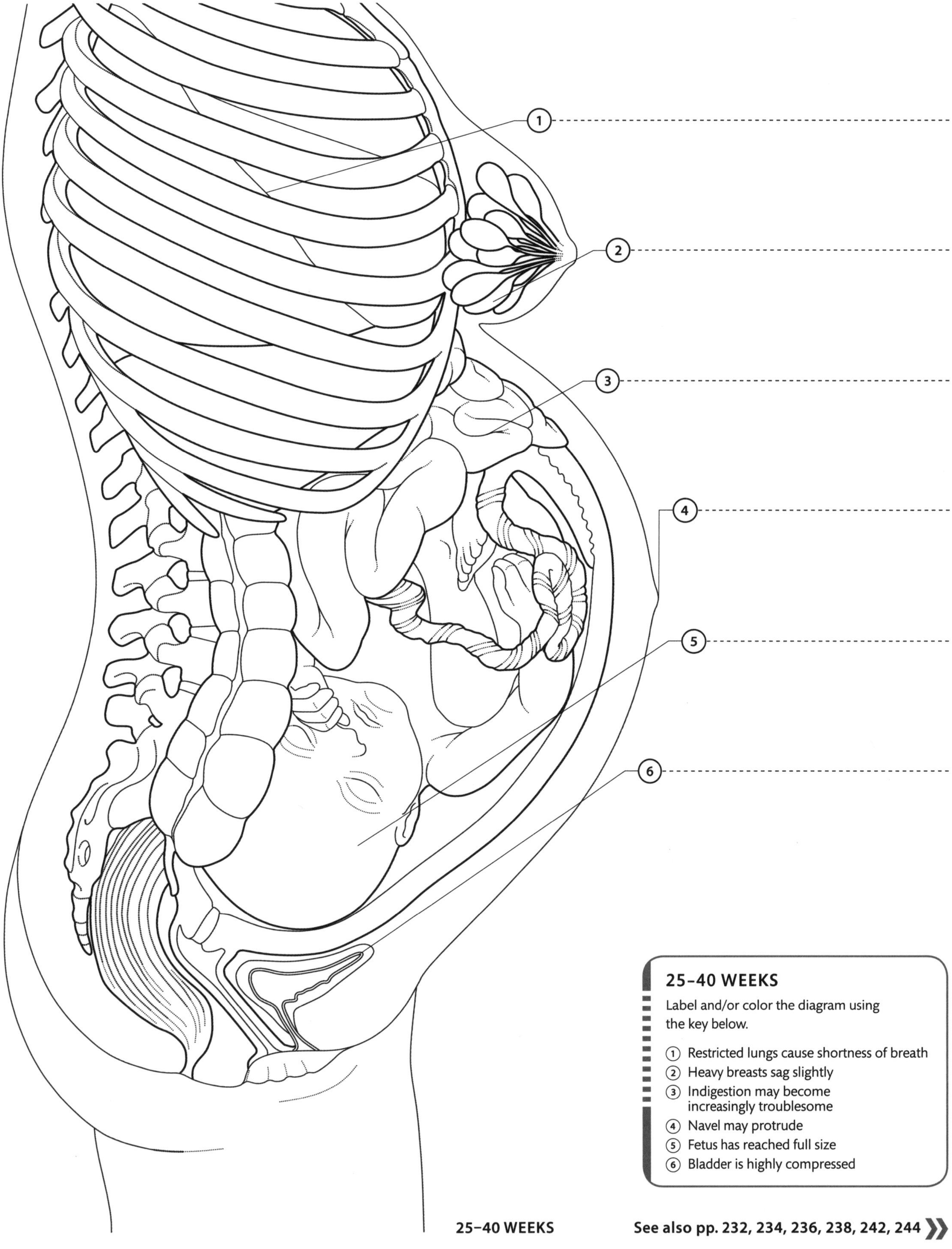

25–40 WEEKS

Label and/or color the diagram using the key below.

1. Restricted lungs cause shortness of breath
2. Heavy breasts sag slightly
3. Indigestion may become increasingly troublesome
4. Navel may protrude
5. Fetus has reached full size
6. Bladder is highly compressed

25–40 WEEKS

See also pp. 232, 234, 236, 238, 242, 244 »

LABOR AND BIRTH

THE TERM "LABOR" USUALLY REFERS TO THE FULL PROCESS OF GIVING BIRTH, FROM THE START OF CERVICAL DILATION TO DELIVERY OF THE PLACENTA.

SIGNS OF LABOR

Generally, there are three signs that labor is starting: a "show," contractions, and the water breaking. Before labor begins, the mucus plug in the cervix is passed as a bloodstained or brownish discharge (the "show"). Contractions start and become stronger and more regular, helping to dilate the cervix. As the contractions increase, the amniotic sac around the baby ruptures, allowing the colorless amniotic fluid to pass out through the birth canal.

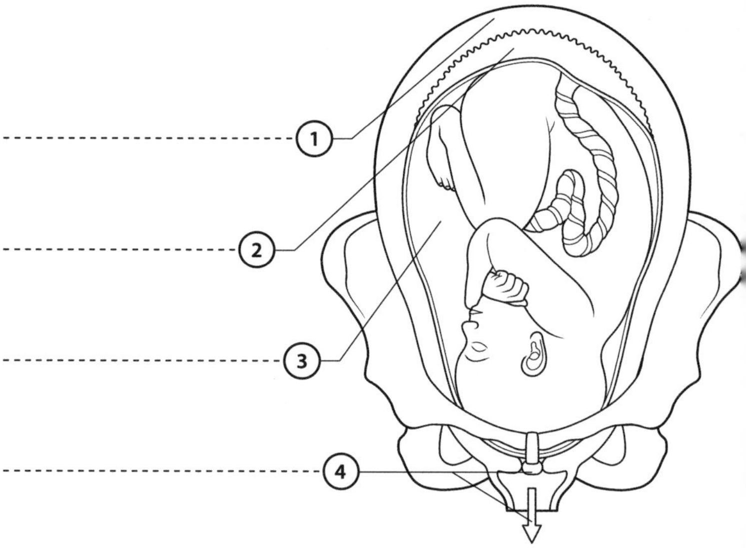

THE "SHOW"

THE "SHOW"

Label and/or color the structures indicated on the diagram using the key below.

1. Uterus
2. Placenta
3. Amniotic fluid
4. Mucus plug ejected

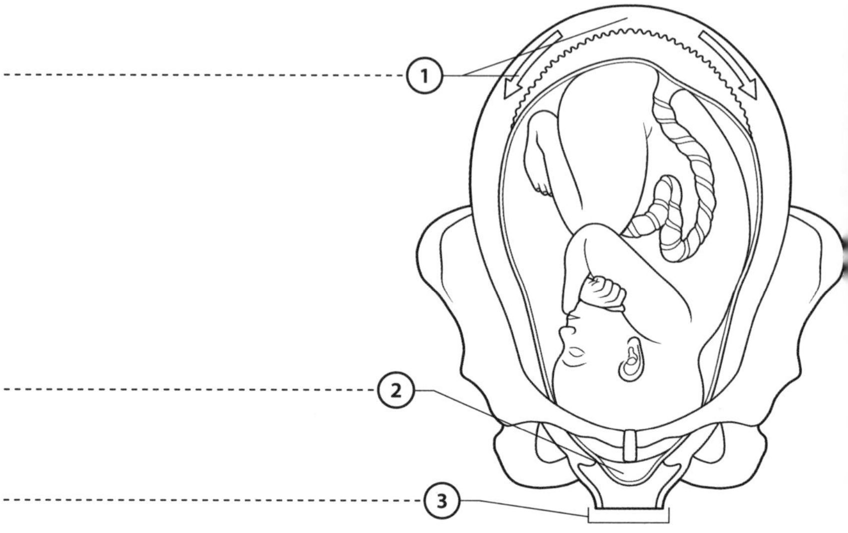

CONTRACTIONS

CONTRACTIONS

Label and/or color the structures indicated on the diagram using the key below.

1. Fundus contracts
2. Bulging amniotic sac
3. Dilating cervix

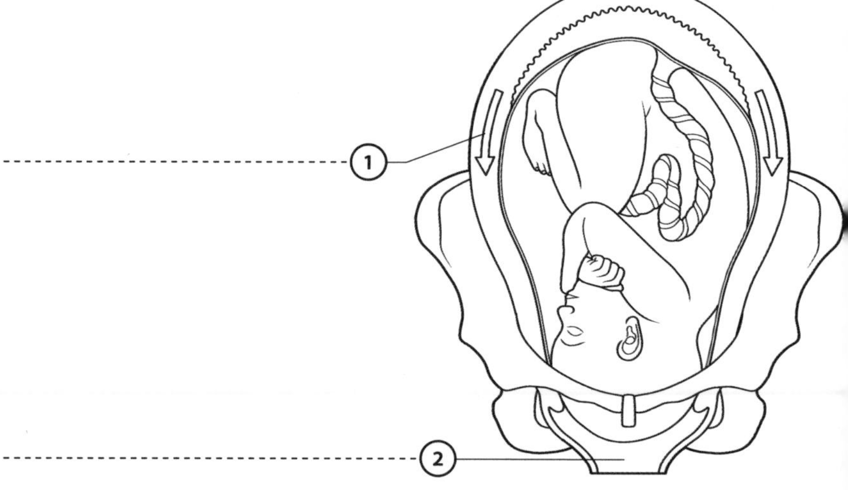

WATER BREAK

WATER BREAK

Label and/or color the structures indicated on the diagram using the key below.

1. Continuing uterine contractions
2. Amniotic fluid drains out through birth canal

STAGES OF LABOR

Labor divides into three main stages, preceded by the latent stage, during which the cervix first starts to dilate. The first stage is defined by dilation of the cervix from $1^1/_2$ in (4 cm) to (4 in) 10 cm. The second stage lasts from full cervical dilation to delivery of the baby. The third stage is the delivery of the placenta. In the first stage, increasingly powerful uterine contractions cause the cervix to dilate. In the second stage, repeated contractions push the baby's head forward through the birth canal, until it becomes visible at the perineum ("crowning"). As the head emerges, the baby turns in the birth canal to allow the shoulders to be delivered; the rest of the baby then slips out easily. In the third stage, the placenta is delivered, usually by pulling on the umbilical cord and pressing on the lower abdominal wall.

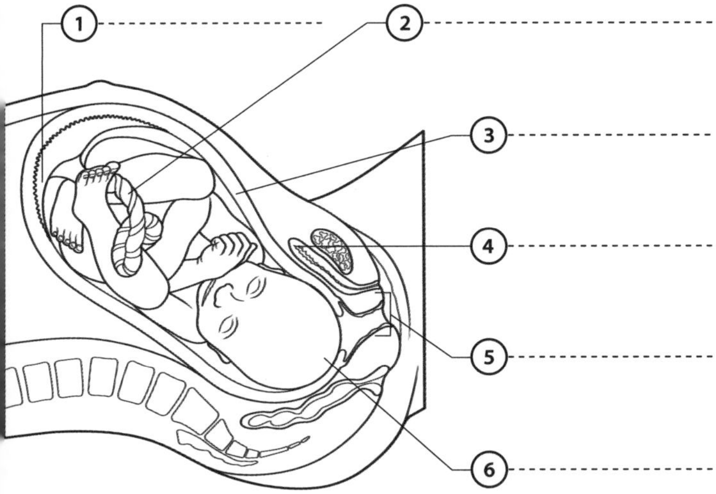

FIRST STAGE

FIRST STAGE

Label and/or color the structures indicated on the diagram using the key below.

1. Placenta
2. Umbilical cord
3. Uterus
4. Bladder
5. Cervix
6. Head

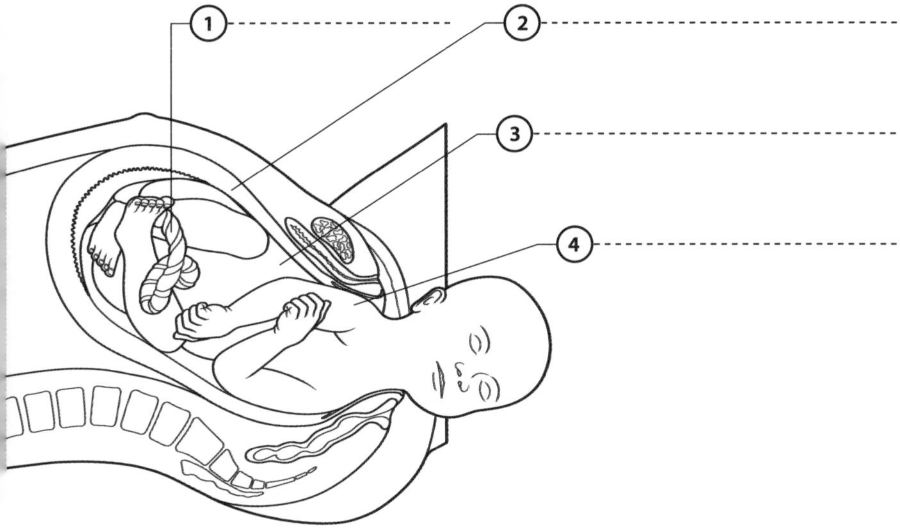

SECOND STAGE

SECOND STAGE

Label and/or color the structures indicated on the diagram using the key below.

1. Umbilical cord
2. Contracting uterus
3. Body
4. Shoulder

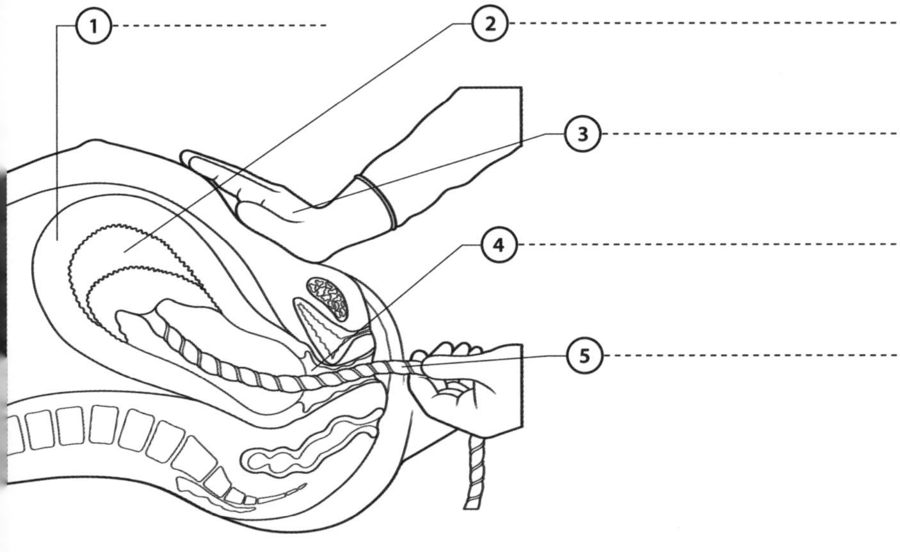

THIRD STAGE

THIRD STAGE

Label and/or color the structures indicated on the diagram using the key below.

1. Uterus
2. Placenta
3. Externally applied abdominal pressure
4. Birth canal
5. Umbilical cord

See also pp. 240, 244 »

THE NEWBORN

IMMEDIATELY AFTER BIRTH THE BABY'S BLOOD CIRCULATION UNDERGOES PROFOUND CHANGES TO MAKE IT SUITABLE FOR BREATHING AIR. A NEWBORN'S ANATOMY ALSO HAS VARIOUS SPECIAL FEATURES TO FACILITATE FURTHER GROWTH AND DEVELOPMENT.

NEWBORN ANATOMY

At birth, a baby has a head that is large in proportion to its body, and often temporarily slightly misshapen due to molding during its passage through the birth canal. The abdomen is relatively large, whereas the chest is bell-shaped and appears small. The breasts may be swollen and sometimes leak a pale, milky fluid. Some newborns have a covering of lanugo, which disappears within weeks or months. A newborn's skeleton is soft and flexible, with bones formed largely of cartilage, and there are flexible fibrous joints (fontanelles) between the bones of the vault of the skull. Although the jaw contains fully formed primary teeth, these do not usually start to erupt until about 6 months. The genitals are large in both sexes, and girls may have a slight vaginal discharge. The liver is relatively large at birth and protrudes below the rib cage. The thymus gland, part of the immune system, is also large because other parts of the immune system are still developing.

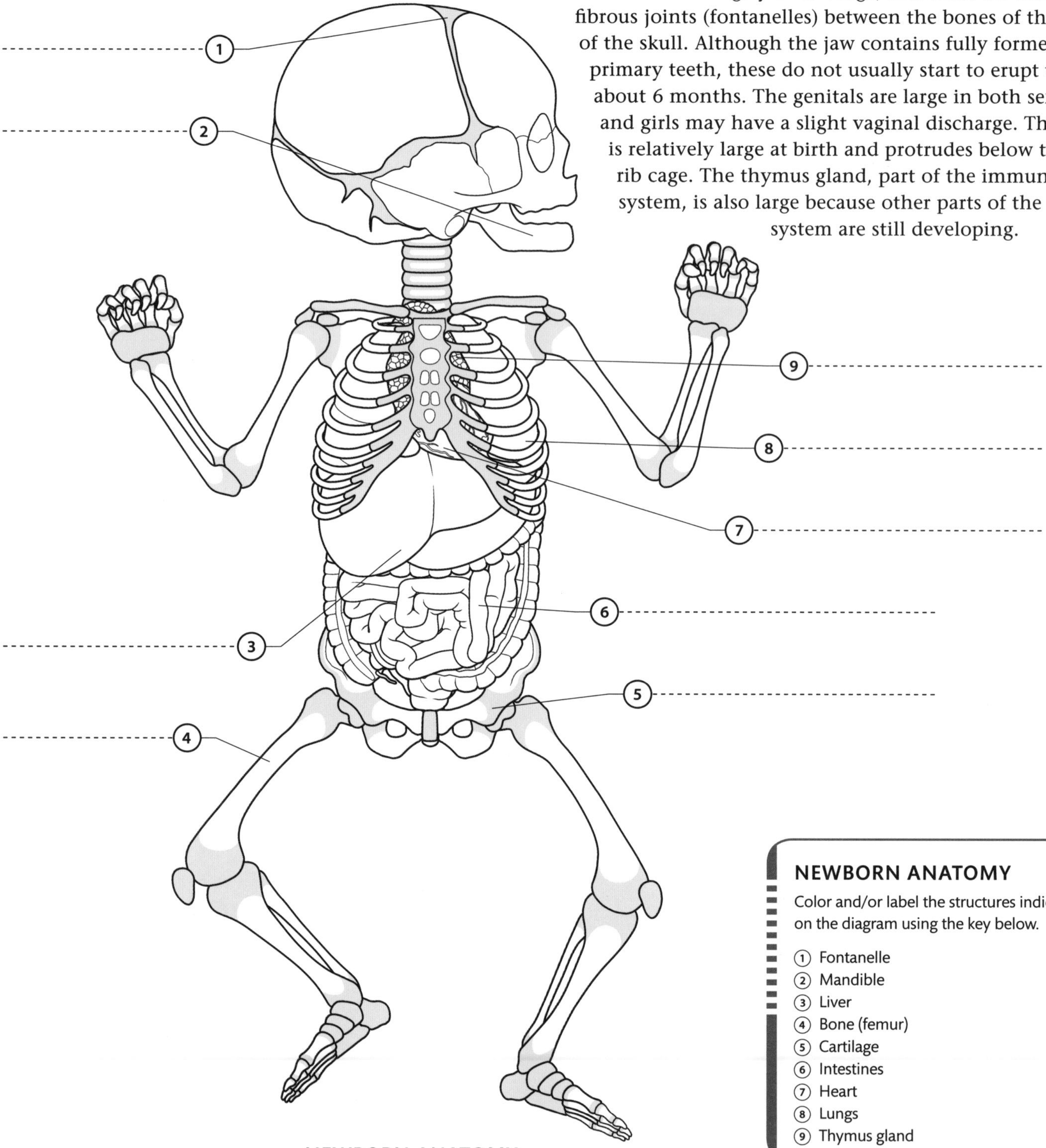

NEWBORN ANATOMY

NEWBORN ANATOMY

Color and/or label the structures indicated on the diagram using the key below.

1. Fontanelle
2. Mandible
3. Liver
4. Bone (femur)
5. Cartilage
6. Intestines
7. Heart
8. Lungs
9. Thymus gland

CIRCULATION CHANGES

In the uterus, the fetus receives nutrients and oxygen and eliminates wastes via the placenta. The fetal circulation therefore has shunts to allow blood to bypass the not-yet-functioning liver and lungs. The ductus venosus shunts incoming blood through the liver to the right atrium, which shunts it through the foramen ovale to the left atrium (mostly bypassing the right ventricle) and onward to the body. Blood that does enter the right ventricle passes into the pulmonary artery but is shunted into the aorta by the ductus arteriosus, bypassing the lungs. With the first breaths, blood travels to the lungs, and the pressure of the blood returning into the left atrium forces shut the foramen ovale. The ductus arteriosus, ductus venosus, and the umbilical vein and arteries close up soon after birth and become ligaments.

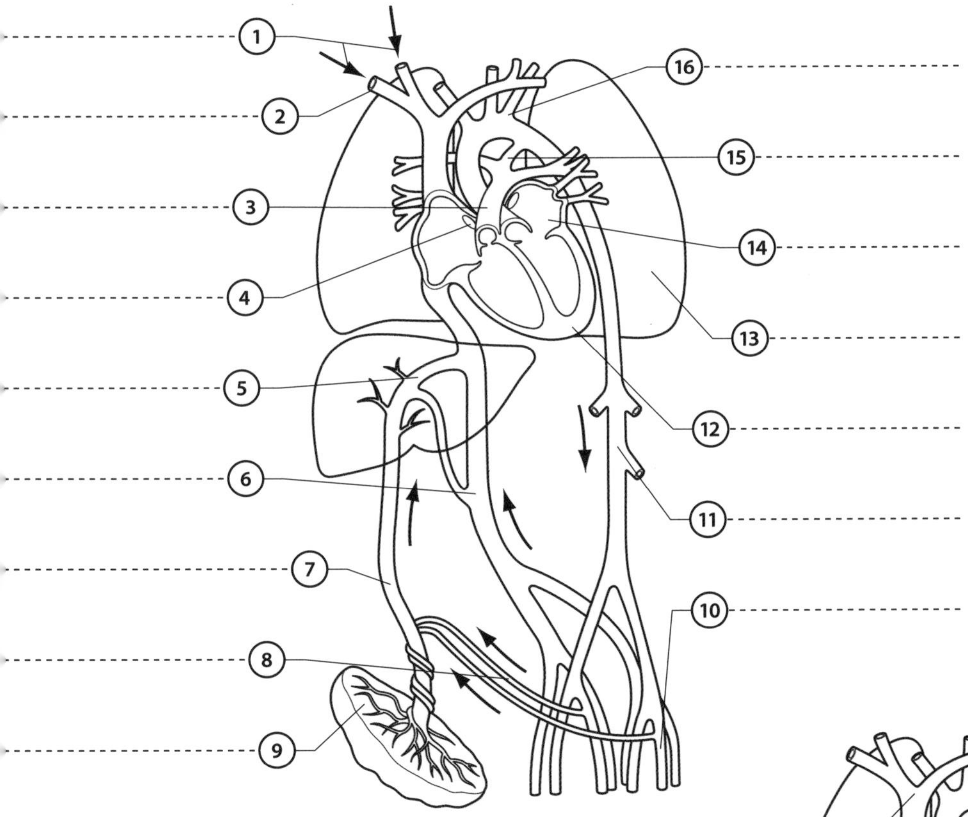

FETAL CIRCULATION

FETAL CIRCULATION

Color and/or label the structures indicated on the diagram using the key below.

1. Direction of blood flow
2. Blood supply from upper body
3. Pulmonary artery
4. Foramen ovale
5. Ductus venosus
6. Inferior vena cava
7. Umbilical vein
8. Umbilical arteries
9. Placenta
10. Blood supply to lower body
11. Descending aorta
12. Heart
13. Left lung
14. Left atrium
15. Ductus arteriosus
16. Blood supply to upper body

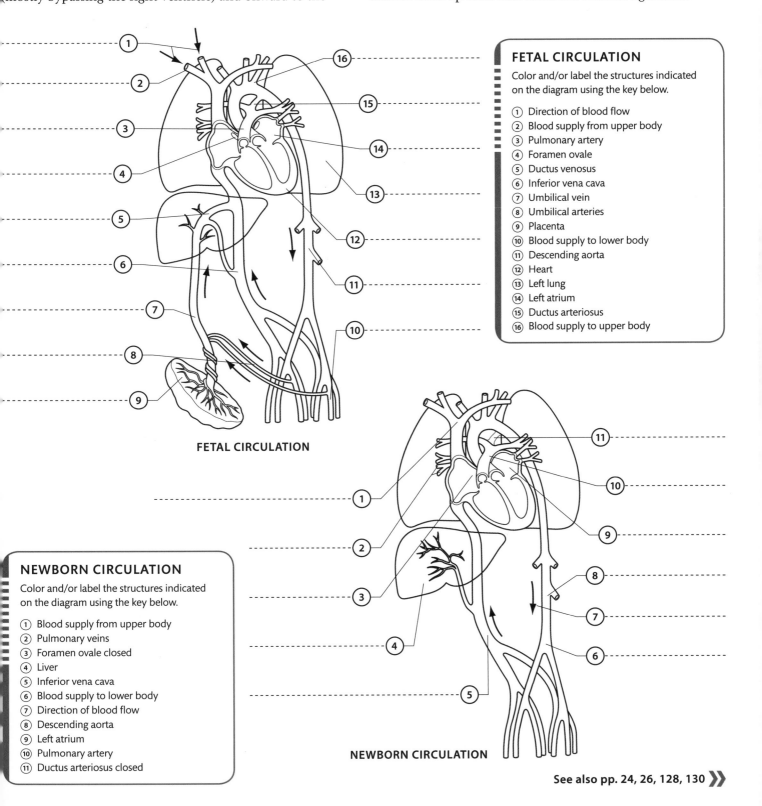

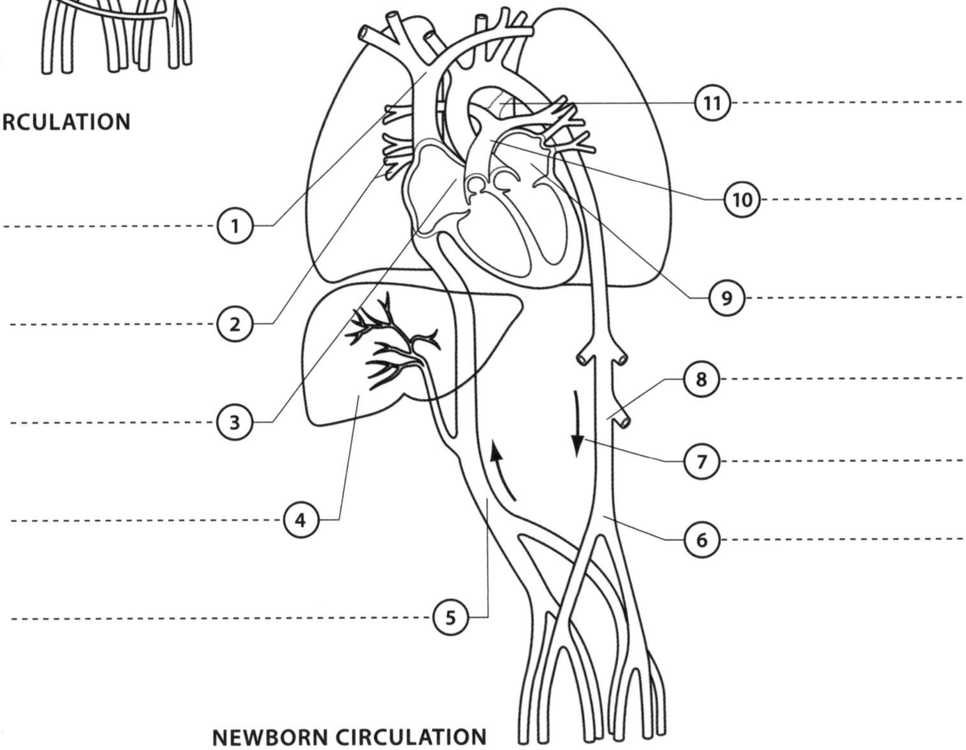

NEWBORN CIRCULATION

NEWBORN CIRCULATION

Color and/or label the structures indicated on the diagram using the key below.

1. Blood supply from upper body
2. Pulmonary veins
3. Foramen ovale closed
4. Liver
5. Inferior vena cava
6. Blood supply to lower body
7. Direction of blood flow
8. Descending aorta
9. Left atrium
10. Pulmonary artery
11. Ductus arteriosus closed

See also pp. 24, 26, 128, 130 »

INDEX

Page numbers in **bold** refer to the color reference pages at the beginning of each chapter.

A

B

C

D

E

I

J

K

L

M

N

O

P

Q

R

S

W

X

Y

Z

ACKNOWLEDGMENTS

Dorling Kindersley would like to thank the following people for their assistance in the preparation of this book:

Ann Baggaley and Laura Palosuo for additional editorial work; Hilary Bird for compiling the index; Katie John for proofreading; Henrietta Gordon, Mayur Murali, Oliver Redfern, Ruth Taylor, and Josephine Wilson for testing the coloring-in artworks.

Dorling Kindersley would also like to thank the following for their kind permission to reproduce their photographs:

(Key: a-above; b-below/bottom; c-centre; f-far; l-left; r-right; t-top)

4 Science Photo Library: Eye of Science (b). **The Wellcome Trust:** Professor Alan Boyde (c). **5 Science Photo Library:** Eye of Science (tr); GJLP (cl); Francis Leroy, Biocosmos (bl); Susumu Nishinaga (crb). **The Wellcome Trust:** EM Unit / Royal Free Med. School (clb); University of Edinburgh (cra). **12 Getty Images:** Dr. Gladden Willis (bc). **Science Photo Library:** Volker Steger (bl). **16 Corbis:** Photo Quest Ltd / Science Photo Library (br). **Science Photo Library:** Biophoto Associates (t); M.I. Walker (bl). **17 Corbis:** Photo Quest Ltd / Science Photo Library (cl). **Getty Images:** Dr. Gladden Willis (cr). **Peter Hurst, University of Otago, NZ:** (tr, br). **Science Photo Library:** Steve Gschmeissner (tl, bl). **22–51 The Wellcome Trust:** Professor Alan Boyde (Chapter bar). **52–79 Science Photo Library:** Eye of Science (Chapter bar). **80–113 Science Photo Library:** Nancy Kedersha. **114–125 Science Photo Library:** GJLP (Chapter bar). **126–153 The Wellcome Trust:** EM Unit / Royal Free Med. School (Chapter bar). **133 Science Photo Library:** Steve Gschmeissner (c, cr); Susumu Nishinaga (cl). **154–167 Science Photo Library:** Francis Leroy, Biocosmos (Chapter bar). **156 Science Photo Library:** Steve Gschmeissner (bl). **158 Science Photo Library:** Steve Gschmeissner (br). **168–185 Science Photo Library:** Eye of Science (Chapter bar). **186–199 The Wellcome Trust:** University of Edinburgh (Chapter bar). **198 Science Photo Library:** Steve Gschmeissner (bl). **203 Getty Images:** PhotoAlto / Teo Lannie (b) **Science Photo Library:** Richard Wehr / Custom Medical Stock Photo (t). **210–245 Science Photo Library:** Susumu Nishinaga (Chapter bar). **213 Alamy Images:** Phototake Inc. (tl). **Science Photo Library:** Steve Gschmeissner (cr). **The Wellcome Trust:** Yorgos Nikas (tr)

For further information see:
www.dkimages.com